KV-026-269

SAC Aberdeen
AC 0026912 5

SOIL ANALYSIS

BOOKS IN SOILS, PLANTS, AND THE ENVIRONMENT

Series Editor
G. STOTZKY
Department of Biology
New York University
New York, New York

Soil Biochemistry, Volume 1, edited by A. D. McLaren and G. H. Peterson

Soil Biochemistry, Volume 2, edited by A. D. McLaren and J. Skujiņš

Soil Biochemistry, Volume 3, edited by E. A. Paul and A. D. McLaren

Soil Biochemistry, Volume 4, edited by E. A. Paul and A. D. McLaren

Soil Biochemistry, Volume 5, edited by E. A. Paul and J. N. Ladd

Soil Biochemistry, Volume 6, edited by Jean-Marc Bollag and G. Stotzky

Organic Chemicals in the Soil Environment, Volume 1, edited by C. A. I. Goring and J. W. Hamaker

Organic Chemicals in the Soil Environment, Volume 2, edited by C. A. I. Goring and J. W. Hamaker

Humic Substances in the Environment, by M. Schnitzer and S. U. Khan

Microbial Life in the Soil: An Introduction, by T. Hattori

Principles of Soil Chemistry, by Kim H. Tan

Soil Analysis: Instrumental Techniques and Related Procedures, edited by Keith A. Smith

Soil Reclamation Processes: Microbiological Analyses and Applications, edited by Robert L. Tate III and Donald A. Klein

Symbiotic Nitrogen Fixation Technology, edited by Gerald H. Elkan

Soil-Water Interactions: Mechanisms and Applications, edited by Shingo Iwata, Toshio Tabuchi, and Benno P. Warkentin

Soil Analysis: Modern Instrumental Techniques, Second Edition, edited by Keith A. Smith

Soil Analysis: Physical Methods, edited by Keith A. Smith and Chris E. Mullins

Growth and Mineral Nutrition of Field Crops, edited by N. K. Fageria, V. C. Baligar, and Charles Allan Jones

Semiarid Lands and Deserts: Soil Resource and Reclamation, edited by J. Skujins

Plant Roots: The Hidden Half, edited by Yaov Waisel, Amram Eshel, and Uzi Kafkafi

Additional Volumes in Preparation

SOIL ANALYSIS

Modern Instrumental Techniques

Second Edition

edited by
KEITH A. SMITH
Edinburgh School of Agriculture
Edinburgh, Scotland

MARCEL DEKKER, INC.   NEW YORK • BASEL • HONG KONG

Library of Congress Cataloging-in-Publication Data

Soil analysis; modern instrumental techniques / edited by Keith A. Smith. -- 2nd ed.
p. cm. -- (Books in soils, plants, and the environment)
Includes bibliographical references and index.
ISBN 0-8247-8355-7 (alk. paper)
1. Soils --Analysis. I. Smith, Keith A. II. Series
S593.S742 1990
631.4'1--dc20 90-13980
CIP

This book is printed on acid-free paper.

MARCEL DEKKER, INC.
270 Madison Avenue, New York, New York 10016

Current printing (last digit):
10 9 8 7 6 5 4 3 2 1

PRINTED IN THE UNITED STATES OF AMERICA

Preface

The theme of this revised and enlarged second edition is the same as that of the first edition, that is, to fill the gap between books covering traditional methods of analysis and specialist monographs on individual instrumental techniques, which are usually not written with soil or plant analysis specifically in mind. The principles of the techniques are combined with discussions of sample preparation and matrix problems, and critical reviews of applications in soil science and related disciplines.

In the 7 years since the Preface to the first edition was written, there have been many developments in the instrumental techniques applied to the analysis of soils and other environmental materials. Some techniques that were not used widely enough to merit inclusion in the first edition have become of much greater significance. For example, inductively coupled plasma (ICP) spectrometry has now become the favored technique for routine multielement analysis in major soil- and plant-testing laboratories. Another new inclusion is the chapter on ion chromatography, which has married the long-established procedures of ion-exchange separation to the technology of high-pressure liquid chromatography to provide a powerful way of determining ionic species in solution, especially anions which were, hitherto, difficult to measure. The last new addition is the chapter on analysis of soil functional groups by NMR spectroscopy. This is a developing area, which seems destined to expand as equipment and awareness of its potential become more widespread.

All 11 chapters of the first edition are retained in revised form; some have undergone a change in authorship. The scale of the revisions varies, depending on the extent of change and development in the subject over the last few years. One general feature of the revisions is the extent to which microprocessor control systems feature in the descriptions of the current generation of instruments. The explosive growth of microprocessors in instrumental control and operation, in all forms of analysis, and concurrent developments in software for data analysis and microcomputers for running that software, are the most significant new features of the field. Not only has the sophistication of what can be done increased greatly, but the cost of doing it has fallen dramatically in real terms. Small academic departments are now able to use techniques that had been restricted, on grounds of cost, to the largest research institutes.

This book is aimed at the researcher working in soil science or a related field who is faced with the problem of making a new determination, or of replacing old analytical equipment to make a routine determination more accurately or more efficiently.

The book will help in evaluating the available techniques so that the optimum choice in terms of speed, cost, or sensitivity may be selected. It will also be useful to teachers and students of postgraduate courses in soil chemistry and soil analysis.

I would like to thank the contributors to this volume for their efforts, and my family, colleagues, and students for their forbearance during the final, frantic phase of editing.

Keith A. Smith

Contents

Contributors

Johnathan R. M. Arah Department of Soil Science, The Edinburgh School of Agriculture, Edinburgh, Scotland

Nicholas T. Basta Department of Agronomy, Iowa State University, Ames, Iowa

John M. Bremner Department of Agronomy, Iowa State University, Ames, Iowa

Peter A. Cawse Harwell Laboratory, Harwell, Oxfordshire, England

David J. Eagle Pesticide Residues Unit, Agricultural Development and Advisory Service, Ministry of Agriculture, Fisheries and Food, Cambridge, England

Edward J. Jewell Analytical Chemistry Department, Agricultural Development and Advisory Service, Ministry of Agriculture, Fisheries and Food, Ashford, Kent, England

Henrik Saaby Johansen Department of Physics, Royal Veterinary and Agricultural University, Copenhagen, Denmark

Angela A. Jones Department of Soil Science, The University of Reading, Reading, Berkshire, England

John L. O. Jones Analytical Chemistry Department, Agricultural Development and Advisory Service, Ministry of Agriculture, Fisheries and Food, Cambridge, England

Victor Middelboe Department of Physics, Royal Veterinary and Agricultural University, Copenhagen, Denmark

Roger P. Paxton Agricultural Development and Advisory Service, Ministry of Agriculture, Fisheries and Food, Cambridge, England

David Robinson Soil-Plant Dynamics Group, Scottish Crop Research Institute, Dundee, Scotland

Leonard Salmon Harwell Laboratory, Harwell, Oxfordshire, England

Albert Scott Central Analytical Laboratory, Scottish Agricultural College, Edinburgh, Scotland

Barry L. Sharp* Analytical Division, The Macaulay Land Use Research Institute, Aberdeen, Scotland

Keith A. Smith Department of Soil Science, The Edinburgh School of Agriculture, Edinburgh, Scotland

M. Ali Tabatabai Department of Agronomy, Iowa State University, Ames, Iowa

Oscar Talibudeen Blackthorn Reach, Aldeburgh, Suffolk, England

Allan M. Ure[†] Analytical Division, The Macaulay Land Use Research Institute, Aberdeen, Scotland

Michael A. Wilson Division of Coal and Energy Technology, CSIRO, North Ryde, New South Wales, Australia

---

Current affiliation:
*Department of Chemistry, Loughborough University of Technology, Loughborough, Leicestershire, England
[†]Department of Pure and Applied Chemistry, Strathclyde University, Glasgow, Scotland

SOIL ANALYSIS

1

Atomic Absorption and Flame Emission Spectrometry

ALLAN M. URE *The Macaulay Land Use Research Institute, Aberdeen, Scotland**

I. INTRODUCTION

Since the independent realization in 1955 by Walsh [1] and by Alkemade and Milatz [2] of the analytical potential of atomic absorption spectrometry in flames, and the pioneering work of Walsh and his colleagues [3,4] in designing a practical atomic absorption instrument, there have been some three decades of research and development. The first applications of the technique were to soils, plant ash, and other agricultural materials [5–7], while the use of solution samples, nebulized into a flame, derives from the earlier use of flame emission spectrometry largely by Lundegårdh [8], again in an agricultural context. The history of these developments and earlier work has been reviewed comprehensively [9]. At present, atomic absorption spectrometry is established as the principal analytical technique for elemental analysis in agricultural and environmental laboratories.

Atomic emission spectrometry using flame photometers with photoelectric detectors was of considerable importance in the 1950s and 1960s, but its use, except for the determination of the alkali metals and the alkaline earths, has to a large extent been superseded by

**Current affiliation*: University of Strathclyde, Glasgow, Scotland.

atomic absorption spectrometry. More recently, however, there has been a revival of interest in atomic emission spectrometry, due to the development of high-temperature sources such as the nitrous oxide/acetylene flame [10–12], various types of high-frequency plasma [13], and, in particular, radio-frequency inductively coupled plasma (ICP) sources [14–18]. This interest stems not only from the high sensitivity of the ICP sources: It is due also to the increasing realization of the advantages of simultaneous multielement analysis offered by atomic emission and atomic fluorescence spectrometry. Conventional atomic absorption spectrometry remains essentially a single-element technique, and although research on multielement atomic absorption techniques using continuum sources has produced practical and sensitive instrumentation, little commercial exploitation of these new developments has occurred [19].

This account will be concerned mainly with atomic absorption spectrometry (AAS) and its use and potential in the analysis of soils and other environmental materials, but reference will be made when appropriate to atomic emission spectrometry (AES) and to atomic fluorescence spectrometry (AFS). In this connection it is important to realize that most commercial atomic absorption spectrometers, using conventional atomic absorption burners, make excellent flame photometers and an emission mode of operation is usually provided. Most instruments can be adapted for AFS but specialized flames or electrothermal atomizers will usually be required. Initial discussion of the principles of AAS will be in terms of flame atomizers, but electrothermal atomization using graphite tube furnace and carbon-rod techniques will also be considered, together with other nonflame methods of analysis.

II. BASIC PRINCIPLES

In its simplest sense atomic absorption spectrometry makes use of the fact that free atoms of an element absorb light at wavelengths characteristic of that element and that the extent of the absorption is a measure of the concentration of these atoms in the light path. The production of atoms from chemical compounds requires a source of energy such as a flame that can vaporize and dissociate the sample compounds into the gaseous elemental state in which *atomic absorption* of radiation can take place. In favorable circumstances, this light energy absorbed by the atoms can subsequently be re-emitted as light of characteristic wavelengths whose intensity can form the basis for analysis by *atomic fluorescence* spectrometry. The function of the flame, therefore, in atomic absorption and fluorescence spectrometry is to produce atoms from the sample compounds, i.e., to act as an *atomizer*. For atomic emission spectrometry, on the other hand, additional energy is required from the flame to excite the emission of atomic spectra, again characteristic of the element, whose intensity is not only a function of the atomic concentration in the flame but also of the temperature of the flame. The energy required for the production of atomic emission spectra increases as the wavelength decreases, and practical chemical flames such as air/acetylene, and even nitrous oxide/acetylene, are of little value for elements whose analytical wavelengths are lower than about 250 nm. Atomic absorption and fluorescence spectrometry, on the other hand, do not have this limitation and are successfully used even at wavelengths below 200 nm. The extremely high temperatures attained by the radio-frequency plasma sources, as distinct from chemical flames, are sufficient to produce emission spectra with excellent sensitivity for elements whose spectral lines lie well below 200 nm.

The basic principles of atomic absorption spectrometry are illustrated in Fig. 1, which shows schematically the basic components of a simple single-beam instrument. A hollow-cathode discharge lamp A, whose cathode is made of the element (copper, for example) that is to be determined, emits a spectrum characteristic of the cathode element (copper). This light is passed through the atomizer in the form of a flame B, into which a fine mist of a solution of the analyte (copper) is sprayed by a pneumatic nebulizer. Atoms of copper are formed from the sample mist by thermal processes in the flame and absorb some of the light from the hollow-cathode lamp at the wavelength of an absorbing (resonance) copper line. The light passed through the flame is received by a monochromator C set to accept and transmit radiation of this wavelength. By means of the adjustable, ganged entrance and exit slits the bandwidth of the monochromator is made narrow enough to reject lines of other wavelengths emitted by the hollow-cathode lamp. The light emerges from the monochromator exit slit and falls on the photocathode of a photomultiplier detector D, and an output current, proportional to the

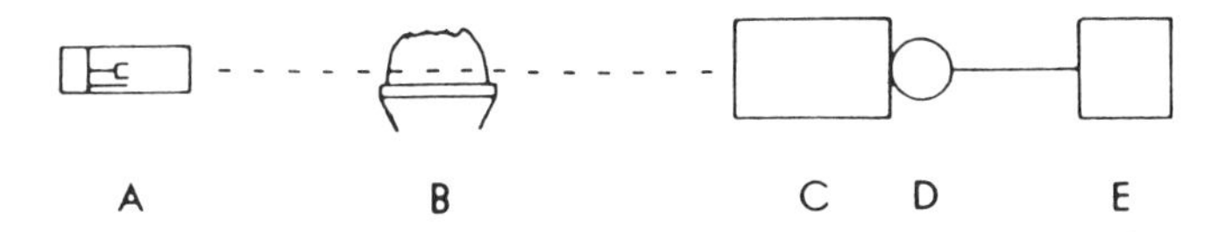

FIG. 1 Schematic diagram of a single-beam atomic absorption spectrometer comprised of a hollow-cathode lamp. A, a burner/nebulizer; B, a monochromator; C, a photomultiplier detector; D, and an output device, E.

incident light intensity, is amplified, processed electronically, and finally presented to a readout device E, such as a moving-coil meter, a pen recorder, or digital display.

If the light intensity measured in the absence of copper atoms in the flame is I_0 and the intensity when copper atoms from a nebulized sample solution are present is I_T, then

$$\%T = \text{percentage transmission} = \frac{100 \times I_T}{I_0} \tag{1}$$

$$\%A = \text{percentage absorption} = \frac{100 \times (I_0 - I_T)}{I_0} \tag{2}$$

$$\text{Absorbance} = \log \frac{100}{\%T} = 2 - \log \%T \tag{3}$$

The atomic concentration in the flame is proportional to the measured absorbance and from this the element concentration in the sample solution can be found by standardization (see Sec. IV.B).

The function of the monochromator in atomic absorption spectrometry is merely to isolate the required line from the remainder of the emitted hollow-cathode lamp spectrum, which is usually simple. The effective bandwidth of the absorption measurement is determined by the line width (ca. 0.001 nm) of the spectral lines emitted by hollow-cathode lamps. It is this excellent monochromation that contributes to atomic absorption spectrometry its remarkable freedom from spectral, superpositional interference.

The situation is quite different in atomic emission spectrometry in which isolation of the desired line from the emitted spectrum is dependent on the monochromator resolution, which cannot approach the equivalent resolution of atomic absorption spectrometry.

III. ATOMIC ABSORPTION SPECTROMETER

A. Introduction

The block diagram of a simple single-beam atomic absorption spectrometer, Fig. 1, serves to illustrate the basic components of the system whose various functions are indicated below. The example of a commercial single-beam instrument (Varian Techtron, AA6) of the previous decade shown in Fig. 2 makes use of an optical bench and consequently well displays these individual components. The modern counterpart in Fig. 3 (Varian Associates Ltd. AA40) conceals most of the constituent parts and highlights the dominant role now played by the built-in computer controlling the instruments and processing the output.

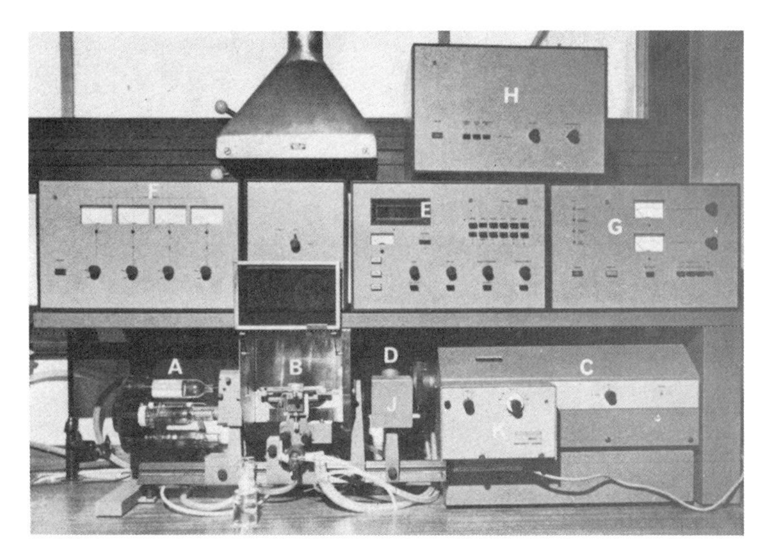

FIG. 2 Commercial single-beam atomic absorption spectrometer (Varian Techtron AA6) showing: hollow-cathode lamp, A; burner/ nebulizer assembly, B; monochromator, C; photomultiplier, D; digital indicator output device, E; hollow-cathode lamp power supply for turret of four lamps, F; automatic gas control unit, G; background correction unit, H. Also shown are the chopper J in position for atomic emission measurements and wavelength scanning attachment K, also for atomic emission spectrometry.

Since the atomic absorption measurement near the detection limit is the measurement of a small reduction in the initially large signal from the hollow-cathode lamp, this detection limit is determined largely by the stability of the lamp output and by variations in the sensitivity of the detector and electronics. One method of optimizing the detection limit is to use averaging output devices such as pen recorders or signal integrators. Another approach is to use a double-beam instrument in which the light from the hollow-cathode lamp is split into two beams, one passing through the flame and the other, the reference beam, passing outside the flame. The atomic absorption measurement made by a common detector is in this case the ratio of the two beam intensities so that lamp and detector noise and instability are reduced considerably. In addition, instrument warmup time, normally about 15 min with a single-beam spectrometer, is virtually eliminated. However, because of the requirement to divide the light beam for double-beam operation, the available light

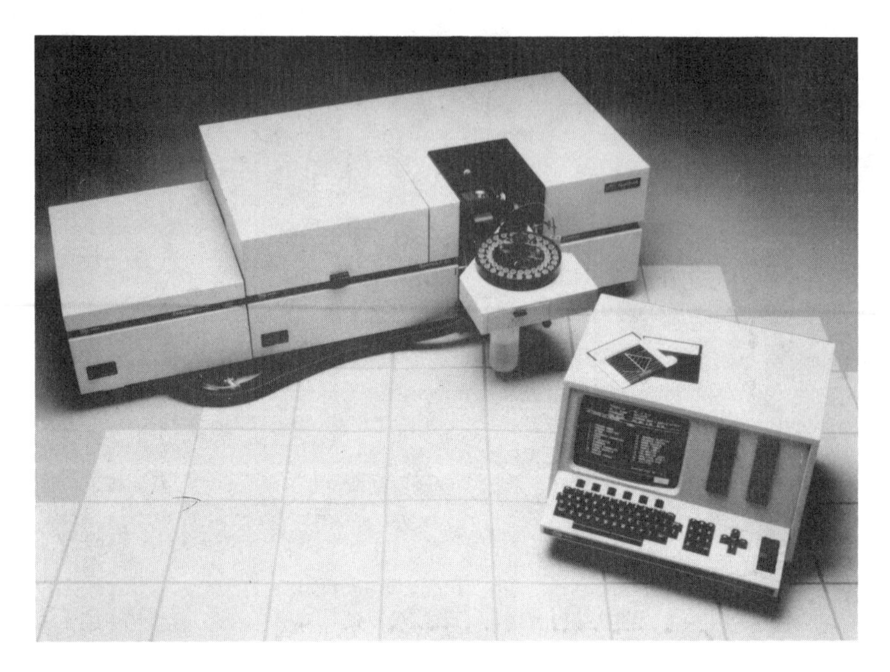

FIG. 3 Modern atomic absorption spectrometer, the Varian AA40, illustrating the dominance of the dedicated microcomputer in controlling the instrument and processing the output. (Courtesy of Varian U.K. Ltd., Walton-on-Thames).

energy at the detector is reduced and this can present some small disadvantage. Double-beam instruments can perhaps more readily incorporate background correction continuum sources (see Sec. VI) but some instruments may sacrifice double-beam operation in the process. Double-beam instruments are necessarily more complicated and their maintenance more difficult, but with modern solid-state electronic technology this disadvantage is a minor one. In practice, similar performance and detection limits are obtainable by both types of instrument, and in choosing an instrument this difference is not necessarily the dominating criterion.

The desirability of attempting to minimize the principal remaining sources of instability, i.e., the production of atoms by the nebulizer/flame combination, has led to the introduction of two-channel double-beam instruments that have a common light source and nebulizer/burner assembly but two monochromators and measuring channels. This not only allows two elements to be determined simultaneously by atomic absorption (or atomic emission), but by using one channel for an internal standard element whose concentration is known or fixed and the other for the element being determined (the analyte), a measurement of the ratio of the absorbance (or measured concentrations) of the two elements can be made. If the two elements are chosen so that the internal standard element and the analyte behave similarly in the flame, this ratio measurement compensates for fluctuations in nebulizer performance and in flame conditions. This internal standard technique, well known in emission spectrometry, requires the incorporation into sample and standard solutions of an internal standard element. The choice of a suitable internal standard [20], however, is often limited in practice, but when this is feasible, considerable gains in precision can be obtained [21].

B. Light Source

Considering, one by one, the components of the atomic absorption spectrometer, we turn first of all to the hollow-cathode discharge lamp (Fig. 1A), almost universally employed as the light source and illustrated in Fig. 2A. The lamp is filled at low pressure with a rare gas, generally argon or neon, whose ions carry an electrical discharge between the wire anode and the hollow cylindrical cathode that is made of the element whose spectrum is required, e.g., of copper for copper analysis. As a result of bombardment of the inner surface of the hollow-cathode by the filler-gas ions, copper, in this example, is sputtered into the discharge where its atoms are excited to emit their characteristic spectrum; this emerges through the glass, UV-transmitting glass, or quartz end window. Hollow-cathode lamps are stable and reliable and have lives often guaranteed by the manufacturer for six months or some 5000 mA hr of operation.

For a few elements, especially the alkali metals, metal-vapor arc discharge lamps are somewhat superior to hollow cathode lamps. They are less convenient to use, partly because they require operation at reduced current to keep the emitted spectral-line widths narrow, but chiefly because they cannot be fitted readily into most commercial instruments. They are only available for the more volatile elements such as the alkali metals, Cd, Hg, and Zn, and have found their chief use in atomic fluorescence in which their high intensity is desirable and line broadening less disadvantageous.

Electrodeless discharge lamps (EDL) excited by microwave-frequency sources have also been developed for atomic absorption and fluorescence spectrometry [22–28] for a wide range of elements. For many elements, however, stable EDL operation requires the use of heated chambers [29] thermostated at temperatures of several hundred degrees Celsius. Their chief virtue lies in the very high intensity, which has given them a principal role in atomic fluorescence spectrometry, where the intensity of the fluorescence signal is proportional to the intensity of the exciting radiation. In commercial instrumentation, however, they have not found much favor as atomic absorption sources. Similar devices operated at radio frequencies [30], more suitable for routine use, are also available from some manufacturers. In atomic absorption, EDLs are mainly of use for the volatile elements arsenic and selenium, for which hollow-cathode lamps are not particularly satisfactory.

C. Atomizer

The flame atomizer illustrated in Fig. 1B is the most common method of producing atoms from the elements in the sample, but in recent years this has been supplemented by a variety of nonflame devices, most of which use electrothermal methods of atomization (see Sec. VII). The flame atomizer consists of a burner in which the flame is produced and a nebulizer to convert the sample solution into a fine mist or aerosol that is fed to the flame. The nebulizer is usually a pneumatic device operating on the principle of a scent or paint spray, although the less convenient ultrasonic nebulizer has been used occasionally because of its greater efficiency. In the flame a complicated process involving desolvation of the solution droplets and dissociation of the chemical compounds of the resultant clotlet results in the production of gaseous atoms of the constituent elements. Although over the years a great variety of burners and nebulizers has been used [31,32], the burners now almost universally employed are of long-slot design burning premixed fuel and oxidant gases and fitted with a pneumatic nebulizer. A typical burner and nebulizer assembly is shown in Fig. 4. Sample solution is aspirated through the sample capillary by the pneumatic nebulizer operated by the oxidant gas and projected, as a spray of fine droplets, into the spray chamber where the larger droplets are intercepted or broken up by the flow spoiler. The surplus solution eventually drains out to waste. The oxidant and fuel gas are mixed in the chamber and carry the sample aerosol to the burner head and into the flame. Although pneumatic nebulizers only succeed in presenting less than 10% of the aspirated sample solution to the flame, alternative, more efficient methods, such as ultrasonic nebulization [33], have not found wide acceptance despite much development work. Heated spray chambers [34,35], or a preheated air supply to the nebulizer [36], have shown improved sensitivity but have not generally been adopted mainly because of memory effects. Automatic sample handling and introduction by flow injection methods offers advantages [37] including economy in the use of releasing

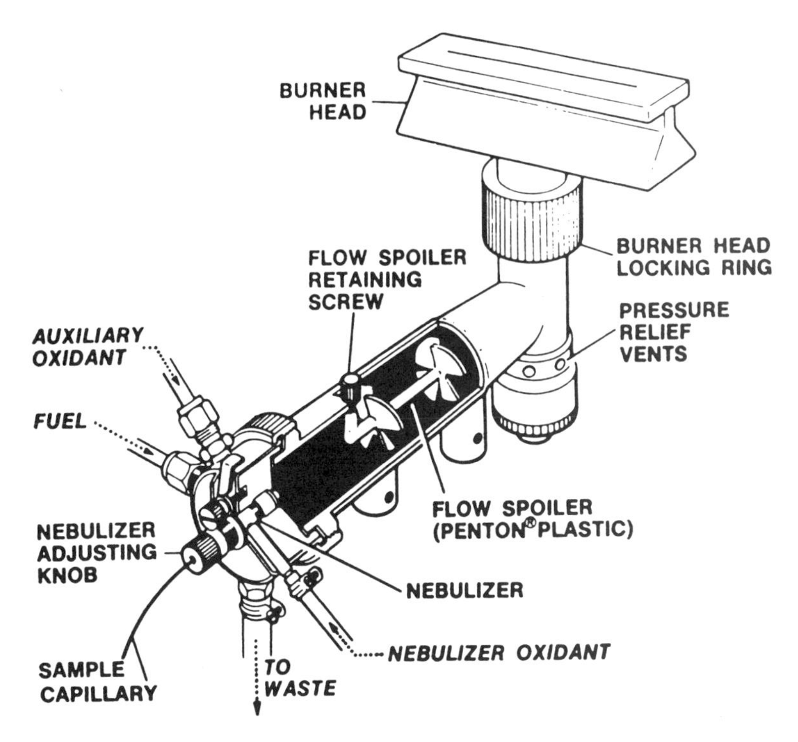

FIG. 4 Diagrammatic illustration of a burner/nebulizer assembly. (From the Perkin Elmer Model 306 Absorption Spectrophotometer, Order No. L333, courtesy of Perkin Elmer Ltd., Buckinghamshire, England.)

agents in the determination of calcium and magnesium [38]. Flow injection procedures are particularly suited to automatic hydride generation methods [39] (see also Sec. VIII.C).

Discrete microsampling techniques that can accept smaller samples and also provide higher sensitivity have been reviewed [40]. Automatic sampling devices are available from most manufacturers that sequentially insert the nebulizer capillary into the samples of a batch. Nebulizers of the Babington [41,42] and Conespray type [43] have been used increasingly because of their ability to handle viscous samples and solutions of high salt content as well as slurries [44].

The type of flame is of critical importance and several criteria govern the choice. If the flame temperature is too low, atomization will be incomplete since the flame cannot supply sufficient energy to dissociate the compounds in the sample. In addition, the degree to which atomization occurs will be a varying function of sample composition, giving rise to matrix, interelement, or anionic interference effects in the atomic absorption measurement. If, on the other hand, the flame temperature is too high, the atoms formed may be ionized to a sufficient extent to reduce the number of atoms present; this number will be dependent on not only the flame temperature but the composition of the sample, since interelement competitive and ionization suppression reactions can occur. The selection of the flame is therefore a compromise between achieving complete atomization (high-temperature flame) and avoiding ionization (low-temperature flame).

For most elements and sample types, the air/acetylene flame represents the best compromise. Even for the easily ionized alkali metal elements, sodium and potassium, the relative freedom from matrix or anion interference effects offered by this flame often makes it the best choice since ionization effects incurred can usually be avoided by the addition of an excess of another easily ionized (alkali) element. For rubidium and cesium, however, lower temperatue flames such as air/propane may be advantageous, provided they are used with standards whose matrix composition is similar to that of the average sample.

For elements such as Al, Ba, Be, Ca, Mo, Si, Ti, etc., which form refractory oxides or other compounds not completely dissociated in the air/acetylene flame, the higher-temperature nitrous oxide/acetylene flame [10] must be used with a fuel-to-oxidant ratio in excess of the stoichiometric proportions, to obtain reducing conditions.

Various other specialized flames are in use, of which only a few can be mentioned here. A multislot burner, the Boling burner [45], is commercially available, which, because the wider cross section of the flame more completely fills the optical path, is much less critical both in its positioning and in its fuel and oxidant settings. It is also less subject to clogging when sample solutions with a high content of dissolved solids, such as EDTA extracts of soils, are aspirated. Separated flames offer considerable improvements in signal-to-noise ratio and detection limits, particularly in atomic emission spectrometry. In these flames the inner combustion zone is separated from the outer diffusion flame by a silica tube [46], plates [47], or by a curtain of argon or nitrogen [48] that prevents the access of atmospheric air necessary for the completion of this secondary combustion. They have also been used routinely in atomic absorption spectrometry for the determination of Cu, Mn, and Zn in EDTA extracts of soil [49]. In addition to their value for emission spectrometry, separated flames seem likely to offer useful atomizers for atomic fluorescence spectrometry [50,51]. The cool argon/hydrogen flame has been used for atomic fluorescence spectrometry

[52] and for the determination of alkali metals [53]. It has useful applications in the determination of arsenic and selenium by the volatile hydride method (see Sec. VIII.B), a method that obviates matrix effects, which are normally large in this flame [53]. The low background of these flames in the ultraviolet is of importance in the determination of arsenic and selenium, for which short wavelength lines are employed. Similar nitrogen/hydrogen flames have been studied [54].

The nitrous oxide/butane (or propane) flame [55] is of limited application, but of considerable value for the determination of calcium in the presence of phosphate. This flame is hot enough almost to eliminate this interference although aluminum interference is only reduced. Because of its docility, this flame requires only the precautions usual for air/acetylene, although normally it is used with 5 mm long high-temperature burners. This is in contrast to the necessity for stringent operational safety procedures with the nitrous oxide/acetylene flame.

Mention might also be made of the use of mixtures of a powdered rock sample and a solid fuel powder [56] that are ignited and burned, without the need for fuel gases, in the optical path of an atomic absorption spectrometer—a system that could find application in portable, prospecting, or field instruments. In general, however, samples for flame atomic absorption, emission, or fluorescence spectrometry must be in solution form and although reasonably successful attempts have been made to aspirate and analyze suspensions or slurries of solid powders in liquids [57,58], matrix interference effects and other difficulties do not usually commend the method. Slurries of soils have been used for the determination of Cu, Fe, and Mg by FAAS [59] and of Cd, Pb by ETAAS [60], but matching of standards and agitation of the slurries while nebulizing will usually be required [61]. Suspensions of carbon have also been used to minimize phosphorus interference in the determination of alkaline earth metals [62,63]. Factors affecting slurry nebulization have been discussed [64,65].

D. Monochromator and Detector

Light from the hollow-cathode lamp is passed through the flame to the entrance slit of the monochromator (Figs. 1C and 2C) by a simple optical system generally consisting of two lenses, one on each side of the flame. As outlined above, the function of the monochromator in atomic absorption is to separate the resonance line used from other lines emitted by the hollow-cathode lamp while the desired wavelength and bandwidth is chosen by an appropriate setting of the monochromator.

Many instruments incorporate a wavelength scanning facility (Fig. 2K), which allows emission spectral lines to be displayed on a recorder together with the background emission for which correction can thus be made.

The monochromator receives light from the hollow-cathode lamp through the flame together with the light emitted from the flame. In the atomic absorption mode, it is almost essential that the signal arising at the detector from flame emission be rejected, and only that originating from the hollow-cathode lamp be accepted. In most commercial instruments this is achieved by modulating the hollow-cathode lamp output intensity so that the signal from the lamp can be separated electronically from the unmodulated flame emission signal and the latter rejected. Strongly emitting species such as the alkali metals will still generate a noise signal at the input of the detector amplifier, and a degradation of the atomic absorption signal-to-noise ratio may occur. The emission-generated noise signal can be minimized by using a narrow monochromator slit to minimize the accepted spectral bandwidth. To permit an instrument incorporating such a modulation system to be used for the measurement of atomic emission from the flame, a mechanical, rotating-sector light-beam chopper (Fig. 2J), synchronized to the detector, must be inserted between the flame and the monochromator to modulate light emitted by the flame.

The detector used almost universally is a photomultiplier tube (Figs. 1D and 2D), whose current output corresponds to the intensity of the light falling on its photocathode. This feeds the amplifiers and output devices which display the measured signals. Although the photomultiplier usually fitted covers a wide band of wavelengths from the ultraviolet to the near infrared, one could be selected to provide maximum sensitivity in the ultraviolet, visible, or red region of the spectrum.

E. Output Devices

The simplest output devices consist of a moving-coil meter or a pen-recorder displaying percentage transmission (%T), from which the absorbance, A, can be calculated. An example of a pen-recording of the transmission for a series of standard solutions for the determination of manganese in EDTA soil extracts is given in Fig. 5. A graph of absorbance versus concentration, i.e., the standard curve, is obtained as shown in Fig. 6. The absorbance versus concentration graph is essentially linear over most of the range with only a slight and quite typical curvature at high concentrations. Most instruments incorporate a logarithmic amplifier, which provides a direct readout of absorbance values, and this, with a provision for curve linearization, forms the basis for displaying outputs directly in concentration terms, using standard solutions for calibration. Provision is usually made for scale expansion,

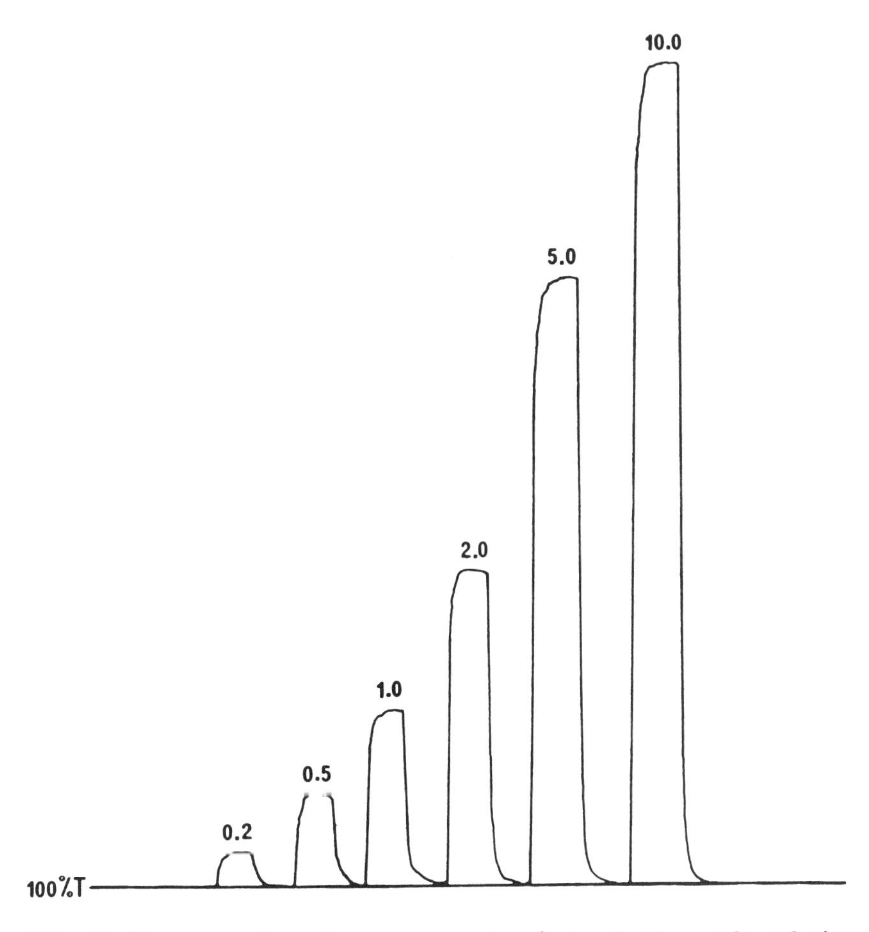

FIG. 5 Pen recording of measurements of standard sample solutions. Manganese, 0-10 μg ml^{-1}, in 0.05 *M* EDTA solution.

which improves the reading accuracy for small absorbance, for integration of the signals and for background correction. For the rapid pulse signals arising from, e.g., electrothermal atomizers, fast peak recorders or cathode-ray tube electronic display units are available together with peak-height and peak-area measuring facilities.

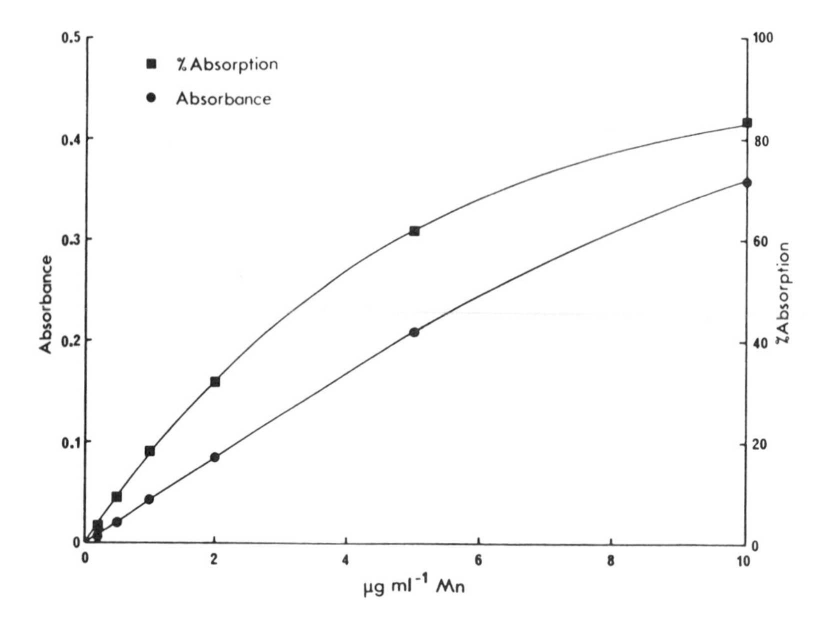

FIG. 6 Standard curve, absorbance versus concentration, for manganese in 0.05 *M* EDTA solution; also shown is the nonlinearity of the graph of % absorption versus concentration.

Most aspects of instrument control, operation, standardization, and data processing or storage are now carried out by a dedicated microcomputer or microprocessor built into the atomic absorption spectrometer or interfaced to it. This development has revolutionized atomic absorption spectrometry, and although the principles of atomic absorption measurement remain the same, few modern instruments make readily available the raw transmittance or absorbance data used, e.g., in Fig. 6 to illustrate standardization procedures. Automated and semiautomated instrumentation is now available commercially in which the choice of element, atomization, and instrumental conditions, output data format, reproducibility and standardization procedures are made under a preselected computer program that also controls the sample and standard solution inputs and changes in conditions for different elements. With such an instrument, illustrated in Fig. 7, up to 16 elements can be determined sequentially in a batch of samples without operator intervention. The use of microcomputer graphics for the display of absorbance/time profiles in electrothermal atomization applications has made a major

FIG. 7 Computer-controlled atomic absorption spectrometer (PU 9400 series) for automatic analysis of up to 16 elements. (Courtesy of Philips Scientific plc., Cambridge, England.)

contribution to the preparation of dry-ash-atomize programs for particular analysis.

Most instruments now provide a print-out of absorbance or concentration data that is also available in digital form for transmission to a central computer for instrument networking schemes that make the automatic laboratory a practical possibility.

IV. OPERATIONAL REQUIREMENTS

A. General Considerations

For each elemental analysis the best wavelength must be selected and the instrumental conditions must be optimized. These include the choice of flame type, fuel and oxidant flow rates, burner height, and slit width. Most manufacturers supply "cookbooks" of methods, and these form a good starting point for establishing operating conditions.

The laboratory itself must be clean and sources of contamination [66] studiously guarded against, especially when trace element analyses are involved. There is always the possibility of contamination by constituents of structural building materials. These can be calcium from plaster work, transition metals from stainless steel and paints, and Ba, Cd, Pb, Sn, and Zn in particular from plastics. This last also applies to the use of plastic laboratory ware; PVC and polystyrene may contain these and other metals as water-soluble stearates. Natural grade (not colored) polyethylene and polypropylene are the preferred plastic materials for containers of standard and sample solutions. Borosilicate glass laboratory ware is equally useful except for the alkalis at low concentrations, but soft glass containers should be avoided. The problem is particularly severe when very sensitive methods, such as electrothermal atomization (ETA), are used [67]. Contamination occurs not only in the laboratory but also at the sampling stage and should be of concern to the analyst since a contaminated sample is not worth analyzing [66].

B. Standardization

The usual method of analysis entails the preparation of an analytical working curve, by measuring the absorbance of a series of standard solutions of the elements of interest and plotting the graph of absorbance versus concentration, as illustrated in Fig. 6. The element concentration in the sample solutions can then be obtained from this graph by interpolation.

The method of standard additions can be used to correct for matrix including ionization, interference effects. This involves one or more incremented additions of analyte in known concentrations to aliquots of the sample solution and measuring the absorbance of the sample, sample plus increment A, sample increment B, etc. Standardization is thus achieved without the preparation of standards in matching matrices, provided the analyte absorbance versus concentration relationship is linear over the range in question and that the interference effect is not dependent on analyte concentration. The method is discussed in detail in most textbooks. It has found limited use in flame applications, but appropriately has been used more often to establish the presence of absence of an interference effect that is thereafter eliminated by other means, e.g., by solvent extraction procedures. In electrothermal atomizer applications the greater incidence of condensed phase and other matrix interferences has entailed a greater use of the standard addition technique. The availability of microcomputers to handle the calculation involved and of autosampler techniques that carry out the additions automatically have made the technique more attractive. It should still be regarded, however, as less than ideal because of the limitations mentioned above and in general is best applied to compensate for residual interference effects that cannot be fully eliminated by the appropriate pretreatment chemistry, matrix modification, or atomization program.

V. ANALYTICAL CAPABILITIES

A. Comparison of Atomic Absorption, Emission, and Fluorescence Spectrometry

As has already been indicated, atomic absorption must be regarded as a single-element technique, and although several multielement instruments have been described and reviewed [68], they have so far had little influence on commercially available equipment. Both atomic emission [69] and atomic fluorescence [70] are capable of simultaneous multielement analysis, and although the development of multielement atomic fluorescence spectrometers has not proceeded far, direct reading spectrometers using dc arc plasma-jet, microwave plasma, and in particular inductively coupled radiofrequency plasma (ICP) sources are available commercially for the simultaneous determination of elements in solution by atomic emission methods. In addition to an extremely high temperature, the ICP source offers detection limits that are often superior to those yielded by atomic absorption, and particularly so for refractory elements such as molybdenum (detection limits of a few nanograms per milliliter). This atomic emission technique has attractive possibilities for the direct analysis of soil extracts, making unnecessary the preconcentration techniques often required for atomic absorption analysis. In view, however, of the complexity of spectra produced by this source, the degree of expertise required will be considerably greater than for atomic absorption spectrometry and a good deal of exploratory work on its application has still to be done before the applied laboratory can make full use of it.

The limits of detection by flame AES, AAS, and AFS have been compared in an excellent refiew of AFS as an analytical technique [71]. AFS shows significant advantages over AAS and AES in detection limit for elements such as Ag, Bi, Cd, Hg, Te, Tl, and Zn. The advantages of AFS, however, are more obvious when nonflame techniques are used, since these avoid the quenching effects of the flame gases and combustion products. The subject of AFS has been treated comprehensively [51]. The place for AFS as a sensitive multielement technique may well have been usurped by the advent of the ICP AES technique and the continued lack of a good range of stable intense excitation sources for AFS.

B. Element Coverage: Sensitivity and Detection Limits

More then 60 elements can be determiend by atomic absorption methods, and the extent to which this wide range of element coverage can be realized in soil analysis depends first of all on the sensitivity of atomic absorption spectrometry for the elements of interest. An indication of the possibilities of soil analysis for different minor and trace elements, judged solely on their detection limits, is given in Table 1. It should be remembered, however, in considering Table 1 that other factors such as matrix or interference effects have a major influence on the viability of a particular analysis and that the quoted detection limits are for single-element solutions under optimum conditions. In many cases the lowest concentration that can usefully be determined in practical sample solutions will be greater, by a factor of 5 or 10, than these detection limits.

The major soil constituents Al, Ca, Fe, K, Mg, Na, and Si have not been included in Table 1, since atomic absorption methods are more than adequately sensitive for their determination as total contents in soils and, with the exception of silicon, in soil extracts. The determination of these elements is discussed in Sec. X.

Table 1, column 2, shows typical, total soil contents [72], together with the contents of the corresponding analysis solutions (column 3) prepared, e.g., by a lithium metaborate fusion/nitric acid dissolution procedure (see Sec. X.A) that typically involves a dilution of the contents from soil (100 mg) to solution (100 ml) of about 1000 times. The dilution will depend on the dissolution procedure but will seldom be much less than 500 times and may often exceed 1000 times. Also listed in Table 1 (columns 4 and 5) are normal content ranges for some elements extractable by 0.05 *M* EDTA and by 0.5 *M* acetic acid, which are in common use for assessing extractable or plant-available contents. These contents are calculated from data in Ref. 72. The last two columns list the detection limits obtained in aqueous solutions by atomic absorption spectrometry using flame and electrothermal atomization (ETA) methods, respectively. Full horizontal lines from element to total or extractable contents in Table 1 indicate that flame atomic absorption offers sufficient sensitivity for a particular element without preconcentration. Dashed lines indicate that the sensitivity is not sufficient for the full range of contents expected, but may be adequate for some samples and some purposes, e.g., detection of excess levels.

The determination of the total soil contents of the minor and trace elements Ba, Cr, Li, Mn, Rb, Sr, Ti, and Zn are indicated in Table 1 to be within the range of flame AAS methods without preconcentration. Examples of the determination of total contents of these elements are given in Sec. XI. The determination of the total contents of Be, Co, Cu, an Ni is marginal, and the remainder of the elements are insufficiently sensitive and require the use of preconcentration techniques, the employment of special methods (see Sec. VIII), or electrothermal atomization techniques. The detection limits for the latter technique (Table 1) are often some two orders of magnitude lower than flame limits.

Table 1 also shows that, on sensitivity criteria, it should be possible by flame atomic absorption spectrometry to determine the extractable contents of Cd, Co, Cr, Cu, Mn, Ni, and Zn in EDTA and acetic acid extracts, Pb, Ti, and V in EDTA and Sr in acetic acid extracts. This also illustrates the fact that the extractant is often chosen not only for its assessment of availability but also to maximize the solution content of the element to bring it within the scope of an analytical technique. For many of the other elements and other extracting solutions, preconcentration or the use of special methods such as ETA will be essential.

The characteristic concentration (often referred to in AAS as *sensitivity* or *reciprocal sensitivity*) is defined as the concentration of an element (measured in micrograms per milliliter) that gives a signal corresponding to 1% absorption, i.e., to an absorbance of 0.0044. It may often be more appropriate, e.g., in ETAAS, to define a characteristic mass (rather than a concentration) as that mass of an element giving a signal corresponding to 1% absorption, i.e., an absorbance of 0.0044. Either of these values used in conjunction with detection limits, allows some assessment of the viability of an analytical technique for a particular application.

The detection limit is defined usually as the element concentration that provides an absorption equal to twice or three times the standard deviation of the background (zero absorption).

C. Accuracy and Precision

For many trace element analyses the obtainable precision of AAS procedures is in the region of ±1–5%, whereas with special procedures and sophisticated instrumentation precisions of better than 1% can be obtained in some cases. The accuracy depends on the extent to which interference effects can be neglected or overcome, and in complex solutions such as those used for the determination of total contents of soils, it may be desirable to match standards and samples for matrix composition if the maximum accuracy is to be achieved. The precision of flame measurements has been discussed recently [73] and it has been shown that for each element a concentration (often some 100 times the characteristic concentration) exists for which the precision is a maximum [74]. The accuracy of AAS methods for soil analysis has been assessed by comparison with neutron activation analysis [75].

TABLE 1 Contents of Various Elements in Soils and Their Detection Limits by AAS

Element	Total content		Extractable content[c] (normal range; μg ml^{-1} in extract)		Detection limit (μg ml^{-1} in solution)	
	Soil: typical conc.[a] (μg g^{-1})	Analysis solution[b] (μg ml^{-1})	0.05 *M* EDTA	0.5 *M* acetic acid	Flame[d]	ETA[e]
Ag	0.1	0.0001			0.002	0.000005
As	6	0.006			0.11	0.00006
B	10	0.01			2.0	
B				0.005—5[f]	2.0	
Ba[g]	1000	1			0.02	0.00004
Be	6	0.006			0.0007	0.000001
Cd	0.5	0.0005			0.0007	0.000004
Cd			<0.002—0.06	<0.0003—0.008		
Co	15	0.015			0.007	0.00003
Co			<0.01—0.8	<0.001—0.05		
Cr	100	0.1			0.005	0.000005
Cr			0.02—0.8	<0.0003—0.03		
Cu	20	0.02			0.002	0.000008
Cu			0.06—2	<0.001—0.08		
Ga	25	0.025			0.038	
Ge	1	0.001			0.038	
Hg	0.1	0.0001			0.16	
Li	50	0.05			0.0015	
Mn	800	0.8			0.002	0.000004
Mn			1—20	0.1—2.5		
Mo	2	0.002			0.03	0.00006
Mo			<0.006—0.2	<0.0001—0.0008		

Ni — — —	50	0.05			0.008	0.000025
Ni ———			0.04—1.0	0.003—0.1		
Pb	20	0.02			0.15	0.00003
Pb ———			0.2—2	<0.00005—0.1		
Rb ———	100	0.1			0.002	
Sc	8	0.008			0.025	
Se	0.5	0.0005			0.25	
Sn	3	0.003			0.031	0.00006
Sn			<0.004—0.2	<0.0005—0.005		
Sr ———	300	0.3			0.002	0.00001
Sr ———				0.005—0.25		
Ti ———	4000	4			0.05	0.00033
Ti ———			0.1—2	<0.003—0.03		
V	100	0.1			0.05	0.00015
V ———			0.04—1	<0.001—0.03		
Y	40	0.04			0.11	
Zn ———	50	0.05			0.001	0.000001
Zn ———			<0.6—4	<0.05—0.75		
Zr	400	0.4			1.0	

[a]*Source*: Ref. 72.
[b]$LiBO_2$ fusion, HNO_3 dissolution, 100 mg soil per 100 ml solution, i.e., 1000 × dilution.
[c]Calculated from data in Ref. 59, EDTA extract, 15 g soil per 75 ml extract, i.e., 5 × dilution; acetic acid extract, 20 g soil per 800 ml extract, i.e., 40 × dilution.
[d]Varian Techtron AA6. Manufacturer's data.
[e]Instrumentation Laboratory 555 Furnace atomizer. Manufacturer's data.
[f]Hot water extraction; 20 g soil per 40 ml extract, i.e., 2 × dilution.
[g]Solid line indicates that determination is possible at "normal" levels on sensitivity criteria; dashed line indicates that determination is possible at "excess" levels on sensitivity criteria.

VI. INTERFERENCE EFFECTS

A. Spectral Effects

The relative freedom of AAS from superpositional spectral interference effects has already been commented on and this largely accounts for the fact that simple, single-element solutions can often be used as standard solutions without rigorous matching of matrix composition.

B. Physical or Matrix Effects

Physical effects such as viscosity of the sample solution, the presence of organic solvents, or high salt contents can also produce interference effects. This can usually be overcome by dilution, by matching the standard solution matrix composition to that of the sample, or by using the method of standard additions [76].

C. Ionization Effects

The extent to which ionization interference occurs depends on the ionization potential of the element being determined and on the flame temperature. An element with a very low ionization potential, such as the alkali metal cesium, can be ionized to the extent of 80% in the air/acetylene flame, whereas lithium, with a much higher ionization potential, is only some 5% ionized. In high-temperature flames, such as that of nitrous oxide/acetylene, cesium is completely ionized [77] and most elements are ionized to some extent. This effect can be countered by the addition of an excess of an easily ionized element such as potassium to suppress the ionization of the element being determined. For very easily ionized elements such as cesium it may be necessary to use a flame of lower temperature.

D. Background Effects

Interference of this type arises in flame atomic absorption whenever the concentration of major elements such as Al, Ca, Mg, and Na is high. The effect is most pronounced at shorter wavelengths. This interference is due to a scattering of light by dried salt particles and results in an apparently enhanced absorption. In some instances molecular absorption can produce a similar interference effect. This type of interference can be removed by separating the element from the interfering species by a technique such as solvent extraction. Alternatively the error can be corrected by repeating the determination at an adjacent wavelength, using a nonabsorbing line, and thus measuring and subtracting the contribution made by scattering or molecular absorption from the original measurement. Most instruments now incorporate a deuterium or hydrogen lamp in a background corrector. These lamps produce a continuous emission, at wavelengths shorter than about 350 nm, which is used to measure the scattering or molecular absorption light loss. This is then automatically subtracted from the original measurement of atomic absorption plus background absorption. Such correctors can correct for a light loss equivalent to an absorbance of about 2, but when the interference is large compared with the atomic absorption signal, the precision of the background-corrected analysis can be significantly degraded.

The use of a background corrector is essential for work with electrothermal atomizers such as the graphite furnace, since the furnace itself retains the smoke and molecular species produced durig atomization of the sample, as well as carbon particles emitted from the rod, all of which can contribute a light-scattering or molecular absorption error. A conventional background-correction system using a H_2 or D_2 continuum source is often sufficient but the higher background absorbances occurring with electrothermal atomization may require the use of a more sophisticated background-correction system using the Zeeman effect produced by a magnetic field [78—86] or the Smith—Hieftje technique [87,88]. Both of these techniques, which are commercially available, can correct in addition errors due to structured background and both obviate the difficulties of alignment that exist with the conventional H_2 and D_2 continuum methods. The Zeeman effect relies on a magnetically induced wavelength displacement of the analytical line that can then be used to measure background absorbance at a wavelength adjacent to that used for the atomic absorption measurement. It is more elaborate instrumentally and more expensive than the Smith—Hieftje system. The latter employs high-current pulsing of the analytical hollow-cathode lamp to broaden the emitted line. The difference between the absorption measured without (i.e., that due to line + background) the current pulse and with it (i.e., that due mainly to the background) is a measure of the background-corrected atomic line absorption. Although there is little evidence of shortened life for the pulsed hollow-cathode lamps, commercial exploitation of the Smith—Hieftje method makes use of specially designed lamps.

E. Chemical Effects

The most frequently encountered interference effects in AAS are chemical effects that alter, and usually reduce, the extent to which atom formation occurs in the flame. Typical examples in flames are the suppressive interference of Al, P, and Si on calcium and the other alkaline earth metals as a result of refractory compound formation. These effects are important in the air/acetylene flame (and in flames of lower temperature) in which they can be ameliorated by the use of releasing agents such as EDTA, or lanthanum and strontium chlorides. EDTA forms a compound with the analyte and inhibits the formation of calcium phosphate, e.g., which is not fully dissociated at the flame temperature. Lanthanum and strontium act by a mass action effect, if they are present in excess, to combine preferentially with the interfering (phosphate) species, leaving the analyte (calcium) unaffected. For greater freedom, indeed almost complete freedom, from many of the common chemical interferences, the higher-temperature flame of nitrous oxide/acetylene is to be recommended, often combined with the use of releasing agents. Most of these common interference effects are well understood and are dealt with in detail in the textbooks (see Sec. XII). A review in depth of interference effects and methods of controlling them has been published [89]. The inductively coupled plasma emission source, with its very high temperature, is relatively free of such interferences that also, of course, occur in atomic emission since the formation of atomic species forms the basis both of atomic absorption and atomic emission spectrometry.

VII. ELECTROTHERMAL ATOMIZATION (ETA)

As has been indicated already, limitations to the sensitivity of atomic absorption spectrometric analysis using flame atomizers, and the need to perform analysis on samples too small for continuous nebulization, have encouraged the use of electrically heated furnace atomizers. Developed largely from the work of L'vov [90–92], Massmann [93, 94], and Woodriff and Ramelow [95], commercial versions are available from many manufacturers. The two principal types are represented by the Heated Graphite Atomizer (Perkin Elmer Ltd.) and the Mini-Massmann atomizer (Varian Techtron) shown in Fig. 8 [96]. A simple carbon-rod atomizer proposed by West and Williams [97] is shown in Fig. 9 in a form modified for atomic fluorescence spectrometric analysis of cadmium [98] in soil extracts and plant material. Koirtyohann and Feldman [99] have compared a graphite tube furnace (Perkin Elmer HGA74) and the Mini-Massmann (Varian Techtron, CRA 63), with the principal conclusions that the former has better relative sensitivity (in micrograms per gram) and precision but greater background absorption difficulties, whereas the latter has a higher absolute (microgram) sensitivity but greater matrix problems. All operate with an inert gas atmosphere, usually of argon, and are heated by passing an electric current through the furnace material.

The technique involves pipetting a small volume (in the range 1–100 μl with different types of apparatus) into the furnace where sequentially the solvent is evaporated at low heat, the organic matrix

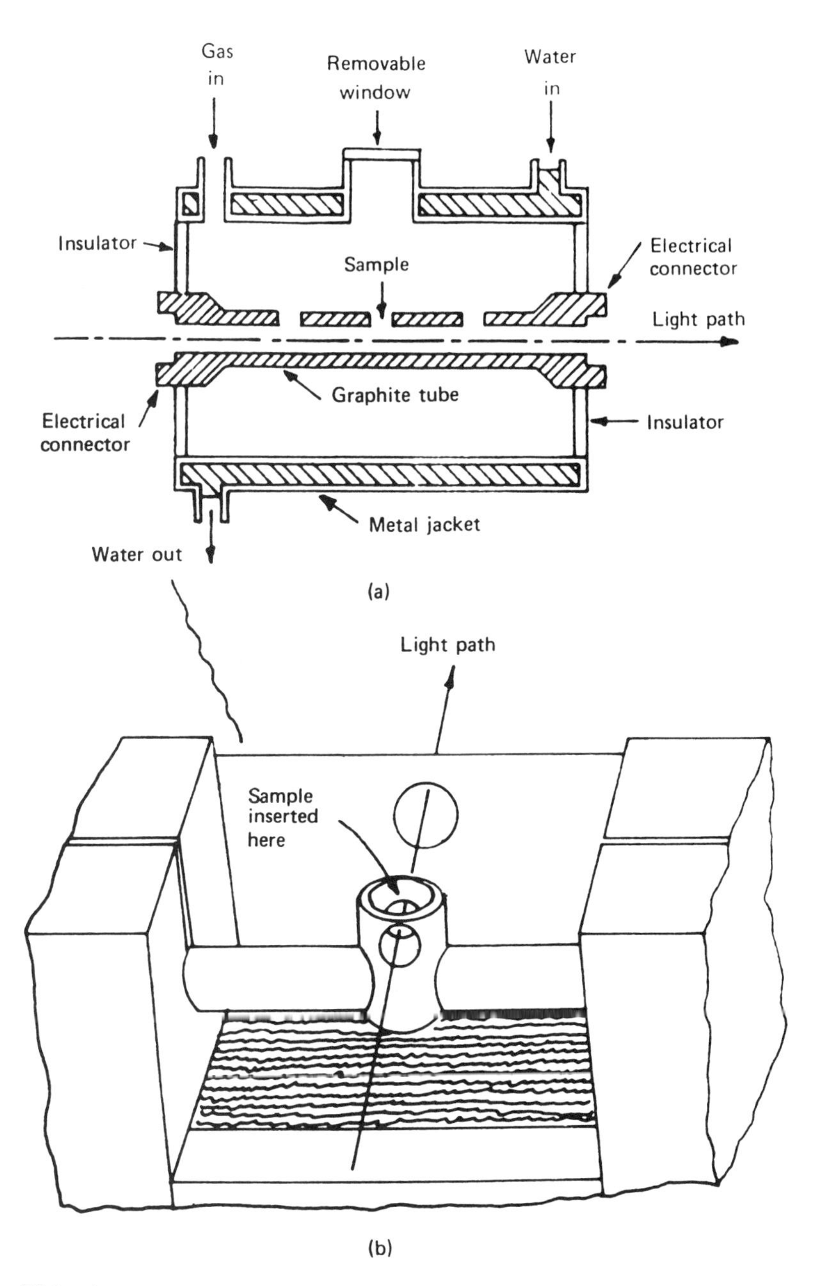

FIG. 8 Heated graphite atomizers: (a) Schematic of the Perkin Elmer furnace and (b) schematic of the Varian Techtron system (the cup shown can be replaced with a small tube). [From Ref. 96, C. Veillon, *Trace Analysis: Spectroscopic Methods for Elements* (J. D. Winefordner, ed.). Copyright © 1976 John Wiley & Sons, Inc. Reprinted by permission of John Wiley & Sons, Inc., New York.]

FIG. 9 Carbon-rod atomizer for the determination of cadmium by atomic fluorescence spectrometry.

ashed at a higher temperature, and the elements pyrolysed or atomized at a still higher temperature for analysis. The furnace is mounted in the optical axis of the atomic absorption instrument in place of the burner assembly. The operation of the atomizer is usually controlled by a time and temperature program, designed to minimize losses of the analyte element during drying and ashing and provide rapid heating at the atomization stage. The atomic absorption signal occurs as a sharp pulse with a rise time of a few tenths of a second, and this requires the use of output devices that have a rapid response for peak measurement or of devices for area measurement. Detailed accounts and reviews have been published on ETA [100–103] and recent developments discussed [104]. Some of these reviews describe other electrothermal atomizers such as the tantalum ribbon atomizer [105] and metal filament atomizers using, e.g., an electrically heated tungsten wire [106–109]. The characteristics of the graphite furnace [110–111] and carbon rod [111] atomizers have been comprehensively reviewed recently. The advantages of graphite furnace and similar electrothermal atomizers compared with flame atomizers can be summarized as follows:

1. Smaller (microliter) samples are required.
2. Higher sensitivity and lower detection limits (see Table 1) are possible as a consequence of longer atomic lifetimes, complete sample consumption (cf. 5–10% for burner nebulizers), and absence of dilution by flame gases.
3. Viscous liquids and even solid samples can be used.
4. The absence of flame background can give a better signal-to-noise ratio and may also allow measurement in the ultraviolet region.
5. For some elements, particularly those forming refractory oxides, the different mechanism of atomization, namely, reduction of oxide to element by hot carbon, gives more effective atomization than a flame.

The principal disadvantage is that the nonspecific light losses caused by smokes and vaporized matrix material are very much greater in furnaces than in flames, and the use of a background-correction facility is almost mandatory. These effects combined with the greater incidence of matrix interference effects at the surface of the electrothermal atomizer and of chemical, vapor-phase interferences have meant that electrothermal methods can appear to be less element-specific than flame methods. The investigation of such effects and the often lengthy development time for the establishment of furnace programs combined with the slower speed of analysis compared to flame methods indicate that for many analyses flame methods are preferred. Electrothermal methods do, however, offer major advantages in those cases where flame methods are insufficiently sensitive, where sample sizes are too small for flame methods, and in those instances where the different atomization mechanism offers freedom from an interference problem occurring in flames. Despite these advantages, the difficulties mentioned above have militated against their use in the soils laboratory. In recent years an improved understanding of the mechanisms of atomization and of interference effects have greatly improved the situation and many analyses at trace and ultra-trade levels can now be carried out effectively. The general principles of accurate analysis using graphite furnace atomization include:

1. The use of isothermal atomization procedures to minimize gas phase interference effects by atomization of the sample into a hot furnace atmosphere. This is achieved by the use of a graphite L'vov platform [112] within the furnace on to which the sample is pipetted. Because the platform is in poor thermal contact with the

furnace walls, it is heated largely by radiative processes, and as a result, atomization is delayed until the furnace is hot. Similar improvements have been achieved by the use of graphite probes [113] for the insertion of the sample into a preheated furnace, by the use of more elaborate designs of a constant temperature furnace [114], or by the use of atomization from a second surface on which the sample is condensed [115].

2. The use of matrix modifiers to transform the analyte to its most temperature-stable form and by this means to facilitate the removal of the more volatile matrix components, e.g., organic matter, by pyrolysis or ashing in the furnace without loss of analyte.

Interferences can occur as a result of carbide formation in the furnace by elements such as Hf, Mo, Nb, Th, W, and Zr, which can be minimized by the use of a tantalum furnace liner [116,117]. In contrast to the flame method, solutions are best prepared, not as chlorides, but as nitrates, because the former can lead to losses of volatile elements as chlorides [118] prior to the atomization stage and to interferences by recombination of analyte atoms to form chloride complexes in the vapor phase [119,120]. Many other interferences have been described [121—128], including the effect of H_2O_2 and various acids [129]. Interference effects may be so severe that it may still be necessary to resort to a solvent extraction separation procedure. It may also be necessary because of complex interference effects to make use of standard addition techniques.

With carbon-rod atomizers, the use of hydrogen to replace argon and form a hydrogen flame around the rod reduces the interference effects [130]. Pyrolytic carbon-coated furnaces have shown not only an improved performance for elements such as molybdenum and vanadium but also a longer life [111]. Totally pyrolytic graphite furnace tubes [131] have been introduced that, because they are less massive, offer potentially higher heating rates, and that do not suffer the progressive change [132] in sensitivity that occurs with coated tubes. Tube lifetimes are also better than for coated tubes.

VIII. SPECIAL TECHNIQUES

A. Boat Technique

Into a small boat of tantalum a small volume (less than 1 ml) of sample solution is pipetted and dried near the atomic absorption spectrometer flame before being introduced into the center of the flame. There the analyte is atomized within a few seconds, producing an atomic absorption signal in the form of a discrete pulse [133]. As the optimum temperature achieved will not exceed 1200°C, this method is only suitable for easily atomized elements such as Ag, As, Bi, Cd,

Hg, Pb, Se, Te, Ti, and Zn, but for these elements improvements in detection limits of 5—50 times can be achieved. The technique can be used with solutions too viscous for pneumatic nebulizers. Compared with the graphite furnace and carbon-rod methods, somewhat larger samples can be used with the boat technique. In a development of the boat technique, the Delves cup method [134], a nickel boat is used with an open tube of nickel or silica placed over the boat and flame to increase atomic residence times and improve the sensitivity. This method has been applied widely to the determination of lead. The use of a background corrector is essential to achieve accuracy with this technique. Comparative detection limits of flame, boat, and Delves techniques are given in Welz [135].

B. Atom-Trapping Atomic Absorption

For a similar group of relatively volatile elements, Ag, As, Au, Bi, Cd, Cu, Mn, Pb, Sb, Se, Tl, and Zn, flame atomic absorption sensitivity can be enhanced by up to 40 times by the use of the atom-trapping technique [136—141]. This employs a water-cooled silica tube in an air-acetylene flame to collect and preconcentrate the analyte from a sample solution nebulized into the flame for a fixed period of time of typically 1 or 2 min. At the end of the collection period the water cooling is switched off, the tube heats up, and the accumulated analyte is released into the flame, and the transient atomic absorption peak measured. The technique uses a simple adaptor for the burner of a conventional atomic absorption spectrometer and provides detection limits intermediate between flame and furnace methods: The detection limits for Pb and Cd, e.g., are 2 μg/l and 0.5 μg/l, respectively. The technique has been applied successfully to waters, soil digests, and soil extracts.

Smaller enhancements of flame atomic absorption sensitivity at least partly attributable to an atom-trapping mechanism have been achieved by the use of mechanically separated flames [142] and by the slotted silica tube device [143].

C. Volatile Hydride Method

The technique makes use of the fact that from elements such as As, Bi, Ge, Sb, Se, Sn, and Te, gaseous covalent hydrides can be formed by reduction in solution and passed into a flame or furnace, where atomization occurs and an atomic absorption measurement is made. The method was introduced by Holak [144] for arsenic, developed by Dalton and Malanoski [145], and optimized in a form made available commercially by Fernandez and Manning [146] and Manning [147]. In addition to Zn/HCl, other reducing agents, including $Mg/TiCl_3$ [148] and $NaBH_4$ [149], have been used and the methods for preparing stable borohydride solutions described [150]. The method has also been applied to elements of special agricultural interest such as selenium [151], but interference effects from copper and nickel, particularly, are likely to be severe in soil and plant matrices [152–154]. The effects of acids used in digestion procedures (for arsenic) have been discussed, together with the effect of the oxidation state of arsenic [155]. The technique has been reviewed recently by Robbins and Caruso [156].

There has been considerable development of automatic equipment for hydride generation atomic absorption spectrometry [157–159] and commercial versions are available. The technique has been applied directly to the speciation of arsenate and arsenite in soil extracts [160]. Hydride generation has been combined with nondispersive AFS [161] and with graphite furnace atomization [162] for the determination of As in soils. Interference effects in hydride methods for As and Se have been reviewed [163].

D. Reduction, Aeration, and Cold-Vapor Methods for Mercury Analysis

Since the detection limit for mercury by normal flame methods is about 0.5 mg $liter^{-1}$, its determination at concentrations at the μg $liter^{-1}$ level is only possible by special methods. Currently, the most important of these is the cold-vapor atomic absorption technique developed by Hatch and Ott [164], based on the formation of atomic mercury in a solution by reduction, followed by aeration of the sample solution to release the mercury vapor [165,166]. A second technique widely used for geological materials volatilizes the mercury by heating the solid sample and determines the evolved mercury directly in the vapor phase or preferably after collection and revolatilization [167,168]. The determination of mercury by such nonflame methods of atomic absorption and fluorescence spectrometry has been reviewed comprehensively [169,170] with applications to plants, rocks, soils, and other sample types. Bibliographies on mercury in the environment are available [171–173].

E. Indirect Methods

Indirect procedures have been described for phosphorus [174–176], arsenic [177], and silicon [178] determination by forming phosphomolybdate, arsenomolybdate, or silicomolybdate and determining the molybdenum content by atomic absorption spectrometry. For the determination of sulfur the interference effects of sulfates and sulfides on the determination by AAS of calcium and iron have been used [179,180], but more commonly sulfate has been precipitated by barium and sulfate estimated by the determination of the residual, or precipitated, barium [181,182]. Fluoride has been determined by making use of the effect on magnesium atomic absorption of fluoride concentration and pH [183].

F. Molecular Emission Methods

Various techniques of molecular emission spectrometry are of potential value for the determination of trace amounts of nonmetals in agricultural samples. These include the use of a cool flame such as the nitrogen/hydrogen diffusion flame for elements such as sulfur [184, 185], phosphorus [186], and the halides [187]. More recently, molecular emission cavity analysis (MECA) [188] has been employed not only for the determination of nonmetals such as sulfur [189] and boron [190], but has been applied in speciation studies for the determination of different sulfur anions in mixtues [191]. A review of the applications of MECA has been published [192].

In the MECA technique, a small (milligram) sample is injected into a cavity at the end of a stainless steel holder and introduced into a cool (ca. 300°C) hydrogen flame, which volatilizes the sample and produces molecular species such as S_2, whose molecular emission is measured by a simple spectrometer. Depending on the combinational form, detection limits of a few milligrams of sulfur can be determined with reproducibilities of 2–3%. While, in common with most molecular emission systems, interference effects occur, procedures for their elimination or reduction have evolved.

G. Miscellaneous Methods

Flame atomic absorption spectrometers have been used as metal-specific detectors in gas chromatography [193,194] and in liquid and high-performance liquid chromatography [194,195]. These, and other techniques used for purposes of speciation, have been reviewed recently [196]. The graphite furnace atomizer has also been combined with gas chromatography for the measurement of volatile alkyl compounds of arsenic and selenium [197].

Various methods of analysis for microsamples by AAS have been reviewed [198]. Useful techniques for the atomization in flames of microsamples have been described [199–204], in most of which a discrete sample is atomized and an atomic absorption signal pulse measured.

Molecular emission techniques of interest in agriculture include (1) the use of the band emission spectrum of boron oxide in an air/hydrogen flame for the determination of boron [205,206], (2) the use of halide bands for halogen determination [207], and (3) the isotopic analysis of nitrogen by using microwave plasma emission sources [208,209].

The use of solid samples instead of solutions for direct atomization in flames has a long history, from the early work of Ramage [210], who introduced a sample in a paper spill into the flame, to that of Govindaraju et al. [211], who used a powder sample in a cup at the end of an iron rod to introduce the sample, both of these workers using atomic emission spectrometry. Various solid sampling systems for use with AAS and AFS have recently been reviewed [212], and many examples using graphite furnace atomizers have been described [213,215]. Although the use of solid samples has its own special problems of matrix effects, sample inhomogeneity, and the difficulties of microsampling, some worthwhile results have been obtained.

IX. PRECONCENTRATION AND SEPARATION

The simplest way to achieve a concentration of the element of interest, and at the same time separate matrix elements that may interfere with the analysis, is to use the method of solvent extraction. The technique is discussed in depth in books by Morrison and Freiser [216] and Cresser [217]. The element is combined with an organic reagent to form a complex that can then be extracted by an organic solvent. Concentration of 10 times or more can readily be achieved and by adjusting the conditions, usually the pH, the extraction can be made selective for the element in question. The most common procedure uses ammonium pyrrolidene dithiocarbamate (APDC) [218] as the chelating or complexing agent and methyl isobutyl ketone (MIBK) as the extracting organic solvent. Other solvents have been used but chlorinated hydrocarbons are undesirable in flames since they inhibit combustion and also form the very poisonous gas phosgene. Chlorinated hydrocarbons such as chloroform can be used safely with furnace atomizers. Silver, Bi, Cd, Co, Cr, Cu, Fe, Hg, In, Mn, Mo, Ni, Pb, Pd, V, and Zn have all been extracted successfully at various pHs and the conditions for extraction for many metals have been studied systematically [219].

Other methods of preconcentration have been used but have not been adopted to any great extent. A useful general treatment of the subject is given by Zolotov [220].

X. APPLICATIONS

A. Total Contents of Soils

The total element content of soils is of less interest agriculturally than the extractable content, but for some purposes this information is of importance. The methods used for soils are similar to those for rocks and other geological materials. For essentially calcareous samples, solutions can usually be prepared by simple digestion with hydrochloric acid. Most soils will, however, require dissolution techniques capable of rendering siliceous material soluble.

Classical silicate dissolution methods using HF/H_2SO_4 have been adapted [221–224] for the determination of the alkalis and alkaline earths by atomic absorption and emission spectrometry. Interference effects due to sulfate and possible losses of alkaline earths and lead as insoluble sulfates make it preferable to substitute HCl, HNO_3, or $HClO_4$ for the H_2SO_4, and such procedures have been described for various elements and groups of elements [225–232]. These and other dissolution techniques for silicate materials are discussed by Angino and Billings [233] and reviewed elsewhere [234]. With some of these HF digestion procedures, soils or sewage sludge samples may require prior oxidation, e.g., by dry-ashing in a muffle furnace at 450°C, to destroy organic matter.

Dissolution methods using HF, aqua regia, or other acids at temperatures of 100–150°C in a sealed polytetrafluoroethylene (PTFE) digestion vessel or "bomb," contained in a stainless steel or aluminum casing, have been developed [235,236]. A comprehensive method of this type has recently been described [237]. After digestion and cooling, the excess HF and insoluble fluorides are complexed in solution by addition of boric acid. This allows the solution to be handled in conventional glassware for periods of at least 1 hr. The method has been widely applied to soils [238], sediments [239], and minerals [240] and up to 18 elements, Al, Ca, Co, Cr, Cu, Fe, K, Li, Mg, Mn, Na, Ni, Pb. Si, Sr, Ti, V, and Zn, have been determined in geological materials by atomic absorption in a single digest [241]. The method retains the silica in solution for analysis but has the disadvantage that commercial versions of the bomb are relatively expensive.

Although Na_2CO_3 fusion procedures have been used [233,242], the consequent strong sodium atomic emission in the visible region can degrade the signal-to-noise ratio of the atomic absorption or emission measurement. The most common fusion procedure is the lithium metaborate method developed by Ingamells [243,244] and suggested for use with atomic absorption analysis [245]. The relative merits of lithium metaborate or tetraborate fusion have also been discussed [246], while a strontium metaborate fusion technique, which avoids the need for addition of strontium (or lanthanum) in the determination of calcium, has been suggested [247,248]. Practical procedures [249,250] involve dissolution in dilute nitric acid of the glass formed by the fusion, and a method of analysis for 10 major and minor elements used in the author's laboratory [251] is outlined in Scheme 1. The nitrous oxide/acetylene flame is used for all the elements except potassium and sodium, which

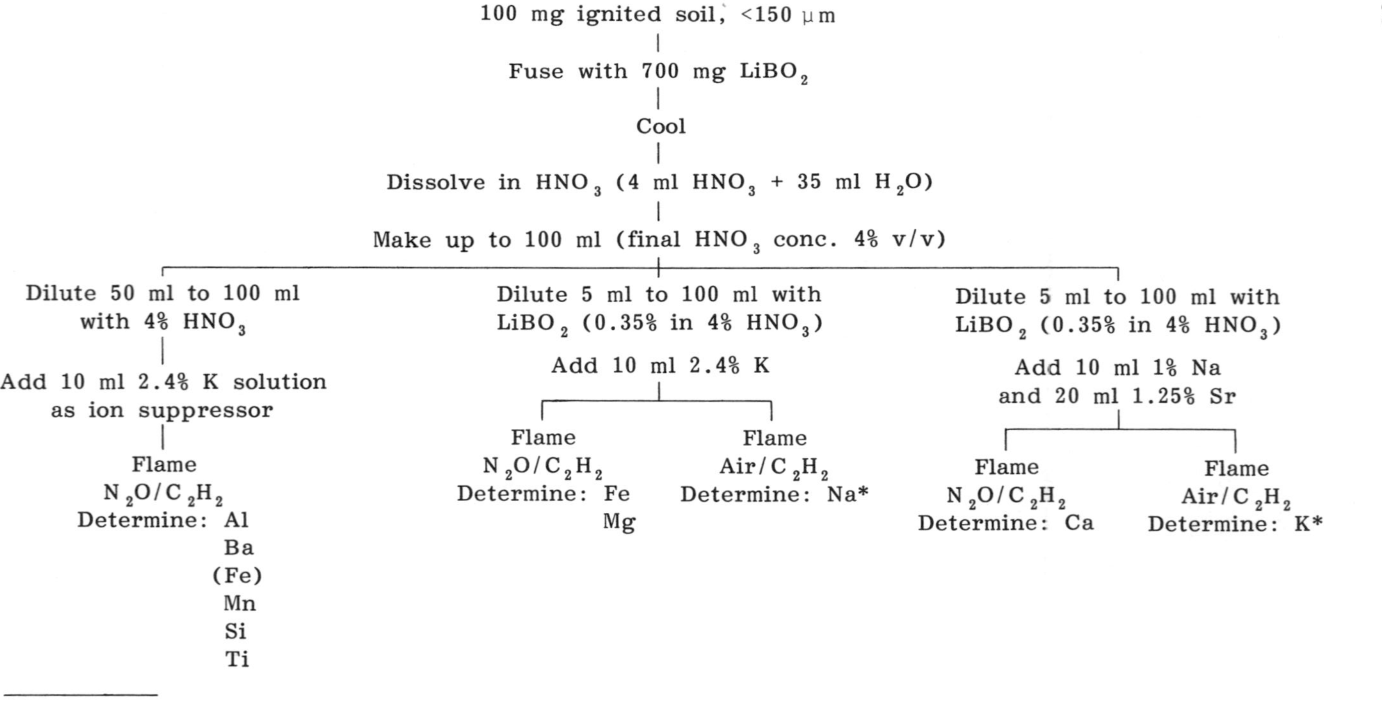

*Na and K by AES, remainder by AAS.
Standard solutions are single element, made up in 0.35% $LiBO_2$, 4% HNO_3 solution.

SCHEME 1 Scheme for the determination of 10 elements in small soil or rock samples by lithium metaborate fusion and nitric acid dissolution using atomic absorption and emission spectrometry.

are determined by emission in the air/acetylene flame. An excess of potassium or sodium is added as the ionization suppressor element, and for the calcium determination strontium is added for the prevention of aluminum or phosphorus interference. Single element standard solutions are used with the same concentrations of $LiBO_2$ and HNO_3, and the same additions of K, Na, or Sr as the sample solutions. Although interference effects are small, because of the high flame temperature and the buffering action of the $LiBO_2$, it is good practice to include a standard rock or soil with each batch of analyses. The chief virtue of the technique is that precise contents, to within ±0.5–1%, can be determined with a smaller soil or mineral sample (ca. 50 mg) than that required for alternative techniques such as X-ray fluorescence spectrometry (XRF). Analysis by XRF is, however, capable of greater precision for most of the major elements and is less tedious when such a large group of elements is involved. However, with some XRF instruments the accuracy for the light elements magnesium and sodium may be inadequate. The technique is discussed fully in Chap. 5.

These dissolution techniques for soils are in general suitable for the determination of major and a few minor elements. Because of the dilution involved, between 400 and 2000 times, most trace elements cannot be determined by flame atomic absorption methods (see Table 1) unless preconcentration techniques are used (see, e.g., Ref. 252).

The determination of the total content of the heavy metals Ag, Cd, Co, Cr, Cu, Fe, Hg, Mn, Mo, Ni, and Zn and also the elements Al, As, B, Bi, Pb, Sb, Se, and Sn in sewage sludges has become recently a matter of agricultural concern, because their disposal on agricultural land as soil conditioners and fertilizers may lead to toxicological problems in plants and animals. These are often assessed as total contents, but complete dissolution of the siliceous material is not always essential since these contaminants are seldom present in silicate-bound form. Various extraction procedures have been used for sludge-treated soils [253–256].

Several dissolution procedures (see Sec. XI) are therefore possible; these include treatment with combinations of HNO_3 and H_2SO_4, digestion with $HClO_4$, aqua regia, HNO_3, and H_2O_2, digestion under pressure [257] and dry ashing at 450°C followed by treatment with acid [258]. This latter technique is not practicable for As, Hg, and Se because of their volatility, but no losses of Cd, Pb, or Zn occur at this temperature [259]. Sensitivity requirements necessitate special methods (see Sec. VIII) for As, Bi, Hg, Sb, Se, and Sn, whereas other elements, such as silver and vanadium, require the use of electrothermal atomization methods (see Sec. VII). As mentioned above, digestion procedures using H_2SO_4 should not be used for lead because of the insolubility of its sulfate, whereas $HClO_4$ procedures entail losses of chromium as volatile oxychloride.

Since the median levels of trace element contents found in a survey [72] of sewage sludges are greater than the typical soil contents in Table 1 by factors of 40 for Cd, 4 for Cr, 35 for Cu, 2 for Ni, 25 for Pb, 50 for Sn, and 40 for Zn, all of these elements are determinable by flame AA methods in most of these digests. Use has been made of the graphite furnace (or sample boat) to ash and atomize a slurry of the sample prepared by homogenization techniques [260,261].

B. Soil Extracts

With the exception of soil sample treatments involving dissolution of siliceous material by hydrofluoric acid or by fusion procedures, few other treatments can be considered capable of providing solutions for the determination of true total trace element contents of a soil. With some vigorous treatment procedures, e.g., nitric acid/perchloric acid digestion, the determined contents will closely approach total contents with some types of sample and for some elements. Between such digestion procedures and much less vigorous extraction procedures designed to simulate plant availability, a whole range of digestion and extraction procedures have grown up in response to local conditions and for a variety of diagnostic purposes. The factors affecting the availability of trace elements in soils have been discussed by Mitchell [262]. The use and justification of the different methods of obtaining soil solutions for trace element analysis cannot, however, be considered in detail here and reference should be made to a detailed treatment [263], to an appropriate soil chemistry textbook [222,264,265], or to references surveying this subject [266—271]. Practical extraction procedures used for soils in the United Kingdom have been described, together with analytical methods that include atomic absorption spectrometry [272,273]. Extraction procedures for clay minerals [274] and sediments [275] have been evaluated recently.

Major nutrients such as Ca, K, Mg, and Na can be assessed in topsoils of noncalcareous nature by 0.5 *M* acetic acid extraction, and their concentration determined by direct analysis of the extract by atomic emission or absorption spectrometry. In calcareous soils, neutral 1 *M* ammonium acetate extracts can be used to determine exchangeable K, Mg, and Na concentrations, but for exchangeable calcium a different procedure, such as leaching with LiCl/Li acetate solution and determining calcium in the leachate [273], is required. Potassium and sodium are determined most readily by flame emission methods, and this allows their simultaneous measurement [276,277]. Calcium, determined by AAS or AES, suffers pronounced depressive interference by aluminum, readily extracted from soils by acetic acid and to a lesser extent by ammonium acetate, and by phosphate, although this is generally of less importance than aluminum in soils. With an air/acetylene flame an addition of lanthanum or strontium is only a palliative for aluminum interference, and for accurate calcium analysis a nitrous oxide/acetylene flame combined with the use of lanthanum or strontium is necessary. Magnesium is also affected, but to a smaller degree, and accurate analysis can be made in an air/acetylene flame by atomic absorption spectrometry, with an excess of lanthanum or strontium added to sample and standard solutions.

In Table 1 it can be seen that the quantity of certain elements in the soil extract (e.g., cadmium in EDTA extract) may exceed the quantity in the solution prepared for a total analysis, as a consequence of the large dilution (ca. 1000 times) involved in the preparation of the latter solutions. In the case of cadmium it is, in fact, possible to determine this element in most soils in EDTA extractable form directly by AAS, whereas the direct determination of total cadmium contents is difficult in uncontaminated soils. The choice of the extractant is usually made for its ability to predict plant uptake, but when this is not the prime criterion, the choice of a different extractant may bring the element within range of the AAS technique. Thus, the cadmium content in acetic acid extracts of many soils will be too low for direct AAS determination, and the higher content of major elements, e.g., aluminum and calcium, in this type of extract entails a greater background absorption interference. This interference, combined with the lower atomic absorption signals for cadmium, introduces greater errors than are found in EDTA extracts for this element. In addition, the use of EDTA for cadmium, like zinc, is likely to be as good as acetic acid in predicting plant availability. For Co, Cr, and Ni, acetic acid is the usual extractant and useful measurements can be made directly on the extract, although for some soils the levels of these elements may be beyond the reach of AAS unless a small concentration is made.

At the Macaulay Institute, Cu, Mn, and Zn in EDTA extracts have been determined using a separated air/acetylene flame [49], with the minimum soil concentrations determined being 0.05, 0.1, and 0.05 $\mu g\ ml^{-1}$, respectively. After long periods of operation (2 hr) with EDTA extracts, the burner slot may become partially salted up and require cleaning before further analysis. Multislot burners of the Boling type [45] are more tolerant of such high dissolved solid contents and can be used to advantage in this situation with no loss of sensitivity. Acetic acid extracts can also be analyzed for these elements in the air/acetylene flame, although in the case of copper the extract may require concentration by a factor of 5–10 times before analysis. This can be performed by evaporating to dryness the 800 ml extract, followed by treatment of the residue with HNO_3 and HCl, before dissolving in 100 ml of 0.05 *M* HCl.

Lead, Ti, and V can all be measured in EDTA extracts. For most of the other elements in Table 1, flame methods are insufficiently sensitive, so preconcentration procedures (Sec. IX), special techniques (Sec. VIII), or electrothermal atomization techniques (Sec. VII) must be followed.

These examples are only for illustration, and for details of these and other analyses in different extracts the references in Sec. XI, the annotated bibliography of applications of AAS in the agricultural field [278], or the general works of reference should be consulted.

XI. APPLICATIONS TO THE DETERMINATION OF TRACE ELEMENTS IN SOILS AND RELATED MATERIALS, AND SOME RECENT CONTRIBUTIONS TO THE METHODOLOGY*

A. Abbreviations Used for Reagents

EDTA Ethylenediaminetetraacetic acid
APDC Ammonium pyrrolidinedithiocarbamate
MIBK Methyl isobutyl ketone
NH_4Ac Ammonium acetate
DTPA Diethylenetriaminepentaacetic acid
NaDDC Sodium diethyldithiocarbamate
TOTP Tri-iso-octyl thiophosphate
HAHDC Hexamethyleneammonium hexamethylenedithiocarbamate

B. Determination of Individual and Groups of Elements

1. *Arsenic*

Arsenic has been determined in $HClO_4$ digests of soils by the volatile hydride (AsH_3) procedure using Ar/H_2 or N_2/H_2 flames [279–281] or electrically heated quartz tube furnace atomizers [282,283]. Various acid digests of soils [284–286] and extracts [287] have been used to prepare samples for hydride generation. Interference effects in the hydride and graphite furnace methods have been examined for arsenic and selenium [288]. The direct vaporization of AsH_3 from soils [289] and slurries [290] has been employed. Other applications of the hydride method include the analysis of soils and

*References to work published after 1971. For earlier work, see Kalra [278].

waters [291] and sewage sludges [292]. Hydride generation has been compared with electrothermal atomization in a graphite furnace for the determination of As [293] and the two methods have been combined for the analysis of soils prepared by pressure digestion [294].

Atomic absorption spectrometry using electrothermal atomization has been applied to As in soil extracts [295,296], and the interference of Al with the As line at 193.7 nm but not at 197.2 nm noted. Speciation of As in soils has been carried out by the use of HPLC and other chromatographic separations in combination with ETAAS [297–299]. The use of Pd as matrix modifier in ETAAS determinations in soils instead of the more usual Cu or Ni has been discussed [300].

2. Antimony

Total antimony has been determined using an air/acetylene flame after fusion with NH_4I and dissolution in HCl [301]. The hydride method has been used for soils following a HNO_3/H_2SO_4 digestion [302], in waters [303] and for sewage sludges [292].

3. Beryllium

A sodium carbonate fusion and HCl dissolution technique followed by a solvent extraction procedure [304] has been used for the determination of beryllium in a N_2O/C_2H_2 flame.

4. Boron

Only a few attempts have been made to determine boron by AAS [305] because of the technique's limited sensitivity, although an extraction procedure developed a few years ago [306] makes it more feasible. The analysis of boron by a variety of methods, including spectrochemical ones, has been reviewed extensively [307]. Band emission (BO) has been applied to the analysis of soils [308].

5. Cadmium

Cadmium has been determined in water, acid digests, or extracts of soils directly by FAAS [286,309–312]. Lower limits of determination and greater freedom from matrix and background effects have been obtained with preconcentration by solvent extraction using dithizone [313], HAHDC/BuOAc [314], APDC/MIBK [315–319], or as iodide complexes [320] and either FAAS or ETAAS determination. When chlorinated organic solvents such as $CHCl_3$ or CCl_4 are used in solvent extraction procedures, the health hazard due to the production of toxic gases (e.g., phosgene, $COCl_2$) in flames can be obviated by back extraction into aqueous solution [321] before analysis.

Electrothermal AAS methods have been applied directly to soils using solid samples [322,323] or slurries [324], as well as for the determinations of Cd volatilized directly from soil [325] and for various soil extracts [320,326]. The carbon-rod atomizer has been used for AAS and AFS determination of cadmium in dithizone/$CHCl_3$ concentrates of acetic acid extracts of soils [98,327,328].

Atom-trapping AAS has found use in the analysis of soil extracts and digests [329,330] and offers considerable promise for samples with difficult matrices such as $CaCl_2$ used, e.g., to predict the availability of Cd (and Pb) to some plants [331].

An AAS method has been compared and found in good agreement with a NAA method [332]. Flame atomic fluorescence spectrometry has been applied for Cd (and Pb) in sewage sludges [333].

Determination of Cd in soils and sludges has been carried out by AAS using procedures designed for a group of heavy metals. See the group analysis section.

The speciation of Cd and other trace metals in a simulated soil solution has been approached by comparing theoretical, calculated ion concentrations with contents measured by AAS [334].

6. *Chromium*

Chromium contents of soils have been determined in the air/acetylene flame by AAS in HF/H_2SO_4 digests [375] and in HNO_3 [336], HCl [337], and NH_4Ac/KCl extracts [338]. The use of $HClO_4$ treatments is inadvisable for chromium because of the volatility of its oxychloride.

7. *Cobalt*

Cobalt can be determined by flame AAS directly in HF [339] or HCl/HNO_3 digests [340] or after concentration by evaporation of acetic acid extracts [341]. Flame AAS has also been applied to the determination of cobalt in DTPA extracts of soils. It has been determined in various extracts by solvent extraction procedures using APDC/MIBK [342,343] or cupferron/MIBK [336] and in MIBK or amyl acetate extracts of soil treated with ammonium hexamethylene dithiocarbamate [344]. Soil extracts concentrated by APDC/MIBK have been analyzed for cobalt by graphite furnace [345]. Electrothermal AAS has been used for HNO_3/HF digests of soils [346] and for the direct determination of cobalt in acetic acid extracts of soils [347].

8. *Copper*

Since copper is to a large extent concentrated in the organic fractions of topsoils, oxidative digestion procedures can give a reasonably good estimate of total copper content in such cases; even dry ashing at 500°C followed by dissolution in HNO_3 has been used. Acid/perchloric acid mixtures have been used widely with flame techniques [315,348–351] and with ICP atomic emission methods [352]. Digestion with HNO_3/HCl followed by treatment with HF has been used for flame determination of Cu (and Zn) [353]. Other procedures for total copper by flame methods include HNO_3/HCl digestion with an APDC/MIBK concentration [354] and HNO_3/H_2SO_4 digestion [355]. Total copper in peat has been determined following $HClO_4$ digestion and concentration by APDC/MIBK [356]. Humic and fulvic acid fractions have also been analyzed [357].

Extractable copper contents have been determined by flame methods in water [358,359], NH_4Ac [273,359], NH_4Ac/KCl [338], LiCl/LiAc [273], EDTA [49,273,360,361], HCl [359,362], HNO_3 [363], and in various other extracts [349,361,364–366], in some cases combined with a subsequent concentration by solvent extraction [315,318,367,368] or by an ion-exchange technique [369]. Copper in the soil solution obtained by centrifugation has been determined by ETAAS [370].

9. *Lead*

Total lead contents have been determined in HNO_3 [371,372], HF/HNO_3 [373,374], HF/$HClO_4$ [375], HNO_3/HCl [354], and $HNO_3/HClO_4$ [315] by air/C_2H_2 flame AAS, but in many cases a concentration/separation procedure is required not only for reasons of sensitivity but also because the short wavelength makes it subject, like cadmium, to background interference. Solvent extraction procedures using APDC/MIBK with citrate masking of iron [342] and HAHDC/BuOAc solvent extraction [314] have all been used. For sediments, HCl/HNO_3 [340] and HF/HCl/$HClO_4$ [376] digestions have been used with the air/C_2H_2 flame. Atom-trapping AAS has been used for soil extracts [329].

Extractable lead has been determined in acetic acid [371,377,378], dilute HCl [379], dilute HNO_3 [336], dilute HCl/H_2SO_4 [326], NH_4Ac [326,380], hydroxylamine hydrochloride in HNO_3 [367], EDTA [368], and water extracts [380,358] by AAS in the air/C_2H_2 flame. Solvent extraction by APDC/MIBK [315,368] for flame AAS, dithizone/$CHCl_3$ extraction with AAS using the carbon-rod atomizer [328], and extraction as iodide complexes [320] have also been employed. Electrothermal AAS methods have been used as slurries of soils [324] and the use of the Zr-coated graphite furnace recommended [381].

10. *Manganese*

Total manganese in soils can be determined in solutions prepared by fusion procedures [242] as already outlined (Sec. X.A), or by HF/

$HClO_4$ [382] and HF/H_2SO_4 [383] digestion methods. It is commonly determined in solutions prepared for major element analysis by similar techniques [384].

Extraction procedures for manganese are legion, as they are designed to extract manganese in a variety of combinational forms, and include extraction with EDTA [49,273,382], NH_4Ac [359,361,385,386], supplemented NH_4Ac [338,387], H_2SO_4 [388], HCl [385], hydroxylamine hydrochloride in HCl or HNO_3 [368,389], water [358], DTPA [366], and many other reagents [70,390–394]. Sensitivity by AAS in the air/acetylene flame is adequate for most samples without concentration procedures, although they have been used [338,395]. Manganese has also been determined in the soil solution [370,396].

11. Mercury

The literature on the determination of mercury by various flameless techniques is so extensive that no attempt will be made here to summarize it. Readers are referred to reviews and bibliographies [100, 169–173] on the subject, to the outline given in Sec. VIII.C and to the few recent references below. Cold-vapor atomic absorption methods continue to be the most popular, and automatic procedures and equipment have been described [397,398] and applied to total and inorganic mercury in sewage sludges. Other applications to sewage sludge include Refs. 399 and 400. Total mercury [401] and total and available mercury in soils [402] have been determined by this method. Graphite furnace AAS has also been applied to soils [403], and methods for the electrostatic [404] and electrolytic [405] deposition of mercury on the graphite furnace for analysis have been described. Mercury has also been volatilized from soils on to gold-coated beads for analysis by cold-vapor AAS [406].

12. Molybdenum

For total molybdenum, soil solutions are prepared by digestion with $HF/HClO_4$ [407] or HF/HCl [408] and will often require the use of a concentration procedure [409] such as an MIBK [407] or Aliquat 336 [410] extraction of the thiocyanate complex, or an extraction of phosphomolybdate by Aliquat 336 in MIBK [411]. Soil extracts of molybdenum by NH_4Ac are seldom within the range of flame AAS, even with solvent extraction procedures. Molybdenum is not easy to determine with the graphite furnace because of the high temperatures required for atomization and the existence of memory effects.

13. Nickel

Nickel in soils has been of less interest agriculturally than cobalt, but with the increasing impact of sewage sludge application to agricultural land and the presence in some of these sludges of potentially toxic amounts of nickel, the demand for its analysis is likely to increase. It has been determined in HNO_3/HCl digestions of sediments [340] in the air/C_2H_2 flame and in acid perchloric digests by ICP AES [352]. Although it has been determined in various soil extracts [412,413], it will often require concentration [338, 344,369,414]. In the analysis of sludges AAS has been compared with dimethylglyoxime methods [415].

14. Rubidium

There have been only a few recent determinations of rubidium in soils by AAS [416,417].

15. Selenium

See hydride method (Sec. VIII.B). This method has been applied to soils and sediments [418]. Electrothermal AAS has been applied to soil digests after complexation with DAN and extraction [419] and also to soils and other materials [420] following combustion in oxygen [421,422]. An AFS method using an As/H_2 flame and evaporation by $La(OH)_3$ coprecipitation have been used for the determination of selenium [423].

16. Silver

Nitric acid digestion followed by solvent extraction concentration with TOTP/MIBK [424], and aqua regia treatment with an APDC/MIBK concentration stage [354] have been applied to the determination of silver in soils. Electrothermal AAS has been used for acid digests of soils [425].

17. Strontium

Although there have been few recent references to strontium analysis in soils, methods for the analysis of soil profiles [416], NH_4Ac extracts [426], and soils reclaimed after contamination with ^{90}Sr [427] have been described.

18. Thallium

Recent interest in the toxicity of thallium has led to the development of electrothermal methods of determining this element in muds, fly ash [428] soil extracts, and digests [381]. In general, the low concentrations involved require the use of solvent-extraction preconcentration [381].

19. *Tin*

Hydride methods [429] have been applied to soils and sediment following preparation by decomposition with NH_4I, sublimation and absorption in tartaric acid, whereas ETAAS has been used for soil digests [430].

20. *Zinc*

Total zinc contents of soils have been determined by AAS in flames following treatment with HF/HNO_3HClO_4 [348,431], $HF/HNO_3/H_2SO_4/HClO_4$ [431], $H_2SO_4/HNO_3/HClO_4$ [433], $HF/H_2SO_4/HNO_3$ [305], and treatments with HF [339] or $HClO_4$ [434]. Other analyses of soils [284,315,390,435,436], sediments [340,384,437,438], and mine spoil [393] have been made using similar treatments. Although in general the sensitivity by flame AAS methods is more than adequate and interferences are controllable, an APDC/MIBK concentration procedure [354], nondispersive AFS [439], AES using an ICP [349], and flameless methods [434] have all been employed.

Extractable zinc can be determined readily with a variety of extracting reagents, including water, with an APDC/MIBK preconcentration stage [434], and without one [358,359]. Also used have been HCl [359,361,362,432,441], HAc [378,442], EDTA [361,432], 0.05 *M* $CaCl_2$ and 2.5% HAc [367], and other or similar treatments [321,336,338,349,368,443]. An argon-separated air/C_2H_2 flame and AFS [70], and a mechanically separated air/C_2H_2 flame, have also been used for extracts of soils [49].

C. Group Analysis

Many applications of AAS to the analysis of soil digests or soil extracts are designed for the determination of groups of several elements, often including major as well as minor and trace elements. These include treatments with water [444–447], H_2O_2 [448], HNO_3/H_2O_2 [448], NH_4Ac [444,445,449–451], dithionite reagents [452–454], sodium pyrophosphate [453,454], NH_4Ac/NH_4F [455], HCl/NH_4F [449], and acid NH_4 oxalate [453,454,456], HCl [457], and other methods [449,458–463] for flame analysis. A flame procedure for Cd, Cu, Mn, Pb, and Zn after dissolution in HF/HNO_3 has been described [312]. Some graphite furnace procedures for groups of elements are given in Refs. 464–467.

Cadmium, Cu, Pb, and Zn have been determined in EDTA soil extracts [462] and Cd, Cu, Cr, Fe, Mn, and Zn in aqua regia digests of soils [468]. The effects of digestion temperature on the seven elements dissolved from soil by HCl [469] and the recoveries of six elements after aqua regia digestion of soils [468] have been discussed. Group extraction procedures for soil analysis by FAAS have been used [470] and continuum-source multielement AAS has been applied to NBS soils SRMs [471]. Digestion methods for sludge analysis have been evaluated [472–474] and compared with dry ashing/dissolution methods [475].

ACKNOWLEDGMENT

The author wishes to acknowledge the helpful advice and constructive criticism of Dr. M. L. Berrow of The Macaulay Land Use Research Institute.

REFERENCES

1. A. Walsh, *Spectrochim. Acta, 7*:108 (1955).
2. C. T. J. Alkemade and J. M. W. Milatz, *J. Opt. Soc. Am., 45*:583 (1955).
3. B. J. Russell, J. P. Shelton, and A. Walsh, *Spectrochim. Acta, 8*:317 (1957).
4. G. F. Box and A. Walsh, *Spectrochim. Acta, 16*:255 (1960).
5. J. E. Allan, *Analyst (Lond.), 83*:466 (1958).
6. D. J. David, *Analyst (Lond.), 86*:730 (1961).
7. W. Slavin, *At. Adsorp. Newsletter*, No. 4 (1962).
8. H. Lundegårdh, *Die Quantitative Spektralanalyse der Elemente*, Vol. I, Gustav Fischer, Jena, 1929.
9. R. L. Mitchell, in *Flame Emission and Atomic Absorption Spectrometry*, Vol. I (J. A. Deans and T. C. Rains, eds.), Marcel Dekker, New York, 1969, Chap. 1.
10. J. B. Willis, *Nature (Lond.), 207*:715 (1965).
11. E. E. Pickett and S. R. Koirtyohann, *Anal. Chem., 41*:28A (1969).
12. G. D. Christian and F. J. Feldman, *Appl. Spectrosc., 25*:660 (1971).
13. B. L. Sharp, in *Selected Annual Reviews of the Analytical Sciences* (L. S. Bark, ed.), Chem. Soc., London, 1976, Chap. 4.
14. V. A. Fassel, in *Coll. Spec. Interna. XVI, 1971, Plenary Lect. Rept.*, Adam Hilger, London, 1972.
15. P. W. J. M. Boumans and F. J. de Boer, *Spectrochim. Acta, 30B*:309 (1975).
16. P. W. J. M. Boumans, F. J. de Boer, F. J. Dahmen, H. Hoelzel, and A. Meier, *Spectrochim. Acta, 30B*:449 (1975).
17. P. W. J. M. Boumans and F. J. de Boer, *Spectrochim. Acta, 31B*:355 (1976).
18. S. Greenfield, H. McD. McGeachin, and P. B. Smith, *Talanta, 23*:1 (1976).

19. T. C. O'Haver, *Analyst (Lond.)*, *109*:211 (1984).
20. T. Takada and K. Nakano, *Spectrochim. Acta*, *36B*:735 (1981).
21. S. B. Smith, Jr., J. A. Blasi, and F. J. Feldman, *Anal. Chem.*, *40*:1525 (1968).
22. J. D. Winefordner and R. A. Staab, *Anal. Chem.*, *36*:1367 (1964).
23. R. M. Dagnall, K. C. Thompson, and T. S. West, *Talanta*, *14*:551 (1967).
24. R. M. Dagnall, K. C. Thompson, and T. S. West, *Talanta*, *14*:557 (1967).
25. R. M. Dagnall, K. C. Thompson, and T. S. West, *Talanta*, *14*:1551 (1967).
26. R. M. Dagnall, K. C. Thompson, and T. S. West, *Talanta*, *14*:1467 (1967).
27. J. M. Hall and C. Woodward, *Spectrosc. Lett.*, *2*:113 (1969).
28. S. R. Goode and D. C. Otto, *Appl. Spectrosc.*, *32*:63 (1978).
29. R. F. Browner and J. D. Winefordner, *Spectrochim. Acta*, *28B*:263 (1973).
30. W. B. Barnett, *At. Absorp. Newsletter*, *12*:142 (1973).
31. R. Mavrodineanu and H. Boiteux, *Flame Spectroscopy*, Wiley, New York, 1965, Chaps. 2—7.
32. R. Herrmann, in *Flame Emission and Atomic Absorption Spectrometry*, Vol. 2 (J. A. Deans and T. C. Rains, eds.), Marcel Dekker, New York, 1971, p. 57.
33. H. C. Hoare, R. A. Mostyn, and B. T. N. Newland, *Anal. Chim. Acta*. *40*:181 (1968).
34. A. A. Venghiattis, *Appl. Opt.*, *7*:1313 (1968).
35. A. Hell, W. F. Ulrich, N. Shifrin, and J. Ramirez-Munoz, *Appl. Opt.*, *7*:1317 (1968).
36. R. A. G. Rawson, *Analyst (Lond.)*, *91*:630 (1966).
37. J. F. Tyson, J. M. H. Appleton, and A. B. Idris, *Analyst (Lond.)*, *108*:153 (1983).
38. S. Zhang, L. Sun, H. Jiang, and Z. Fang, *Gangpuxue Yu Guangpu Fenxi*, *4*:42 (1984).
39. J. W. Carnahan, *Am. Lab.*, *15*:31 (1983).
40. M. S. Cresser, *Prog. Anal. At. Spectrosc.*, *4*:219 (1981).
41. B. Thelin, *Analyst (Lond.)*, *106*:54 (1981).
42. N. Mohamed and R. C. Fry, *Anal. Chem.*, *53*:450 (1981).
43. B. L. Sharp, "Method of and Apparatus for Nebulisation of Liquids and Liquid Suspensions, UK Patent Assignment 8432338, 1984.
44. M. D. Wickman, R. C. Fry, and N. Mohamed, *Appl. Spectrosc.*, *37*:254 (1983).
45. E. A. Boling, *Spectrochim. Acta*, *22*:425 (1966).
46. D. N. Hingle, G. F. Kirkbright, M. Sargent, and T. S. West, *Lab. Practice*, *18*:1069 (1969).

47. A. M. Ure, *Proc. Soc. Anal. Chem.*, *7*:192 (1970).
48. G. F. Kirkbright, M. Sargent, and T. S. West, *Talanta, 16*: 1467 (1969).
49. A. M. Ure and M. L. Berrow, *Anal. Chim. Acta, 52*:247 (1970).
50. R. M. Dagnall, G. F. Kirkbright, T. S. West, and R. Wood, *Anal. Chim. Acta, 47*:407 (1969).
51. V. Sychra, V. Svoboda, and I. Rubeska, *Atomic Fluorescence Spectroscopy*, Van Nostrand, London, 1975.
52. C. Veillon, J. M. Mansfield, M. L. Parsons, and J. D. Winefordner, *Anal. Chem.*, *38*:204 (1966).
53. H. L. Kahn and J. E. Schallis, *At. Absorp. Newsletter, 7*:5 (1968).
54. R. M. Dagnall, K. C. Thompson, and T. S. West, *At. Absorp. Newsletter, 6*:117 (1967).
55. L. R. P. Butler and A. Fulton, *Appl. Opt.*, *7*:2131 (1968).
56. A. A. Venghiattis, *Spectrochim. Acta, 23B*:67 (1967).
57. M. Kashiki and S. Oshima, *Anal. Chim. Acta, 51*:387 (1970).
58. W. W. Harrison and P. O. Juliano, *Anal. Chem.*, *43*:248 (1971).
59. J. Stupar and R. Ajlec, *Analyst (Lond.)*, *107*:144 (1982).
60. K. W. Jackson and A. P. Newman, *Analyst (Lond.)*, *108*:261 (1983).
61. C. W. Fuller, R. C. Hutton, and B. Preston, *Analyst (Lond.)*, *106*:913 (1981).
62. T. Kono and T. Kojima, *Bunseki Kagaku 32*:327 (1983).
63. D. M. Marinescue, *Spectrosc. Lett.*, *16*:583 (1983).
64. R. Ajlec and J. Stupar, *Vestn. Slov. Kem. Drus.*, *29*:295 (1982).
65. F. Figeoli, S. Landi, and G. Lucci, *Ann. Chim. (Rome)*, *72*: 63 (1982).
66. R. O. Scott and A. M. Ure, *Proc. Soc. Anal. Chem.*, *9*:288 (1972).
67. T. Harada, K. Fujiwara, and K. Fuwa, *Bunseki Kagaku*, *26*: 877 (1977).
68. K. W. Busch and G. H. Morrison, *Anal. Chem.*, *45*:712A (1973).
69. B. L. Vallee and M. Margoshes, *Anal. Chem.*, *28*:175 (1956).
70. R. M. Dagnall, G. F. Kirkbright, T. S. West, and R. Wood, *Anal. Chem.*, *43*:1765 (1971).
71. R. F. Browner, *Analyst (Lond.)*, *99*:617 (1974).
72. M. L. Berrow and J. C. Burridge, in *Inorganic Pollution in Agriculture*, MAFF Tech. Bull. GFM 3, HMSO, London, 1979.
73. N. W. Bower and J. D. Ingle, Jr., *Anal. Chem.*, *49*:574 (1977).
74. J. T. H. Roos, *Spectrochim. Acta, 28B*:407 (1973).
75. R. L. Brunelle, C. M. Hoffman, K. B. Snow, and M. J. Pro, *J. Assoc. Off. Anal. Chem.*, *52*:911 (1969).
76. J. Ramirez-Munoz, *Atomic Absorption Spectroscopy*, Elsevier, Amsterdam, 1968, p. 328.

77. A. M. Ure and R. L. Mitchell, in *Flame Emission and Atomic Absorption Spectroscopy*, Vol. 3 (J. A. Deans and T. C. Rains, eds.), Marcel Dekker, New York, 1975, p. 1.
78. H. Prugger and R. Torge, German Patent 1,964,469, filed December 23, 1969.
79. Y. Uchida and S. Hattori, *Oyo Butsuri, 44*:852 (1975).
80. S. D. Brown, *Anal. Chem., 49*:1269A (1977).
81. E. Grassam, J. B. Dawson, and D. J. Ellis, *Analyst (Lond.), 102*:804 (1977).
82. K. Yasudo, H. Koizumi, K. Onishi, and T. Noda, *Prog. Anal. Atom. Spectrosc., 3*:299 (1980).
83. H. Koizumi, H. Yamada, K. Yasuda, K. Uchino, and K. Oishi, *Spectrochim. Acta, B36*:603 (1981).
84. D. A. Hull, M. Muhammed, J. G. Lanese, S. D. Reish, and T. T. Finkelstein, *J. Pharm. Sci., 70*:500 (1981).
85. Y. Ma and X. Feng, *Hua. Hsueh Tung Pao, 2*:90 (1981).
86. F. J. Fernandez, W. Bohler, M. M. Beaty, and W. B. Barnett, *At. Spectrosc., 2*:73 (1981).
87. S. B. Smith and G. M. Hieftje, *Appl. Spectrosc., 37*:419 (1983).
88. L. De Galan and M. T. C. De Loos-Vollebregt, *Spectrochim. Acta, B39*:1011 (1984).
89. I. Rubeska and J. Musil, *Prog. Anal. Atom. Spectrosc., 2*: 309 (1979).
90 B. V. L'vov, *Ing. Fiz. Zhur., 11*:44 (1959).
91. B. V. L'vov, *Spectrochim. Acta, 17*:761 (1961).
92. B. V. L'vov, *Atomic Absorption Spectrochemical Analysis*, Adam Hilger, London, 1970.
93. H. Massmann, *Z. Anal. Chem., 225*:203 (1967).
94. H. Massmann, *Spectrochim. Acta, 23B*:215 (1968).
95. R. Woodriff and G. Ramelow, *Spectrochim. Acta, 23B*:665 (1968).
96. C. Veillon, in *Trace Analysis, Spectroscopic Methods for Elements* (J. D. Winefordner, ed.), Wiley, New York 1976, p. 155.
97. T. S. West and X. K. Williams, *Anal. Chim. Acta, 45*:27 (1969).
98. A. M. Ure and M. P. Hernandez-Artiga, *Anal. Chim. Acta, 94*:195 (1977).
99. S. R. Koirtyohann and C. Feldman, in *Developments in Applied Spectroscopy*, Vol, 3, Plenum Press, New York, 1964, p. 180.
100. G. F. Kirkbright, *Analyst (Lond.), 96*:609 (1971).
101. A. Syty, *Crit. Rev. Anal. Chem., 4*:155 (1974).
102. R. A. Woodriff, *Appl. Spectrosc., 28*:413 (1974).
103. C. W. Fuller, *Electrothermal Atomization for Atomic Absorption Spectrophotometry*, Chem. Soc., London, 1977.

104. G. M. Hieftje, T. R. Copeland, and D. R. de Olivares, *Anal. Chem.*, *48*:142R (1976).
105. J. Y. Hwang, C. J. Mokeler, and P. A. Ullucci, *Anal. Chem.*, *44*:2018 (1972).
106. J. E. Cantle and T. S. West, *Talanta*, *20*:459 (1973).
107. N. T. Faithfull, *Lab. Practice*, *26*:946 (1977).
108. N. T. Faithfull, *Lab. Practice*, *27*:25 (1978).
109. N. T. Faithfull, *Lab. Practice*, *27*:26 (1978).
110. R. E. Sturgeon, *Anal. Chem.*, *49*:1255A (1977).
111. R. E. Sturgeon and C. L. Chakrabarti, *Prog. Anal. Atomic Spectrosc.*, *1*:5 (1978).
112. B. V. L'vov, *Spectrochim. Acta*, *33B*:153 (1978).
113. J. Marshall, S. K. Giri, D. Littlejohn, and J. M. Ottaway, *Anal. Chim. Acta*, *147*:173 (1983).
114. W. Frech and S. Jonsson, *Spectrochim. Acta*, *B37*:1021 (1982).
115. J. A. Holcombe and M. T. Sheehan, *Appl. Spectrosc.*, *36*:631 (1982).
116. G. D. Renshaw, C. A. Pounds, and E. F. Pearson, *At. Absorp. Newsletter*, *12*:55 (1973).
117. B. V. L'vov and L. A. Pelieva, *Can. J. Spectrosc.*, *23*:1 (1978).
118. D. A. Segar and J. G. Gonzalez, *Anal. Chim. Acta*, *58*:7 (1972).
119. M. Binnewies and H. Schafer, *Z. Anorg. Allg. Chem.*, *395*:77 (1973).
120. A. Katz and N. Taitel, *Talanta*, *24*:132 (1977).
121. J. M. Ottaway, *Proc. Anal. Div. Chem. Soc.*, *13*:185 (1976).
122. W. Frech and A. Cedergren, *Anal. Chim. Acta*, *82*:83 (1976).
123. W. Frech and A. Cedergren, *Anal. Chim. Acta*, *82*:93 (1976).
124. W. Frech and A. Cedergren, *Anal. Chim. Acta*, *88*:57 (1977).
125. J. A. Persson, W. Frech, and A. Cedergren, *Anal. Chim. Acta*, *92*:85 (1977).
126. J. A. Persson, W. Frech, and A. Cedergren, *Anal. Chim. Acta*, *92*:95 (1977).
127. K. Johansson, W. Frech, and A. Cedergren, *Anal. Chim. Acta*, *94*:245 (1977).
128. C. W. Fuller, *At. Absorp. Newsletter*, *16*:106 (1977).
129. J. F. Alder and D. A. Hickman, *At. Absorp. Newsletter*, *16*:110 (1977).
130. M. D. Amos, P. A. Bennett, K. G. Brodie, P. W. Y. Lung, and J. P. Matousek, *Anal. Chem.*, *43*:211 (1971).
131. T. C. Dymott, M. P. Wassall, and P. J. Whiteside, *Analyst (Lond.)*, *110*:467 (1985).
132. C. A. Shand and A. M. Ure, *J. Anal. At. Spectrosc.*, *2*:143 (1987).

133. H. L. Kahn, G. E. Peterson, and J. E. Schallis, *At. Absorp. Newsletter, 7*:35 (1968)
134. H. T. Delves, *Analyst (Lond.), 95*:431 (1970).
135. B. Welz, *Atomic Absorption Spectroscopy*, English ed., Verlag Chemie, Weinheim, 1976, p. 42.
136. C. Lau, A. Held, and R. Stephens, *Can. J. Spectrosc., 21*: 100 (1976).
137. J. Khalighie, A. M. Ure, and T. S. West, *Anal. Chim. Acta, 107*:191 (1979).
138. C. M. Lau, A. M. Ure, and T. S. West, *Anal. Chim. Acta. 146*:171 (1983).
139. C. M. Lau, A. M. Ure, and T. S. West, *Anal. Proc., 20*:114 (1983).
140. S. M. Fraser, A. M. Ure, M. C. Mitchell, and T. S. West, *J. Anal. Atom. Spectrom., 1*:19 (1988).
141. C. Hallam and K. C. Thompson, *Analyst (Lond.), 110*:497 (1985).
142. A. M. Ure and M. L. Berrow, *Anal. Chim. Acta, 52*:247 (1970).
143. R. J. Watling, *Anal. Chim. Acta, 94*:181 (1977).
144. W. Holak, *Anal. Chem., 41*:1712 (1969).
145. E. F. Dalton and A. J. Malanoski, *At. Absorp. Newsletter, 10*:92 (1971).
146. F. J. Fernandez and D. C. Manning, *At. Absorp. Newsletter, 10*:86 (1971).
147. D. C. Manning, *At. Absorp. Newsletter, 10*:123 (1971).
148. E. N. Pollock and S. J. West, *At. Absorp. Newsletter 11*:104 (1972).
149. F. J. Schmidt and J. L. Royer, *Anal. Lett., 6*:17 (1973).
150. J. R. Knechtel and J. L. Fraser, *Analyst (Lond.), 103*:104 (1978).
151. Y. Yamamoto, T. Kumamaru, Y. Hayashi, and M. Kanke, *Anal. Lett., 5*:717 (1972).
152. M. Bedard and J. D. Kerbyson, *Can. J. Spectrosc., 21*:64 (1975).
153. J. Guimont, M. Pichette, and N. Rheaume, *At. Absorp. Newsletter, 16*:53 (1977).
154. G. F. Kirkbright and M. Taddia, *Anal. Chim. Acta, 100*:145 (1978).
155. H. K. Kang and J. L. Valentine, *Anal. Chem., 49*:1829 (1977).
156. W. B. Robbins and J. A. Caruso, *Anal. Chem., 51*:889A (1979).
157. K. S. Subramanian, *Fres. Z. Anal. Chem., 305*:390 (1981).
158. A. L. Dennis and D. G. Porter, *J. Automatic Chem., 2*:134 (1980).
159. D. Astrom, *Anal. Chem., 54*:190 (1982).

160. D. Tsalev and I. Petrov, *Dokl. Bolg. Akad. Nauk*, *34*:1413 (1981).
161. K. Braun, W. Slavin, and A. Walsh, *Spectrochim. Acta*, *B37*: 721 (1982).
162. K. Kuga and K. Tsuju, *Anal. Lett.*, *15*:47 (1982).
163. A. W. Gunn, Tech. Rept. TR169, W. R. C. Environmental Protection, Marlow, England, Oct. 1981.
164. W. R. Hatch and W. L. Ott, *Anal. Chem.*, *40*:2085 (1968).
165. Y. Kimura and V. L. Miller, *Anal. Chim. Acta*, *27*:325 (1962).
166. N. S. Poluektov and R. A. Vitkun, *Zh. Anal. Khim.*, *18*:37 (1963).
167. W. W. Vaughan and J. H. McCarthy, Jr., *U.S. Geol. Surv. Prof. Paper 510D*, 123 (1964).
168. S. H. Williston and M. H. Morris, U.S. Patent No. 3,173,016 (1965).
169. S. Chilov, *Talanta*, *22*:205 (1975).
170. A. M. Ure, *Anal. Chim. Acta*, *76*:1 (1975).
171. M. E. Barnes (ed.), *Mercury Contamination in the Natural Environment*, PB 192910, U.S. Dept. of the Interior, Washington, D.C., 1970.
172. D. Taylor, *Mercury as an Environmental Pollutant, A Bibliography*, 3rd ed., ICI Brixham Lab., England, 1973.
173. E. A. Jenne and W. Sanders, *J. Water Pollut. Contr. Fed.*, *45*:1952 (1973).
174. W. S. Zaugg and R. J. Knox, *Anal. Chem.*, *38*:1759 (1966).
175. W. S. Zaugg and R. J. Knox, *Anal. Biochem.*, *20*:282 (1967).
176. T. R. Hurford and D. F. Boltz, *Anal. Chem.*, *40*:379 (1968).
177. S. S. Michael, *Anal. Chem.*, *49*:451 (1977).
178. C. Riddle and A. Turek, *Anal. Chim. Acta*, *92*:49 (1977).
179. G. D. Christian and F. J. Feldman, *Anal. Chim. Acta*, *40*:173 (1968).
180. M. Kunishi and S. Ohno, *At. Absorp. Newsletter*, *13*:29 (1974).
181. R. H. Loeppert and H. L. Breland, *Proc. Soil Crop. Sci. Soc. Florida*, *32*:145 (1973).
182. O. C. Bataglia, *Cienc. Cult. (Sao Paulo)*, *28*:672 (1976).
183. C. C. Fongi and C. O. Huber, *Spectrochim. Acta*, *31B*:113 (1976).
184. R. M. Dagnall, K. C. Thompson, and T. S. West, *Analyst (Lond.)*, *92*:506 (1967).
185. K. M. Aldous, R. M. Dagnall, and T. S. West, *Analyst (Lond.)*, *95*:417 (1970).
186. R. M. Dagnall, K. C. Thompson, and T. S. West, *Analyst (Lond.)*, *93*:72 (1968).
187. R. M. Dagnall, K. C. Thompson, and T. S. West, *Analyst (Lond.)*, *94*:643 (1969).

188. R. Belcher, *Z. Anal. Chem.*, *263*:257 (1973).
189. R. Belcher, S. L. Bogdanski, and A. Townshend, *Anal. Chim. Acta*, *67*:1 (1973).
190. R. Belcher, S. A. Ghonaim, and A. Townshend, *Anal. Chim. Acta*, *71*:255 (1974).
191. M. Q. Al-Abachi, R. Belcher, S. L. Bogdanski, and A. Townshend, *Anal. Chim. Acta*, *86*:139 (1976).
192. A. Townshend, *Proc. Anal. Div. Chem. Soc.*, *13*:64 (1976).
193. B. Kolb, G. Kemmner, F. H. Schleser, and E. Wiedeking, *Z. Anal. Chem.*, *221*:166 (1966).
194. J. C. Van Loon, B. Radziuk, N. Kahn, J. Lichwa, F. J. Fernandez, and J. D. Kerber, *At. Absorp. Newsletter*, *16*: 79 (1977).
195. S. E. Manahan and D. R. Jones IV, *Anal. Lett.*, *6*:745 (1973).
196. F. J. Fernandez, *At. Absorp. Newsletter*, *16*:33 (1977).
197. G. E. Parris, W. R. Blair, and F. G. Brinckman, *Anal. Chem.*, *49*:378 (1977).
198. W. J. Price, *Chem. in Britain*, *14*:140 (1978).
199. W. Slavin, *Atomic Absorption Spectroscopy*, Wiley Interscience, New York, 1968, p. 16.
200. M. S. Cresser, *Anal. Chim. Acta*, *80*:170 (1975).
201. D. C. Manning, *At. Absorp. Newsletter*, *14*:99 (1975).
202. P. D. Goulden, *At. Absorp. Newsletter*, *16*:121 (1977).
203. G. A. Eagle and M. J. Orren, *At. Absorp. Newsletter*, *16*: 151 (1977).
204. W. J. Simmons and L. A. Plues-Foster, *Aust. J. Soil Res.*, *15*:171 (1977).
205. E. E. Pickett and M. L. Franklin, *J. Assoc. Off. Anal. Chem.*, *60*:1164 (1977).
206. D. D. Siemer, *Anal. Chem.*, *54*:1321 (1982).
207. D. Marquardt, R. Stösser, and G. Henrion, *Spectrochim. Acta*, *B36*:943 (1981).
208. C. P. Lloyd-Jones, G. A. Hudd, and D. G. Hill-Cottingham, *Analyst (Lond.)*, *99*:580 (1974).
209. E. J. Skerrett and C. P. Lloyd-Jones, *Analyst (Lond.)*, *102*: 969 (1977).
210. H. Ramage, *Nature (Lond.)*, *123*:601 (1929).
211. K. Govindaraju, J. Morel, and N. L'Homel, *Geostandards Newsletter*, *1*:137 (1977).
212. V. A. Razumov, *Zh. Prickl. Spek*, *24*:1117 (1976).
213. F. J. Langmyhr, R. Solberg, and Y. Thomassen, *Anal. Chim. Acta*, *92*:105 (1977).
214. K. Fujiwara, Y. Umezawa, Y. Numata, F. Fuwa, and S. Fujiwara, *Bunseki Kagaku*, *26*:735 (1977).
215. H. U. Meisch and W. Reinle, *Mikrochim. Acta*, *1*:505 (1977).

216. G. H. Morrison and H. Freiser, *Solvent Extraction in Analytical Chemistry*, Wiley, New York, 1957.
217. M. S. Cresser, *Solvent Extraction in Flame Spectroscopic Analysis*, Butterworths, London, 1978.
218. H. Malissa and E. Schoffman, *Mikrochim. Acta, 1*:187 (1955).
219. E. Lakanen, *At. Absorp. Newsletter 5*:17 (1966).
220. Yu. A. Zolotov, *Pure Appl. Chem., 50*:129 (1978).
221. L. Shapiro and W. W. Brannock, *Rapid Analysis of Silicate Rocks*, U.S. Geol. Survey Bull. 1036-C, 1956.
222. M. L. Jackson, *Soil Chemical Analysis*, Prentice-Hall, Englewood Cliffs, N.J., 1958.
223. D. J. Trent and W. Slavin, *At. Absorp. Newsletter 3*:118 (1964).
224. D. J. Trent and W. Slavin, *At. Absorp. Newsletter*, No. *19*:1 (1964).
225. C. B. Belt, Jr., *Econ. Geol., 59*:240 (1964).
226. I. Rubeska and B. Moldan, *Acta. Chim. Acad. Sci. Hung., 44*:367 (1965).
227. L. R. Hossner and L. W. Ferrara, *At. Absorp. Newsletter, 6*:71 (1967).
228. R. A. Isaac and J. D. Kerber, in *Instrumental Methods for the Analysis of Soils and Plant Tissue* (L. M. Walsh, ed.), Soil Sci. Soc. Am., Madison, Wis., 1971, p. 17.
229. E. J. Perkins, J. R. S. Gilchrist, O. J. Abbot, and W. Halcrow, *Mar. Pollut. Bull., 4*:59 (1973).
230. J. Bolton, *J. Sci. Fd. Agric., 24*:727 (1973).
231. H. Erlenkeuser, E. Suess, and H. Willkomm, *Geochim. Cosmochim. Acta, 38*:823 (1974).
232. V. I. Shcherbakov, A. V. Karyakin, L. N. Bannykh, and V. I. Lebedev, *Zavod. Lab., 43*:957 (1977).
233. E. E. Angino and G. K. Billings, *Atomic Absorption Spectrometry in Geology*, Elsevier, Amsterdam, 1067, p. 0C.
234. I. A. Voinovitch, *Bull. Liaison Lab. Points Chaussees, 79*:81 (1975).
235. B. Bernas, *Anal. Chem., 40*:1682 (1968).
236. F. J. Langmyhr and P. E. Paus, *Anal. Chim. Acta, 43*:397 (1968).
237. W. J. Price and P. J. Whiteside, *Analyst (Lond.), 102*:664 (1977).
238. L. A. Lerner, L. P. Orlova, and D. N. Ivanov, *Agrokhimiya, 5*:138 (1971).
239. H. Agemian and A. S. Y. Chau, *Anal. Chim. Acta, 80*:61 (1975).
240. R. J. Guest, D. R. MacPherson, and R. J. Pugliese, *Anal. Chim. Acta, 96*:185 (1978).

241. D. E. Buckley and R. E. Cranston, *Chem. Geol.*, *7*:273 (1971).
242. L. A. Lerner, E. I. Tikhomirova, and L. F. Plotnikova, *Pochvovedenie*, *1*:122 (1975).
243. C. O. Ingamells, *Talanta*, *11*:665 (1964).
244. C. O. Ingamells, *Anal. Chem.*, *38*:1228 (1966).
245. N. H. Suhr and C. O. Ingamells, *Anal. Chem.*, *38*:730 (1966).
246. H. Bennett, *Analyst (Lond.)*, *102*:153 (1977).
247. E. Jeanroy, *Analusis*, *2*:703 (1973-74).
248. R. Righi and F. De Coninck, *Geoderma*, *19*:339 (1977).
249. J. C. Van Loon and C. M. Parissis, *Analyst (Lond.)*, *94*: 1057 (1969).
250. P. L. Boar and L. K. Ingram, *Analyst (Lond.)*, *95*:124 (1970).
251. A. A. Verbeek, M. C. Mitchell, and A. M. Ure, *Anal. Chim. Acta*, *135*:215 (1982).
252. P. Hannaker and T. C. Hughes, *Anal. Chem.*, *49*:1485 (1977).
253. Department of the Environment, *Analysis of Raw, Potable and Waste Waters*, HMSO, London, 1972.
254. L. Wiklander and K. Vahtras, *Geoderma*, *19*:123 (1977).
255. A. Rosopulo and W. Scholl, *Landwirtsch. Forsch.*, *31*:74 (1978).
256. R. Knechtel, K. Conn, and J. Fraser, *The Analysis of Chemical Digester Sludges for Metals by Several Laboratory Groups*, Tech. Dev. Rept. EPS4-WP-78-1, Water Pollution Control Directorate, Environmental Protection Service, Ottawa, 1978.
257. P. J. Ballinger, *Science & Industry (Philips)*, *8*:12 (1976).
258. J. C. Van Loon, J. Lichwa, D. Ruttan, and J. Kinrade, *Water Air Soil Pollut.*, *2*:473 (1973).
259. J. C. Van Loon, *Heavy Metals in Agricultural Lands Receiving Chemical Sewage, Sludges IV, Analytical Methods for Sewage Sludge Analysis*, Environmental Protection Service, Ottawa, 1976.
260. J. N. Lester, R. M. Harrison, and R. Perry, *Sci. Total Environ.*, *8*:153 (1977).
261. S. Stoveland, M. Astruc, R. Perry, and J. N. Lester, *Sci. Total Environ.*, *9*:263 (1978).
262. R. L. Mitchell, *Geol. Soc. Am. Bull.*, *83*:1069 (1972).
263. F. R. Cox and E. J. Kamprath, in *Micronutrients in Agriculture* (J. J. Morvedt, P. M. Giordano, and W. L. Lindsay, eds.), Soil Sci. Soc. Am., Madison, Wis., 1972, p. 289.
264. C. S. Piper, *Soil and Plant Analysis*, Wiley-Interscience, New York, 1944.
265. P. R. Hesse, *A Textbook of Soil Chemical Analysis*, Murray, London, 1971.

266. H. D. Chapman (ed.), *Diagnostic Criteria for Plants and Soils*, Univ. Calif., Div. of Agric. Sciences, 1966.
267. S. G. Dolar and D. R. Keeney, *J. Sci. Fd. Agric., 22*:273 (1971).
268. S. G. Dolar and D. R. Keeney, *J. Sci. Fd. Agric., 22*:279 (1971).
269. S. G. Dolar, D. R. Keeney, and L. M. Walsh, *J. Sci. Fd. Agric., 22*:282 (1971).
270. O. K. Borggaard, *Acta. Agric. Scand., 26*:144 (1976).
271. W. L. Lindsay and W. A. Norvell, *Soil Sci. Soc. Am. J., 42*: 421 (1978).
272. R. O. Scott, R. L. Mitchell, D. Purves, and R. C. Voss, *Spectrochemical Methods for the Analysis of Soils, Plants and Other Agricultural Materials*, Consultative Committee for the Development of Spectrochemical Work (The Macaulay Institute for Soil Research, Aberdeen), Bulletin 2, 1971.
273. Ministry of Agriculture, Fisheries and Food, *The Analysis of Agricultural Materials*, Tech. Bull. 27, HMSO, London, 1973.
274. R. W. Lahann, *J. Environ. Sci. Health, A11*:639 (1976).
275. H. Agemian and A. S. Y. Chau, *Analyst (Lond.), 101*:761 (1976).
276. R. L. Mitchell, *Spectrochim. Acta, 4*:62 (1950).
277. A. M. Ure, *Brit. Commun. Electronics, 5*:846 (1958).
278. Y. P. Kalra, *Applications of Atomic Absorption Spectrophotometry in Soil, Plant, Water and Fertilizer Analysis: An Annotated Bibliography*, Part I, File Rept. 6935-2, Northern Forest Res. Centre, Canadian Forest Service, Alberta, 1977.
279. Y. Nakamura, H. Nagai, D. Kubota, and S. Himeno, *Bunseki Kagaku, 22*:1543 (1973).
280. M. L. Kokot, *At. Absorp. Newsletter, 15*:105 (1976).
281. L. Ebdon, J. R. Wilkinson, and K. W. Jackson, *Anal. Chim. Acta, 136*:191 (1982).
282. P. N. Vijan, A. C. Rayner, D. Sturgis, and G. R. Wood, *Anal. Chim. Acta, 82*:329 (1976).
283. R. G. Smith, J. C. Van Loon, J. R. Knechtel, J. L. Fraser, A. E. Pitts, and A. E. Hodges, *Anal. Chim. Acta, 93*:61 (1977).
284. J. Hubert, R. M. Candelaria, and H. G. Applegate, *At. Spectrosc., 1*:90 (1980).
285. H. Agemian and E. Bedek, *Anal. Chim. Acta, 119*:323 (1980).
286. H. Bruene and R. Ellinghaus, *Landwirtsch. Forsch., Sonderh., 38*:338 (1982).
287. T. J. Forehand, A. E. Dupuy, and H. Tai, *Anal. Chem., 48*: 999 (1976).
288. F. D. Pierce and H. R. Brown, *Anal. Chem., 49*:1417 (1977).

289. J. R. Melton, W. L. Hoover, J. L. Ayers, and P. A. Howard, *Soil Sci. Soc. Am. Proc.*, *37*:558 (1973).
290. R. M. Brown, C. J. Pickford, and W. L. Davison, *Int. J. Environ. Anal. Chem.*, *18*:135 (1984).
291. E. Bolibrzuch, Mater. Konwersatorium Spektrom. At. Emisyjenj, Absorpc., Spektrom. Mas. 120th 157 (1983).
292. S. Kempton, R. M. Sterritt, and J. N. Lester, *Talanta*, *29*: 675 (1982).
293. K. G. Brodie, *International Lab.*, Sept./Oct., 65 (1977).
294. I. Petrov and D. Tsalev, *Pochvozn. Agrokhim.*, *14*:20 (1979).
295. S. Constantini, R. Giordano, and P. Ravagnan, *Ann. 1st Super Sanita*, *16*:287 (1980).
296. M. A. Lovell and J. G. Farmer, *Int. J. Environ. Anal. Chem.*, *14*:181 (1983).
297. E. A. Woolson, N. Akaronson, and R. Iadevaia, *J. Agric. Fd. Chem.*, *30*:580 (1982).
298. T. Takamatsu, H. Aoiki, and T. Yoshida, *Soil Sci.*, *133*:239 (1982).
299. E. A. Woolson, *Proc. Int. 5th Congr. Pestic. Chem.*, Vol. 4, p. 79, 1983.
300. X. Shan, Z. Ni, and L. Zhang, *Anal. Chim. Acta*, *151*:179 (1983).
301. D. J. Nicolas, *Anal. Chim. Acta*, *55*:59 (1971).
302. P. N. Vijan, *Am. Lab.*, No. *8*:32 (1979).
303. B. J. A. Haring, W. Van Delft, and C. M. Bom, *Eres. Z. Anal. Chem.*, *310*:217 (1982).
304. T. Asam, *Nippon Dojo-Hiryogaku Zasshi*, *46*:421 (1975).
305. M. M. Ashry, *Geochim. Cosmochim. Acta*, *37*:2449 (1973).
306. E. E. Pickett and J. C. M. Pau, *J. Ass. Off. Anal. Chem.*, *56*:151 (1973).
307. F. Thevenot and J. Cueilleron, *Analusis*, *5*:105 (1977).
308. S. Inoue and M. Sasaki, *Bunseki Kagaku*, *31*:E127 (1982).
309. M. K. John, H. H. Chuah, and C. J. Van Laerhoven, *Environ. Sci. Technol.*, *6*:555 (1972).
310. S. Yamasaki, *Nippon Dojo-Hiryogaku Zasshi*, *44*:383 (1973).
311. G. Tyler, B. Mornsjo, and B. Nilsson, *Plant Soil*, *40*:237 (1974).
312. N. I. Ward, R. R. Brooks, and E. Roberts, *N.Z. J. Sci.*, *20*:413 (1977).
313. C. H. Williams and D. J. David, *Soil Sci.*, *121*:86 (1976).
314. V. M. Byr'ko and V. S. Gorbatov, *Vestn. Mosk. Univ. Ser. 17 Pochvoved*, *3*:82 (1981).
315. Y. Takijima and F. Katsumi, *Soil Sci. Plant Nutr.* (Tokyo), *19*:29 (1973).
316. M. J. Dudas, *At. Absorp. Newsletter*, *13*:109 (1974).
317. W. Schmidt and F. Dietl, *Z. Anal. Chem.*, *295*:110 (1979).

318. J. Wen and T. Geng, *Fenxi Huaxue 9*:565 (1981).
319. A. Brashnarova and L. Stanislavova, *Pchvozn. Agrokhim.*, *18*:10 (1983).
320. A. M. Aziz-Alrahaman, M. A. Al-Hajjaji, and I. Z. Al-Zamil, *Int. J. Environ. Anal. Chem.*, *15*:9 (1983).
321. A. H. C. Roberts, M. A. Turner, and J. K. Syers, *Analyst (Lond.)*, *101*:574 (1976).
322. V. N. Oreshkin, Y. I. Belyaev, G. L. Vnukovskaya, and Y. G. Tatzil, *Pochvovedenie*, No. *5*:109 (1979).
323. I. Petrov and D. Isalev, *Dokl. Bolg. Akad. Nauk.*, *35*:467 (1982).
324. K. W. Jackson and A. P. Newman, *Analyst (Lond.)*, *108*:261 (1983).
325. A. Weitz, G. Fuchs, and K. Bachman, *Fres. Z. Anal. Chem.*, *313*:38 (1982).
326. H. L. Kahn, F. J. Fernandez, and S. Slavin, *At. Absorp. Newsletter*, *11*:42 (1972).
327. A. M. Ure and M. C. Mitchell, *Anal. Chim. Acta*, *87*:283 (1976).
328. A. M. Ure, M. P. Hernandez-Artiga, and M. C. Mitchell, *Anal. Chim. Acta*, *97*:37 (1978).
329. C. M. Lau, A. M. Ure, and T. S. West, *Anal. Chim. Acta*, *146*:171 (1983).
330. C. M. Lau, A. M. Ure, and T. S. West, *Anal. Proc.*, *20*:114 (1983).
331. D. R. Sauerbeck and P. Styperek, "Processing and Use of Sewage Sludge," *Proc. 3rd Int. Symp.*, Brighton, 1983, D. Reidel, Dordrecht, 1984, p. 431.
332. E. Steinnes, *Talanta*, *24*:121 (1977).
333. C. J. Ritter, *Int. Lab.*, Oct. 30 (1982).
334. S. P. McGrath, *Analyst (Lond.)*, *111*:459 (1986).
335. L. A. Lerner, L. F. Plotnikova, and E. I. Tikhomirova, *Pochvovedenie*, *1*:126 (1976).
336. N. Baumslag and P. Keen, *Arch. Environ. Health*, *25*:23 (1972).
337. R. Nakagawa and Y. Ohyagi, *Nippon Kagaku Kaishi*, *12*:2331 (1974).
338. K. Aoba and K. Sekiya, *Engei Shikenjo Hokoku, Ser. A*, *12*: 79 (1973).
339. R. C. P. Sinha and B. K. Banerjee, *Technol. (Sindri, India)*, *11*:263 (1974).
340. B. G. Oliver, *Environ. Sci. Technol.*, *7*:135 (1973).
341. A. M. Ure and R. L. Mitchell, *Spectrochim. Acta*, *23B*:79 (1967).
342. A. Sapek, *Chem. Anal. (Warsaw)*, *19*:687 (1974).
343. M. C. Lucas and E. M. De Sequeira, *Agron. Lusit.*, *41*:113 (1981).

344. A. I. Busev, V. M. Byrko, L. A. Lerner, and V. I. Migunova, *Zh. Analit. Khim.*, *27*:607 (1972).
345. M. J. Dudas, *At. Absorp. Newsletter, 13*:67 (1974).
346. E. D. Prudnikov, *Zh. Prikl. Spektrosk.*, *33*:804 (1980).
347. M. C. Mitchell, M. L. Berrow, and C. A. Shand, *J. Anal. Atom. Spectrom.*, *2*:261 (1987).
348. J. Masui, S. Shoji, and K. Minami, *Soil Sci. Plant Nutr. (Tokyo)*, *18*:31 (1972).
349. F. D. Macias, *Soil Sci.*, *115*:276 (1973).
350. H. Karim, J. E. Sedberry, and B. J. Miller, *Commun. Soil Sci. Plant Anal.*, *7*:437 (1976).
351. D. E. Pride, *J. Geochem. Explor.*, *7*:361 (1977).
352. R. H. Scott and M. L. Kokot, *Anal. Chim. Acta, 75*:257 (1975).
353. L. M. Shuman, *Soil Sci.*, *127*:10 (1979).
354. R. Horton and J. J. Lynch, Geol. Surv. Can., Paper 75-1, Pt. A, 1975, p. 213.
355. B. D. Balraadjsing, *Inst. Bodemvruchtbaarheid Rapp.*, No. 12 (1972).
356. A. Sapek and B. Sapek, *Chem. Anal. (Warsaw)*, *17*:339 (1972).
357. R. E. Truitt and J. H. Weber, *Anal. Chem.*, *53*:337 (1981).
358. W. H. Ko and F. K. Hora, *Soil Sci.*, *113*:42 (1972).
359. S. J. Locascio, J. G. A. Fiskell, and F. G. Martin, *J. Am. Soc. Hort. Sci.*, *97*:119 (1972).
360. A. Oeien and K. Gjerdingen, *Acta. Agric. Scand.*, *22*:173 (1972).
361. M. Facchinetti, R. L. Grassi, and A. L. Diez, *Agrochimica*, *17*:413 (1973).
362. B. R. Singh and K. Steenberg, *Plant Soil, 40*:637 (1974).
363. M. E. Varju and E. Elek, *At. Absorp. Newsletter 10*:128 (1971).
364. A. Olumuyina, E. E. Osiname, F. E. Schulte, and R. B. Corey, *J. Sci. Fd. Agric.*, *24*:1341 (1973).
365. R. G. McLaren and D. V. Crawford, *J. Soil Sci.*, *24*:172 (1973).
366. G. W. Wallingford, L. S. Murphy, W. L. Powers, and H. L. Manges, *Soil Sci. Soc. Am. Proc.*, *39*:482 (1975).
367. S. S. Iyengar, D. C. Martens, and W. P. Miller, *Soil Sci.*, *131*:95 (1981).
368. T. T. Chao and R. F. Sanzolone, *J. Res. U.S. Geol. Surv.*, *1*:681 (1973).
369. Zs. Horvath, K. Falb, and M. Varju, *At. Absorp. Newsletter*, *16*:152 (1977).
370. D. J. Linehan, A. H. Sinclair, and M. C. Mitchell, *Plant Soil*, *86*:147 (1985).
371. J. L. Seeley, D. Dick, J. H. Arvik, R. L. Zimdahl, and R. K. Skogerboe, *Appl. Spectrosc.*, *26*:456 (1972).

372. D. D. Hemphill, C. J. Marienfeld, R. S. Reddy, and J. O. Pierce, *Arch. Environ. Health*, *28*:190 (1974).
373. N. I. Ward, R. D. Reeves, and R. R. Brooks, *N.Z. J. Sci.*, *18*:261 (1975).
374. N. I. Ward, R. R. Brooks, and R. D. Reeves, *N.Z. J. Sci.*, *19*:81 (1976).
375. L. Faitondzhiev and A. Brashnarova, *Landwirtsch. Forsch.*, *32*:1 (1979).
376. E. A. Crecelius and D. Z. Piper, *Environ. Sci. Technol.*, *7*: 1053 (1973).
377. D. Briggs, *Nature (Lond.)*, *238*:166 (1972).
378. F. Beavington, *Aust. J. Soil Res.*, *11*:27 (1973).
379. L. D. Jordan and D. J. Hogan, *N.Z. J. Sci.*, *18*:253 (1975).
380. J. V. Lagerwerff and D. L. Brower, *Soil Sci. Soc. Am. Proc.*, *37*:11 (1973).
381. W. Schmidt and F. Dietl, *Fres'. Z. Anal. Chem.*, *315*:687 (1983).
382. S. Ravikovitch and J. Navrot, *Soil Sci.*, *121*:25 (1976).
383. D. H, Yaalon, C. Jungreis, and H. Koyumdjisky, *Geoderma*, *7*:71 (1972).
384. M. S. Cresser and R. Hargitt, *Anal. Chim. Acta*, *82*:203 (1976).
385. R. J. Bartlett and C. J. Picarelli, *Soil Sci.*, *116*:77 (1973).
386. A. Siman, F. W. Cradock, and A. W. Hudson, *Plant Soil*, *41*:129 (1974).
387. R. F. Farley and A. P. Draycott, *Plant Soil*, *38*:235 (1973).
388. O. G. Oniani and N. V. Egorashvili, *Agrokhimiya*, *9*:137 (1974).
389. L. B. Grass, A. J. Mackenzie, B. D. Meek, and W. F. Spencer, *Soil Sci. Soc. Am. Proc.*, *37*:14 (1973).
390. A. Kadow, E. Rabban, and I. Rubeska, Inst. Appl. Res. Natural Resource Abu-Ghraib Iraq Tech. Rept. 2, 1970.
391. D. S. Cronan, *Can. J. Earth Sci.*, *9*:319 (1972).
392. F. J. Dewis, A. A. Levinson, and P. Bagliss, *Geochim. Cosmochim. Acta*, *36*:1359 (1972).
393. P. B. Hoyt and M. D. Webber, *Can. J. Soil Sci.*, *54*:53 (1974).
394. U. K. Misra, R. K. Blanchar, and W. J. Upchurch, *Soil Sci. Soc. Am. Proc.*, *38*:897 (1974).
395. R. J. Gibbs, *Sci.*, *180*:73 (1973).
396. S. K. Yoon, J. T. Gilmour, and B. R. Wells, *Soil Sci. Soc. Am. Proc.*, *39*:685 (1975).
397. K. G. Van Ettekoven, H_2O, *13*:326 (1980).
398. M. Goto, T. Shibakawa, T. Arita, and D. Ishii, *Anal. Chem. Acta*, *140*:179 (1982).
399. H. Besson, A. Bernard, and C. Ducauze, *C.R. Seances Acad. Agric. Fr.*, *67*:51 (1981).

400. J. Welsh and J. Frost, *Anal. Newsletter*, *1*:22 (1982).
401. K. Graeser and K. Staeger, *Z. Gesamte Hyg. Ihre. Grenzgeb.*, *29*:734 (1983).
402. H. Lang, Y. Liu, and Z. Cheng, *Huanjing Kexue*, *3*:61 (1983).
403. Z. Ni and F. Yang, *Huanjung Huoxue*, *1*:83 (1982).
404. G. Tosi, E. Desimoni, F. Palmisano, and L. Sabbatini, *Analyst (Lond.)*, *107*:96 (1982).
405. D. A. Frick and D. F. Tallman, *Anal. Chem.*, *54*:1217 (1982).
406. R. D. Rogers, *Soil Sci. Soc. Am. J.*, *43*:289 (1979).
407. C. H. Kim, C. M. Owens, and L. E. Smythe, *Talanta*, *21*:445 (1974).
408. J. C. Van Loon, *At. Absorp. Newsletter*, *11*:60 (1972).
409. D. Hutchinson, *Analyst (Lond.)*, *97*:118 (1972).
410. C. H. Kim, P. W. Alexander, and L. E. Smythe, *Talanta*, *23*:229 (1976).
411. P. Dharma Rao, *At. Absorp. Newsletter*, *10*:118 (1971).
412. H. F. Massey, *Soil Sci.*, *114*:217 (1972).
413. A. J. Anderson, D. R. Meyer, and F. K. Mayer, *Aust. J. Agric. Res.*, *24*:557 (1973).
414. A. R. Memon, S. I. To, and M. Yatazawa, *Soil Sci. Plant Nutr.*, *26*:271 (1980).
415. E. Haenni and R. C. Daniel, *Mitt. Geb. Lebensmittelunters. Hyg.*, *73*:94 (1982).
416. L. W. Murdock and C. I. Rich, *Soil Sci. Soc. Am. Proc.*, *36*:167 (1972).
417. A. A. R. Hafey and P. P. Stout, *Soil Sci. Soc. Am. Proc.*, *37*:572 (1973).
418. H. Agemian and E. Bedek, *Anal. Chim. Acta*, *119*:223 (1980).
419. X. Q. Shan, L. Z. Jin, and Z. M. Ni, *At. Spectros.*, *3*:41 (1982).
420. C. A. Shand and A. M. Ure, *J. Anal. Atom. Spectrom.*, *2*:143 (1987).
421. H. B. Han, G. Kaiser, and G. Tolg, *Anal. Chim. Acta*, *126*:9 (1981).
422. G. Knapp, S. E. Raptis, G. Kaiser, G. Tolg, P. Schranell, and B. Feiser, *Res. Z. Chem.*, *308*:197 (1981).
423. J. Azad, G. F. Kirkbright, and R. D. Snook, *Analyst (Lond.)*, *104*:232 (1979).
424. T. T. Chao, J. W. Ball, and J. M. Nakagawa, *Anal. Chim. Acta*, *54*:77 (1971).
425. K. C. Jones, P. J. Peterson, and B. E. Davies, *Geoderma*, *33*:157 (1984).
426. J. M. Soileau, *Agron. J.*, *65*:625 (1973).
427. J. V. Lagerwerff and W. D. Kemper, *Soil Sci. Soc. Am. Proc.*, *39*:1077 (1975).
428. H. Han and Z. Ni, *Huanjing Huaxue*, *2*:44 (1983).

429. D. Gladwell, M. Thompson, and S. J. Wood, *J. Geochem. Explor.*, *16*:41 (1981).
430. L. Jin and Z. Ni, *Huanjung Huaxue*, *1*:281 (1982).
431. M. K. John, *Soil Sci.*, *113*:222 (1972).
432. M. L. Harimho and K. Igue, *Agron. J.*, *64*:3 (1972).
433. B. D. Balraadjsing, Inst. Bodemvruchtbaarheid Rapp., No. 11, 1972.
434. D. D. Siemer and R. W. Stone, *Appl. Spectrosc.*, *29*:240 (1975).
435. D. D. Warncke and S. Barbir, *Soil Sci. Soc. Am. Proc.*, *36*:39 (1972).
436. M. P. Ireland, *J. Soil Sci.*, *26*:313 (1975).
437. A. S. G. Jones, *Mar. Geol.*, *14*:M1 (1973).
438. G. W. Bryan and L. G. Hummerstone, *J. Mar. Biol. Ass. U.K.*, *53*:839 (1973).
439. J. D. Norris and T. S. West, *Anal. Chim. Acta*, *71*:289 (1974).
440. K. G. Tiller, M. P. C. De Vries, and J. L. Honeysett, *Aust. J. Soil Res.*, *10*:151 (1972).
441. G. A. Selevtsova, L. A. Volodina, M. A. Knyazeva, and I. A. Yanchuk, *Khim. Sel'sk. Khoz.*, *13*:755 (1975).
442. D. Purves and E. J. Mackenzie, *Plant Soil*, *39*:361 (1973).
443. W. H. McKee, *Soil Sci. Soc. Am. J.*, *40*:586 (1976).
444. H. Nishita and R. M. Haug, *Soil Sci.*, *118*:421 (1974).
445. J. G. A. Fiskell and D. V. Calvert, *Soil Sci.*, *120*:132 (1975).
446. U. K. Misra, W. J. Upchurch, and C. E. Marshall, *Soil Sci.*, *121*:323 (1976).
447. M. L. Tsap, Z. V. Proskura, S. P. Sheredko, and N. S. Gedz, *Agrokhimiya*, 5:150 (1976).
448. L. F. Elliott, T. A. Travis, and T. M. McCalla, *Soil Sci. Soc. Am. J.*, *40*:513 (1976).
449. J. R. Jorgenson and C. G. Wells, *Plant Soil*, [illegible] ([illegible]).
450. S. Payette and P. Morisset, *Soil Sci.*, *117*:352 (1974).
451. R. M. Carlson and J. R. Buchanan, *Soil Sci. Soc. Am. Proc.*, *38*:113 (1974).
452. E. A. Jenne, J. W. Ball, and C. Simpson, *J. Environ. Qual.*, *3*:281 (1974).
453. P. S. Sidhu, J. L. Seghal, M. K. Sinha, and N. S. Randhawa, *Geoderma*, *18*:241 (1977).
454. M. D. Webber, P. B. Hoyt, M. Nyborg, and D. Corneau, *Can. J. Soil Sci.*, *57*:361 (1977).
455. J. L. McIntosh and K. E. Varney, *Agron. J.*, *65*:629 (1973).
456. P. L. Searle and B. K. Daley, *Geoderma*, *19*:1 (1977).
457. M. J. Dudas and S. Pawluk, *Can. J. Soil Sci.*, *57*:329 (1977).

458. S. Lian and A. Tanaka, *Soil Sci. Plant Nutr.*, Tokyo, *18*:15 (1972).
459. D. J. Carmody, J. B. Pearce, and W. E. Yasso, *Mar. Pollut. Bull.*, *4*:132 (1973).
460. J. A. C. Fortescue, S. A. Curtis, and E. Gawron, *Can. J. Spectrosc.*, *18*:23 (1973).
461. D. E. Baker, *Soil Sci. Soc. Am. Proc.*, *37*:537 (1973).
462. S. M. Griffith and M. Schnitzer, *Soil Sci.*, *120*:126 (1975).
463. E. A. Forbes, *N.Z. J. Agric. Res.*, *19*:153 (1976).
464. J. Yan, *Ti (Ch'iu) Hua Hsueh*, No. *4*:291 (1975).
465. S. Yamasaki, A. Yoshino, and A. Kishita, *Soil Sci. Plant Nutr. (Tokyo)*, *21*:63 (1975).
466. K. L. Liu, I. D. Pulford, and H. J. Duncan, *Anal. Chim. Acta*, *106*:319 (1979).
467. Z. Wu and Y. Ma, *Huanjung Huaxue*, *1*:228 (1982).
468. M. L. Berrow and W. M. Stein, *Analyst (Lond.)*, *108*:277 (1983).
469. S. V. Raju and K. Sundhaker, *Geonews (Secunderabad, India)*, *10*:357 (1982).
470. G. Henrion, K. Bode, and J. Pelzer, *Z. Chem.*, *23*:424 (1983).
471. J. M. Harnly, J. S. Kane, and N. J. Miller, *Appl. Spectrosc.*, *36*:637 (1982).
472. D. Geyer and P. Martin, *Abwasertechnik*, *32*:17 (1981).
473. S. Kempton, R. M. Sterritt, and J. N. Lester, *Talanta*, *29*:675 (1982).
474. R. Smith, *Water SA*, *9*:31 (1983).
475. S. A. Katz, S. W. Jeniss, T. Mount, R. E. Tout, and A. Chatt, *Int. J. Environ. Anal. Chem.*, *9*:209 (1981).

2

Inductively Coupled Plasma Spectrometry

BARRY L. SHARP *The Macaulay Land Use Research Institute, Aberdeen, Scotland*

I. INTRODUCTION

The development of inductively coupled plasma atomic emission spectrometry (ICP-AES) has arisen in response to the need for a solution analysis technique combining the sensitivity and precision of atomic absorption spectrometry with a capability for simultaneous multielement determinations. Although high-temperature arc and spark plasmas had been used for the multielement analysis of solid materials many years before the advent of the ICP, their rodlike structure made them unsuitable for the injection of liquid samples, even when these were finely dispersed as aerosols.

In 1961, Reed [1,2] described the first ICP operating at atmospheric pressure. This plasma did not require electrodes, and its robustness for sample introduction was demonstrated by its use for high-temperature crystal growing. Greenfield et al. [3,4] were the first to recognize the analytical potential of the ICP operated in an annular configuration. Wendt and Fassel [5] also noted Reed's work and experimented with a laminar flow "teardrop" shaped ICP, but later Fassel et al. [6] described the medium power (1–3 kW) 18 mm annular plasma now favored in modern analytical instruments. The importance of employing the correct plasma operating conditions was demonstrated by Dickinson and Fassel in 1969 [7], and this theme was further investigated in two important papers by Boumans et al. [8,9]. Although much of the early work was aimed at the development of the ICP source, Greenfield's group employed their plasma

**Current affiliation:* Loughborough University of Technology, Loughborough, Leicestershire, England

aspects that have a direct bearing on the analytical use of the ICP. for industrial analysis from the outset and reviewed their only experiences with the technique in three papers published in 1975–76 [10,11,12].

This pioneering work demonstrated the analytical advantages of the technique, and the subsequent development of commercial instruments has led to the ICP becoming the preferred source for the multielement analysis of samples in solution.

II. PLASMA EMISSION SPECTROMETRY

A. Formation of a Plasma

An ICP is formed by coupling the energy from a radiofrequency magnetic field (1–3 kW power at 27–50 MHz) to free electrons in a suitable gas. The gas, usually argon, is contained in a plasma "torch" constructed from a high-temperature resistant material, e.g., quartz, that is transparent to the rf radiation, and the magnetic field is produced from a 2- or 3-turn water-cooled copper coil placed around the upper part of the torch. The initial electron "seeding" of the neutral gas is provided by a spark discharge.

The electrons are accelerated by the magnetic field oscillation and describe reversing elliptical paths around the field lines. There is efficient power transfer to the electrons, because the electron collision frequency (ca. 10^{10} s^{-1}) is much higher than the applied field frequency (ca. 10^{7} s^{-1}) typically used for ICPs, and therefore, the electrons undergo many collisions during each field cycle. Eventually, some electrons attain energies equivalent to the ionization potential of the gas, at which point they may undergo inelastic collisions and cause further ionization. Rapidly, an equilibrium is reached in which the rate of electron production is balanced by losses due to recombination and diffusion and a stable plasma is formed.

The plasma is effectively a conductor and is heated by the flow of current induced by the radiofrequency field. The largest current flow and heat dissipation occur in the periphery of the conductor. This so-called "skin" effect, coupled with a suitable gas flow geometry, produces an annular or doughnut-shaped plasma. Electrically, the coil and plasma form a transformer with the plasma acting as a one-turn secondary coil of finite resistance.

B. Plasma Structure and Spectral Emission

The temperature distribution in a typical annular ICP is shown in Fig. 1 [13], and it is this that confers on this source almost ideal properties for the vaporization/atomization and excitation of solution samples. A sample aerosol particle, given sufficient velocity, will penetrate the base of the plasma and be constrained by the high-temperature plasma ring to pass through the central channel. The

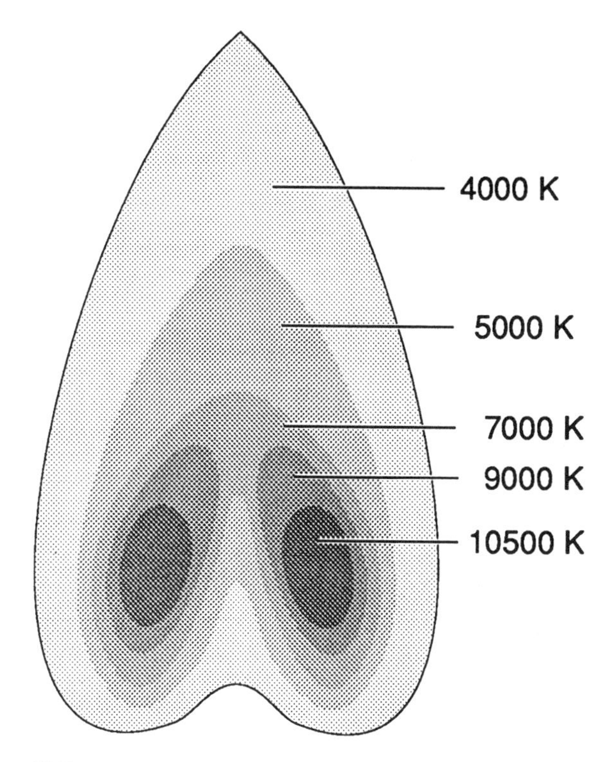

FIG. 1 Approximate temperature (kelvin) map of an analytical ICP.

injection velocity is typically ca. 7 m s^{-1}, and therefore, the particle will be exposed to a temperature of ca. 6000–8000 K for a period of a few milliseconds. Experiments have shown that this is sufficient to provide 100% atomization efficiency for particles having diameters <10 μm.

Different zones within the central channel of the plasma (Fig. 2) give rise to different types of emission. In the initial radiation zone (IRZ), atomic emission of low-excitation potential lines takes place. This gives way to a strong ionic and high-excitation potential atomic emission in the hotter normal analytical zone (NAZ), with low-energy atomic emission again occurring in the cooler regions of the tail flame. The actual positions of the different zones are determined by the operating conditions, particularly the applied power and sample carrier gas flow, the quantity and particle size distribution of the sample aerosol, the solvent vapor loading, and the matrix components of the sample.

The principles of emission spectrometry in general [14] and plasma spectrometry in particular [15,16] have been covered extensively in the literature, and here we need only consider those

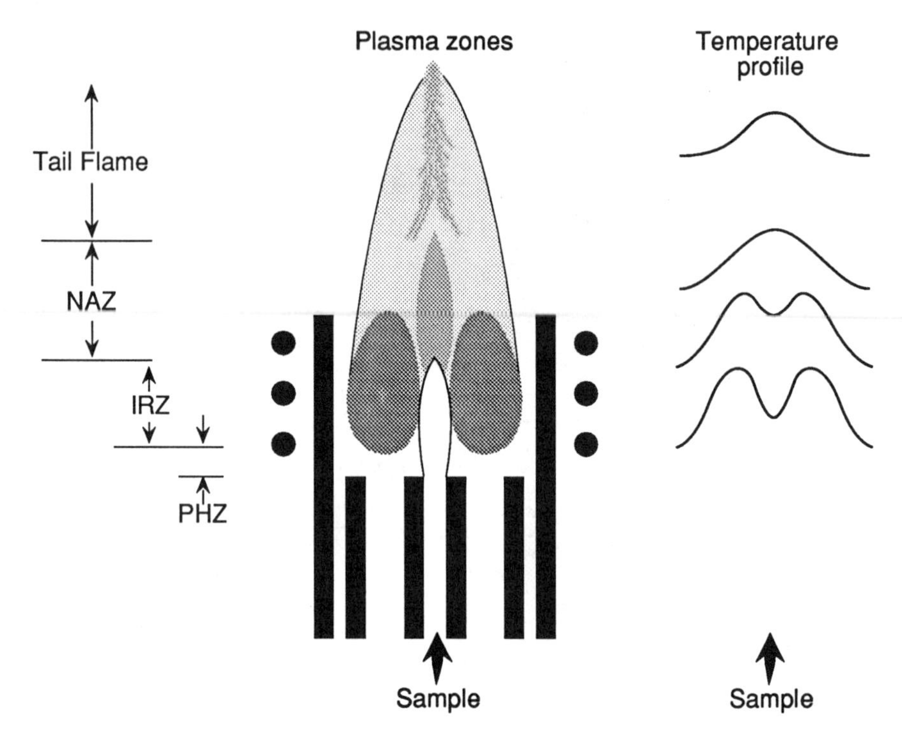

FIG. 2 Axial channel emission zone structure of an ICP. PHZ, preheating zone; IRZ, initial radiation zone; NAZ, normal analytical zone.

The equation describing the spectral radiance B of an assembly of atoms contained in a source in local thermal equilibrium (LTE) having a temperature T is

$$B = \frac{1}{4\pi} h\nu_0 \frac{NL}{Z(t)} g_k A_{ki} \exp(-E_k/kT) \tag{1}$$

where

h = Planck's constant
ν_0 = the frequency of the emitted photons
N = the number of atoms per unit volume
$Z(t)$ = the partition function
g_k = the statistical weight of the kth state

A_{ki} = the Einstein transition probability for spontaneous emission
L = the optical depth of the source
E_k = the excitation energy of the kth state
k = the Boltzmann constant

This equation expresses the linear relationship that exists between the spectral radiance and the concentration of free atoms in the source that underpins calibration for analytical purposes. Equation (1), however, is based on the assumption that all photons emitted escape from the source and are available for detection, i.e., the source is said to be optically thin. In fact, this assumption is only valid when the product of the number density N and the optical depth L are low. As the product (NL) increases, there is a greater probability that emitted photons will be absorbed by unexcited atoms. This process of "self-absorption" degrades the proportionality between emission intensity and atomic concentration, producing a characteristic curvature of the calibration graph toward the concentration axis (Fig. 3). A point is eventually reached where the emission intensity at the line center ceases to be dependent on the atomic concentration, and the source approximates, over this narrow spectral region, to a black body whose radiance may be described by Planck's law and is a function of temperature only. It may be shown [15] that in this region the intensity is proportional to $(NL)^{1/2}$, resulting in another linear section in the plot of log B against log N, but with a slope equal to 0.5. Such a curve is obtained only if the spectral bandpass of the measuring instrument can accommodate

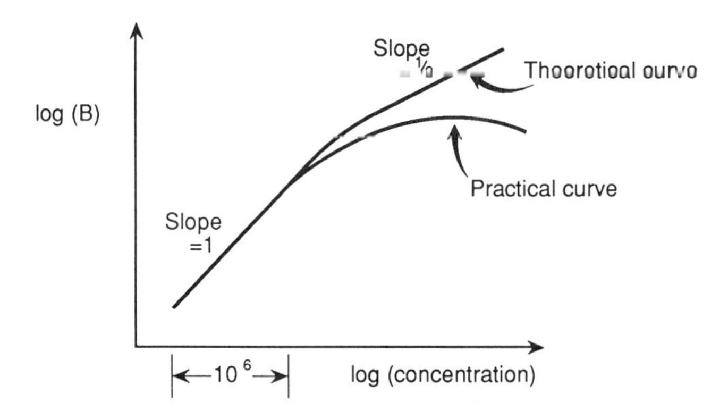

FIG. 3 Relationship between emission intensity (B) and concentration of analyte atoms.

the very broad spectral lines that may occur at high analyte concentrations. Normally, this is not the case and the calibration curvature is more severe. Matrix effects may also cause calibration curvature.

The analyte atoms are confined to a narrow central channel through the plasma, thereby minimizing the optical depth L and in the NAZ (from where the analyte emission is normally viewed), the transverse temperature profile is reasonably constant. The result is that the linear portion of the calibration curve extends typically over six orders of magnitude from the detection limit.

The temperature distribution along the central channel of the ICP is reflected in the types of emission that are observed. This has important implications for the use of the ICP for spectrometric analysis in establishing optimum operating parameters and for the selection of spectral lines. Two concepts are useful here: that of the "norm temperature" and a division of spectral lines into "hard" or "soft" lines as proposed by Boumans [17].

Equation (1) contains two terms that are temperature-dependent, the partition function $Z(t)$ and the exponential Boltzmann term. The partition function describes the change in the relative population of the excited states with temperature. Initially, the exponential term increases at the greater rate and therefore the intensity of the spectral line rises, but as the excited states become more populated, the increase in $Z(t)$ accelerates and the intensity of the transition begins to decrease. For elements of low ionization potential, there will also be a significant population of the excited states of the first ionic state. Thus, there is a certain temperature at which the emission intensity of a particular spectral line reaches a maximum and this is known as the norm temperature. The Boumans' classification divides lines according to their difficulty of excitation, with lines having excitation potentials below ca. 4.5 eV being termed soft lines and those above hard lines.

For soft lines it has been shown that low in the plasma, in the vicinity of the IRZ, the peak in the spatial emission profile correlates with the norm temperature [18]. Thus, in this region of the central channel, the excitation is essentially thermal in nature. A consequence of this is that both the position and magnitude of the spatial emission peaks for the soft lines are strongly dependent on the operating conditions and the matrix components of the sample. These dependencies may be summarized as follows:

1. An increase in applied power enhances the emission and shifts the peak lower in the plasma.
2. An increase in the sample carrier flow rate reduces the emission intensity and shifts the peak higher in the plasma.
3. The presence of an increasing concentration of an easily ionizable element enhances the emission and shifts the peak lower in the plasma, but higher up in the vicinity of the NAZ, the enhancement is much less [19]. This is not an ionization interference in the classical

sense and the normal technique of buffering the source to reduce the problem is inappropriate.

For hard lines, the position of the peak of the spatial emission profile is remarkably constant, even for lines having widely different excitation characteristics. Thus, low wavelength (< 300 nm) atomic lines and all ionic lines peak at approximately the same position in the plasma in the region of the NAZ. This stability of the spatial emission pattern is reflected in the effects of parametric changes and the composition of the matrix. These will be summarized as follows:

1. An increase in the applied power produces an increase in the emission intensity, but the position of the peak changes very little.

2. An increase in the sample carrier gas flow produces a small but significant upward shift in the spatial emission peak and a reduction in the intensity.

3. An increasing concentration of an easily ionizable element causes a depression in emission intensity in the vicinity of the spatial emission peak, but an enhancement lower in the plasma, with the result that there is a cross-over region where the effect of the interfering elements is minimized.

These characteristics point to an essentially nonthermal excitation environment in the NAZ. For more information on excitation mechanisms, the reader is referred to the work by Blades [20].

III. INSTRUMENTATION

The instrumentation required for ICP-AES is shown in Fig. 4 and comprises three basic units: the source, a spectrometer, and a computer for control and data analysis. The following sections are devoted, in turn, to a discussion of these components.

A. Plasma Source

The source unit includes the rf generator, the plasma torch and gas flow control system, and the sample introduction system.

1. RF Generators

A number of different designs for the rf generator are available, and they are categorized according to whether they use a crystal oscillator and power amplifier to drive the load coil, or whether the load coil is part of a free-running rf oscillator. The type used is not particularly important in terms of analytical performance, except that the crystal controlled oscillators must have automatic tuning if organic solvents or hydride generation are to be used, or if the frequent ingress of small quantities of air to the sample introduction system is anticipated. Failure to provide automatic retuning under

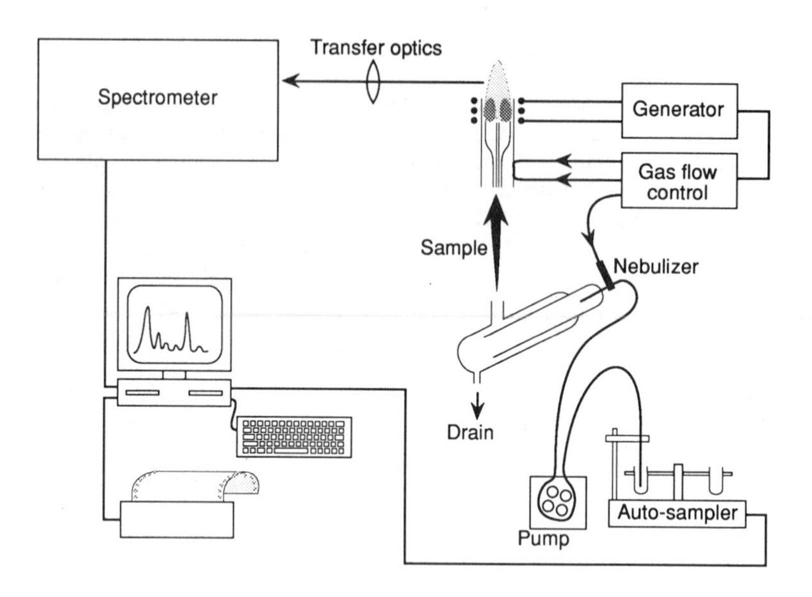

FIG. 4 Schematic diagram of an inductively coupled plasma optical emission spectrometer.

these circumstances will cause the plasma to be extinguished. Most commercial generators operate at 27.12 MHz, but it has been shown that plasmas operated at 40–50 MHz provide a higher signal-to-background ratio (SBR) in the emission spectrum [21], and therefore, some manufacturers offer units operating at this frequency.

An ICP power output of up to 2 kW is desirable for normal operating conditions. The precision of the ICP-AES technique is of the order of 1%, and a typical emission line might show an intensity response to a power variation of 1–5% per percentage change in power; therefore, power stability and regulation of the power output with respect to the supply voltage are important. A stability figure of 0.1%, sustainable over a typical working day, is needed if frequent recalibration is to be avoided. The power unit should be well screened, at least to meet the regulations of the country of use, and well enough to prevent interference with other equipment. Experience suggests that few generators meet the latter requirements.

2. *Torches*

The plasma is sustained in a plasma torch that in essence consists of three concentric quartz tubes arranged to provide a suitable gas flow geometry. Most modern instruments use a torch similar to that described by Scott et al. (Fig. 5a) [6] that has an 18 mm ID outer

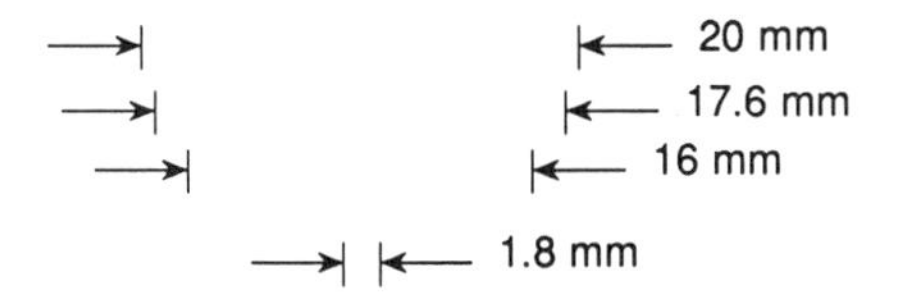

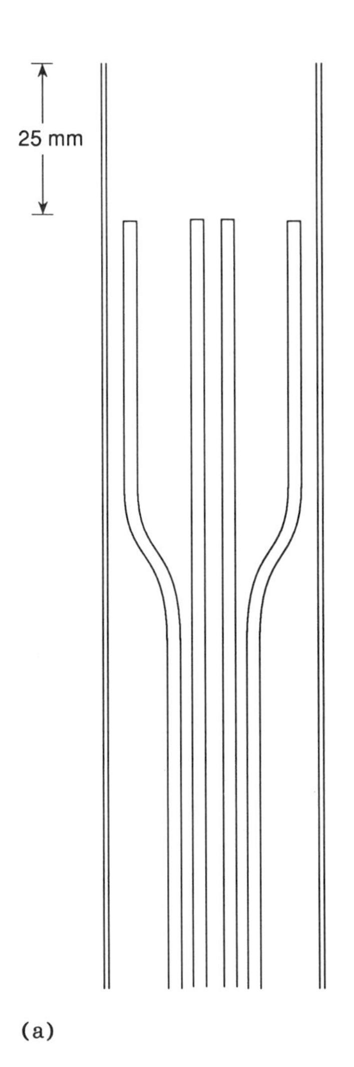

(a)

FIG. 5 Torches for inductively coupled plasma atomic emission spectrometry. (a) Scott–Fassel-type torch.

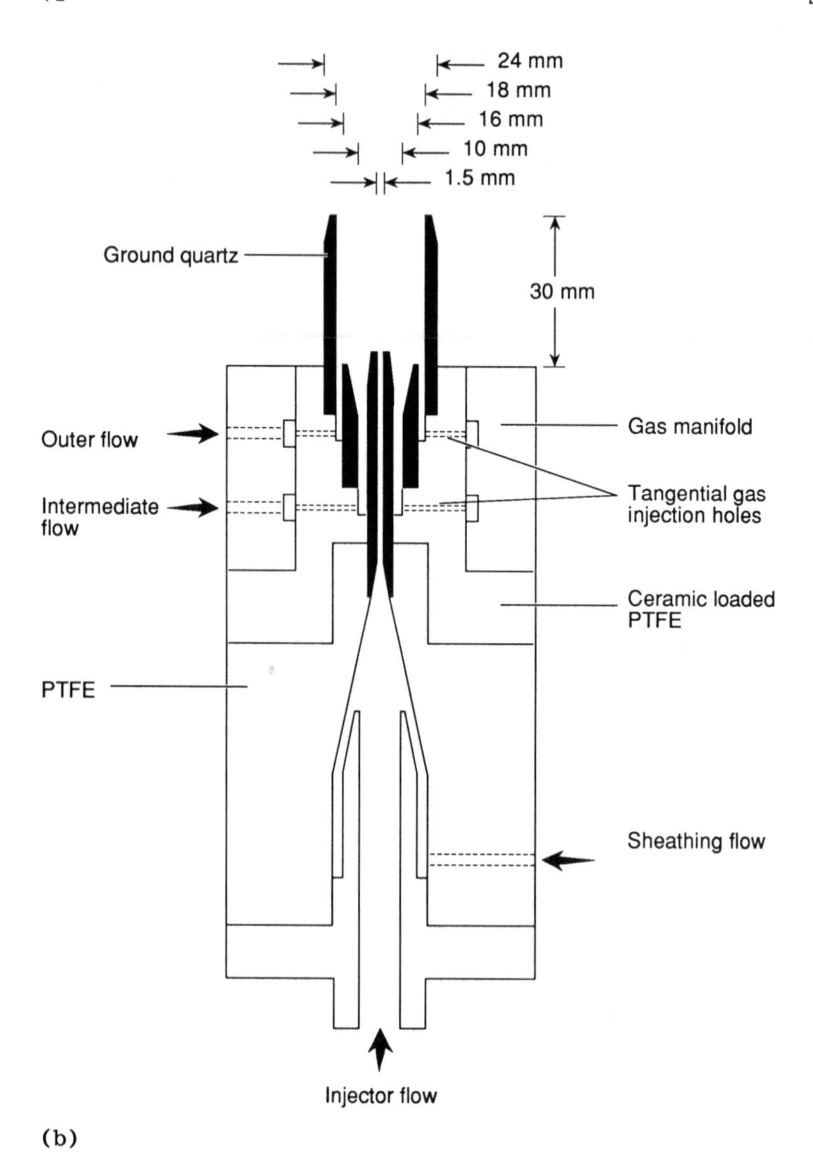

(b)

FIG. 5 (Continued) (b) Demountable torch as used at the Macaulay Land Use Research Institute.

tube, employs a 1 mm gap between the tulip-shaped intermediate tube and the outer tube, and has a 1.8 mm ID injector tube. Such torches will run at optimum performance at powers of ca. 1 kW, with

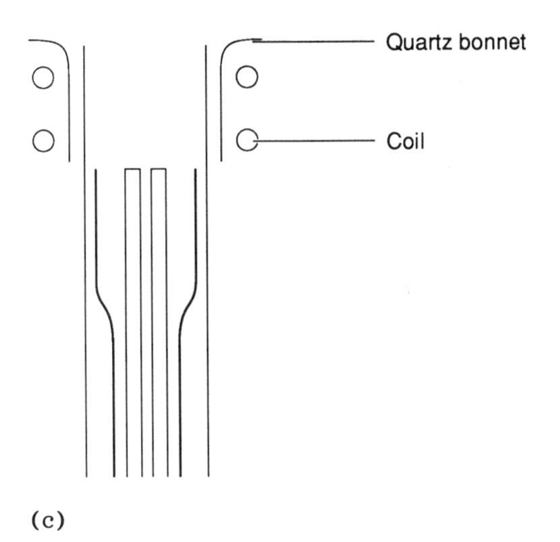

(c)

FIG. 5 (Continued) (c) Torch with quartz bonnet.

an outer Ar flow of ca. 10 l min^{-1} and an injector flow rate of 1 l min^{-1}. The Scott–Fassel torch requires a high degree of concentricity and most are fabricated in one piece from fused silica. A problem with this form of construction is that erosion of the injector tube tip, or of the outer tube in the vicinity of the plasma, may necessitate replacement of the whole torch due to relatively minor local damage. This problem is circumvented by the use of a demountable torch; and Fig. 5b shows one of this type that has been in use at the Macaulay Institute since 1983. The body supporting the outer quartz tubes is made of ceramic-loaded PTFE to provide thermal stability, and the sample introduction tube and injector mounting are made of normal PTFE. Gas is introduced tangentially through an array of 12 × 0.5 mm diameter holes for the outer flow. A very important feature of this torch is the use of 3 mm thick ground quartz tubes that are shaped to produce smooth gas flow patterns. The thick quartz is very robust, so injector and intermediate tubes appear to last almost indefinitely. The outer tube has a useful lifetime of approximately one year and may be replaced simply by plugging in a new one without dismantling the torch or removing it from the load coil. An additional feature of this torch is that it incorporates a sheathing flow for the injector flow, as first proposed by Mermet and Trassy [22]. This helps to minimize deposition on the tip of the injector when introducing solutions containing a high solids content. It has also been found to improve the sensitivity of determination for the alkali metals and, when the flow is doped with H_2, to offer potential improvements in the decomposition of the sample aerosol [23].

Although there is a general consensus on the overall design of the torch, there is still a variation in the type of injector used. Figures 6a through 6c show three commonly used designs. The device shown in Figure 6a provides a turbulent aerosol stream, whereas that in Fig. 6c produces laminar flow, and that in Fig. 6b has characteristics intermediate between the other two. Figure 6d is a schematic of the injector tube used in the author's laboratory; it incorporates two further features, a 20° (full angle) angle inlet taper and a tapered injector tip. This design offers several advantages. In particular, the tapered tip reduces salt deposition and also prevents the buildup of carbon deposits when organic solvents are used [24]; and the low-angle inlet taper prevents excessive deposition of aerosol particles at the inlet.

Torches are best left undisturbed unless there is a noticeable deterioration in analytical performance. Cleaning involves an overnight soak in concentrated nitric acid, and some brushing of the

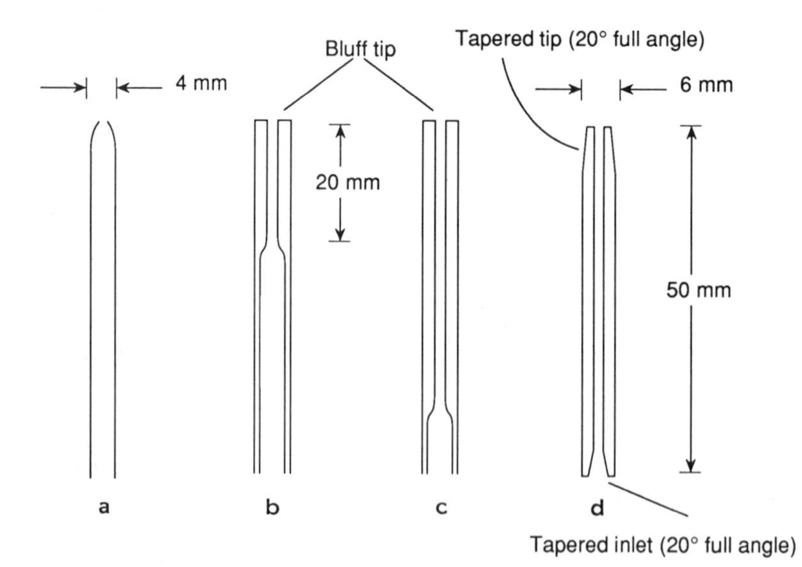

FIG. 6 Injector tubes used in torches for ICP-AES. (a) Turbulent constricted injector tip. (b) Intermediate laminar/turbulent capillary injector tip. (c) Fully laminar capillary injector. (d) Streamlined capillary injector for demountable torch.

injector tip or the outer wall may be necessary to remove solid deposits. Ablation and devitrification of the outer tube may eventually lead to arcing of the plasma through the wall onto the coil. This may be avoided by the addition of a quartz "bonnet" (see Fig. 5c) between the coil and the torch.

The gas consumption and power requirement for the ICP has been shown by Angleys and Mermet [25] to be dependent on the annular gap between the intermediate and outer tube. For example, a torch having an annular gap of 0.3 mm can be operated successfully at an outer flow of 3 l min^{-1} with an applied power of 0.3 kW. The requirement for good concentricity, however, places severe dimensional constraints on such designs and the practical limit appears to be about 0.4 mm. A high-efficiency torch using ca. 6 l min^{-1} Ar has appeared on the market [26], but there is evidence that the structure and excitation conditions may be different to those obtaining in conventional designs. The alternative approach to reducing gas and power consumption is to reduce the torch size, and one manufacturer is offering a mini-torch design on a commercial instrument [27].

B. Sample Introduction Systems

The ICP can accept samples in vapor, liquid, or solid form, and systems employing each mode of sample introduction have been described. The robustness of the ICP for sample introduction is one of its great strengths, and modifications to the method of sample introduction are the simplest way of enhancing the performance of the technique for specific applications.

1. Nebulization

The vast majority of analyses are carried out using pneumatic nebulization for sample introduction. Sharp [28] has discussed comprehensively this topic, and therefore, the information given here is limited to practical matters relating to the principal types of nebulizer in current use.

Figure 7 shows the three principal types supplied with commercial instruments. The concentric and cross-flow nebulizers are self-priming, whereas the Babington nebulizer requires a pump to deliver the solution. The glass concentric nebulizer is probably the most widely used and with care provides stable and trouble-free operation. Its principal disadvantages are its fragility and the fineness of the liquid-carrying capillary (ID ~ 0.3 mm) and of the annular gas orifice (thickness ~ 0.02 mm). Blockages are commonly due to the accumulation of salt particles in the annular gas orifice and to physical obstruction of the liquid capillary, usually

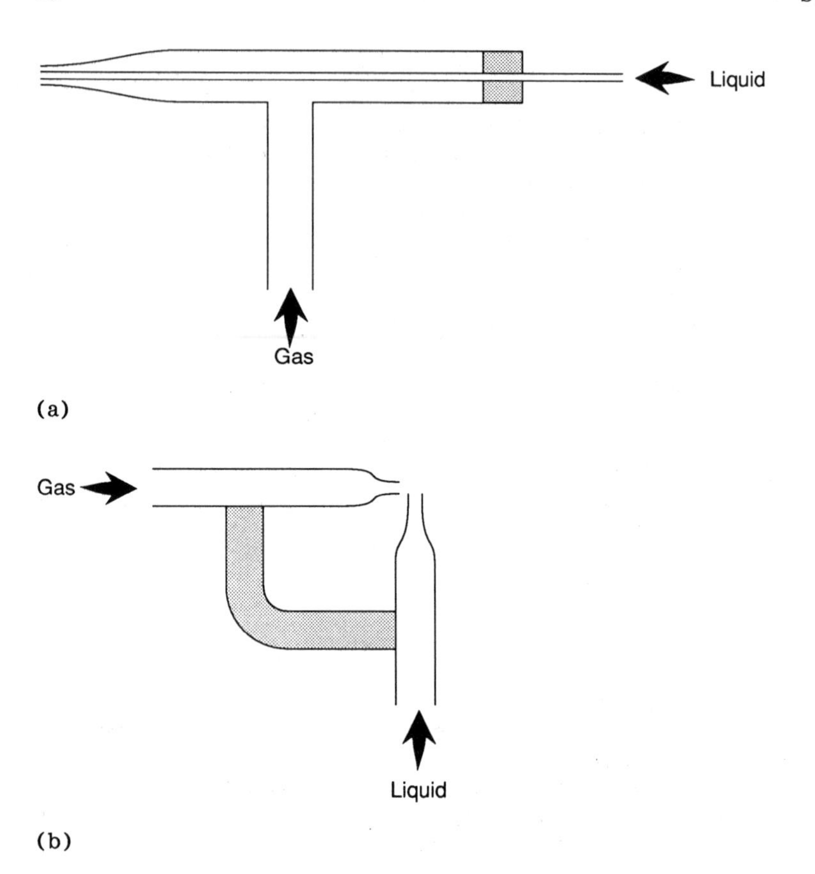

FIG. 7 Commonly used types of pneumatic nebulizer. (a) Concentric-flow. (b) Cross-flow.

by fibers attaching themselves to the wall at the tube exit. Recessed-tip versions of the concentric nebulizer have been reported to enable highly concentrated salt solutions to be handled without salting-up of the gas orifice [29]. Other necessary precautions are to ensure an adequate period of washing of the nebulizer between samples and to prevent excessive air ingress that will promote drying of the tip and crystal formation. The use of argon saturated with water vapor also helps prevent salt formation, and in extreme cases periodic injections of water into the gas line clean the nebulizer without the need to interrupt the plasma. A blocked nebulizer can be cleaned by soaking, back-flushing (either orifice) with a syringe, or ultrasonic cleaning in acid and/or a surfactant.

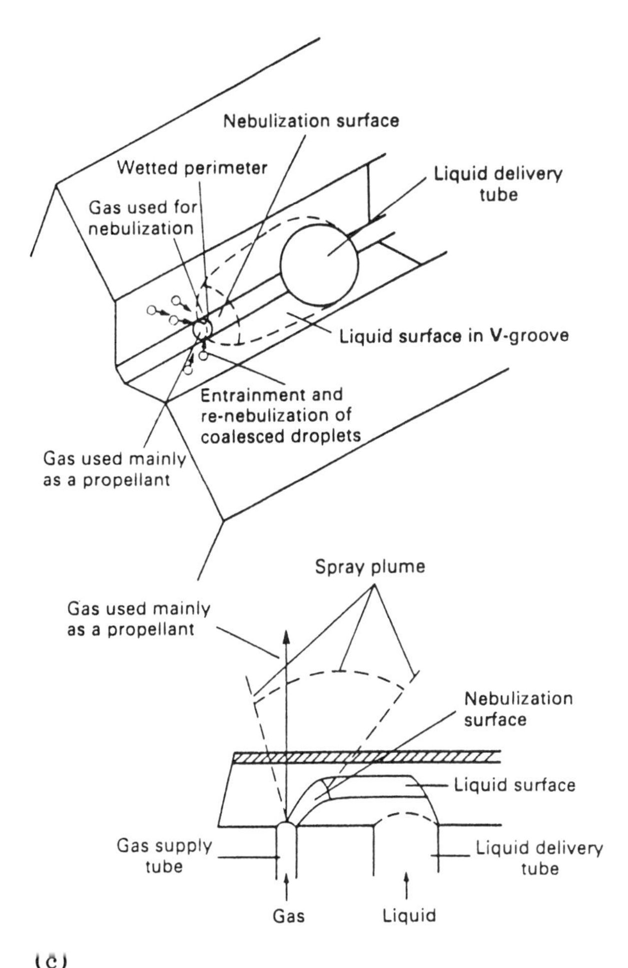

(c)

FIG. 7 (Continued) (c) Babington v-groove type.

Mechanical cleaning should only be used as a last resort and then only soft materials should be used, e.g., a camel hair brush or nylon fishing line. The glass construction precludes the use of HF, solutions containing high concentrations of fluoride ion, or strong caustic solutions.

The cross-flow nebulizer (Fig. 7b) offers similar analytical performance to the concentric flow design and requires the same basic care. Various materials are used for its construction including sapphire, glass, and inert plastics such as Ryton. The most critical mechanical feature is the relative alignment of the liquid-carrying capillary and the gas orifice, and therefore, a rigid design is essential. The liquid-carrying capillary generally has an internal diameter not exceeding 0.5 mm and that of the gas orifice is usually 0.1–0.2 mm. As a general rule, liquid capillaries of <0.3 mm ID should be avoided as they are prone to blockage. The cross-flow nebulizer is perhaps slightly less prone to blockage than concentric types. Most cross-flow nebulizers operate at relatively low pressure (20–40 psig), but an all-glass high-pressure (200 psig) design is available and appears to offer some advantages in terms of efficiency and analytical precision [30]. A drawback of this device is that it needs a separate gas control system because most commercial instruments are designed to work from liquid gas storage tanks with a maximum supply pressure of ca. 0.7 MPa (100 psig).

Very concentrated solutions, or those containing significant quantities of suspended solids, require the use of a nebulizer based on the Babington principle [31], in which the liquid is not confined to a narrow capillary but is conducted to the gas orifice along an open slot or v-groove (Fig. 7c) [28,32–36]. The key problem of achieving satisfactory gas-liquid mixing has been addressed in the linear conespray nebulizer that employs a slotted gas jet fed with liquid on both sides from a flat-topped weir [28]. Most commercial designs are not good in this respect, and as a result, generally achieve lower analyte transport efficiencies. This is readily overcome, however, by increasing the sample feed rate, provided that an adequate sample is available. The current generation of Babington nebulizers are devices optimized for their sample-handling capacity, but it is probable that future improvements in design will ultimately result in their becoming the preferred choice for most applications.

The spray produced by the nebulizer passes into a spray chamber to remove large aerosol droplets (>10 μm). Design of the spray chamber should ensure a minimum loss of the small droplets, permit rapid changes from one sample to another without memory effects, and provide for free drainage of the waste solution without causing pressure pulsations that will be reflected in noise on the optical signal [33]. Temperature control of the spray chamber can result in improved precision, and cooling is commonly employed to control the solvent loading of the plasma when organic solvents are used [37].

The use of a peristaltic pump to feed the nebulizer has a number of advantages. It minimizes variations in the feed rate; the pressure it generates will often clear blockages in the capillary; and the pump may be reversed to backflush the nebulizer for cleaning. Furthermore, the pump enables the sample flow rate to be controlled independently of the nebulizer gas flow rate. Most nebulizers provide improved efficiency at lower liquid flow rates, and this may also be accompanied by improved signal-to-noise ratio leading to lower detection limits. To avoid surges in the flow that are reflected in the optical signal, the pump should be of the multiroller type (e.g., 10 rollers) run close to its maximum speed, with narrow-bore pump tubing. The important consideration is the ratio of the period of signal variation to the measurement period.

2. *Chemical Vapor Generation*

The direct introduction of the analyte to the ICP in vapor form offers the potential for achieving 100% transport efficiency and therefore improved sensitivity, a separation of the analyte from the matrix and a consequent reduction in interferences, and avoids the problems associated with nebulization. The techniques used are modifications of those previously developed for use in AAS, namely, the cold vapor technique for Hg, hydride generation for As, Pb, Sb, Se, and Sn, and the generation of volatile chlorides and organometallic compounds. The necessity for modification arises because the ICP is not tolerant of high levels of molecular gases, e.g., H_2, H_2O, and CO_2, as these quench the plasma and may cause an impedance mismatch with the generator, causing the plasma to be extinguished. Higher power levels (2–2.5 kW) are usually employed to accommodate greater levels of vapor input, but the essential step is to limit the production of vapor to a suitable level. This is usually accomplished by generating the hydrides in a continuous-flow apparatus (with a facility for rapid exchange between samples and blanks) that produces a continuous flow of molecular gas, allowing the generator to be correctly matched to the plasma and providing a stable background emission. Commercial systems are available from several manufacturers.

Optimization studies have been carried out by Parker et al. [38] and Pyen et al. [39], and a mathematical model of hydride generation that describes the factors affecting the shape and magnitude of the analytical signal has been reported by Barnes and coworkers [40]. Readers interested in the mechanism of hydride generation should consult the continuing series of papers by Welz and coworkers [41] and Agterdenbos and coworkers [42].

3. *Electrothermal Vaporization (ETV)*

The use of electrothermal vaporization, ETV [43], for sample introduction provides approximately an order-of-magnitude improvement in the detection limit (for elements that are readily vaporized) compared with nebulization. This technique involves evaporation of a solid or liquid sample directly into the plasma from an electrically heated graphite cuvette or metal filament.

The signal from an ETV sample introduction is transient, which complicates the use of automatic background correction by wavelength scanning. Uniform heating of the furnace is necessary to avoid multiple peaks in the emission signal and a close coupling of the furnace with the torch minimizes transport losses.

Laboratories possessing automated electrothermal atomization AA equipment may not consider ETV-ICP to be a worthwhile investment. But when the number of elements to be determined at the ultratrace level exceeds 5–10 and the sample size is limited, the technique has a lot to offer.

C. Spectrometers

The most important features of a spectrometer for ICP-AES are its resolution, light throughput, stability, and stray light performance. Unfortunately, high resolution is not compatible with high light throughput (for dispersive spectrometers) and stability, and a compromise is necessary. It must be stressed that this compromise is determined entirely by the type of samples that are to be analyzed. Soil, plant, and other environmental samples generally only require moderate resolution, but do need high light throughput and stability, and good stray light characteristics, to ensure that light from strong matrix lines is not present in the spectral bandpass when weak analyte lines are being measured.

The largest contribution to the physical line width of spectral lines emitted from an ICP is Doppler broadening, and since the Doppler half-width is inversely proportional to the square root of the atomic mass, there is a progressive reduction in line width for heavier elements. Thus, e.g., the light element boron has a line width of 5.0 pm (B I 249.773 nm), whereas that for molybdenum is 1.2 pm (Mo II 202.030 nm) [44]. This simple picture is complicated by the fact that many spectral lines exhibit hyperfine structure, due to either the effect of nuclear spin or the presence of isotopic components. The hyperfine structure is not normally resolved by analytical instruments and can result in the occurrence of quite broad spectral lines. This effect has been observed for lines of Bi, Co, Eu, Ho, La, Ln, Lu, Mn, Nb, Pb, Pr, Pt, Re, Sb, Ta, Tb, and V [44]. The practical resolving power of an ICP is always determined by the slit-width. A narrow slit (e.g., 3 μm) set to provide diffraction-limited performance will pass such a small light flux that the attainable signal-to-noise ratio (SNR) and detection limit will be limited by the shot noise. Opening the slit will increase the light flux, but the noise will increase only as the square root of the light flux, with a consequent gain in the SNR and detection limit. The limit to this improvement is set by the increasing contribution from flicker noise (which is proportional to the signal intensity) that for the ICP limits the attainable SNR to ca. 200. Typical slit widths used vary between 20 and 50 μm, depending on the type of instrument and the particular application.

The majority of spectrometers used in commercial ICP systems are either of the single-channel scanning type, or of the multichannel fixed-wavelength type. Figure 8 is a schematic diagram of a typical scanning spectrometer using the Czerny-Turner mounting. Figure 9 shows the optical configuration of a typical multi-channel spectrometer using the Paschen-Runge mounting. The detailed designs of these spectrometers have been described elsewhere [16,45], and here discussion will be limited to their operational characteristics. The basic requirements for any spectrometer for ICP-AES are that the wavelength coverage extends from 170-800 nm and that the detection system be linear over a range of 10^6, which covers the operating range of most photomultiplier tubes (anode currents 10^{-9} -10^{-3} amp) and encompasses the linear dynamic range of the ICP (~10^5). A number of elements, notably As, Hg, P, and S, have their best analytical lines below 200 nm and for these an evacuated spectrometer is required.

Several factors must be considered before an objective choice can be made between a scanning and multichannel system. A list of the operating characteristics of each type follows.

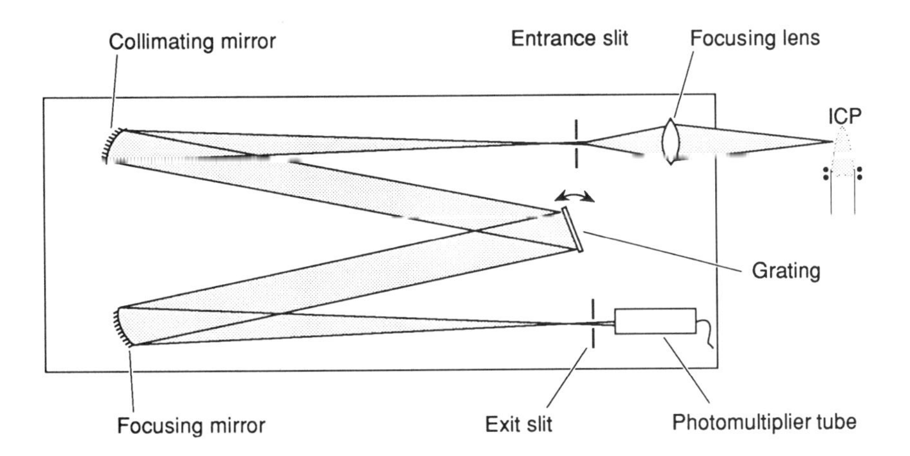

FIG. 8 Single-channel scanning spectrometer based on the Czerny-Turner optical configuration.

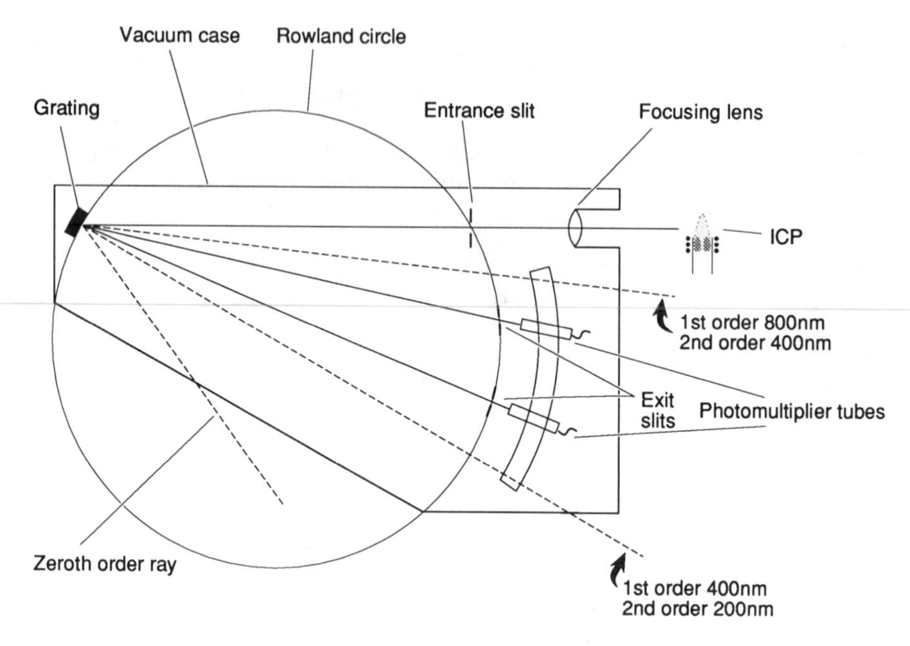

FIG. 9 Multichannel simultaneous spectrometer using the Paschen–Runge optical configuration.

1. *Multichannel Spectrometers*

Advantages may be summarized as follows:

1. *Speed of analysis.* Because all the analyte lines are measured simultaneously, the speed of analysis is limited mainly by the sample-changeover time.

2. *Minimal sample consumption.* Multielement determinations can be performed on a few ml of sample, or on microsamples if ETV or HPLC is used for sample introduction.

3. *Accuracy and stability of wavelength setting.* The use of fixed slits and a modern thermally stable construction ensure that the correct wavelength is measured. The stability of the wavelength setting provides good analytical precision (0.5–1.0% RSD) for routine measurements.

4. *Flexibility of background correction mode.* Most modern multichannel spectrometers incorporate a scanned entrance slit or spectrum shifter to enable simultaneous scanning of all the programmed spectral lines. This feature, together with the ability to quantify simultaneously elements causing spectral interference, permits both automatic off-peak and on-peak matrix background corrections.

5. *Ease of applying matrix corrections.* Rotational interferences caused by matrix interferences can be compensated without time penalty by the simultaneous determination of the concentrations of the interfering elements and the application of suitable corrections.

6. *Ease of implementing internal standardization.* The simultaneous measurement of reference lines can in specific circumstances compensate for changes in instrument performance (see Sec. VI).

Disadvantages are:

1. *High cost* relative to single-channel spectrometers.

2. *Inflexibility.* The suite of lines to be used must be chosen in advance of installation and subsequent modification is difficult and expensive. There is no facility for moving to another line if a new sample type renders a line unusable because of either its sensitivity, or the occurrence of spectral interferences.

3. *Moderate resolution.* The requirement to maintain good optical stability imposes limitations on the slit width that can be used in a multichannel spectrometer. Typically, entrance slits of 20 μm and exit slits of 50 μm are used that, for a spectrometer having a reciprocal linear dispersion of 1.0 nm mm^{-1} in the first order, yields a bandpass of 0.05 nm. This imposes limitations on the use of automatic background correction and the ability of the instrument to minimize spectral interferences.

For these reasons, multichannel spectrometers are normally employed when high sample throughput is required and the elements to be determined are known in advance of purchase. A particular ability of multichannel instruments is the provision of multielement determinations on microsamples, provided that an appropriate sample introduction procedure is used.

2. *Single-Channel Scanning Spectrometers*

Advantages include:

1. *Flexibility.* All the spectral lines for each element are available for selection to meet the requirements of a particular application.

2. *High resolution.* Because wavelength selection is achieved dynamically, narrower slits, e.g., 20 μm, can be used, leading to improved resolution. The consequences of this are reduced spectral interferences and improved detection limits.

3. *Ease of automatic background correction.* The total signal and background signal are measured separately and the net analyte signal calculated by difference. Background correction is therefore automatic and can be extended to work in the presence of spectral interference, either by implementation of a suitable software algorithm, or by intervention of the analyst following the graphical display of the data.

4. *Lower cost* relative to multichannel spectrometers.

5. *Survey analysis.* It is possible to scan the entire spectrum of the ICP and using software-based reference tables to produce a "total analysis" on a semiquantitative basis.

Disadvantages are:

1. *Speed of analysis.* The sequential measurement of spectral lines imposes a time penalty in proportion to the number of elements determined. Modern instruments have extremely rapid wavelength drives and can perform rapid survey analyses, but at the expense of precision.

2. *Higher sample consumption.* This is a direct consequence of (1) above and occurs in proportion to the number of elements determined.

3. *Reduced ability to perform matrix correction.* Correction for either spectral or matrix interferences requires the measurement of additional spectral lines that can only be accomplished with a time penalty on a sequential instrument.

4. *Inability to implement internal standardization.*

Scanning spectrometers have therefore found application in situations where sample throughput is not the principal criterion, where the range of elements determined varies widely from sample to sample, and in some applications where there has been a paramount need to minimize spectral interferences.

An instrument combining the merits of scanning and multichannel spectrometers would have obvious advantages and a number of such systems have appered on the market. The simplest approach is to add a separate scanning monochromator system viewing the plasma along a different axis. The advantages are that the best features of each type of spectrometer are available, but such "two-box" solutions are expensive. A simpler alternative is to provide a separate scanning or "(N+1)" channel within the frame of the multichannel spectrometer. The resolution of such systems is inevitably poorer than that achieved by stand-alone scanning spectrometers, but the engineering is more compact and the standard electronics of the multichannel spectrometer can be used to process the signal from the roving channel.

IV. ANALYTICAL CAPABILITY

A. Element Coverage

The ICP is suitable for the determination of most elements in the periodic table except the halides, the inert gases and atmospheric gases, although by excluding air from the plasma, nitrogen may be determined [46,47].

B. Detection Limits, Precision, and Dynamic Range

The detection limit C_L is defined by

$$C_L = \frac{2\sigma_B}{S} \quad \cdots \tag{2}$$

where σ_B is the standard deviation of the background signal and S is the sensitivity, i.e., the slope of the linear calibration curve relating signal strength to concentration. For a given concentration C, the equivalent net line intensity is I_C, and hence, the sensitivity is given by

$$S = \frac{I_C}{C} \quad \cdots \tag{3}$$

and if we introduce the background intensity I_B, Eq. (2) becomes

$$C_L = 2\sigma_B \frac{I_B}{I_B} \frac{C}{I_C} \quad \cdots \tag{4}$$

or

$$C_L = \frac{2\sigma_B}{I_B} \frac{C}{\text{SBR}} \quad \cdots \tag{5}$$

where SBR is the signal-to-background ratio. An ICP operating under flicker noise limited conditions should achieve a relative standard deviation of the background signal of 0.01, so that Eq. (5) becomes

$$C_L = 0.02 \frac{C}{\text{SBR}} \quad \cdots \tag{6}$$

From Eq. (6) it can be seen that the detection limit should correspond to an SBR of 0.02, but in practice this will vary, depending on whether the background is true background or is structured because of the presence of spectral lines or bands [44].

For small signals (SBR<1), the standard deviation of the net line signal remains fairly constant, being determined by the deviation in the background signal. However, at concentrations above the background equivalent concentration (BEC corresponding to the concentration required to yield a SBR of unity), the standard deviation increases in proportion to the net signal so that the relative standard deviation becomes constant at a value of ca. 0.01 (Fig. 10).

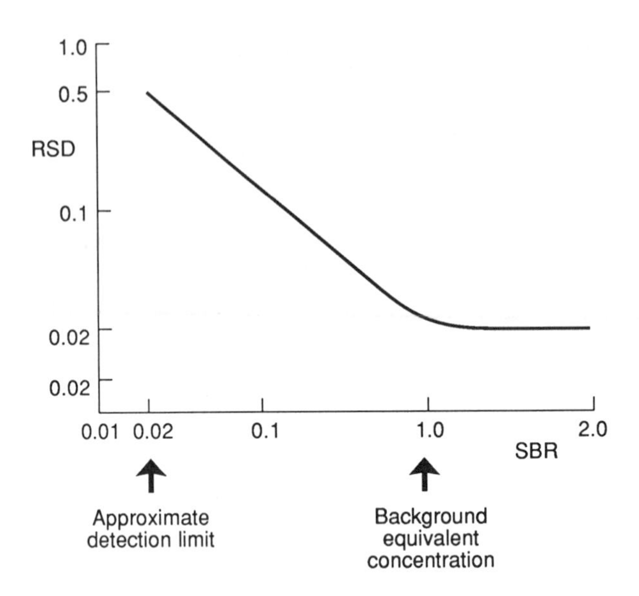

FIG. 10 Variation in relative standard deviation with SBR (concentration) for ICP-AES.

Detection limits for ICP-AES using pneumatic nebulization for sample introduction vary from ca. 0.1 ng ml^{-1} for elements such as Ca, Ba, Mg, Mn, and Sr up to ca. 100 ng ml^{-1} for elements such as As, K, P, Pb, Se, and U. An order-of-magnitude improvement is available for many elements using ETV and a similar improvement is achieved by the use of hydride generation for the elements As, Bi, Pb, Sb, Se, and Sn. A list of detection limits for ICP-AES (and ICP-MS) for lines that have been found to be appropriate for environmental samples is given in Table 1. Note that these are the best attainable values on a particular instrument using aqueous standards and may be severely degraded for real samples. The linear dynamic range of ICP-AES is usually 10^3-10^6 and, for most elements, extends from ca. 1000 μg ml^{-1} down to the detection limit.

V. INTERFERENCES

The increasing popularity of ICP-AES for elemental analysis owes much to the remarkable degree of freedom from interferences that the technique provides. Nevertheless, interferences do occur and they are divided into two classes, translational (or additive) and rotational (or multiplicative) interferences. Translational interferences are those in which a given concentration of the interferent produces a corresponding addition to the analyte signal, independent of the analyte concentration. Rotational interferences occur when the effect of a given concentration of the interferent is to multiply the analyte signal by a constant factor, and thus to change the slope of the response curve. The factor is independent, to a fist approximation, of the concentration of the analyte. Translational interferences are caused by spectral overlaps or by stray light in the spectrometer, whereas rotational interferences are caused by more general matrix effects.

A. Spectral Interferences

Spectral interferences can be divided conveniently into three classes. These are:

1. Line overlap, in which the interfering line shares a common portion of the spectrum with the analyte line.
2. Line interference, in which the interfering line intrudes into the spectral bandpass of the spectrometer.
3. Background interference, in which there is a general change in the intensity of the background continuum, or in which the wing of an intense neighboring spectral line intrudes into the bandpass of the spectrometer.

The practical significance of line overlap is that its effect cannot be completely removed by improved instrumentation, i.e., higher resolution. The severity of the interference depends on the relative intensity of the two lines and on the degree of overlap. But note that most spectrometers operate with a spectral bandpass 3–4 times the physical line width, so that even partially overlapped lines will be completely unresolved.

The simplest way of dealing with line overlap is to choose another line, but when this is inappropriate, the total apparent analyte signal must be corrected for the contribution of the interfering line. Clearly, the success of this approach depends on the relative magnitude of the correction and ideally it should be less than 10% of the total intensity. Any correction must decrease the relative precision and have an adverse effect on the detection limit. Church [48] concluded that the greatest error arises from the uncertainty in the correction factor, which is typically of the order of 5%, and Thompson et al. [49] found that this translates to an increase in the detection limit of ca. 10% of the magnitude of background correction.

TABLE 1 Detection Limits for ICP-AES and ICP-MS for a 27-MHz Plasma

	ICP-AES			ICP-MS
Element	Ionization state	Wavelength (nm)	Det. limit (ng ml^{-1})	Det. limit (ng ml^{-1})
Ag	I	328.068	4.0	0.02
Al	I	308.215	18.0	0.04
	I	396.152	11.0	
As	I	193.759	50.0	0.12
B	I	182.59		
	I	249.678	6.0	0.1
Ba	II	455.404	1.0	
Ca	II	317.933	9.4	
	II	393.366	0.18	
Cd	II	226.502	2.8	0.04
Co	II	228.616	5.1	0.008
	II	238.892	5.7	
Cr	II	267.716	5.1	0.028
Cu	I	324.754	2.3	0.026
Fe	II	259.940	1.7	0.7
Hg	I	184.960	—	0.04
	II	194.227	25.0	
K	I	766.491	80.0	
Li	I	670.780	1.9	0.024
Mg	II	279.079	0.03	0.027
Mn	II	257.610	1.0	0.028
Mo	II	202.030	7.4	0.07
	I	379.830	5.3	
	I	313.259	11.0	
Na	I	588.995	20.0	0.062
Ni	II	231.604	7.1	0.085

(continued)

TABLE 1 (Continued)

Element	ICP-AES Ionization state	ICP-AES Wavelength (nm)	ICP-AES Det. limit (ng ml^{-1})	ICP-MS Det. limit (ng ml^{-1})
P	I	178.287	100.0	2.0
	I	213.618	73.0	
Pb	II	220.350	40.0	0.016
S	I	182.036	30.0	1.0
Se	I	196.026	71.0	1.4
Si	I	251.611	9.1	10
	I	288.158	18.0	
Sn	II	189.926	25.0	0.093
Ti	II	337.280	3.7	0.02
V	II	290.882	6.8	0.01
Zn	I	213.856	1.7	0.077

Two methods are available for making corrections, either blank subtraction, or cross-calibration of the contribution of the interfering element at the analyte wavelength (sometimes referred to as "on-peak" correction). Blank subtraction has only limited application when the samples have a known and constant concentration of the interfering element. The cross-calibration method involves preparing a calibration curve for the interferent at the analyte wavelength and then at analysis time, using an interference-free line, measuring the concentration of the interfering element. Clearly, this procedure is more efficiently carried out on a multichannel spectrometer.

Line interference is caused by the inability of the spectrometer to resolve neighboring spectral lines. Increasing the resolution will therefore both decrease the severity of the interference and potentially provide a dramatic improvement in the detection limit. The ability to control resolution is limited to scanning spectrometers, where the options of using narrower slits, or a higher-order spectrum, are available.

Two methods of background correction may be used to reduce the effects of line interference. The cross-calibration procedure, described previously, is most appropriate for multichannel spectrometers, but for scanning spectrometers, automatic background correction (sometimes referred to as "off-peak" correction) may be possible. This is the process of scanning the spectrum and extracting the net analyte signal intensity from the data. Its success depends on the ratio of the intensities of the analyte and interfering line and on the degree of resolution achieved by the spectrometer. When the resultant overlap is slight, measurement of the background on the side of the spectral line away from the interfering line may be perfectly adequate. The difficulty of making accurate background correction increases substantially when the interfering line makes a significant contribution at the central wavelength of the analyte line.

Computer-based techniques are available for the resolution of overlapping spectra. However, considerable care is necessary in developing algorithms for peak detection and background correction. A common effect of a poorly resolved spectral interference is to shift the peak of the resultant spectral profile. Unless the peak detection algorithm is sufficiently rigorous, it may simply fit a single profile to the shifted peak and the interference will pass undetected.

Background interference may be caused by the wings of strong spectral lines, e.g., Ca II 396.847 nm on Al I 396.152 nm; the presence of molecular band-heads generated from sample components, e.g, CN, NO, OH bands; increases in the intensity of this recombination continuum, e.g., from the presence of Al in the spectral region below 220 nm; or stray light, e.g., from high levels of Ca or Mg. So varied are these causes that background-correction techniques depending on prior knowledge of the existence and source of the interference are not applicable. Automatic background correction works extremely well and is only complicated by the occasional necessity to allow for a sloping baseline. The relative magnitude of the interference is the most important consideration and when the factor by which the background intensity changes exceeds an order of magnitude, there will be a significant loss in detection power.

B. Matrix Interferences

The rotational interferences that occur due to the presence of concomitant elements (the matrix) in the analyte solution are derived from two sources: the effect that the matrix has on the structure and excitation conditions of the plasma and its effect on the sample introduction process.

Interferences generally occur as suppressions of the emission intensity and increase with the total excitation potential (ionization + excitation) of the analytical line. In the case of some lines of low-excitation potential, e.g., Li 670.78 nm, the interference may be positive, but the magnitude is small, generally less than 5%.

The magnitude of the interference caused by a particular matrix correlates with the sum of its ionization energy plus its dissociation energy. Ramsey and Thompson [50] termed this the "matrix energy demand." Serious interferences (>10%) must be anticipated when elements of high-excitation potential are to be determined in matrices of high-energy demand.

A considerable effort has been devoted to studying the mechanism of matrix interferences in the ICP, particularly the effects of easily ionizable elements, but a definitive explanation is not yet available. It has been established, however, that the observed interferences due to the presence of easily ionizable elements are not attributable to simple shifts in the ionization equilibria as occur in flame spectroscopy. Ionization buffering is not therefore appropriate for use in ICP-AES.

C. Physical Effects

For self-priming nebulizers the sample flow rate is determined by the nebulizer suction, by the difference in static head between the nebulizer and sample, by the hydrodynamic resistance of the transport tubing, and by the effective viscosity of the sample solution. Variation in any of these quantities will affect the instrumental response. The use of a peristaltic pump, however, will provide a constant flow of sample more or less independent of the sample properties and the nebulizer.

The important properties influencing nebulization itself are the solution viscosity and surface tension, and there will be secondary effects caused by the presence of particles and fibers. Efficient nebulization is promoted by solvents with low surface tension and low viscosity; however, surfactants do not necessarily improve the process because the fragmentation of the liquid occurs on a timescale that is short compared with the migration of the active molecules to the surface layer.

The principal factors affecting transport of the aerosol produced by the nebulizer to the plasma are the solvent volatility and evaporation factor [51] that determine the total solvent loading and the ratio of liquid aerosol flux to vapor flux. These factors are important for organic liquids for which cooling of the spray chamber, and/or the use of a condenser, may be necessary to control the solvent loading.

The practical consequences of these observations are that it is necessary to produce the samples and standards in similar solvents (e.g., acid) approximately matched in molarity [52]. Further, when the samples contain high concentrations of solutes that might affect the bulk solution properties, e.g., brines, surfactants, organic liquids, dissolved polymeric materials, the standards must be constituted similarly.

VI. DATA PROCESSING

Software for automatic calculation of analyte concentrations and archiving of operational and analytical data is standard on commercial ICP systems. Varying degrees of computer control of the instrument are available and some manufacturers offer optimization algorithms to assist in method development. However, the information content in the signals produced is far greater than is actually used, and there will undoubtedly be significant advances in extracting this information in the next few years.

A. Calibration

Careful studies of the calibration curves derived from ICP-AES have shown them to be truly linear over a concentration range of five to six orders of magnitude. As previously indicated, this range generally extends from about 1000 $\mu g\ ml^{-1}$ down to the detection limit, although curvature may be evident at lower concentrations for the very sensitive lines of elements such as Ca (Ca II 393.366 nm) and Mg (Mg II 279.553 nm). Most curve-fitting algorithms use the least-squares procedure for finding the "best" line, but some care must be exercised in using this approach, in which the computer will attempt to minimize the absolute sum of the squares of the differences of the data points and the fitted line. In an ICP calibration curve, the standard deviation remains approximately constant (equal to the standard deviation of the background signal) for the first two orders-of-magnitude increase in concentration above the detection limit. Least-squares fitting will therefore give equal weighting to the data points in this concentration range and produce a satisfactory calibration line. However, above the background equivalent concentration (BEC), the standard deviation increases in proportion to the signal. Least-squares fitting, therefore, will give undue weight to the data point of highest concentration and largely ignore the data points at the low concentration end. This problem may be overcome by weighting the fit, or by transforming the data to logarithmic scales.

Multielement instrument calibration is time-consuming, and it is impractical to repeat the entire process as a means of correcting for instrumental drift. Drift arises from a variety of causes, but manifests itself as either a translation or rotation of the calibration curve. A commonly used drift correction procedure involves recording a low and high signal for every analytical line using a minimum number of multielement calibration solutions immediately after the full calibration is completed, and automatically calculating the response curve. The solutions used may be standards, but the procedure does not depend on knowing the concentration of the analyte elements. The same solutions are rerun each time the curve needs to be updated. The preparation of multielement calibration solutions is not a trivial problem and the two main factors that must be taken into account are:

1. The presence of other analytes in the material used to prepare initial single-element stock solutions prior to mixing. This is a particular concern if multielement standards are made by attempting to mix single-element commercial standards sold for AA analysis.
2. The compatibility of the analytes and their solvent in the mixed solution (e.g., Ag in HCl, Pb in CrO_4^{2-} solution).

Commercial multielement standards are now available, but the cost is high. Such standards can be made in the laboratory by starting with "Spec-Pure" chemicals and mixing the single-element stock solutions in the appropriate ratios. A cheaper alternative for a working standard is to mix commercial AA stock solutions, or solutions made in the laboratory from Analar reagents, and then obtain the actual concentration of the multielement solution by analysis against an accurately prepared standard.

B. Optimization

The performance of an ICP system, both in terms of the detection limit and the freedom from interferences, is strongly influenced by the operating parameters. The effects of individual parametric changes, however, are not independent; the optimum power determined at fixed levels of the other parameters will vary each time one or all of them are changed. The most important operating parameters for the ICP are the power, observation height, and injector flow rate, with the outer flow, intermediate flow, and sample uptake rate being of secondary importance. To map the complex multivariate response surface associated with these parameters would take many hundreds of measurements and is impractical for routine optimization.

Often, instruments operated under so-called "optimized conditions" are, in fact, operated under arbitrary conditions that happen to meet a defined level of performance. The establishment of a true optimum both improves the data quality and provides an unambiguous statement of "best" performance, and therefore enables comparisons of procedures or instruments even though they are different.

The most widely used optimization technique is based on implementation of the "variable-step simplex" algorithm described by Nelder and Mead [53]. A simplex is simply a geometric figure in factor space having $N + 1$ vertices, where N is the number of factors to be optimized. The response of the instrument is recorded at the parametric settings corresponding to each vertex and then these responses are ranked. The procedure, in essence, involves reflecting the simplex through its centroid away from the vertex of worst response toward the factor space of best response. This produces a new vertex that is used to generate a new simplex and the old worst response is discarded. The procedure is terminated when additional simplexes produce no significant improvement in performance. The mathematics of simplex optimization are simple, and the procedure is implemented by following a set of well-defined rules. It is readily programmed and commercial software is available.

The first step in optimization is to define the objective variable by which performance will be estimated. The two most commonly used variables are the signal-to-background ratio (SBR) [54–57] and the relative freedom from interference. For a single analytical line, convergence is usually quite rapid, occurring in 20–25 vertices. Figure 11 shows the results in terms of the best and worst response at each step recorded for the optimization of the Ca II 393.36 nm line for minimum interference in the presence of a synthetic plant matrix. Note that the initial conditions fortuitously produced some results with relatively low levels of interference, ~3%, but other vertices represented unacceptably high levels. Gradually, convergence occurs and the level of interference falls to ~1%.

Optimization becomes more difficult when conditions suitable for simultaneous multielement analysis are sought. In practice, it is found that optimization of the SBR, or interference performance, of a single or group of hard lines (e.g., Cd II 226.5 nm, Zn II 202.1 nm) will produce the most suitable conditions. Inspection will then show that the plasma is being viewed in the normal analytical zone (NAZ) where efficient excitation and maximum freedom from interference occur.

C. Internal Standardization

Departures from the ideal in ICP performance can occur because of changes in the performance of the nebulizer, drift in the values of the operating parameters, and effects induced by the matrix. When

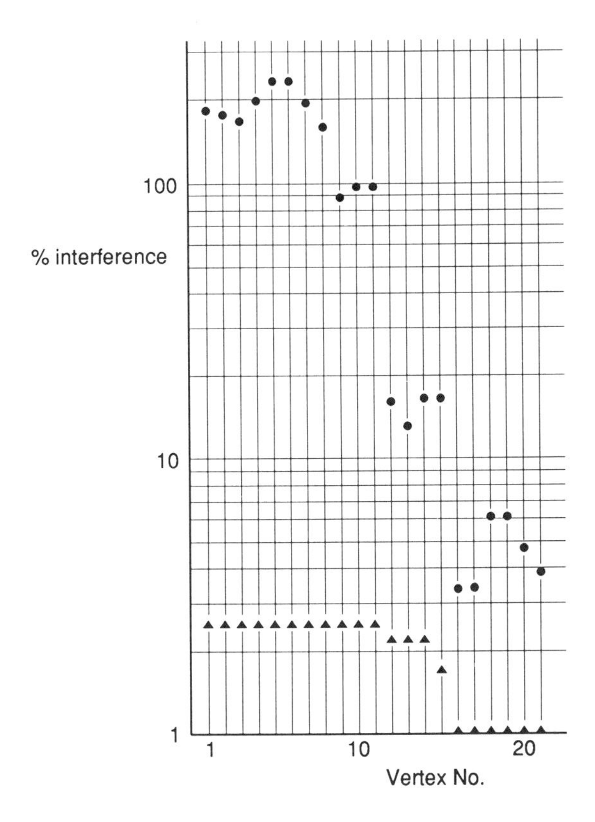

FIG. 11 Optimization of an ICP for minimum interference by use of the variable-step simplex algorithm. The best and worst response vertices are mapped on the response axis for each step of the optimization

these departures result in a significant loss in precision or accuracy, internal standardization can be of benefit.

The essential criterion in selecting internal standards is that the variances in their signals show a positive correlation with the variances of the analyte signals. In conventional arc spectrometry this was achieved by selecting lines with similar excitation characteristics, and such lines were termed homologous pairs. For ICP-AES it has been found more effective to choose internal reference lines that reflect changes in the parametric settings of the source that affect most analyte lines in a similar sense, although in varying magnitude. Two schemes have been particularly successful in this respect: Myers–Tracy [58,59] signal compensation that is commercially available and the parameter related internal standard method (PRISM) of Ramsey and Thompson [60,61]. In the former, about 5% of the plasma radiation incident on the main spectrometer is split from the beam and passed to two filter photometers, one monitoring the Sc II line at 424.638 nm and the other, the argon background emission at 450 nm. The internal reference signals derived from these observations have been shown to compensate both random noise, resulting in improved precision, and drift caused by changes in the performance of the nebulizer. In the PRISM method, two internal standard lines are chosen, one of low-excitation potential, e.g., Li 670.78 nm or Rb 780.02 nm, whose variance has been shown to correlate with variations in the sample introduction process, and a high-excitation potential line, e.g., Zn II 202.55 nm or Cd II 226.50 nm, whose variance correlates with changes in the power coupled to the plasma. Correction coefficients are calculated for each analyte line and it has been shown that both random and systematic errors are compensated. The most important benefit from a technique such as PRISM is not the improvement in precision obtained (typically, a factor of 2), but its ability to reduce matrix effects that can produce errors of >10%.

An alternative to deriving software correction factors from internal standards is to use them for interactive instrument control, and preliminary experiments of this type have been reported [62].

VII. INDUCTIVELY COUPLED PLASMA MASS SPECTROMETRY

A. Instrumentation

The concept of using mass spectrometric rather than optical detection with plasma sources was pioneered by Gray in the early 1970s [63] and ultimately resulted in the introduction of two commercial ICP-MS instruments in 1983.

A schematic diagram of an ICP-MS system is shown in Fig. 12. The tail flame of the ICP impinges on a water-cooled nickel cone having a 0.5–1.0 mm sampling orifice. Ionized gas from the central channel of the ICP flows through the cone and expands supersonically to the vacuum in the first stage that is maintained at a pressure of about 1 torr by a rotary pump. A second skimmer cone with a 1.0 mm aperture is mounted a few mm behind the sampling orifice and samples a portion of the supersonic jet into the second vacuum stage that is at a pressure of 10^{-4} torr. This is sustained by a diffusion pump or cryogenic pump, depending on the particular instrument.

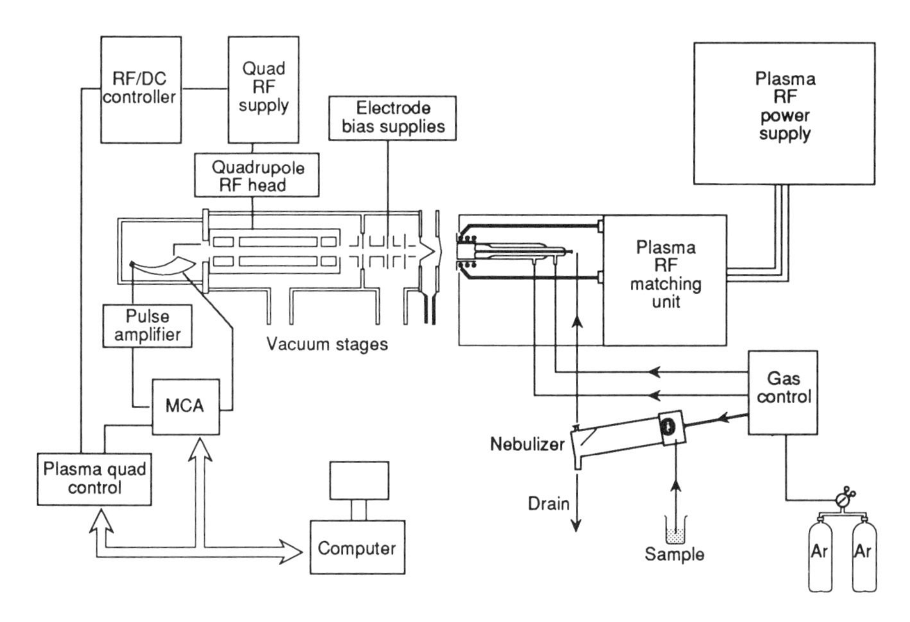

FIG. 12 Schematic diagram of an ICP-MS instrument.

The ions passing through the skimmer are accelerated by a negatively charged cylindrical extraction electrode and passed to a lens stack; this focusses the ion beam through a differential aperture into the quadrupole mass analyzer, which operates at a pressure of 10^{-6} torr. A photon stop at the beginning of the lens stack prevents light from traversing the quadrupole and reaching the detector. Ions transmitted by the quadrupole mass filter are detected by a channel electron multiplier that is normally operated in pulse counting mode.

B. Analytical Characteristics

Reports describing analytical applications of ICP-MS are beginning to appear in the literature and the numbers are likely to increase rapidly as more commercial instruments are purchased. The merits of the technique may be summarized as follows:

1. High sensitivity for most elements in the periodic table covering the range 0.001 ng ml^{-1} (e.g., for Bi, Co) to >10 ng ml^{-1} (e.g., for Ca, K, S) [64] (see Table 1).

2. The ability to carry out isotopic and isotope ratio determinations and to use isotope dilution internal standardization for high-accuracy determinations [65].
3. Greatly simplified spectra compared with the optical technique, with the entire periodic table being covered by 211 isotopic lines.
4. The ability to carry out rapid semiquantitative analyses for the majority of elements in the periodic table.

The disadvantages are:

1. Reduced detection limits for some elements, e.g., Ca and Fe, because of isobaric interferences from the plasma gas.
2. Restricted use of solvents and acids because of interferences produced from molecular ions derived from the solvent. Nitric acid is the normally preferred medium.
3. The total dissolved solids content of the sample solution must be kept below ~1000 $\mu g\ ml^{-1}$ to reduce clogging of the sampling orifice.
4. Greater sensitivity to matrix interferences, particularly those caused by easily ionizable materials, than the optical technique.
5. Less robust in terms of sample handling than the optical technique, e.g., water loading of the plasma must be controlled to reduce the effects of stable oxide formation.
6. Higher cost per analysis than the optical technique.

The high cost of ICP-MS will largely preclude it from routine soil analysis, but in the related geochemical and biochemical sciences, it will undoubtedly have a major impact. The high sensitivity for some difficult elements, notably As, Co, I, Mo, and Pb, is valuable; the simplified mass spectra facilitate analyses for the rare earth elements at high sensitivity; and perhaps most important, it enables stable isotope tracers to be used in metabolic studies and to be supported by a high sample-throughput technique (in comparison with, e.g., thermal ionization mass spectrometry).

VIII. APPLICATIONS

The applications of ICP-AES for the analysis of soils are many and varied, and it is gradually replacing atomic absorption as the preferred technique for the multielement analysis of solution samples. Some general observations on the use of ICP-AES for soil analysis are made, together with selected examples illustrating the potential of the technique and a classified list of key references.

A. General Aspects

ICP-AES may be used for the determination of major, minor, and trace elements in whole soils and soil extracts with the following limitations:

1. The detection limits for some elements, notably As, Cd, Co, Hg, Mo, Pb, and Se, are inadequate and separation/preconcentration methods are required.
2. Matrix interferences will be encountered from the major elements, i.e., Al, Ca, Fe, K, Mg, and Si, or from the presence of high levels of fluxing agents such as $LiBO_2$. Spectral interferences occur, notably from Fe [e.g., in the determination of B (249.77 nm), Co (238.89 nm), and Mo (202.15 nm)], Ca [e.g., in the determination of Al (396.15 nm)], and Al (for analytes having lines below 220 nm).
3. Soil extracts in, e.g., water, acetic acid, EDTA, or aqua regia may be analyzed directly, but when strong salt solutions are used for the extraction, e.g., molar NH_4Cl or $CaCl_2$, problems with sample introduction may be encountered, necessitating the dilution of the sample or addition of a surfactant (e.g., Triton X-100) to improve the flow characteristics of the material.
4. Sample preparation procedures should be applicable to a variety of elements to take advantage of the multielement capability of the technique. Procedures developed for AAS are readily adapted for ICP-AES, and the freedom from chemical interference enables a wider range of reagents to be used (e.g., phosphatic reagents can be used even if Ca is an analyte). Total dissolved solids contents are best kept below 10 mg ml^{-1}, if possible.
5. The slurry technique may be used as a method of minimizing the sample preparation stage, but the accuracy may suffer if the matrix is of variable composition [66].

B. Examples of Applications

Table 2 contains a list of references classified into the following categories: general soil analysis, sample preparation, interferences, separations and preconcentration procedures, hydride generation, slurry nebulization, and organic solvents. The capacity and versatility of ICP-AES are well illustrated by the following examples.

1. Determination of Mo in Soil Extracts

The availability of Mo in Scottish soils is determined by extraction with neutral 1 *M* ammonium acetate (50 g:400 ml, shaken at 20°C

TABLE 2 Applications of ICP-AES to Soil Analysis: References to Applications Areas

Soil analysis, general	Sample preparation	Interferences	Separations/preconcentration	Hydride/vapor generation	Organic solvents
45	112–115	29	49	38	26
48		50	120	39	37
77–111		60		40	126
		61		41	127
		62		42	128
		116		46	
		117		67	
		118		68	
		119		69	
				70	
				71	
				72	
				83	
				74	
				121	
				122	
				123	
				124	
				125	

for 16 hr). Extractable contents vary widely from <0.01–0.8 mg kg^{-1}, but for advisory purposes 0.01 mg kg^{-1} is considered the lowest level required to retain satisfactory growth in brassica crops, whereas levels above 0.08 mg kg^{-1} give rise to toxic effects in cattle. At 16-fold dilution, these concentrations are too low to be determined directly even by graphite-furnace atomic absorption, and in any case, the ammonium acetate causes rapid degradation of the graphite tube surface.

The presence of the ammonium acetate, co-extracted major cations (ca. 200–4000 mg kg^{-1}; Na 10–150 mg kg^{-1}; Fe 1–50 mg kg^{-1}), and organic matter preclude the use of direct preconcentration by reduction of volume, and a two-step process is necessary. The procedure used in this laboratory involves evaporation of the ammonium acetate solution to dryness followed by oxidation with 4 × 5 ml aliquots of redistilled nitric acid, taking to dryness on each occasion. The residue is then treated with 2 × 10 ml aliquots of 6 *M* HCl, which improves its solubility, and dissolved finally in 10 ml of 0.05 *M* HCl. The pH of this solution is adjusted to 1.5 with NH_3. This step is advantageous because Mo in acid solutions may exist as the MoO_2^+ cation or as heteropoly anions. The higher pH favors formation of the simple MoO_4^{2-} anion and improves the efficiency of removal of the major cations by ion-exchange chromatography. Thus, a 10 ml column of strong cation exchange resin is prepared in NH_4^+ form. Three ml of the solution are injected onto the column and pumped through using an aqueous carrier stream at a flow rate of 0.5 ml min^{-1}. On-line detection using ICP-AES indicated that the Mo is eluted between 2 and 10 ml in conjunction with the soluble organic fraction. In routine analysis, the eluents are collected off-line and the 8 ml aliquots analyzed in the normal manner. The column treatment reduces the levels of major cations in synthetic solutions to less than 0.1 $\mu g\ ml^{-1}$; although with soil extracts, a few $\mu g\ ml^{-1}$ are carried through the column, these levels are insufficient to cause interference. The final solution, when introduced into the plasma, causes a strong red coloration of the plasma due to emission from cyanogen bands. The extensive band emission precludes the use of some of the most common Mo lines, but the spectrum in the region of 290 nm is clear, permitting use of the Mo 292.339 nm line that yields a detection limit of ~0.015 $\mu g\ ml^{-1}$. The dilution in the exchange step is × 2.7, but the initial stage involves a × 5 concentration so that the method has a detection limit of 0.008 mg kg^{-1}. The results by this procedure agree well with those obtained by a chemical-concentration-dc arc method [67].

Molybdenum in herbage samples may often be determined directly by ICP-AES, provided care is taken to assess and correct for interference from Fe that occurs on most of the useful Mo lines (e.g., Mo 202.03 nm, Mo 379.83 nm). However, sensitivity is a problem at the lowest levels (0.02 mg kg^{-1}) and a preconcentration/separation procedure such as that described above may be necessary.

2. *Determination of B in Soil Extracts*

The determination of boron presents similar problems to those encountered with Mo in that the colorimetric methods lack sensitivity and are prone to interference, and the element is difficult to atomize in flame or furnace atomic absorption. The detection limit by ICP-AES is of the order of 0.006 $\mu g\ ml^{-1}$ (B 249.67 nm) and is therefore adequate for determinations in soil extracts or plant materials, provided that the potential spectral interference from Fe is assessed and if necessary corrected.

Boron availability in topsoils is estimated by boiling with distilled water under reflux for 10 min, using a soil:water ratio of 1:2 (w/v), and filtering immediately under vacuum. Pyrex glassware can be conditioned for this determination by carrying out blank runs, but silica apparatus is preferred. The extractable levels typically cover the range 0.3–3.5 mg kg^{-1} (the range for satisfactory plant growth is 0.7–3.0 mg kg^{-1}). Boron can be determined directly in the aqueous extract, and as the co-extracted levels of Fe are low (<10 μg ml^{-1}), there are no significant interferences.

Boron may also be determined in plant materials, blood plasma, urine, and diet samples. For plants, 1 g of dried milled sample is taken to dryness with 10 ml of saturated $Ca(OH)_2$, which prevents the loss of volatile boric acid. The sample is then ashed for a minimum of 3 hr at 450–500°C. The residue is dissolved and taken to dryness with two 8 ml aliquots of 6 *M* HCl, then dissolved in 5 ml of 1.5 *M* HCl, and the final volume adjusted to 50 ml, using deionized water. For diet or tissue samples, a wet oxidative digestion is necessary and the most convenient method of achieving this and avoiding the use of perchloric acid is to carry out a bomb digestion with nitric acid in a microwave oven. The digestion is extremely rapid and the final solution can be dried and redissolved to give convenient acid strength and concentration ranges.

3. *Determination of Hydride-Forming Elements (As, Bi, Ge, Pb, Sb, Se, Sn, Te)*

The detection limits by ICP-AES for this important group of elements are too high to allow their direct determination in soils or soil extracts. The problem is overcome by conversion of the elements to their volatile hydrides, using the well-known $NaBH_4$ reduction procedure.

The use of continuous-flow hydride generation has been studied extensively by Thompson and coworkers [45,68], who have developed methods for a range of geological [69,70] and environmental materials including soils and sediments [71,72], waters [73], and herbage [74]. It was found that As, Sb, Bi, Se, and Te are best reduced in strong acid solution (5 *M* HCl), whereas Ge and Sn are most efficiently reduced at much lower acidities, e.g., in 0.1 *M* HCl or in weak organic acids such as 1% w/v tartaric acid. Additionally, hydride generation efficiency is reduced for the high oxidation states of As(V), Sb(V), and Se(VI). Prior reduction of As and Sb to the (III) state is accomplished readily using an aqueous KI solution, but reduction using the bromide ion is preferred for Se(VI) to avoid direct reduction to Se(0). In the hydride generation manifold described by Goulter [75] (Fig. 13), a flow of 6% w/v H_2O_2 is incorporated prior to the addition of $NaBH_4$, which both reduces the generation of hydrogen and increases the sensitivity for Pb.

Mutual interferences between the hydride forming elements is minimal, but some transition elements, notably Cu^{2+}, Ni^{2+} and to

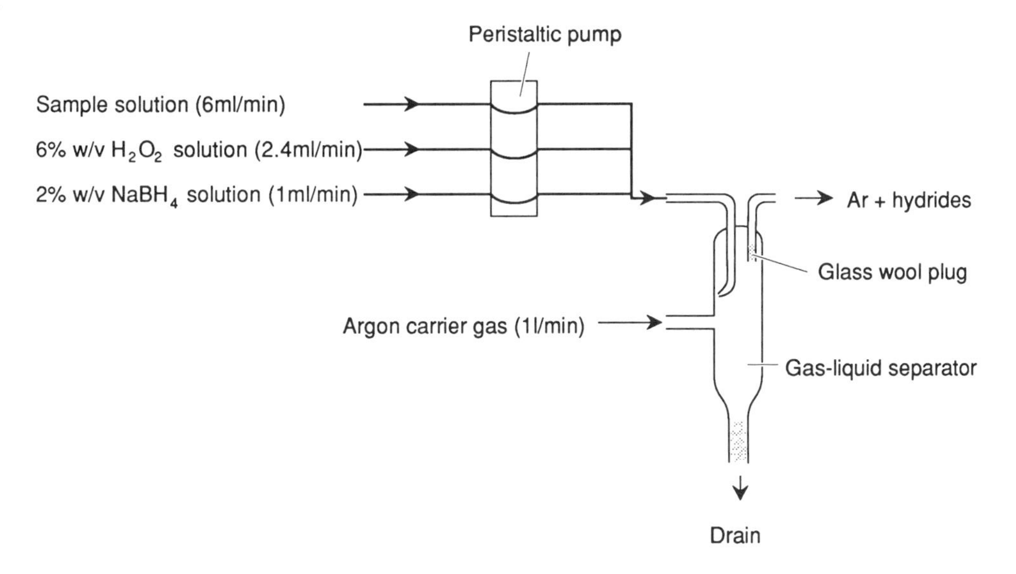

FIG. 13 Schematic diagram of the ARL hydride generation system.

a lesser extent Fe^{3+}, cause serious interference. For Ge and Sn, the use of tartaric acid to control the pH also complexes the transition elements and masks their effect, but for the remainder a separation is necessary. The preferred method is that described by Bedard and Kerbyson [76], which involves co-precipitation of the analytes with $La(OH)_3$ and redissolution in 5 *M* HCl. Following the dissolution and the addition of La, the pH is raised with ammonia, causing precipitation to occur, with the copper being retained in solution as the $Cu(NH_3)_4{}^{2+}$ complex. Unfortunately, iron is also precipitated in this procedure so that its interference effect is not entirely removed.

For the determination of As, Sb, and Bi in soils and sediments [72], the sample is digested with concentrated hydrochloric acid (0.25 g:5 ml concentrated HCl) at 150°C in capped test tubes in a heating block. The prior reduction is effected with 5 ml aqueous KI solution. To determine Se in soils, an oxidative digestion is necessary to release Se from the soil organic matter [72]. Thus, 0.5 g of soil is first subjected to mild oxidation using 2 ml of concentrated nitric acid at 50°C overnight. To the cooled solution, 1 ml of 72% w/w perchloric acid is added and the oxidation carried out in two stages at 100°C until the emission of brown fumes ceases, and then at 170 ± 1°C for 1 hr until the sample is nearly dry. The residue is dissolved in a minimal quantity of concentrated HCl and made up to 5 ml using deionized water. The co-precipitation procedure is then employed to remove Cu and Ni and the prior reduction carried out using 10 ml of 5% w/v KBr solution.

REFERENCES

1. T. B. Reed, *J. Appl. Phys.*, *32*:821 (1961).
2. T. B. Reed, *Int. Sci. Technol.*, *42*:42 (1962).
3. S. Greenfield, I. L. Jones, and C. T. Berry, *Analyst*, *89*:713 (1964).
4. S. Greenfield and H. McD. McGeachin, *Chem. Brit.*, *16*:653 (1980).
5. R. H. Wendt and V. A. Fassel, *Anal. Chem.*, *37*:920 (1965).
6. R. H. Scott, V. A. Fassel, R. N. Kniseley, and D. E. Nixon, *Anal. Chem.*, *46*:76 (1974).
7. G. W. Dickinson and V. A. Fassel, *Anal. Chem.*, *41*:1021 (1969).
8. P. W. J. M. Boumans and F. J. de Boer, *Spectrochim. Acta*, *27B*:391 (1972).
9. P. W. J. M. Boumans and F. J. de Boer, *Spectrochim. Acta*, *30B*:309 (1975).
10. S. Greenfield, H. McD. McGeachin, and P. B. Smith, *Talanta*, *22*:1 (1975).
11. S. Greenfield, H. McD. McGeachin, and P. B. Smith, *Talanta*, *22*:553 (1975).
12. S. Greenfield, H. McD. McGeachin, and P. B. Smith, *Talanta*, *23*:1 (1976).
13. M. I. Boulos and R. M. Barnes, in *Developments in Atomic Plasma Spectrochemical Analysis* (R. M. Barnes, ed.), Heyden, Philadelphia, Pa., 1981, p. 20.
14. P. W. J. M. Boumans, *Theory of Spectrochemical Excitation*, Hilger and Watts, London, 1966.
15. B. L. Sharp, *Selected Annaul Reviews of the Analytical Sciences*, Vol. 4, Royal Society of Chemistry, 1976.
16. P. W. J. M. Boumans, ed., *Inductively Coupled Plasma Emission Spectrometry*, Parts I and II, Wiley, New York, 1987.
17. P. W. J. M. Boumans, in "Second ICP Conference, Noordwijk Aan Zee, Instrumentation—Sample Introduction and Plasma Torches" (J. M. Mermet, ed.), *ICP Info. Newsletter*, *4*:89 (1978).
18. M. W. Blades and G. Horlick, *Spectrochim. Acta*, *36B*:861 (1981).
19. M. W. Blades and G. Horlick, *Spectrochim. Acta*, *36B*:881 (1981).

20. M. W. Blades, in *Inductively Coupled Plasma Emission Spectrometry* (P. W. J. M. Boumans, ed.), Wiley, New York, 1987.
21. B. Capelle, J. M. Mermet, and J. Robin, *Appl. Spectrosc.*, *36*:102 (1982).
22. J. M. Mermet, C. Trassy, and P. Ripoche, in *Developments in Atomic Plasma Spectrochemical Analysis*, Heyden, Philadelphia, Pa., 1981, p. 245.
23. B. L. Sharp, N. W. Barnett, J. C. Burridge, and J. F. Tyson, *J. Anal. At. Spectrom.*, *2*:177R (1987); reported from M. Muszynski, paper presented at the 1987 Winter Conference on Plasma and Laser Spectrochemistry, Lyon, France, Jan. 6–12, 1987.
24. P. W. J. M. Boumans and M. C. Lux-Steiner, *Spectrochim. Acta*, *37B*:97 (1982).
25. G. Angleys and J. M. Mermet, *Appl. Spectrosc.*, *38*:647 (1984).
26. R. C. Ng, H. Kaiser, and B. Meddings, *Spectrochim. Acta*, *40B*:63 (1985).
27. M. W. Tikkanen, S. D. Arellano, J. E. Goulter, and M. W. Routh, *Spectrosc. (Springfield, Ore.)*, *1*:30 (1986).
28. B. L. Sharp, *J. Anal. At. Spectrom.*, *3*:613; *3*:939 (1988).
29. B. R. Baginski and J. E. Meinhard, *Appl. Spectrosc.*, *38*:568 (1984).
30. H. Anderson, H. Kaiser, and B. Meddings, in *Developments in Atomic Plasma Spectrochemical Analysis* (R. M. Barnes, ed.), Heyden, Philadelphia, Pa., 1981.
31. R. S. Babington, U.S. Pats. 3,421,692; 3,421,699; 3,425,058; 3,425,059, and 3,504,859.
32. R. F. Suddendorf and K. W. Boyer, *Anal. Chem.*, *50*:1769 (1978).
33. P. A. M. Ripson and L. de Galan, *Spectrochim. Acta*, *36B*:71 (1981)
34. J. F. Wolcott and C. B. Sobel, *Appl. Spectrosc.*, *30*.005 (1982).
35. L. Ebdon and M. Cave, *Analyst*, *107*:172 (1982).
36. G. L. Moore, P. J. Humphries-Cuff, and A. E. Watson, *ICP Inf. Newsletter*, *9*:763 (1984).
37. F. J. M. J. Maessen, G. Krunning, and J. Balke, *Spectrochim. Acta*, *41B*:3 (1986).
38. L. R. Parker, Jr., N. H. Tioh, and R. M. Barnes, *Appl. Spectrosc.*, *39*:45 (1985).
39. G. S. Pyen, S. E. Long, and R. F. Browner, *Appl. Spectrosc.*, *40*:246 (1986).
40. X. Wang and R. M. Barnes, *Spectrochim. Acta*, *41*:967 (1986).
41. B. Welz and M. Schubert-Jacobs, *J. Anal. At. Spectrom.*, *1*:23 (1986).

42. J. Agterdenbos, J. P. M. van Noort, F. F. Peters, and D. Bax, *Spectrochim. Acta, 41B*:283 (1986).
43. H. Matusiewicz, *J. Anal. At. Spectrom., 1*:171 (1986).
44. P. W. J. M. Boumans and J. J. A. M. Vrakking, *J. Anal. At. Spectrom., 2*:513 (1987).
45. M. Thompson and J. N. Walsh, *A Handbook of Inductively Coupled Plasma Spectrometry*, Blackie, Glasgow and London, 1983.
46. J. F. Alder, A. M. Gunn, and G. F. Kirkbright, *Anal. Chim. Acta, 92*:43 (1977).
47. J. A. C. Broekaert and P. B. Zeeman, *Spectrochim. Acta, 39*: 851 (1984).
48. S. E. Church, *Geostand. Newsletter, 2*:133 (1981).
49. M. Thompson, M. H. Ramsey, and B. Pahlavanpour, *Analyst, 107*:1330 (1982).
50. M. H. Ramsey and M. Thompson, *J. Anal. At. Spectrom., 1*: 185 (1986).
51. G. Kreuning and F. J. M. J. Maessen, *Spectrochim. Acta, 42B*: 677 (1987).
52. S. Greenfield, H. McD. McGeachin, and F. A. Chalmers, *ICP Inf. Newsletter, 3*:117 (1977).
53. J. A. Nelder and R. Mead, *Computer J., 7*:308 (1965).
54. S. Greenfield and D. T. Burns, *Spectrochim Acta, 34B*:423 (1979).
55. L. A. Yarbro and S. N. Deming, *Anal. Chim. Acta, 73*:391 (1974).
56. L. Ebdon, M. R. Cave, and D. J. Mowthorpe, *Anal. Chim. Acta, 115*:179 (1980).
57. S. P. Terblanche, K. Visser, and P. B. Zeeman, *Spectrochim. Acta, 36B*:293 (1981).
58. S. A. Myers and D. H. Tracy, *Spectrochim. Acta, 39B*:1227 (1983).
59. J. Marshall, G. Rodgers, and W. C. Campbell, *J. Anal. At. Spectrom., 3*:241 (1988).
60. M. H. Ramsey and M. Thompson, *Analyst, 110*:519 (1985).
61. M. H. Ramsey and M. Thompson, *J. Anal. At. Spectrom., 2*: 497 (1987).
62. M. Thompson, M. H. Ramsey, B. J. Coles, and C. M. Du, *J. Anal. At. Spectrom., 2*:185 (1987).
63. A. L. Gray, *Analyst, 100*:289 (1975).
64. A. L. Gary and A. R. Date, *Analyst, 108*:1033 (1983).
65. D. Beauchemin, J. W. McLaren, A. P. Myktiuk, and S. S. Berman, *J. Anal. At. Spectrom., 3*:305 (1988).
66. L. Ebdon and J. R. Wilkinson, *J. Anal. At. Spectrom., 2*:39 (1987).

67. R. L. Mitchell, *Tech. Commun. Commonwealth Bur. Soil Sci.*, *44A* (1964).
68. M. Thompson, B. Pahlavanpour, S. J. Walton, and G. F. Kirkbright, *Analyst*, *103*:568 (1978).
69. M. Thompson and B. Pahlavanpour, *Anal. Chim. Acta*, *109*:251 (1979).
70. B. Pahlavanpour, M. Thompson, and S. J. Walton, *J. Geochem. Explor.*, *12*:45 (1979).
71. B. Pahlavanpour, J. H. Pullen, and M. Thompson, *Analyst*, *105*:274 (1980).
72. B. Pahlavanpour, M. Thompson, and L. T. Thorne, *Analyst*, *105*:756 (1980).
73. M. Thompson, B. Pahlavanpour, and L. T. Thorne, *Water Res.*, *15*:407 (1981).
74. B. Pahlavanpour, M. Thompson, and L. T. Thorne, *Analyst*, *106*:467 (1981).
75. J. E. Goulter, Paper 129, Conference on Analytical Chemistry and Applied Spectroscopy, Pittsburgh, Pa., 1981.
76. M. Bedard and J. D. Kerbyson, *Can. J. Spectrosc.*, *21*:64 (1976).
77. J. B. Jones, Jr., "Analysis of Soil Analyses and Plant Tissue Ash by ICAP Emission Spectroscopy, in *Applications of Inductively Coupled Plasmas to Emission Spectroscopy* (R. M. Barnes, ed.), Franklin Institute Press, Philadelphia, Pa., 1978.
78. J. L. Hern, "Elemental Analysis in Agriculture Using Inductively Coupled Plasma-Atomic Emission Spectroscopy, in *Applications of Plasma Emission Spectrochemistry* (R. M. Barnes, ed.), Heyden, Philadelphia, Pa., 1979.
79. J. B. Jones, Jr., "Analysis of Soil Extracts, Plant Tissue and Other Biological Substances," in *Developments in Atomic Plasma Spectrochemical Analysis* (R. M. Barnes, ed.), Heyden, Philadelphia, Pa., 1981.
80. R. C. Munter and R. A. Grande, "Plant Tissue and Soil Extract Analysis by ICP-AES," in *Developments in Atomic Plasma Spectrochemical Analysis* (R. M. Barnes, ed.), Heyden, Philadelphia, Pa., 1981.
81. P. N. Soltanpour and S. M. Workman, "Use of Inductively-Coupled Plasma Spectrometry for the Simultaneous Determination of Macro- and Micronutrients in NH_4HCO_3-DTPA Extracts of Soils," in *Developments in Atomic Plasma Spectrochemical Analysis* (R. M. Barnes, ed.), Heyden, Philadelphia, Pa., 1981.
82. B. L. Sharp, "Applications: Agriculture and Food," in *Inductively Coupled Plasma Emission Spectrometry* (P. W. J. M. Boumans, ed.), Wiley, New York, 1987.

83. R. H. Scott and M. L. Kokot, *Anal. Chim. Acta, 75*:257 (1975).
84. J. B. Jones, *Commun. Soil Sci. Plant Anal., 8*:349 (1977).
85. R. L. Dahlquist and J. W. Knoll, *Appl. Spectrosc., 32*:1 (1978).
86. N. R. McQuaker, P. D. Kluckner, and G. N. Chang, *Anal. Chem., 51*:888 (1979).
87. J. L. Manzoori, *Talanta, 27*:682 (1980).
88. P. N. Soltanpour, S. M. Workman, and A. P. Schwab, *Soil Sci. Am. J., 43*:75 (1979).
89. S. Yamasaki, *Kagaku no Ryoiki, Zokan, 127*:187 (1980).
90. D. W. Hoult, M. M. Beaty, and G. F. Wallace, *At. Spectrosc., 1*:157 (1980).
91. P. J. Lechler, W. R. Roy, and R. K. Leininger, *Soil Sci., 130*:238 (1980).
92. W. D. Gestring and P. N. Soltanpour, *Comm. Soil Sci. Plant Anal., 12*:733 (1981).
93. C. Riandey, P. Alphonse, R. Gavinelli, and M. Pinta, *Analusis, 10*:323 (1982).
94. J. B. Jones, *J. Assoc. Off. Anal. Chem., 65*:781 (1982).
95. C. F. Aten, J. B. Bourke, and J. C. Walton, *J. Assoc. Off. Anal. Chem., 66*:766 (1983).
96. P. Schramel, L. Xu, A. Wolf, and A. Hasse, *Fresenius Z. Anal. Chem., 313*:213 (1982).
97. J. Benton-Jones, *Spectrochim. Acta, 38B*:271 (1983).
98. R. A. Isaac and W. C. Johnson, *Spectrochim. Acta, 38B*:277 (1983).
99. G. A. Spiers, M. J. Dudas, and L. W. Hodgins, *Comm. Soil Sci. Plant Anal., 14*:629 (1983).
100. M. Thompson, *Appl. Environ. Geochem.*, 75 1983, Ed. I. Thornton, Academic, London.
101. E. Janssens and R. Dams, *4th Heavy Met. Environ., Int. Conf., 1*:226 (1983).
102. M. W. Pritchard and J. Lee, *Anal. Chim. Acta, 157*:313 (1984).
103. J. A. C. Broekaert and P. B. Zeeman, *Spectrochim. Acta, 39B*:851 (1984).
104. A. Brzezinska, A. Balicki, and J. Van Loon, *Water, Air, Soil Pollut., 21*:323 (1984).
105. D. J. Lyons and P. J. Lynch, *Commun. Soil Plant Anal., 16*:15 (1985).
106. I. R. Willett and B. A. Zarcinas, *Commun. Soil Plant Anal., 17*:183 (1986).
107. Brzezinska-Paudyn, J. Van Loon, and R. Hancock, *At. Spectrosc., 7*:72 (1986).

108. M. Baucells, G. Lacort, and M. Roura, *Analyst, 110*:1423 (1985).
109. E. E. Cary, D. L. Grunes, V. R. Bohman, and C. Sanchirico, *Agron. J., 78*:933 (1986).
110. I. Novozamsky, R. Van Eck, V. J. G. Houba, and J. J. Van der Lee, *Neth. J. Agric. Sci., 34*:185 (1986).
111. R. M. Awadallah, M. K. Sherif, H. A. Amrallah, and F. Grass, *J. Radioanal. Nucl. Chem., 98*:235 (1986).
112. D. R. Parker and H. H. Gardner, *Comm. Soil Sci. Plant Anal., 12*:1311 (1981).
113. S. M. Combs and R. H. Dowdy, *Comm. Soil Sci. Plant Anal., 13*:87 (1982).
114. L. D. Tyler and M. B. McBride, *Soil Sci., 134*:198 (1982).
115. P. J. Lamothe, T. L. Fries, and J. J. Consul, *Anal. Chem., 58*:1881 (1986).
116. M. H. Ramsey, M. Thompson, and S. J. Walton, *J. Anal. At. Spectrom., 2*:33 (1987).
117. F. J. M. J. Maessen, H. Balke, and J. L. M. de Boer, *ICP Inf. Newsletter, 8*:21 (1982).
118. M. Thomson and M. H. Ramsey, *Analyst, 110*:1413 (1985).
119. H. Tao, Y. Iwata, T. Hasegawa, Y. Nojiri, H. Haraguchi, and K. Fuwa, *Bull. Chem. Soc. Jpn., 56*:1074 (1983).
120. A. F. Ward, L. F. Marciello, L. Carrara, and V. J. Luciano, *Spectrosc. Lett., 13*:803 (1980).
121. R. Roehl and H. J. Hoffman, *Fresenius Z. Anal. Chem., 322*:439 (1985).
122. G. Vajucic, J. D. Steiner, M. Siroki, and M. Herak, *J. Anal. At. Spectrom., 1*:171 (1986).
123. D. L. Miles and K. Lewin, "Determination of Trace Levels of Sulphide in Natural Waters by ICP-OES," Paper presented at the Colloquium Spectroscopicum Internationale XXIV, Garmisch-Partenkirchen, FRG, Sept. 15–20, 1985.
124. A. Sanz-Medal, J. E. Sanchez-Uria, and S. Arribas Jimeno, *Analyst, 110*:563 (1985).
125. S. M. Workman and P. N. Soltanpour, *Soil Sci. Soc. Am. J., 44*:1331 (1980).
126. W. Nisamaneepong, D. L. Haas, and J. A. Caruso, *Spectrochim. Acta, 40B*:3 (1985).
127. T. Brotherton, B. Barnes, N. Vela, and J. Caruso, *J. Anal. At. Spectrom., 2*:389 (1987).
128. M. Thompson and L. Zao, *Analyst, 110*:229 (1985).

3

Ion-Selective Electrodes

OSCAR TALIBUDEEN *Blackthorn Reach, Aldeburgh, Suffolk, England*

I. INTRODUCTION

The first ion-selective electrode (ISE) was invented in the first decade of this century. The active component was a glass membrane sensitive to hydrogen ions, and the electrode forms the basis of the modern pH glass electrode. Such electrodes were first used in soil and plant science more than 50 years ago. About 20 years later, glass membranes sensitive to Na^+ and K^+ ions were devised by modifying the composition of the glass, although interference by hydrogen ions has not been eliminated. In the intervening period, clay mineral membranes were extensively tested for cation analyses in solutions, but these were slow in response, not very ion-specific, and too delicate for use in routine work. Since then much attention has been given to crystalline membranes and liquid membranes sensitive to a wide range of cations and anions. This has resulted in the availability of membranes of high ion specificity and robustness under appropriate environmental conditions.

This chapter is mainly concerned with the use of ISEs in the analysis of soil extracts, but also considers briefly their use in plant analysis and in vivo plant studies. The reader is referred to the voluminous literature on the subject for details of the pH glass electrode.

II. BASIC PRINCIPLES

In a modern ISE, a passive membrane separates the internal standard and test solutions of the ions. Electrons, simple ions, or charged or neutral complexes of the test ion are transported across the membrane interfaces to extents that are proportional to the compositions of the solutions on either side of the membrane. The electrostatic potential difference (EMF) thus developed across the membrane can be measured by coupling the membrane half-cell with a standard reference electrode half-cell. This potential difference E is given by the Nernst equation

$$E = E_0 + \frac{RT}{z_i F} \ln a_i \qquad (1)$$

where z_i is the valence and $a_i = \gamma_i c_i$, the activity of the ion, with γ and c the ionic activity coefficient and concentration, respectively. The theoretical slope of the E:ln a_i relationship at 25°C is 59.12/z_i mV for ions of valence z_i, the slope being positive for cations and negative for anions.

A. Types of Membranes

Membranes are essentially of three types, further subdivisions of which are shown in Table 1, with their principal parameters in Table 2 (see Sec. II.B).

1. *Solid Ion-Exchange or Fixed-Site Membranes*

These principally consist of inorganic ionic crystals in which their manner of preparation induces specific preference for the cation or the anion, and ion selectivity depends not only on the ion-exchange constant but also on the relative mobility of the specific and interfering ions in the membrane matrix, much more so than in liquid membranes. Examples of these are the Ag halide-sulfide, the LaF_3, and the glass electrodes (for H_3O^+, Li^+, Na^+, and K^+).

Recently, Ebdon [3] described a new type of solid-state membrane in which ion-selective sensor groups are attached to a tough and inert polymer by covalent bonds. These ISEs have not as yet been universally or commercially evaluated so they are excluded from Table 1. Robustness, long operational and shelf life, improved ion selectivity, reproducible and stable response, low cost, chemical stability toward physically and chemically hostile environments, and applicability to an unlimited number of ions are the potential advantages over commonly used ISEs claimed by the authors. Sensors for calcium (diallylphenyl phosphonate groups) and nitrate (quaternary ammonium salts with allyl substituents) ions were covalently

TABLE 1 Types of Ion-Selective Membranes Commonly Used

Type of membrane		Composition
1. Without sites	(a) Nonporous solid	Organic crystals; plastic film impregnated neutral carrier
	(b) Nonporous liquid	Organic liquid containing neutral carrier
2. With fixed sites	(a) Nonporous solid	Glass; defect inorganic crystals; high cross-link exchange resins
	(b) Porous solid	Compacted membranes of very small particles of nonporous solids
3. With mobile sites	(a) Liquid	Ion exchangers; hydrophobic salts with dissociation and ion pairing in organic solvent
	(b) Solid	Pure crystalline halides and sulfides; lanthanum fluoride

Source: Adapted from Ref. 1.

bonded in these ISEs to a triblock elastomer, poly-(styrene-*b*-butadiene-*b*-styrene). For these, a few supporting experimental data are given by the authors.

2. *Liquid Ion-Exchange Membranes*

Like the liquid "neutral-carrier" membranes descibed below, these were until recently truly liquid in that the ion-complexing exchanger was dissolved in an inert solvent and physically restrained in an inert membrane, at least temporarily, and replenished at intervals. In modern practice, the dissolved exchanger is impregnated in thermosetting plastic materials so that the so-called liquid membrane is cast at ambient temperatures as a thin film in which the solution is held for much longer periods. Such membranes are much more robust but cannot be recharged or reconditioned after the impregnated solution has been lost. The replacement film costs more but is considered to be much more reliable and long-lasting. The interfaces of a cation-selective ion-exchange membrane must be permeable to

TABLE 2 Composite Selectivity Parameters for the Various Types of Membranes

Ion-exchange membrane	Composite selectivity parameter	Comment
Solid	$\frac{u_j}{u_i} \cdot K_{ij}$	Ion mobility opposes equilibrium selectivity
Liquid		
(1) Dissociated	$\frac{u_j}{u_i} \cdot \frac{k_j}{k_i}$	u and k depend only on the solvent
(2) Associated; poorly mobile sites	$\frac{u_j}{u_i} \cdot \frac{k_j}{k_i}$	$u_i, u_j \gg u_s$
(3) Associated; highly mobile sites	$\frac{u_{js}}{u_{is}} \cdot K_{ij}$	
Neutral carriers	$\frac{K_j}{K_i}$	$u_{js} = u_{is}$

Notes

1. u refers to mobility in the membrane.
2. Subscripts i, j, s, is, and js refer to specific ion, interfering ion, neutral carrier, and the complexed ions.
3. K_{ij} (= K_j/K_i) is the equilibrium constant of the reaction

$$J^+ \ (aq) + IS^+ \ (\text{membrane}) \rightleftharpoons I^+ \ (aq) + JS^+ \ (\text{membrane})$$

4. K_i is the formation constant of the reaction

$$I^+ + X^- + S \rightleftharpoons IS^+ + X^-$$

5. k_i, k_j are the partition coefficients of the dissociated ions i, j between water and the solvent of the liquid membrane.

Source: Adapted from Ref. 2.

the cation but not to its co-anion, nor to the cation-selective anion dissolved in the inert solvent within the membrane (Fig. 1 [4]). The cation complex formed within the membrane is uncharged and freely mobile, in which case the selectivity constant of the complexing ion for the specific and interfering cations determines the EMF response of the electrode to cation concentration. If, however, the

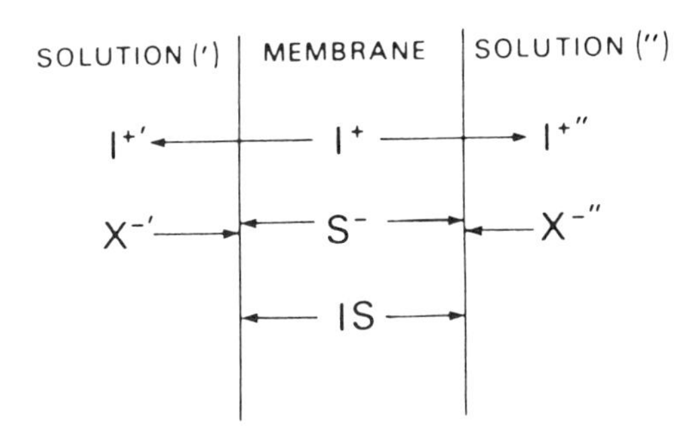

FIG. 1 Ion transfer in a system incorporating a liquid ion-exchanger membrane. I^+, S^-, and X^- refer to the counter ions, the ion-exchanging site, and co-ions, respectively. (From Ref. 2.)

cation complex dissociates fully, then the partition coefficients of the two cations between water and the solvent in the membrane are implicated.

3. Liquid Neutral Carrier Membranes

The ion-selective components in such membranes are macrocyclic antibiotics and cyclic polyethers in low or high dielectric constant solvents, depending on whether they are thin or thick membranes. They form a high concentration of a complex with the same charge as the complexed ion. The mechanism of the development of the potential difference is represented diagrammatically in Fig. 2 and is controlled by the equilibrium constant K_{is} of the reaction

$$I^+ + X^- + (S^*) \rightleftharpoons (IS^*)^+ + (X^*)^-$$

Neutral carriers can be tailored to bind a particular cation and hence provide a simple mechanism of high specificity.

4. Ion-Selective Field Effect Transistors (ISFETs)

Since 1970, further miniaturization of the ISE has become possible by developments in the technology of solid-state integrated circuits, which have produced the *ion-selective field effect transistor*, or ISFET. In this device, the gate metal in the *metal-oxide-semiconductor field effect transistor*, or MOSFET, is replaced by an ion-selective membrane coupled to an adjacent reference electrode. An all-inclusive descriptive name for the device is "IGFET" (insulated gate field effect transistor), but ion selectivity is implicit in the term ISFET so it is preferred here.

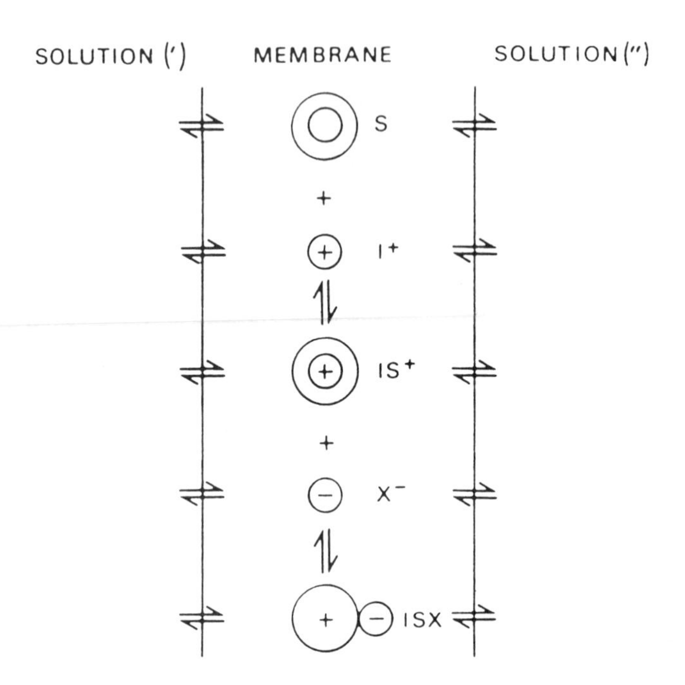

FIG. 2 Ion transfer in a system incorporating a neutral carrier membrane. S, I^+, $(IS)^+$, X^-, and ISX refer to the neutral molecular carrier, free cation, complexed cation, free anion, and neutral complex, respectively. (From Ref. 2.)

No applications have been reported so far to soil science and its ancillary disciplines, but only to animal and human physiology and to environmental science. However, the increased interest in ion fluxes in soils, plants, and across the soil-plant interface suggests that the advantages of ISFETs merit the attention of research scientists in this area with a view to future applications. These advantages are abstracted below from selected papers that also provide accounts of the historical development, the underlying theory, and a bibliography [5,6]. Chemically sensitive field effect transistors (CHEMFETs) have also been reported, designed to detect and measure neutral molecules like H_2 and O_2 and also organophosphorus compounds [7], but their performance is difficult to assess from that publication.

A schematic diagram of an ISFET is reproduced in Fig. 3, which shows that its components are a drain (1), source (2), substrate (3), insulator (5), metal lead (6), and encapsulant (10). The

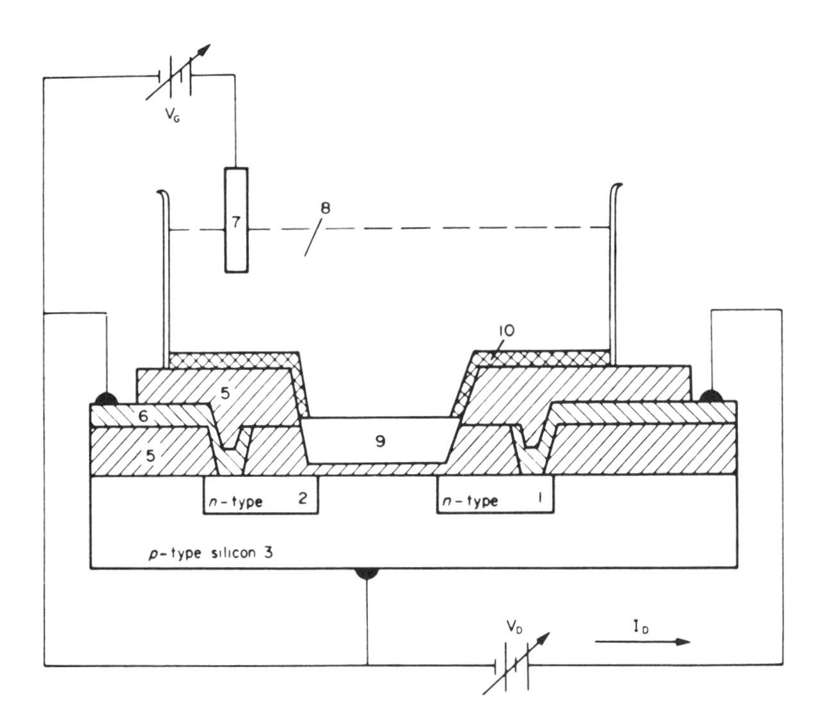

FIG. 3 Schematic diagram of ISFET. (1) Drain, (2) source, (3) substrate, (5) insulator, (6) metal lead, (7) reference electrode, (8) solution, (9) membrane, (10) encapsulant. (From Ref. 5.)

metal gate of the MOSFET is replaced by the reference electrode (7), test solution (8), and ion-selective membrane (9). The gate voltage V_G relates to (7), the drain voltage V_D and the drain current I_D to (9). Its basic difference from the conventional ISE setup is that its ion-selective membrane is integrated fully with the solid-state amplifier, eliminating the need for a conducting lead. This makes the detector assembly both very small and also a low-impedance device, greatly increasing the sensitivity of the signal and decreasing the cost of the ancillary electronics, compared with the high-impedance ISE.

Figure 4 is reproduced here to illustrate the construction and size of an ISFET device in an application to human physiology [5]. The figure shows an ISFET chip (2), dimensions 1.28 × 2.16 mm, mounted on a dual-lumen catheter (1), a section of the upper lumen being cut off 6 mm from the end to expose the ISFET to view.

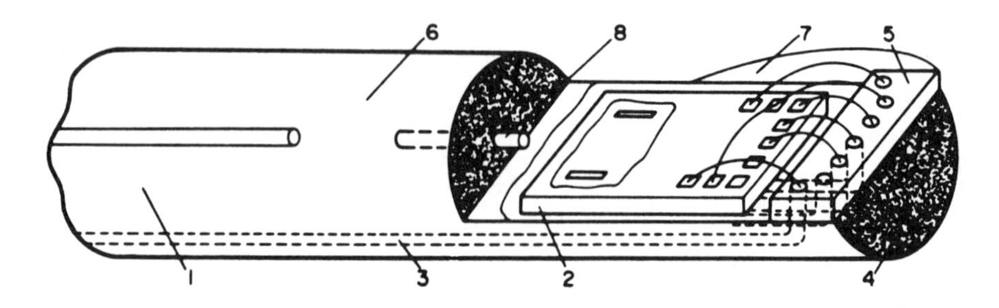

FIG. 4 ISFET chip mounted on dual-lumen 6F catheter (see text for description). (From Ref. 5.)

Lacquered copper leads (3) are threaded through the lower lumen (4) anchored with a layer of epoxy resin (5). The upper lumen (6) houses an Ag/AgCl wire (7, not shown) and a glass capillary (8) for a liquid junction, all comprising a reference electrode. The whole assembly is set in an epoxy resin encapsulant. In the ISFET chip (Fig. 5), Q2 and Q4 are conventional metal gate IGFETs for monitoring the parameters of the basic device independently of the CHEMFETs Q1 and Q3 that are made without the metal gate. The latter two can be used as a multisensor or a reference electrode for measuring standard conditions. The regions 1–9 are metal contact pads to which wires are bonded for connections to the external circuitry. The CHEMFET gates are well removed from these pads to facilitate the deposition of the ion-selective membrane, and also from the area of the chip containing the electrical connections that are coated with an insulating material.

The elongated *n*-type areas A, B, C, and D (Fig. 5a) have appreciable resistance and are the "source" and "drain" of the two CHEMFET gates, carrying current to the bonding pads for the two CHEMFET devices (see Fig. 3). Point 5 in the circuit diagram (Fig. 5b) is the connection point to the substrate silicon.

In their review of chemically sensitive field effect transistors, Janata and Huber [5] list various kinds of ISFETs for different ions: (1) *H_3O^+-sensitive*, the ion-selective "membrane" is Si_3N_4 and ca. 0.04 μm thick, 54–60 mV/pH, pH 1–14; (2) *Na^+-sensitive*, aluminosilicate and borosilicate glass, 55 mV/p(Na), p(Na) 0–3, sensitivity decreasing at lower Na concentrations; (3) *K^+-sensitive*, 80–150 μm thick valinomycin membrane, near-theoretical sensitivity down to p(K) 4.5; (4) *Ca^{2+}-sensitive*, p(1,1,3,3-tetramethyl-butylphenyl) phosphoric acid membrane on polymer base, 80–150 μm thick, near-theoretical sensitivity down to p(Ca^{2+}) 6.5; (5) *halide-sensitive*,

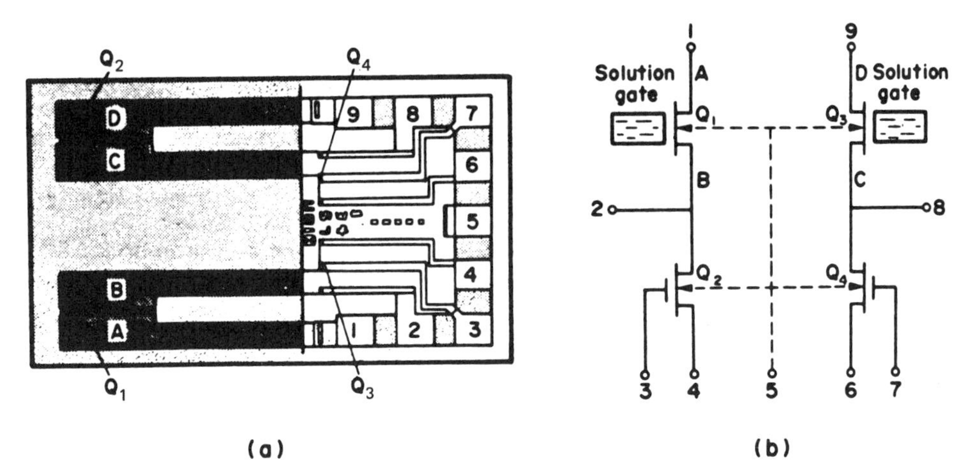

FIG. 5 (a) Photomicrograph and (b) circuit schematic of a multi-device ISFET chip. (From Ref. 5.)

AgCl or AgCl/Ag_2S in elastomer membranes, near-theoretical sensitivities for Cl^-, I^-, and CN^- down to $p(A^-)$ 5 or lower.

The response times and measurable ranges of ion concentrations of ISFETs are similar to those of ISEs, except perhaps for NO_3^-, the lower concentration limit for which is stated to be only 10^{-1} *M*. Reference has already been made to the fact that the ancillary electronics for ISFETs are much simpler and cheaper than those required for ISEs. Greater sensitivity than with ISEs is also attained sometimes from the more compact and much smaller assembly, resulting in an improved signal-to-noise ratio. Intermittent refilling of internal buffers and ion-selective liquids is eliminated. Multi-ion sensors also become more feasible. All these features offer greater advantages for continuous monitoring in confined spaces.

B. Factors Affecting Performance

1. *Response Time*

The response time of an ISE is essentially a property of the physical structure of the electrode that might affect the time taken by the ion-sensitive part of the membrane to come to equilibrium with the solution interface, and the time taken by the ion in bulk solution to come to equilibrium with the membrane, determined by the thickness of the unstirred boundary layer between the bulk solution and the membrane surface. The rate-controlling step is then the diffusion of the ion across this layer, giving an exponential time relationship for ion-exchange membranes and a square-root time relationship for neutral carrier membranes.

Interfering ions can delay the formation of the transported species that are necessary to develop a Nernst potential difference across the ISE membrane and thus affect the response time. In experiments where the kinetics of changes in ion concentration are being measured, a knowledge of the response time of an ISE is essential. Morf et al. [8] calculated theoretical response times by including a term in the Nernst relationship that was a function of all the experimental variables that could affect the development of the Nernst potential difference, e.g., rate of stirring of the bulk solution. Equation (1) was thus rewritten

$$E = E_{i,0} + E' + \frac{RT}{z_i F} \ln a_i \tag{2}$$

where E' is the potential difference contribution produced by diffusion layers in the aqueous sample, a_i the activity of the ion in the boundary layer, and z_i its valence. From the equation subsequently derived to express the time dependence of a_i and hence the factors affecting response time, these authors drew the following conclusions. With ion-exchange membranes, the boundary layer, and thus the response time, can be drastically reduced by (1) efficient stirring (or faster flow rates in continuous flow work), (2) avoiding pores and impurities on the membrane surface, or by using microelectrodes with minimal membrane surface, and (3) changing from a lower to a higher activity in the sample solution during sequential measurement. In neutral carrier membranes, diffusion of dissolved electrolyte (ligand and complex), resulting from the attainment of a steady-state concentration gradient across the membrane, is the main factor governing the response time. Thus, the ion extraction capacity, the resistance to diffusion, the thickness of the membrane, and the polar nature of the solvents in the membrane are all implicated.

2. *Sensitivity of Measurements*

The detection limit of an ISE is normally defined as the lowest concentration of the ion at which a finite potential difference is developed across the membrane. (More rigorously, this could be the lowest concentration at which a Nernstian, or near-Nernstian, response to concentration is measured.) Detection limits are set by the nature of the ISE, of which two examples are given below.

a. ISE: Types 2 and 3b (Table 1). These types consist of ionic crystal membranes in which the electron transfer is through interstitial ions or vacancies in the ionic lattice, so that the detection limit is governed by the solubility product of the ionic solid, or by the product of the ionic defects at the surface of the crystal, whichever is the larger [9]. For an AgCl membrane, for instance, the solubility product $[a_{Ag}a_{Cl}]$ sets the detection limit $a_{Ag,0}$ of the silver ion such that a Nernstian response occurs only when $a_{Ag,0}$ is appreciably higher than $[a_{Ag}a_{Cl}]^{1/2}$. By contrast, for an AgI Membrane at whose surface the concentration of Ag defects greatly exceeds $[a_{Ag}a_{I}]^{1/2}$, then the detection limit of Ag is much higher than this defect Ag concentration.

b. ISE: Types 1a, 1b, and 3a (Table 1). In these types the active agent, or ligand (dissolved in an inert solvent in the membrane), forms a charged, or an uncharged, complex with the ion in solution. The transport of this complex within the membrane gives rise to the potential difference across it and hence the detection and measurement of the activity of the ion in the external solution. The steps leading to such transport are the partition of ion i between the external and internal solutions and the membrane solvent, the formation of a complex between the ligand and the ion, and the solubility of this complex in the membrane solvent. These steps also apply to liquid membranes physically immobilized in a porous matrix, but when the inert solvent is replaced by an inert solid matrix, the partition and solubility steps are inapplicable. Thus, the thermodynamic constant for the formation of the complex between ion i and the ligand essentially controls the detection limit.

For most ISEs, the detection limit is around 10^{-5} *M*. In favorable environmental conditions in solution, ISEs of type 2 (Table 1) can give a Nernstian response down to 10^{-7} *M*, and some ISEs of type 2 with favorable formation constants, to 10^{-6} *M*. These limits are affected adversely, however, by interfering ions and molecules. (A larger mole fraction of water molecules in concentrated aqueous solutions of ions could also prevent access of the ion to the hydrophobic ISE membrane and thus affect adversely the detection limit.)

3. *Selectivity*

In the presence of an interfering ion j, therefore, the selectivity coefficient for ion i, K_{ij}, or the partition coefficient ratio k_i/k_j, is a more general expression of the formation constant K_i of the ligand complex with ion i. So it is convenient and adequate to consider the significance of the detection limit as a special case of the amendment of Nernstian response [Eq. (1)] by the selectivity coefficient K_{ij}. Equations of varying complexities have been proposed and described to express this, in which terms of ion selectivity and mobility (in the membrane) are incorporated.

For solid-state membranes measuring the activity of anion Y^{z-}, typical equations for an AgX (silver halide) membrane [Eqs. (3a) and (3b)] or a mixed AgX + AgY membrane [Eq. (4)], are given by Morf et al. [9]

$$E = E_{y,0} - \frac{RT}{zF} \ln a_y \tag{3a}$$

which describes the measured EMF values of a_x given by

$$a_x < K_{x,y}^{(\mathrm{pot})} a_y^{1/z}$$

where the potentiometric selectivity coefficient $K_{x,y}^{(\mathrm{pot})} = L_{\mathrm{AgX}}/L_{\mathrm{Ag}_z\mathrm{Y}}^{1/z}$, L representing the solubility products. The close agreement between calculated and experimental values of $K_{x,y}$ for AgX membranes (Fig. 6) confirms the validity of these relationships. At higher values of a_x, the membrane EMF is only sensitive to a_x [6] and is given by

$$E = E_{x,0} - RT \ln a_x \tag{3b}$$

The EMF of mixed-phase solid-state electrodes is sensitive to a_x and a_y and is given by

$$E = E_{x,0} - \ln \left(a_x + K_{x,y} a_y^{1/z} \right) \tag{4}$$

denoting that the activity of one ion can be measured at a constant activity of the other (for the Cl, Br system, see Ref. 10).

The AgX membranes can also be used to assay cations that react with AgX, e.g., the mercuric Hg^{2+} ion [8] for which the EMF is given by

$$E = E_{\mathrm{Ag},0} + \frac{RT}{F} \ln (n a_{\mathrm{Hg}} + \alpha) \tag{5}$$

where $\bar{n}$ is the average number of X^- ions coordinated to the Hg^{2+} ion at the phase boundary and α the defect activity of Ag^+ at the membrane surface that is governed by the conditions of its preparation.

In solid-state ion-exchange membranes, the low ion mobility within the membrane does not influence ion selectivity. In liquid membranes, where the complex formed with the ion within the membrane is neutral, the complex-forming ionizable species is an aliphatic amine or acid dissolved in a water-immiscible solvent. Thus, ionic mobility in the solvent, the partition coefficient of the ion between solvent and water, and the formation constant of the neutral complex are all invoked in producing the EMF.

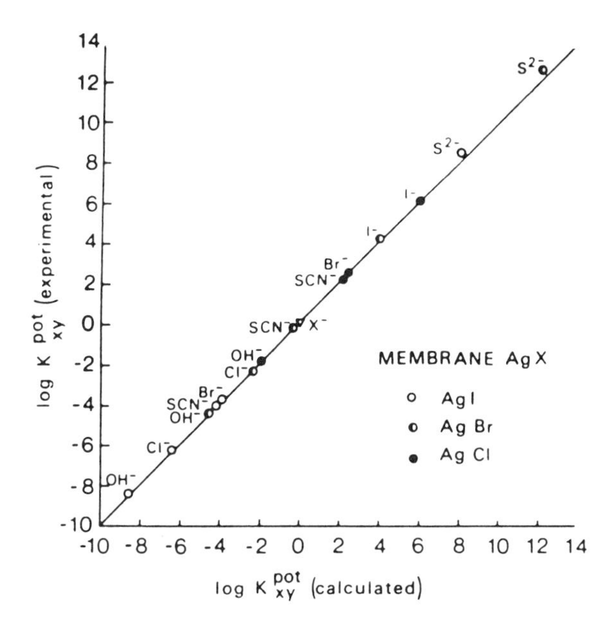

FIG. 6 Comparison of the experimental and calculated values of anion selectivity of different silver halide membrane electrodes. [From Ref. 9, reprinted with permission from W. E. Morf, G. Kahr, and W. Simon, Theoretical treatment of the selectivity and detection limit of silver compound membrane electrodes, *Analytical Chemistry, 46*: 1541 (1974). Copyright © by the American Chemical Society, Washington, D.C.]

For neutral carrier liquid membranes, measuring the activity a_i of ion i in the presence of an interfering ion activity a_j in the simplest case of both ions with the same valence z, Morf and Simon [11] give the EMF of the ISE according to the equation

$$E = E_{i,0} + \frac{RT}{zF} \ln\left[a_i + K_{i,j}^{(\text{pot})} a_j\right] \tag{6}$$

which is a simpler version of that for the liquid ion-exchange membrane, which also incorportates terms for the valences, mobilities, and transference numbers of ions i and j.

The potentimetric selectivity coefficient $K_{i,j}^{(\text{pot})}$ is given by

$$K_{i,j}^{(\text{pot})} = \frac{u_i}{u_j} K_{ij} \tag{7}$$

where u is the mobility of the ionic form in the membrane and $K_{i,j}$ the equilibrium constant of the reaction

$$[I]^{z_i} \text{ (membrane)} + [J]^{z_i} \; (aq) = [J]^{z_i} \text{ (membrane)} + [I]^{z_i} \; (aq) \qquad (8)$$

C. Reference Electrodes

As with pH measurements using the glass electrode, ion activity measurements require the ISE to be coupled with an external reference electrode. The composition of the reference solution in the latter must, however, be chosen such that it does not affect the concentration of the specific ion in the sample or introduce significant concentrations of interfering ions. It is also essential that the liquid junction potential (LJP) between the reference electrode and the calibrating and test solutions of the specific ion be invariant. The LJP depends on the nature of the junction and the interdiffusion of the external and internal solutions. Principally, three devices are popular for reducing LJP. The first is a filling solution in a *single junction electrode* (Fig. 7a) in which cation and anion have the same mobility, e.g., a cation-anion pair chosen from Cs^+ (67.5), Tl^+ (65.3), K^+ (64.2), NH_4^+ (64.3), $\frac{1}{2}Pb^{2+}$ (61.0), and $\frac{1}{2}SO_4^{2-}$ (68.5), Br^- (67.3), I^- (66.3), ClO_4^- (64.0), $\frac{1}{2}C_2O_4^{2-}$ (63.0), NO_3^- (61.6), $\frac{1}{2}CO_3^{2-}$ (60.0), where the values in parentheses represent ionic mobility or conductance. The second is a *double junction electrode* (Fig. 7b) in which the composition and concentration of the filling solution in the outer chamber are chosen such that the LJP with the test solution (Table 3) are minimal. The third involves addition of an *inert* electrolyte to the solutions in both electrode compartments in direct contact (single junction), or through a salt bridge of the inert electrolyte at the same concentrations (double junction); this effectively serves to decrease the relative concentration of the potential-determining ion and also to maintain its activity coefficient invariant. A reference electrode must be stable, have a high exchange current to maintain its constant potential under high-current demand, and be reproducible so that the temperature and concentration dependence of its EMF is not subject to hysteresis.

ISEs also require an internal reference electrode, in which the reference solution has a composition unrelated to the requirements of the test and standard solutions but is primarily concerned with the ion-selectivity mechanism of the ISE membrane.

Most of the reference electrodes in current practice are anion-responsive. The few cation-responsive electrodes in use are for special applications. The external reference electrode of the saturated KCl-calomel variety, because of its fixed potential, is used extensively, but the Ag/AgCl/saturated KCl electrode has been adopted increasingly for general use principally because the $Hg/Hg_2Cl_2/KCl$ electrode is a toxic hazard if damaged. A saturated

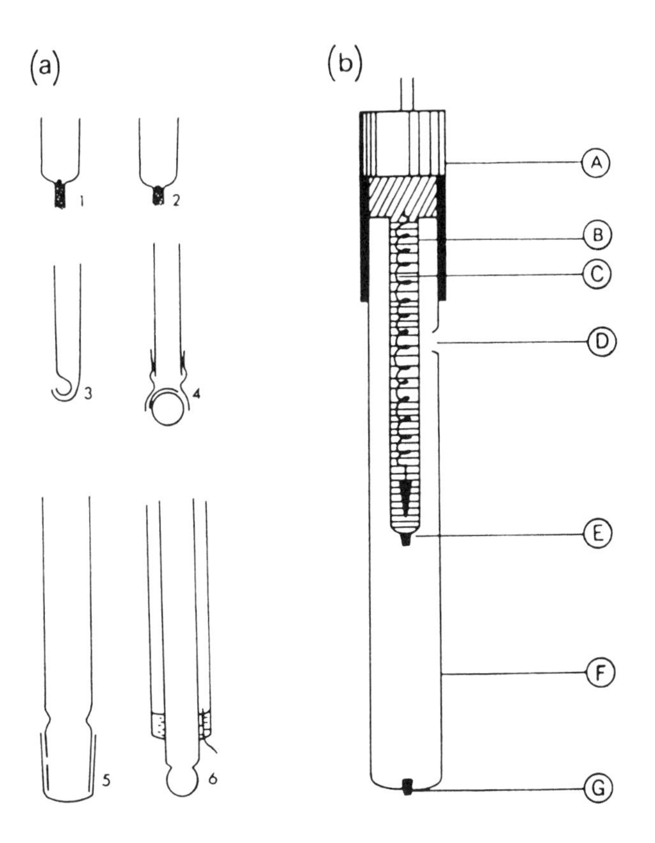

FIG. 7 Types of liquid junctions in reference electrodes: (a) *Single junction* with restrained flow: (1) palladium annulus; (2) ceramic plug; (3) junction to overcome density gradient; (4) cellophane membrane; (5) ground glass sleeve; (6) fiber wick in combination glass electrode. (From Ref. 12.) (b) *Double junction*: (A) cap with air vent; (B) removable reference element; (C) reference element chamber containing concentrated KCl solution; (D) filling hole closed with teat during use; (E) ceramic diaphragm; (F) chamber containing bridging electrolyte; (G) ceramic diaphragm. (Courtesy of Corning Medical Ltd., Halstead, Essex, England.)

KCl solution deposits crystals that clog the electrode junction when the temperature fluctuates, hence, undersaturated solutions, 3.5 or 4.0 *M* KCl, are used increasingly. The insoluble chlorides, especially AgCl, are more soluble in concentrated KCl solutions, which results in drifts of the fixed potential of the reference electrode. In some applications, where continuous monitoring of solutions containing dilute electrolyte solutions is required, the use of

TABLE 3 Filling Solutions Recommended for Single (S) and Double (D) Junction Reference Electrode Half-Cells, Used with Various ISEs

Specific ion	Type of junction	Filling solution	Source[a]
Ba^{2+}	S	Equitransferent	O
	D	0.1 *M* $NaNO_3$	C
Ca^{2+}	S	Equitransferent	O
	S	4 *M* KCl (Ag)	O
	D	1 *M* KNO_3	O
	D	1 *M* KCl+sucrose	O
	D	0.1 *M* KNO_3	C
Cu^{2+}	S	Equitransferent	O
	D	1 *M* KNO_3	O,C
K^+	S	CCl_3COOLi	O
	S	0.06 *M* NaCl	O
	D	0.1 *M* Et_4NCl	C
Na^+	S	CCl_3COOLi	O
	D	1 *M* KNO_3	O
	D	0.1 *M* $CaCl_2$	C
Br^-	S	Equitransferent	O
	D	1 *M* KNO_3	O,C
Cl^-	S	Equitransferent	O
	D	1 *M* KNO_3	O,C
	D	1 *M* KCl+sucrose	O
CN^-	S	Equitransferent	O
	D	1 *M* KNO_3	O,C
F^-	S	Equitransferent	O
	D	1 *M* KNO_3	C
BF_4^-	D	0.1 *M* NH_4F	O
I^-	S	Equitransferent	O
	D	1 *M* KNO_3	C
NO_3^-	D	0.1 *M* KCl	O
	D	0.02 or 0.04 *M* $(NH_4)_2SO_4$	O
	D	0.1 *M* NaF	C
S^{2-}	S	1 *M* KNO_3 or $NaNO_3$	O,C
	S	4 *M* KCl (Ag)	O

[a]O: *Analytical Methods Guide*, Orion Research Inc., East Grinstead, West Sussex, England. C: N. Walton, Corning Medical Ltd., Halstead, Essex, England.

dilute KCl, KNO_3, K_2SO_4, NH_4NO_3, $(NH_4)_2SO_4$, and NH_4F solutions is sometimes preferred for specific applications (Table 3). The anion chosen is tailored to the specific ion under test; the simplest and most obvious example is the use of NO_3^- or SO_4^{2-} filling solutions in a chloride ion assay.

Other variations are necessitated by the use of extracting solutions in analytical practice, or test solutions containing appreciable concentrations of other chemicals, for recovering the specific ion from an adsorbed phase. Direct ISE measurements on the extract necessitate the use of the same concentration of the extracting agent, or foreign chemical compound, in the salt bridge of a double junction electrode. Similarly, with test solutions containing nonaqueous solvents, the salt bridge has to include a solvent, usually ethanol, that is miscible with water and the nonaqueous solvent.

D. Calibration Methods

The method adopted for calibrating an ISE/reference electrode combination depends on (1) the versatility of the meter used [13]; (2) the magnitude of the variations in the ionic strength, or the concentrations of foreign chemical compounds, or the pH, of the test solutions; (3) the deviation of the ISE from the theoretical Nernstian response because of temperature changes, interfering ions, etc.

The most direct procedure is the calibration curve method relating the measured EMF to the concentration of the specific ion, taking steps normally necessary to account for item (2) above, or the activity of the ion if its activity coefficient can be calculated in the standard and test solutions. The method is accurate but laborious. For a small number of observations, greater precision is possible by using two standard solutions of concentrations, first above and then below that of the test solution, whose concentration can then be interpolated, independent of deviations from Nernstian slope.

With modern meters, the more usual practice is to use two solutions, whose specific ion concentration (or activity) ratio is known accurately, to calibrate the ISE/reference electrode/meter system. This concentration ratio is conventionally 10 and the range spans that in which the concentrations of the test solutions occur. If the ISE does not have a true Nernstian EMF response because of factors stated in item (3) above, such meters incorporate a scale-expansion facility that modifies the experimental slope to the theoretical one, and also a temperature compensating adjustment.

For solutions in which variations in ionic strength make accurate measurements difficult, or where hydrogen ion activity interferes with the specific ion measurement, or where the specific ion needs to be converted to an ion pair or ion complex, the use of ionic strength adjusters, pH buffers, and releasing agents added to standard and test solutions alike is recommended. Such combination buffer solutions are known as *total ionic strength adjusters* (ISA or TISAB). The composition of such "cocktails" can be tailored to the specific needs of a particular experimental technique [14]. TISAB contains 114 ml glacial acetic acid, 116 g NaCl, and 0.6 g Na citrate per liter H_2O titrated to pH 5.0–5.5 with 5 *M* NaOH. The addition of 1,2-cyclohexylenediamine-tetraacetic acid to complex any Al ions in solution is also recommended (Orion Research Inc. Applications Guide No. 5).

The most precise and rapid way of preparing calibration standards is the series addition method in which the standard (incorporated in the TISAB or ISA appropriate to the application) at the high-concentration end of the range in which the test solutions lie is successively diluted with the appropriate TISAB or ISA to give a full range of standards.

E. Methods of Analysis

The direct methods of measurement involve the use of a calibration curve, or the differential standards procedure for small numbers of samples in which two standards of concentrations close to the test solution are used; both have been indicated in the previous section. The precision with which the concentrations (or activities) of the specific ion is assayed depends on (1) the sensitivity of the meter (i.e., its response in millivolts to increments in concentration), its stability, and its precision; (2) the steps taken to ensure that the ionic strength, pH, concentrations of interfering ions, and presence of other compounds are normalized in calibration standards and test solutions; (3) temperature control; and (4) the extent to which measurements on standards and test solutions are replicated.

Several indirect methods are in general use, for which greater precision than the direct methods is claimed by their originators. However, these are only suitable for rapid routine processing of large numbers of samples, if we assume preprogrammed microprocessor-operated meters are available [9] or simple computer programs are used. These methods are summarized below.

1. Standard Solution Addition to Test Solution

A volume v_i of the specific ion i at concentration c_i, to be determined, is spiked with a known increment of the ion (volume v_s of a standard solution of concentration c_s where $c_s \sim 100c_i$). The change in the measured EMF (positive for a cation and negative for an anion) is given by Thomas [15]

$$E = E_{(i,s)} - E_i$$

$$= S \log \frac{\nu_i c_i + \nu_s c_s}{c_i (\nu_i + \nu_s)} \tag{9}$$

which follows from

$$E_{(i,s)} = E_0 + S \log \left[x_{(i,s)} f_{(i,s)} \frac{\nu_i c_i + \nu_s c_s}{\nu_i + \nu_s} \right] \tag{10}$$

[where S is the calibration slope and $x_{(i,s)}$ and $f_{(i,s)}$ are the fractions of uncomplexed ions and the ion activity coefficient in the spike solution]. Equation (9) is obtained by assuming that x and f do not change on spiking, a simplification that follows if the conditions specified in item (2) are fulfilled. This simplified equation can be transformed to calculate c_i; thus,

$$c_i = \frac{c_s}{10^{\Delta E/S}(1 + \nu_i/\nu_s) - \nu_i/\nu_s} \tag{11}$$

If the spike volume ν_s is <1% of the test solution volume, this simplifies to

$$c_i = \frac{c_s \nu_s}{(10^{\Delta E/S} - 1)\, \nu_i} \tag{12}$$

Equation (10) can be further extended by measuring $E_{(i,s)}$ for two successive spikes, and by iterative calculation of c_i and S starting from $c_i = 0$, the true convergent value of c_i is obtained.

2. Test Solution Addition to Standard Solution

Here, a volume ν_i of the test solution is added to a known volume ν_s of the standard solution under the same equilibrium conditions as for the case above except that $c_s < c_i$; then,

$$c_i = c_s \left[10^{\Delta E/S} \left(1 + \frac{\nu_s}{\nu_i} \right) - \frac{\nu_s}{\nu_i} \right] \tag{13}$$

3. Subtraction of Specific Ion from Test Solution

This is achieved by adding a known amount of a complexing or precipitating agent such that a stoichiometric amount of the specific ion is removed. The EMF of this solution is then given by

$$E_{(i,c)} = \text{constant} + S \log\left[x_{(i,c)} f_{(i,c)} \frac{\nu_i c_i - \nu_{(i,c)} c_c}{\nu_i + \nu_{(i,c)}}\right] \tag{14}$$

where the subscript i,c refers to the test and complexing solutions mixture and c to the complexing solution.

4. *Titration Methods*

In these, a standard titrant is added in small increments to the test solution such that the specific ion is precipitated or chelated and the EMF is measured with an ISE/reference cell combination. Alternatively, an excess of the standard titrant is added and this excess determined by back-titration with a standard solution of the specific ion. The ISE used could be sensitive to the specific ion or to the reacting ion in the standard titrant. Potentiometric titrations are considered to be more precise than the direct methods described above, but require more time for experimentation and data processing to obtain the desired result than even the indirect methods. Their essential principle is the measurement of the change of EMF with titrant addition, which is maximum at the equivalence point, is considerably more than the response slope of the ISE (Secs. II.B and II.D), and is affected little by liquid junction potentials and activity coefficients. Sharper breaks in the measured EMF at the end point, greater stoichiometry of the titration reaction, and a higher reaction velocity all lead to greater precision.

Potentiometric titration curves (EMF versus titrant volume) are sigmoidal (Fig. 8a), from which the equivalence point is most conveniently obtained graphically by plotting the first and second derivatives (Fig. 8b and 8c) of the sigmoidal relationship. Figure 8d is another variant—the reciprocal of the first derivative versus the titrant volume, in which the two arms intersect the volume axis at the equivalence point.

An even more laborious and more precise procedure is to determine the equivalence point potential by the direct method first, then add the titrant until this EMF is attained, thus giving the exact titration value of the amount of the specific ion in the test solution.

5. *Gran Plot*

In very dilute solutions of the specific ion, the inflexion point in the sigmoidal curve is ill-defined and difficult to evaluate (Fig. 9), when the EMF values measured on adding increments of the titrant to the test solution are plotted against titrant volume, or indeed by any of the indirect methods described above. For acid–base titrations, Sorensen [17] proposed that the antilogarithm of the pH should be plotted against titrant volume, whose final value substantially exceeds the equivalence point. Gran [18] extended this to

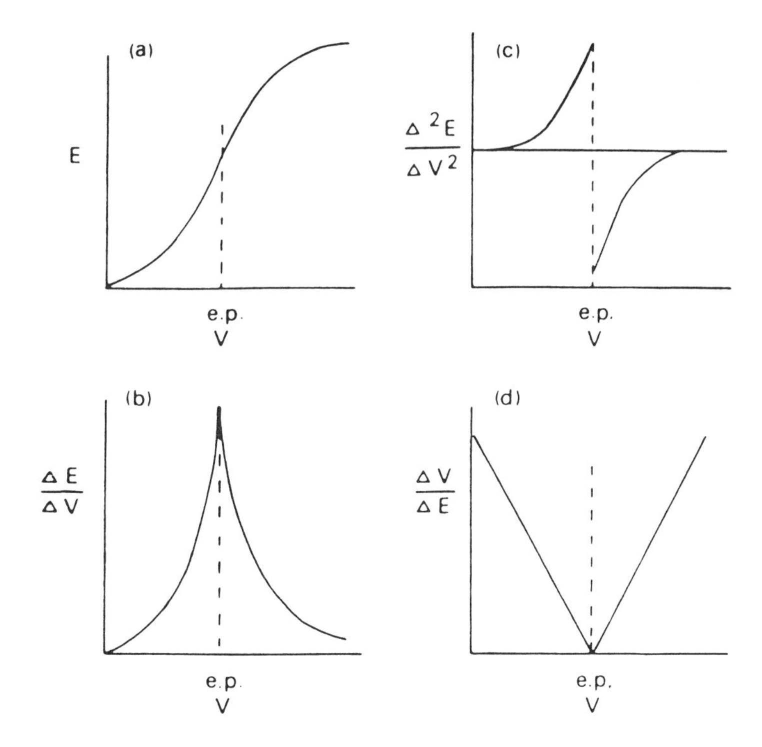

FIG. 8 Titration curves. (a) EMF versus titration volume; (b) first derivative curve; (c) second derivative curve; (d) extrapolated $\Delta V/\Delta E{:}v$ curve. e.p. = equivalence point. (From Ref. 16.)

precipitation, complexation, and oxidation-reduction reactions, incorporating ions of different valencies and also including a correction factor for increasing the volume of the mixture when more titrant was added. This converts a curvilinear antilog plot to a linear form, making possible a precise extrapolation to the equivalence point (Fig. 10).

In infinitely dilute solutions (where the activity coefficient f_i of ion $i = 1$) or in solutions where f_i remains unchanged on adding titrant, and in the absence of complexation in the test solution of foreign ions (i.e., $x = 1$), the theoretical basis of the Gran plot can be seen by rewriting Eq. (10) in the antilogarithmic form; thus,

$$10^{(E_{i,s}-E_0)/S} = \frac{v_i c_i + v_s c_s}{v_i + v_s} \tag{15}$$

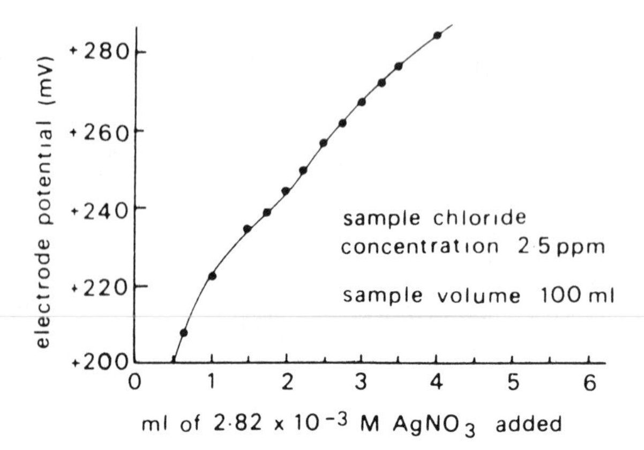

FIG. 9 Conventional plot of titration data for low chloride concentrations. (From *Orion Analytical Methods Guide*, 9th ed., Orion Research Inc., East Grinstead, West Sussex, England, Dec. 1978, p. 34.)

In practice, the appropriate volume correction factor can be tabulated for application to the antilogarithm of the measured EMF values of the control, or blank, and the test solutions.

A simpler expedient is to plot the measured EMF values on an antilogarithmic scale, as the ordinates, against the added titrant

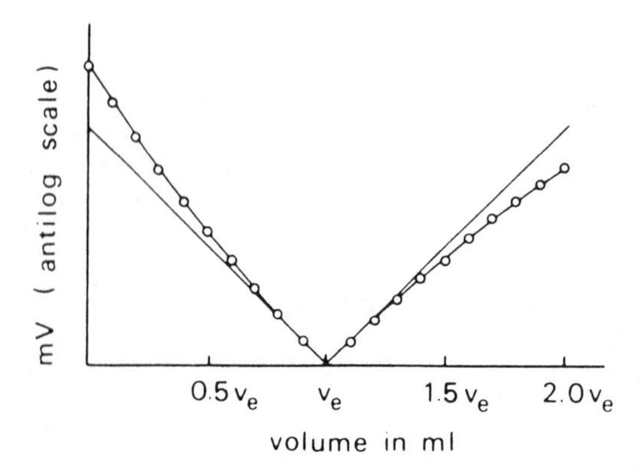

FIG. 10 Titration of 40 ml 0.025 *M* HCl with 0.1 *M* NaOH; —o—o—o, Sorenson plot; —, Gran plot. (From Ref. 18.)

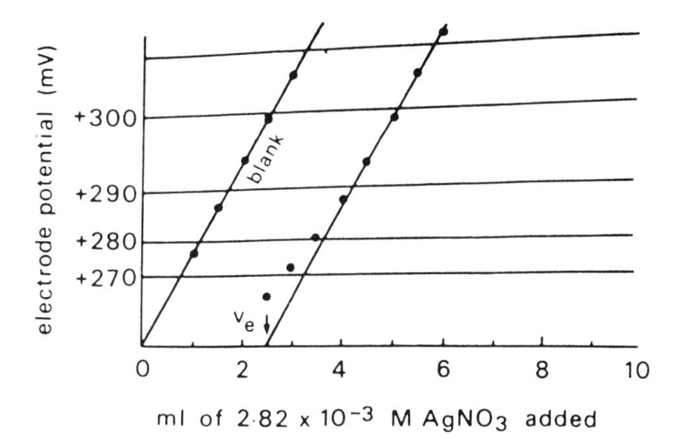

FIG. 11 Gran plot of titration data for low chloride concentrations (from Fig. 6) using an AgS electrode. (From *Orion Analytical Methods Guide*, 9th ed., Orion Research Inc., East Grinstead, West Sussex, England, Dec. 1978, p. 34.)

volumes on a linear scale, as the abscissas (Fig. 11). The volume correction is made by giving a finite slope to the abscissa lines (above the x axis) corresponding to the percentage increase in the initial solution volume ν_i when a predetermined excess volume of titrant, $(\nu_s)_f$, has been added finally, i.e., $100(\nu_s)_f/\nu_i$. The EMF at the origin is that measured on the control or blank before adding the titrant and forms the base line to which the EMF values of the titrated test solution are extrapolated. The scale on the antilog axis is increasingly positive for cation assay and negative for anion assay. The titrant volume obtained by extrapolating to the base line the linear part of the EMF:titrant volume curve at high titrant volumes furthest away from the equivalence point gives the equivalence point in volume units ν_e of added titrant.

6. *Differential Potentiometry*

The principle is illustrated in the determination of chloride ions with a cell of the type

$$\text{Ag, AgCl} \,/\, \underset{\text{solution}}{\overset{\text{A}}{\text{standard Cl}^-}} \,//\, \underset{\text{membrane}}{\text{selective}} \,//\, \underset{\text{solution}}{\overset{\text{B}}{\text{test Cl}^-}} \,/\, \text{AgCl, Ag}$$

in which the reference electrodes can be replaced by matched chloride ISEs [19]. The cell is calibrated by measuring the EMF values of a series of standard solutions of the specific ions put in the half-cell B to produce a calibration curve, from which the concentration of the test solution can be interpolated.

III. COMMONLY AVAILABLE ELECTRODES

Earlier (Sec. II.A and Table 1) a summary list was given of the types of ISEs that are feasible in principle. All of these have been prepared and tested in research laboratories, and most are commercially available and have acceptably high sensitivity, reproducibility, and ruggedness. Their basic construction and performance characteristics are described in this section. Data sheets supplied by the manufacturers at the time of writing and compilation have been used to prepare the appropriate tables and diagrams. The information given thus is intended as a guide to potential users in soil and plant science so that they can select the ISE that best fits their environmental conditions. However, the reader is strongly advised to obtain from the manufacturer the up-to-date characteristics of the ISE selected before planning an experiment and buying the ISE.

A. Solid-State and Liquid Membranes

Solid-state membranes consist of (1) crystalline pure or mixed halides of silver and copper, or a halide-sulfide mixture, for Cu^{2+}, Br^-, Cl^-, I^-, CN^-, and S^{2-}; (2) lanthanum fluoride for F^-; (3) glass for Na^+ and K^+; (4) a silver phosphate-chalcogenide for phosphate; (5) a mixture of silver and copper sulfate and sulfide for SO_4^{2-}. Details of types (1–3) are given in Tables 4–8 and are discussed briefly in the succeeding sections. The phosphate and sulfate electrodes (types 4 and 5) have not as yet been produced commercially but are of sufficient importance to soil and plant science to merit special mention here.

The phosphate-sensitive electrode [20] gives a linear response in the concentration range 10^{-8}–10^{-3} *M* with a sensitivity of 14–15 mV per decade change in PO_4^{3-} concentration instead of the theoretical 19.38 mV at 20°C. The response time of the electrode is <30 sec and the EMF stability is better than 0.5 mV day^{-1}. The SO_4^{2-} and NO_3^- ions do not interfere, but K_{ij} for the chloride ion is 10^2 (see Sec. III.C). The limiting pH at which the electrode will function is thus given by

$$pH = 11.566 - \frac{1}{3}\, p(PO_4) \qquad (16)$$

because of the complementary hydrolytic relationship between Ag_3PO_4 and AgOH.

The sulfate-sensitive electrode [21] of composition $PbSO_4$-PbS-Ag_2S-Cu_2S, has a linear reponse between 10^{-4} and 10^{-1} *M*, with a slope "only slightly less than the theoretical" at 18–20°C that remains unchanged over 2 months. Preliminary measurements with less satisfactory homogeneous $BaSO_4$ electrodes suggest that Cl^- and NO_3^- ions interfere only slightly but CO_3^{2-} and phosphate ions do so seriously. However, K_{ij} values for the heterogeneous composition electrodes are not given, nor is the effect of pH on electrode performance.

Most commonly available ISEs are of the *liquid membrane* type (Table 4) of various constructions (Sec. III.B) and basic principles (Secs. II.A and II.B) involving charged sites (i.e., ionizing organic salts), or neutral complexing carriers dissolved in a water-immiscible solvent, or impregnated in solid solution with an inert carrier.

B. Constructional Details

The three basic types of *solid-state membranes* are shown in Fig. 12. The glass electrode (Fig. 12a) is essentially the same as the pH electrode except that the membrane is less robust and of a composition that is sensitive to the specific ion. The membrane in the two crystal electrodes (Fig. 12b and c) has the same construction in that the small crystals of the active solid are pelleted into a disk to form a continuous ionic matrix in an inert cementing material of fine porosity. Electrical continuity with the external lead is through an internal reference electrode and a suitable filling solution (Fig. 12b), as with the glass electrode, or by cementing the internal conducting wire (mostly Ag or Pt) directly to the crystalline pellet (Fig. 12c). The filling solution can be replenished in a fully or partly demountable container (for easy cleaning), or be in a sealed unit, the latter considered by the authors to be much more satisfactory than the former.

Liquid-state membranes also occur in three forms, all with an internal reference electrode and filling solution. In the earlier, and now obsolescent form, the ion-specific liquid is restrained within the electrode by an inert porous membrane that can be replaced cheaply and quickly if contaminated or damaged. The small leakage rate (0.1–0.2 ml day^{-1}) can be minimized greatly by sealing the open end of the membrane with a thin impermeable and inert film when not in use. However, contamination of the internal electrode by the ion-specific liquid is a potential hazard to the long-term stability of the electrode. This and the possible leakage of the membrane liquid have led to the introduction of two alternative constructions. In one (Fig. 13), the

TABLE 4 Commonly Available Ion-Selective Electrodes

Ion		Type of electrode membrane	Manufacturers[a]	Applications (refs.)			
Direct	Indirect			Soil pastes and suspensions	Soil extracts	Plant extracts	Others
Cations							
Ammonium		liquid (in polymer)	P	[45]	—	—	—
Barium	sulfate	liquid	C, O	—	—	—	—
Cadmium		solid-state	P	—	—	—	—
Calcium		liquid (in polymer)	C, O, K, P	[25]	[39,40]	[108]	—
Calcium + magnesium		liquid (in polymer)	C, O	[25,41,109]	—	—	—
Copper		solid-state (crystal)	C, O, P	—	—	—	[84]
Lithium		plastic	P	—	—	—	—
Potassium		solid-state (glass)	B, K	[30,31,33, 35,62]	—	—	—

		liquid (in polymer)	C, O	—	[38]	—	—
		plastic	P	—	—	—	—
Sodium		solid-state (glass)	B, C, K	[27,29,32, 36,37]	[22—24,26]	—	[107]
		solid-state (crystal)	O	—	—	—	—
		liquid (in polymer)	B	—	—	—	—
Anions							
Bromide		solid-state (crystal)	C, K, O, P	—	[73,74]	[73]	—
Chloride		solid-state (crystal)	C, K, O, P	[70]	[66—69]	[116]	—
Cyanide		solid-state (crystal)	C, K, O, P	—	—	—	—
Fluoride	aluminum	solid-state (crystal)	C, O, P	—	[75,76]	[76,113,114]	[115]
	lanthanum	solid-state (crystal)	K	—	—	—	—
Fluoroborate	boron	liquid (free flowing)	O, P	—	—	—	—
Iodide		solid-state (crystal)	C, K, O, P	—	—	—	—
Nitrate		liquid (in polymer)	C, K, O, P	[46,56,60, 62—64]	[46—55]	[54,56,59, 110—112]	—
Sulfide		solid-state (crystal)	C, K, O, P	—	[82]	—	—

[a] B, Beckman; C, Corning; K, Kent Industrial Measurements; O, Orion; P, Philips.

TABLE 5 Cation-Selective Electrodes: Performance Characteristics

Electrode	Manufacturer (see Table 4)	Membrane composition	Range (M)[a]
NH_4^+	P	—	10^{-6}—0.1
Ba^{2+}	C	Antarox	5×10^{-5}—1
	O	—	10^{-6}—1
Ca^{2+}	C	$Ca(Me_4 \cdot Bu \cdot Ph \cdot PO_4)$	5×10^{-6}—1
	K	—	5×10^{-6}—1
	O	—	5×10^{-7}—1
	P	—	10^{-6}—1
(Ca + Mg) and water hardness	C	$Ca(Me_4 \cdot Bu \cdot Ph \cdot PO_4)$	$>10^{-4}$
	O	—	6×10^{-6}—1
Cd^{2+}	P	solid-state	10^{-6}—1
Cu^{2+}	C	$Cu_xS + Ag_2S$	$>10^{-6}$
	O	—	10^{-8}—satd.
	P	—	10^{-6}—1
Li^+	P	plastic[c]	10^{-5}—1
K^+	B	glass	10^{-6}—1
	C	valinomycin	5×10^{-5}—1
	K	—	—
	O	—	10^{-6}—1
	P	—	10^{-6}—1
Na^+	B	glass	10^{-6}—1
	C	glass	10^{-6}—1
	K	glass	5×10^{-9}—10^{-1}
	K	glass	5×10^{-5}—10^{-1}
	O	glass	10^{-8}—1
	P	plastic[c]	10^{-5}—1

[a]The lower limit is taken to indicate also the detection limit.
[b]Usually at specific ion concentrations $>10^{-3}$ *M*; larger times for more dilute solutions.
[c]"Plastic" signifies a plastic membrane wetted with an ion-selective liquid (e.g., Fig. 14). (—) indicates performance characteristics are not readily available.

Sensitivity (mV per decade molarity change)	Response time (sec)[b]	Operational life (months)	pH range	Temperature range (°C)
56 ± 3	<30	6	4—10	0—50
—	<90	6	2—11	—
23 ± 2	—	6	3—11	0—50
—	<90	6	4—9	—
—	—	—	4—10	0—50
24	—	—	2.5—11	0—40
27 ± 2	<30	6	3—10	0—50
—	<90	6	4.5—8	0—50
24	—	—	—	0—50
27	—	—	—	—
—	<90	3	3—7	0—80
26	—	—	—	0—80
27 ± 2	<20	—	1—14	0—50
56 ± 3	—	—	—	—
—	—	—	—	—
—	<90	6	4—9	—
—	—	—	—	—
54	—	—	2—12	0—40
56 ± 3	<30	6	2—12	0—50
—	—	—	—	—
—	—	>12	—	—
—	15—90	—	>p(Na)+3	0—80
—	—	—	>p(Na)+3	0—80
55	<90	>12	>p(Na)+3	0—80
55 ± 3	<30	6	2—12	0—50

TABLE 6 Cation-Selective Electrodes: Interferences by Ions Commonly

Electrode	Manufacturer (see Table 4)	Selectivity ratio				
		H^+	NH_4^+	Na^+	K^+	Mg^{2+}
NH_4^+	P	3×10^{-2}	1	2×10^{-3}	2×10^{-1}	10^{-5}
Ba^{2+}	C	—	—	4×10^{-4}	9×10^{-3}	—
	O	6×10^{-2}	4×10^{-3}	3×10^{-3}	2×10^{-2}	8×10^{-6}
Ca^{2+}	C	—	—	1.5×10^{-4}	$<10^{-6}$	2.5×10^{-4}
	K	(< pH4)	—	8.3×10^{-4}	—	4.9×10^{-4}
	O[a]	—	—	3×10^{-1}	—	1.0
	P	—	10^{-5}	3×10^{-4}	2×10^{-4}	4×10^{-4}
(Ca + Mg) and water hardness	C	—	—	2×10^{-2}	—	—
	O(10^{-3} *M*)[a]	—	—	3×10^{-2}	—	—
Cd^{2+}	P	2.4	—	—	—	—
Cu^{2+}	C	—	—	—	—	—
	O[a]	—	—	—	—	—
	P	—	—	—	—	—
Li^+	P	1	5×10^{-2}	5×10^{-2}	7×10^{-3}	—
K^+	B	—	—	—	—	—
	C	—	3×10^{-1}	2.6×10^{-3}	1.0	1.9×10^{-3}
	K	—	—	—	—	—
	O(10^{-3} *M*)[a]	10^{-2}	6×10^{-3}	2.0	1.0	—
	P	10^{-5}	10^{-2}	3×10^{-6}	—	6×10^{-6}
Na^+	B	—	—	—	—	—
	C	(pH≤pNa+3)	2×10^{-2}	1.0	3×10^{-2}	—
	K	—	—	—	—	—
	O	—	—	—	—	—
	P	5×10^{-1}	2×10^{-1}	1.0	5×10^{-1}	8×10^{-4}

[a]For Orion ISEs: Maximum molarity of interfering ion giving 10% error at specific ion molarity given in brackets. (—) indicates performance characteristics are not readily available.

membrane liquid is immobilized for up to 3 months in a plastic sponge, providing a continuous pathway for charge transfer. The same effect is also achieved by incorporating films of the ion-specific liquid randomly in a thin plastic film that is inert to the charge transfer process but has sufficient chemical affinity for the membrane liquid to form a pseudo-solid solution that is physically homogeneous and stable (Fig. 14).

Found in Soil Water

$K_{ij} = a_i/a_j$					
Ca^{2+}	Zn^{2+}	Cu^{2+}	Pb^{2+}	Fe^{2+}	Others
3×10^{-5}	—	—	—	—	Ba^{2+}: 3×10^{-5}
2.5×10^{-3}	—	—	2.4	—	—
2×10^{-4}	—	—	—	—	—
1.0	1.0	—	0.1	0.8	—
1.0	3.7	1.7×10^{-4}	1.5×10^{-2}	—	Ba^{2+}: 3.7×10^{-2}
—	1.0	4×10^{-2}	1×10^{-5}	2×10^{-2}	Ba^{2+}: 3×10^{-2}
1.0	—	—	—	—	Ba^{2+}: 4×10^{-4}
—	—	—	—	3×10^{-3}	—
—	3×10^{-5}	3×10^{-5}	—	6×10^{-5}	Ba^{2+}: 6×10^{-4}
—	—	—	6	196	Mn^{2+}: 2.7; Al^{3+}: 0.13
1×10^{-4}	—	1.0	$5 \times 10^{+3}$	—	Mn^{2+}: 6.4×10^{-5}
—	—	—	—	—	$S^{2-} \leq 10^{-7}$ *M*;
—	—	—	—	—	high Cl^-
—	—	—	—	—	Cs^+: 3×10^{-3}
—	—	—	—	—	—
2.5×10^{-3}	—	—	—	—	—
—	—	—	—	—	—
—	—	—	—	—	—
3×10^{-5}	—	—	—	—	Ba^{2+}: 4×10^{-4}
—	—	—	—	—	—
—	—	—	—	—	—
—	—	—	—	—	—
—	—	—	—	—	—
2×10^{-3}	—	—	—	—	Ba^{2+}: 5×10^{-3}

C. Performance Characteristics and Interferences

Table 4 summarizes ISEs that are currently available commercially and includes references to various types of applications in soil and plant science. Some of the more important of these applications are described in Sec. IV. Tables 5 to 8 provide a ready reference to the composition of the membranes, operational range of the specific

TABLE 7 Anion-Selective Electrodes: Performance Characteristics

Electrode	Manufacturer (see Table 4)	Membrane composition	Range (M)[a]	Sensitivity (mV per decade molarity change)	Response time (sec)	Operational life (months)	pH range	Temperature range (°C)
Br^-	B	$AgBr/Ag_2S$	$>10^{-7}$	—	—	—	0—14	—
	C	$AgBr/Ag_2S$	5×10^{-6}—1	—	<90	<36	2—12	—
	K	$AgBr/Ag_2S$	5×10^{-7}—1	—	—	—	0—14	0—80
	O	—	5×10^{-6}—1	57	—	—	2—12	0—80
	P	—	10^{-6}—1	56 ± 3	<30	12	1—11	0—50
Cl^-	B	$AgCl/Ag_2S$	$>5 \times 10^{-5}$	—	—	—	0—14	—
	C	$AgCl/Ag_2S$	$>5 \times 10^{-5}$	—	<90	<36	3—10	—
	K	—	5×10^{-5}—1	58	40	—	2—11	0—100
	O	—	$>8 \times 10^{-6}$(liquid); $>10^{-5}$(solid)	57	—	—	2—11	0—80
	P	—	10^{-5}—10^{-1}	56 ± 3	>60	12	1—10	0—50
CN^-	K	AgI	$>10^{-6}$	—	<90	<36	—	—
	K	AgI	5×10^{-7}—10^{-2}	—	—	>2	0—14(opt 13)	0—80
	O	—	10^{-6}—10^{-2}	54	—	—	11—13	0—80
	P	—	10^{-7}—10^{-1}	56 ± 3	<30	12	1—12	0—50

F^-	B	—	$>10^{-6}$	—	—	—	4—13	—
	C	LaF_3	$>5 \times 10^{-7}$	—	<90	<36	5—8	—
	K	LaF_3	10^{-6}—1	—	60—120 at 10 *M*	—	5—7 at 10^{-6} *M* 6—11 at 10^{-1} *M*	0—60
	O	—	10^{-6}—satd.	56	—	—	5—8	0—80
	P	—	10^{-6}—10^{-1}	56 ± 3	<60	12	4—8	0—50
BF_6^{3-}	O	—	3×10^{-6}—satd.	56	—	—	3—10	0—50
	P							
I^-	B	AgI	$>10^{-}$[illegible]	—	—	—	0—14	—
	C	AgI	$>10^{-}$[illegible]	—	<90	<36	3—12	—
	O	—	7×10^{-6}—1	55	—	—	2.5—11	0—40
	P	—	10^{-7}—1	56 ± 3	<20	12	1—12	0—50
NO_3^-	C	$(C_8H_{17})_4NNO_3$	$>10^{-}$[illegible]	—	<90	6	4—11	5—40
	K	—	10^{-6}—10^{-1}	—	—	>12	3—11	5—40
	O	—	6×10^{-6}—1	55	—	—	3—10	0—50
	P	—	10^{-5}—1	56 ± 3	<30	6	1—9	0—50
S^{2-}	B	Ag_2S	$>10^{-}$[illegible]	—	—	—	>12	—
	C	Ag_2S	$>10^{-}$[illegible]	—	<90	<36	0—13/14	—
	K	Ag_2S	10^{-7}—1	—	—	—	0—14	0—100
	O	—	10^{-7}—1	28	—	—	0—13/14	0—80
	P	—	10^{-6}—10^{-1}	27 ± 2	<15	12	1—14	0—50

[a]The lower limit is taken to indicate the detection limit also.

TABLE 8 Anion-Selective Electrodes: Interference by Ions Commonly

Electrode	Manufacturer (see Table 4)	Selectivity ratio OH⁻	F⁻	Cl⁻	Br⁻	I⁻
Br^-	C	3×10^{-5}	—	2.5×10^{-3}	1.0	$5 \times 10^{+3}$
	O[b]	—	—	—	—	(2×10^{-4})
	P	1×10^{-3}	—	6×10^{-3}	—	$2.0 \times 10^{+1}$
Cl^-	C	1.3×10^{-2}	—	1.0	$3 \times 10^{+2}$	$2 \times 10^{+6}$
	K	1×10^{-2}	—	1.0	$3 \times 10^{+2}$	$2 \times 10^{+6}$
	O[a]	(10^{-4})	(7×10^{-4})	1.0	(4×10^{-5})	(8×10^{-6})
	P	2.4×10^{-2}	—	1.0	1.2	$2 \times 10^{+1}$
CN^-	C	10^{-8}	—	10^{-6}	2×10^{-4}	very large
	K	—	—	—	—	—
	O[b]	—	—	(10^{+6})	$(<5 \times 10^{+3})$	(10^{-1})
	P	—	—	2×10^{-5}	2×10^{-4}	3
F^-	C	10^{-1}	1.0	—	—	—
	O[b]	>pH7	1.0	—	—	—
	P	10^{-1}	—	—	—	—
BF_6^{3-}	O[a]	(2×10^{-1})	(2×10^{-1})	(2×10^{-1})	(2×10^{-1})	—
I^-	C	10^{-8}	—	10^{-6}	2×10^{-4}	1.0
	O[a]	—	—	—	—	1.0
	P	—	—	6.6×10^{-6}	6.5×10^{-5}	1.0
NO_3^-	C	—	10^{-6}	9.4×10^{-3}	5×10^{-2}	4.1
	K	—	—	4×10^{-3}	1.3×10^{-1}	11
	O[a]	—	—	(4×10^{-2})	(8×10^{-4})	(6×10^{-6})
	P	1×10^{-3}	5×10^{-3}	10^{-2}	3×10^{-1}	10
S^{2-}	C	—	—	—	—	—
	K	—	—	—	—	—
	O[a]	—	—	—	—	—
	P	—	—	—	—	—

[a]Orion ISEs only: Value given is the maximum molarity m_j of the interfering ion that gives a 10% error in the measured concentration of the specific ion $(m_i)_{apparent}$ when its true concentration is 10^{-3} *M*.
[b]Orion ISEs only: Value given is the ratio m_j/m_i. (—) indicates performance characteristics not readily available.

Found in Soil Water

$K_{ij} = a_i/a_j$					
CN^-	SO_4^{2-}	S^{2-}	PO_4^{3-}	NO_2^-	Others
$1.2 \times 10^{+4}$	—	very large	—	—	NH_4^+: 3×10^{-1}
—	—	$(<10^{-7})$	—	—	—
$2.5 \times 10^{+1}$	—	—	—	—	CO_3^{2-}: 2.3×10^{-3}
$5 \times 10^{+6}$	—	very large	—	—	NH_4^+: 8.3
large	—	large	—	—	SCN^-: 2×10^2; $S_2O_3^{2-}$ interferes
—	(3×10^{-5})	—	—	—	(NO_3^-: 3×10^{-5}), (HO_3^-; acetate$^-$: 4×10^{-4})
$4 \times 10^{+2}$	—	—	—	—	CO_3^{2-}: 3×10^{-3}
1.0	—	very large	—	—	—
—	—	—	—	—	—
1.0	—	$(<10^{+7})$	—	—	—
1.0	—	—	—	—	CO_3^{2-}: 3.6×10^{-4}
—	—	—	—	—	—
—	—	—	—	—	—
—	—	—	—	—	—
—	(2×10^{-1})	—	—	—	(NO_3^-: 2×10^{-2})
2.5	—	very large	3×10^{-10}	—	AsO_4^{3-}: 3×10^{-10}
—	—	$(<10^{-7})$	—	—	—
3.4×10^{-1}	—	—	—	—	—
—	3.5×10^{-3}	—	10^{-6}	3×10^{-2}	—
—	7.4×10^{-5}	—	2.5×10^{-5}	4×10^{-2}	HCO_3^-; acetate$^-$: $< 3 \times 10^{-4}$
(2×10^{-4})	—	(HS^-: 4×10^{-3})	—	(8×10^{-4})	HCO_3^-; acetate$^-$: 3×10^{-2}
—	—	—	—	5×10^{-2}	HCO_3^-: 10; MnO_4^-: 63
—	—	1.0	—	—	Hg^{2+}: very large
—	—	—	—	—	—
—	—	1.0	—	—	Hg^{2+}: $<10^{-7}$
4^-	—	1.0	—	—	—

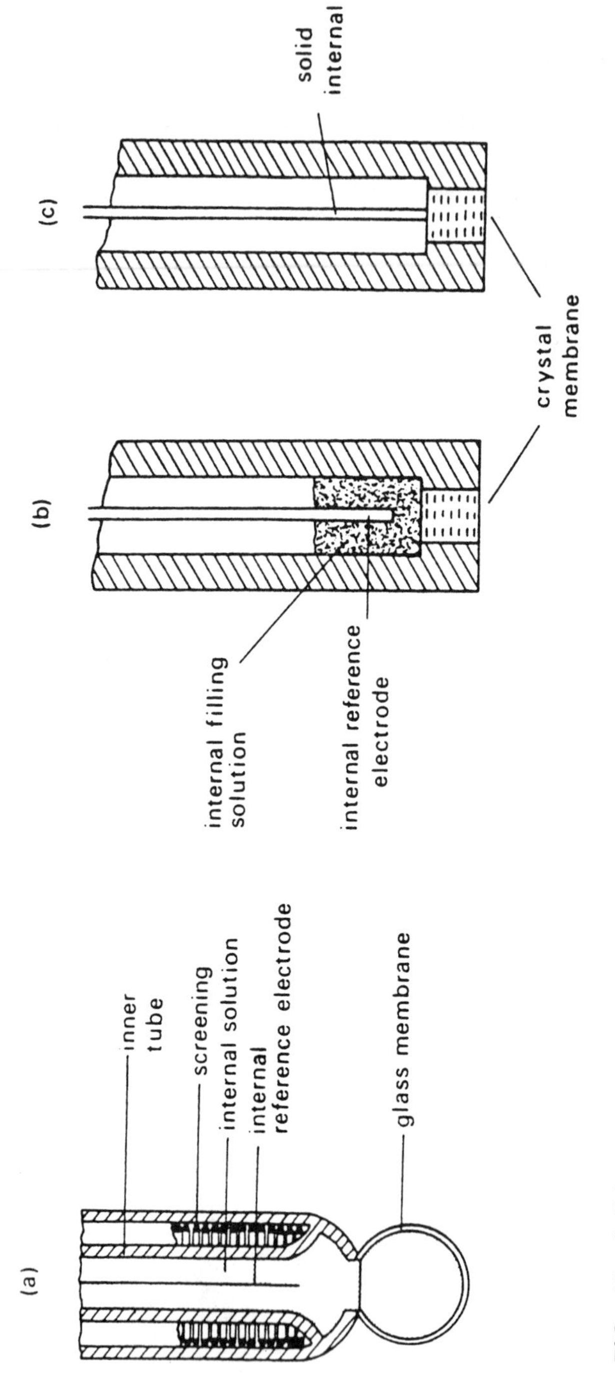

FIG. 12 Solid-state membranes. (a) Glass; (b) crystalline, with internal reference solution; (c) crystalline, with internal solid connection. (From Ref. 115.)

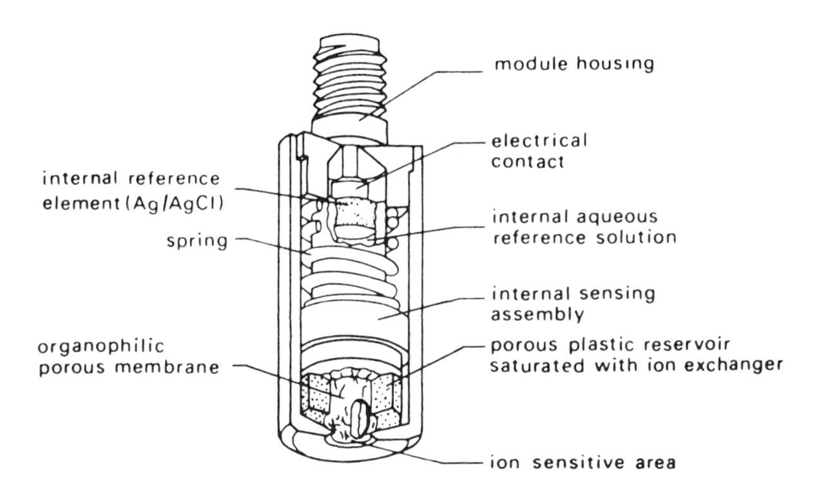

FIG. 13 Free-flowing liquid membrane: liquid in porous plastic reservoir. [From *Orion Analytical Methods Guide*, data sheet on liquid membrane electrodes (electrode now obsolete). Orion Research Inc., Cambridge, Mass., 1978.]

ion concentration, sensitivity of response (i.e., millivolts per decade change in the specific ion concentration), response time, the pH and temperature ranges in which each electrode functions, and the interfering effects of foreign ions most likely to occur in the soil and plant environment. These data should be read in conjunction with Sec. II.C to understand their full import. The reader's attention is drawn to the manner in which ion interference is expressed in Tables 6 and 8 for Orion ISEs (see footnote to the tables).

The lower limit of the operational concentration range in Tables 5 and 7 is taken as the detection limit of the specific ion, although the latter could well be a tenth of the lower limit under ideal conditions, e.g., good temperature stability, absence of interfering ions, and high stability of the measuring device. Response times are also affected adversely by these factors and could be as little as 1–10 sec under ideal conditions.

D. Gas-Sensing Electrodes

Strictly, these are not sensitive to the specific ion, the sensing device being a pH glass electrode immersed in a film, directly connected to a small reservoir of a pH buffer solution within the electrode, between the glass membrane and an inert gas-permeable

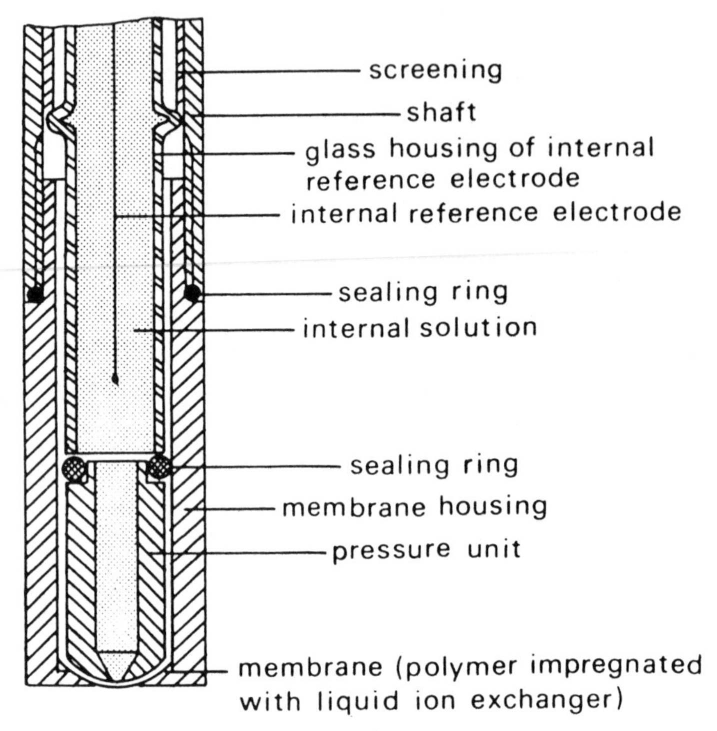

FIG. 14 Liquid membrane immobilized in polymer. (From Philips Bulletin on Analytical Equipment, Oct. 1970, courtesy of Philips-Pye-Unicam Ltd., Cambridge, England.)

membrane (Fig. 15). The appropriate gas is generated from the specific ion solution by adjusting its pH (Table 9) in an enclosed microcell incorporating the gas-permeable membrane very close to the generating solution.

IV. APPLICATIONS IN SOIL SCIENCE AND RELATED DISCIPLINES

A. Introduction

Ion-selective electrodes offer a rapid method of determining the activities of a range of ions in soil extracts and in soil pastes and suspensions. In extracts they are a cheap alternative to spectrophotometric and colorimetric techniques, but the facility to measure the activity of an ion directly in a solution in equilibrium and in contact with the soil is unique. The extension of this to the determination

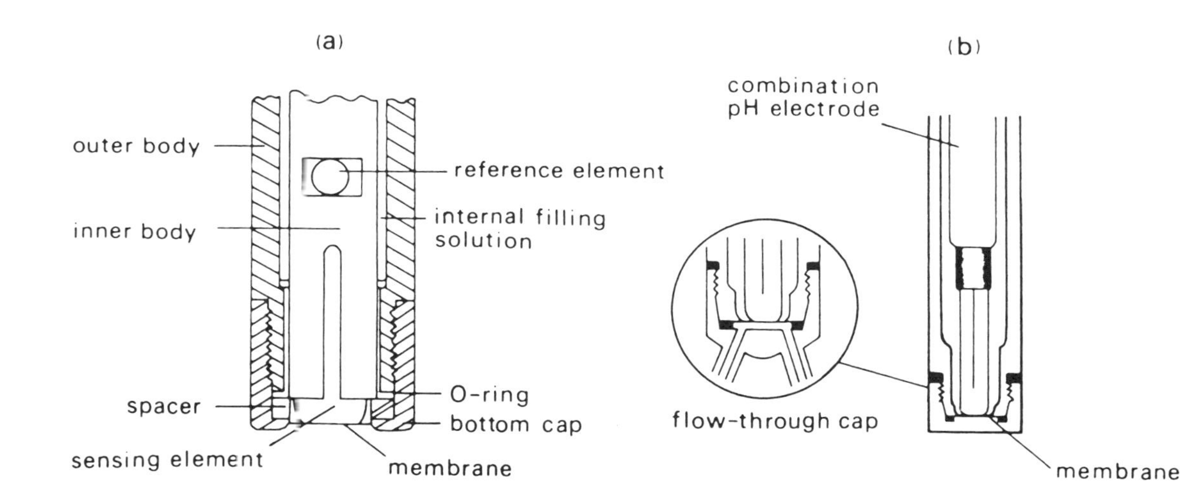

FIG. 15 Gas-sensing electrodes. (a) Ammonia electrode. (From *Orion Analytical Methods Guide*, Orion Research Inc., Cambridge, Mass. 1978.) (b) Sulfur dioxide probe. (Courtesy of Kent Industrial Measurements, Ltd., Chertsey, Surrey, England.) The sensing element is a pH glass electrode.

TABLE 9 Gas-Sensing Probes for Dissolved Gases Incorporating a pH Glass Electrode

Gas	Manufacturer (see Table 4)	Detection limit (M)	Operational pH	Response time (sec)	Interferences	Ref. (soils)
NH_3	K	5×10^{-6}–5×10^{-2}	≥ 11	< 60	Volatile amines	[96–98]
	O	10^{-6}–1	≥ 11	< 60		
CO_2	O	10^{-4}–10^{-2}	5	< 60	Volatile weak acids	
$NO + NO_2$	O	5×10^{-7}–10^{-2}	0–2	< 60	Volatile weak acids	[100]
SO_2	K	5×10^{-5}–5×10^{-2}	1.7	< 600	HF, HCl, acetic acid	

activity of nutrients in the soil in situ is perhaps one of the ultimate goals in soil chemistry and plant nutrition studies.

Although ion-selective electrodes have been in use in soil science for nearly 20 years, only rarely have they been established in routine analytical techniques. Their main limitation is imperfect selectivity for the measured ion, so that the presence of other ions (as is usually the case in soils) causes errors, although the selectivity of electrode membranes has been improved progressively. This imperfect selectivity has necessitated careful selection of extracting agents and of reference electrode salt bridges. With a calomel reference electrode, saturated KCl is usually used but is obviously not ideal when measuring a halide, ammonium or nitrate ion and out of the question for K^+ and Cl^- determinations (see Sec. II.C).

That ISEs measure activities can be a great advantage, but often total concentrations are required. They can be calculated from activities but generally the problem is overcome by equalizing the ionic strength of the standards and the samples.

The following is a review of how ISEs have been employed in soil science.

B. Cation-Sensitive Glass Electrodes

1. *Sodium*

Bower [22] realized the value of glass sodium-selective electrodes in soil salinity investigations. He first used a Beckman sodium alumino-silicate glass electrode for determining sodium in soil water extracts. This glass, however, was equally sensitive to K^+, which consequently had to be precipitated with tetraphenyl boron. A later Beckman sodium electrode, made with lithium aluminosilicate glass, was much less sensitive to potassium. With this, Bower [23] measured Na^+ in soil water extracts mixed 1:1 with 1 *M* magnesium acetate to give a constant activity coefficient and to buffer the solution to pH >6, thus eliminating interference from H^+ ions. He also determined exchangeable sodium in the soil by measuring Na^+ in a 0.5 *M* $MgAc_2$ extract with the electrode and similarly the cation exchange capacity (CEC) by saturating the exchange sites with sodium, then displacing it with 0.5 *M* $MgAc_2$. He found good agreement between his electrode measurements and measurements made on the same extracts by flame photometry. This observation was confirmed by Fehrenbacher et al. [24], who obtained a correlation coefficient of 0.9982 between electrode and flame photometer measurements of sodium in 126 soil water samples obtained from field piezometers and percolation studies (Fig. 16). However, El-Swaify and Gazdar [25] found that the Beckman sodium electrode gave poor response slopes in 0.2, 0.5, and 1.0 *M* ammonium acetate and also in

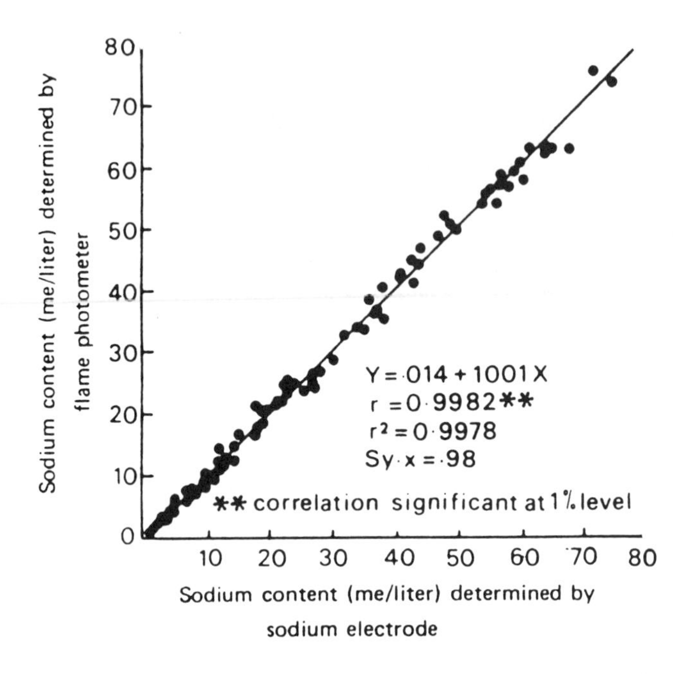

FIG. 16 Relationship between the sodium contents of soil water as determined by the ISE and flame photometry. $(Na)_{photometry} = 0.014 + 1.001(Na)_{ISE}$; $r = 0.9982$**. [Reproduced from Ref. 24, *Soil Sci. Soc. Am. Proc.*, *27*(2):153 (March–April 1963) by permission of the Soil Science Society of America.]

0.1 and 0.25 *M* $MgAc_2$, although in 0.5 *M* $MgAc_2$ the response was close to the theoretical (see Sec. II.B). Using an electrode in a soil suspension or paste may produce errors caused by the surface charge on soil particles, e.g., the suspension effect with the sodium glass electrode [26].

Quite large errors may be involved in electrode measurements in soil pastes. Susini et al. [27] obtained 25–30% errors in sodium determinations in saturated soil pastes with a sodium glass electrode. They stated, though, that in salinity investigations a high degree of accuracy is not always necessary and the lack of it is compensated for by the speed of measurement. Nevertheless, over the wide concentration range that they measured (13–10,000 μg Na g^{-1}), the correlations between the electrode measurements and measurements made on the supernatant solutions were very good ($r = 0.9986$). The regression coefficient ($b = 0.75$) was rather small, but this may have been in part due to the difference between activity measured by the electrode and concentration by the flame photometer.

Orlov and Alzubaidi [28] measured sodium activities in 1:5 soil: water suspensions with a glass electrode to distinguish solonetz and saline soils. They found that these values correlated well with exchangeable sodium; solonetz soils had pNa values of 1.5 or less, and nonalkaline soils pNa values greater than 3.0. Alzubaidi [29] used the same method to measure the degree of dissociation of the exchangeable sodium in soils expressed as

$$\text{Dissociation percentage} = \frac{a_{\text{Na}} \text{ in suspension}}{a_{\text{Na}} + \text{exch. Na}} \times 100$$

Using a technique described briefly in the next section on potassium, Krupskiy et al. [30–32] measured sodium ion activities in soil pastes with a glass electrode that could tolerate a 100:1 excess of Ca, 4:1 of K, and pH down to 1.6. They also made measurements in the field and found that field soils at 15–20% moisture gave similar results to the same soils in the laboratory.

2. *Potassium*

Mortland [33] determined exchangeable potassium in 30 soils using the early Beckman cation-selective electrode, with which Bower [22, 23] had first measured sodium. This could only be achieved in soils containing little sodium and lithium because the glass membrane was equally sensitive to those ions. He eliminated H^+ interference by extracting the exchangeable K^+ with 0.5 *M* $MgAc_2$ in a 1:10 soil-solution ratio as Bower had done for Na^+, measuring K^+ in the suspension with the electrode and in the supernatant by flame photometry. The correlation between the two sets of values was satisfactory ($r = 0.902$). The standard curve for the electrode in KCl solutions, made up in 0.5 *M* $MgAc_2$, had a response slope of 55.3 mV per 10-fold change in a_{K^+} between 10^{-4} and 10^{-1} *M*. Mortland also compared the electrode measurements of exchangeable K with those made by flame photometry on *M* NH_4Ac extractions ($r = 0.825$). Surprisingly, a saturated KCl calomel reference electrode was used in these measurements.

Farrell and Scott [34] found the valinomycin ISE to be far superior in determining exchangeable K in soils, extracted with *M* NH_4OAc or 0.5 *M* $BaCl_2$, compared with the K-sensitive glass electrode. Exchangeable K values agreed to within ±5% of values determined by conventional atomic absorption analysis.

Orlov and Alzubaidi [35] used a K^+ glass electrode to determine K^+ activities in soil suspensions (1:2, soil to water) and compared the results with flame photometer measurements made in the supernatant liquid. In forest soils the two determinations agreed well, but in a chestnut soil the electrode readings tended to be higher than those of the flame photometer, probably because of interference by sodium.

Very thorough investigations into the use of ion-selective electrodes in soil pastes were made by Krupskiy et al. [30,31]. Measuring K^+ with a glass electrode, they found that the $p(K^+)$ value measured (= -log [K]) decreased linearly as the logarithm of the dilution factor of the paste increased. The regression coefficients were characteristic of individual soils. It was therefore necessary to select a standard moisture content in which to make measurements. These workers selected the lower liquid limit as determined by the Vasil'yev balance cone, at which the paste consistency is such that the cone sinks in it to a certain mark. Another interesting feature of their work was the use of lithium acetate in agar as a second salt bridge. They found lithium only interfered in the electrode measurement when the Li:K ion ratio was greater than 5:1. The acetate ion has a mobility similar to that of the Li^+ ion because of the considerable hydration capacity of Li^+.

The K^+ electrode was also sensitive to sodium and could not be used where the K:Na ion value was less than 1:1. Figure 17 shows that the electrode calibration curve was linear down to quite low K^+ concentrations, i.e., high $p(K^+)$ values. These critical $p(K^+)$ values increased with decreasing Na^+ concentrations. By fitting second-degree polynomials to a series of calibration curves, obtained in the presence of a range of sodium concentrations, Krupskiy et al. [30] were able to derive one equation to elucidate the effect on the EMF:K activity relationship at any given concentration of sodium.

3. *Cation Exchange*

The measurement of ion activities in suspensions lends itself to application in cation exchange studies. Kennedy and Brown [36] measured the rate of exchange of calcium and magnesium for sodium with a sodium electrode on sand-sized stream sediment and on kaolinite, montmorillonite, and illite. They added $CaCl_2$ solution to a suspension of clay or sand particles and measured the increase in Na^+ activity in the free solution with the electrode. The results were plotted on a chart recorder attached to the millivoltmeter. However, they found that a second method, in which the Na saturated clay or sand was added to a 0.085 *M* $MgCl_2$ solution, gave smoother curves. It proved possible to measure the rates of exchange up to 90% Na saturation. Beyond this point the rate of electrode response for the small change in activity was too slow.

Malcolm and Kennedy [37] also used this second method to measure rates of Na^+ and K^+ exchange with Mg^{2+}, Ca^{2+}, and Ba^{2+} on

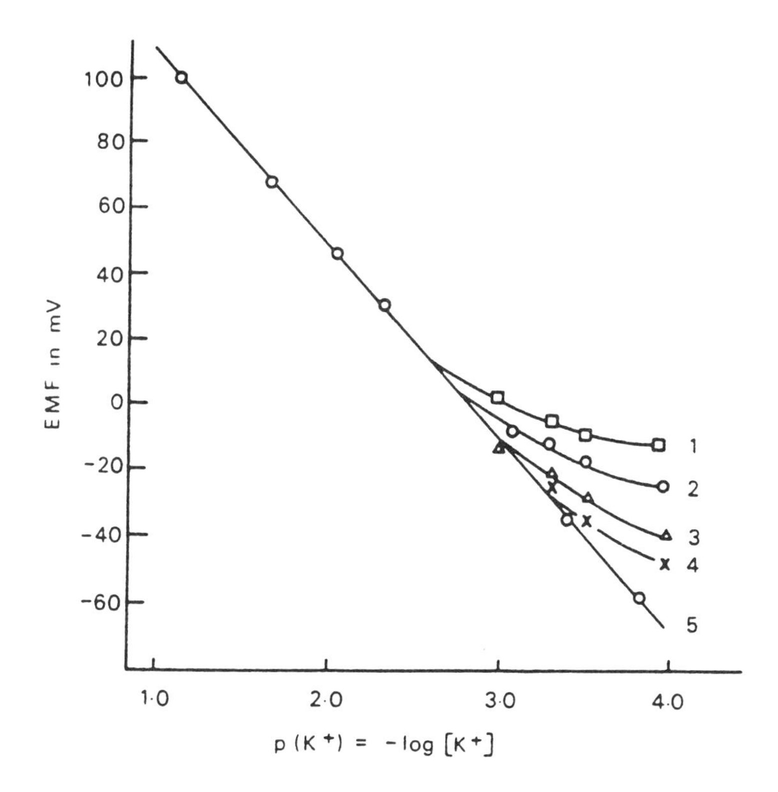

FIG. 17 Dependence of the EMF of a potassium ISE on sodium activity in solution. p(Na) values: (1) 2.29; (2) 2.42; (3) 3.32; (4) 3.06; (5) no sodium. (From Ref. 30, reprinted with permission of Scripta Publishing Co., Silver Spring, Md.)

several standard clay minerals using Beckman electrodes for K^+ and Na^+. They added the K^+ or Na^+ saturated clay suspensions rapidly into 0.1 *M* Ca, Ba, or Mg chloride solutions and then monitored the release of Na^+ or K^+ into the solution with the electrode. However, all exchanges except K^+ for Ba^{2+} on vermiculite were as fast as, or faster than, the response rate of the electrode and hence could not be measured. Exchange on the illite and the kaolinite appeared to be even faster than the electrode response in the blank. The authors suggested this may be because suspensions of these two minerals were made more concentrated to compensate for their low exchange capacities and so were perhaps not well mixed. The electrode would then have measured artificially high local concentrations. They discounted any suspension effect because the percentage change in conductance in the liquid, produced by addition of the clay, was only of the order of 0.5%. The CECs of the clays, taken as the total K released into the solution after 15 min, agreed within 3% with values obtained using ^{85}Sr exchange.

C. Cation-Sensitive Liquid Membranes

1. *Potassium, Calcium, and Magnesium*

With the introduction of macrocyclic antibiotics as selective membranes, the problem of sodium interference in the measurement of K^+ was greatly reduced. Banin and Shaked [38] used a valinomycin-in-polymer K electrode with a calomel reference electrode and a lithium trichloroacetate salt bridge to measure potassium activity in aqueous extracts of 16 soils, with soil:water ratios ranging from 1:10 to a saturated paste. The measurements were made in the filtered liquid. They obtained a high correlation between K concentrations measured by the ISE and the flame photometer but a low regression coefficient ($r = 0.92$, $b = 0.65$), i.e., 35% of the potassium in the extract was not measured by the electrode. Ionic strength effects were accounted for in the calculation of concentration from the activity measurements. The authors suggested that the potassium in the soil solution is bound by ligands and thus the concentration measured is determined by the dissociation constant of these ligands. However, we have often observed that submicron particles in filtrates from soil suspensions cause erroneously high flame photometer readings (Talibudeen et al., unpublished work).

Calcium-selective liquid membrane electrodes have been employed to determine soil calcium by several workers. El Swaify and Gazdar [25] tested Orion Ca^{2+} and $Ca^{2+} + Mg^{2+}$ electrodes with extracting solutions for exchangeable cations. They found that the electrodes' response slopes generally decreased with increasing sodium or ammonium acetate concentrations up to 1 *M* and that only the calcium electrode in 0.2 *M* or 0.5 *M* NaAc gave sufficient response to be of practical value in determining exchangeable calcium. The calcium electrode was sensitive to magnesium to such an extent that Mg could not be determined by the difference between the Ca^{2+} and $Ca^{2+} + Mg^{2+}$ electrode measurements.

Woolson et al. [39] compared electrode and atomic absorption determinations of calcium in 0.5 *M* NaAc extracts of soil diluted 10-fold. The correlation and regression coefficients for 12 soils were both 0.99. Similarly, Doiron and Chamberland [40] obtained, for 93 soils, good correlation between calcium concentrations measured by the Orion electrode and by atomic absorption in 0.5 *M* KCl soil extracts ($r = 0.91$). However, the regression coefficient was only 0.85 and they also suggested that the 15% Ca^{2+} not measured by the electrode might be complexed by ligands.

McLean and Snyder [41] investigated pH-dependent charges on illite and bentonite with the aid of ion-selective electrodes. Activities of Ca^{2+} and Rb^{+} were measured in mixed suspensions of H^{+}, Ca^{2+}, and Rb^{+} saturated clay, Rb^{+} with a Beckman cation glass electrode, and Ca^{2+} with the Orion Ca^{2+} + Mg^{2+} electrode. The Beckman glass electrode is sensitive to all monovalent cations, so a correction had to be made for hydrogen.

Wang and Yu [42] used a novel device to eliminate errors in measuring [42] the lime potential of soils (pH - 0.5 pCa) caused by (1) the inability to calculate precisely Ca^{2+} activity, especially at high ionic strengths, (2) the "suspension effect" resulting from the junction between the reference electrodes (coupled with the pH glass and Ca^{2+} electrodes), and (3) the uncertainties arising from the salt bridges for these reference electrodes in their different liquid junction potentials with the soil suspension. They measured the potential difference $E_{Ca} - E_{H}$ between the pH glass electrode and the Ca-ISE inserted together in the soil suspension, thus avoiding the use of reference electrodes. The lime potential was then given by

$$\mathrm{pH} - 0.5\ \mathrm{pCa} = \frac{(E_{Ca} - E_{H}) - (E^{0}{}_{Ca} - E^{0}{}_{H})}{S_{H}} + 0.5\ \mathrm{pCa}\left(\frac{S_{Ca}}{S_{H}} - 1\right) \tag{17}$$

where $E^{0}{}_{Ca}$ and $E^{0}{}_{H}$ are the respective electrode potentials in their standard states, and the theoretical value for both S coefficients would be 58.17 mV at 20°C, realizable in practice for the glass electrode but deviating slightly for the Ca-ISE.

Wang and Yu [43] used their 1981 method, employing the Ca^{2+} ISE and the pH glass electrode jointly, to show that the lime potential (pH-1/2 pCa) curve of a soil is "more distinct" than its pH curve. The lime potentials of ferralsols were found to be higher (5.0–7.0) than those of acrisols and luvisols (<5.0).

Parra and Torrent [44] designed a continuous procedure using the Orion potassium ISE (no. 93-19) coupled with a double junction reference electrode containing 0.02 M $CaCl_2$ for obtaining the quantity-intensity relationship for soil potassium. Increasing amounts of KCl in $CaCl_2$ were added successively to the soil suspended in 0.01 M $CaCl_2$ to give final K concentrations in the soil solution between 5×10^{-5} and 5×10^{-3} M potassium. After each addition and stirring for 6 min, the K concentration in the suspension was measured with the ISE and the Q/I relationship obtained in the normal way. The "suspension effect" was said to be negligible and the procedure more rapid than the conventional "batch" method. The results obtained were the same as with the latter method and so the continuous procedure was considered more suitable for routine analysis.

2. Ammonium

The Philips ammonium electrode has been used to measure ammonium activities in soil pastes [45] but is not ideally suited for use with soil suspensions because of the low NH_4^+ concentrations normally present and because of its relatively high selectivity for potassium (see Sec. III.C).

D. Anion-Sensitive Liquid Membranes

1. Nitrate

The measurement of nitrate has probably been the most important application of ion-selective electrodes in soil. Only rarely is nitrate adsorbed by the soil and so its total concentration in the soil is measureable in the soil solution. However, soils with some positively charged surfaces will adsorb nitrate; then a salt solution, usually a sulfate, is used to displace the nitrate. Except for chloride in saline soils, interfering ions are not usually present in sufficient concentration to cause serious errors in the electrode measurement of nitrate, hence its popularity in use.

Several extracting solutions have been used to remove nitrate from the soil for measurement with the electrode. Myers and Paul [46] obtained 94% recovery of added nitrate, measured by the electrode in clear water extracts of soil (1:5 and 1:10 soil to water, shaken for 30 min). Electrometric determinations correlated well with measurements made on the same extracts by the phenoldisulfonic acid colorimetric method ($r = 0.991$ for 75 soils) and the relationship between the two determinations was almost 1:1 ($b = 0.98$). Extractants containing $CuSO_4$ (with Ag_2SO_4 added to precipitate chloride) and saturated $CaSO_4$ were also useful. Other workers used 0.01 *M* $CuSO_4$ [47], 0.005 *M* $CuSO_4$ + Ag_2SO_4 [48], saturated $Ca(OH)_2$ [49], 0.25 *M* K_2SO_4, saturated $CaSO_4$ [50], and $KAl(SO_4)_2$ [51] as extractants for nitrate.

Soil:solution ratios varied from 1:5 to 1:20. Too great an excess of solution could give too small a concentration of nitrate. Shaking periods used were generally between 15 min and an hour. These methods mainly correlated well with determinations by other recognized techniques, including phenoldisulfonic acid colorimetry and steam distillation.

Black and Waring [44] extracted NO_3^- from an oxisol as efficiently with 0.05 *M* K_2SO_4 as with 4 *M* KCl (oxisols have positively charged sites that adsorb anions). Measurements made with the Orion electrode in the supernatant 0.05 *M* K_2SO_4 extract gave a correlation coefficient of 0.99 with the steam distillation determinations made on the same solutions and the regression coefficient was 0.97 (Fig. 18). Except in such soils, the extracting solutions serve mainly to keep

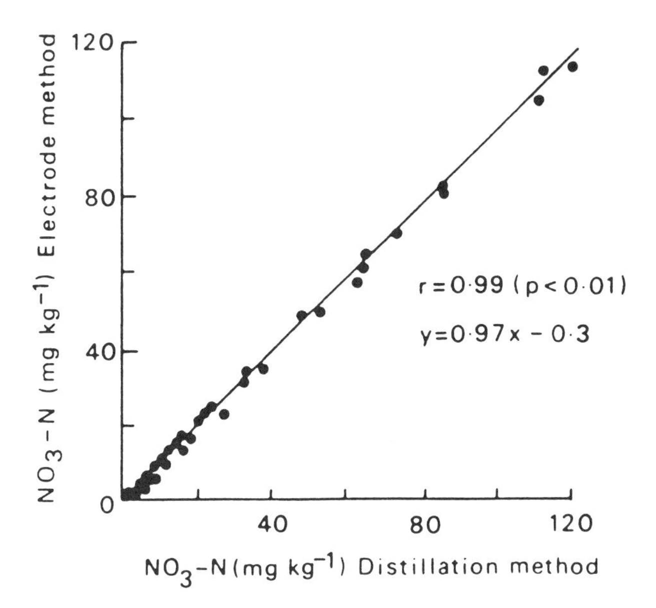

FIG. 18 Nitrate levels in field samples determined by the nitrate ISE and by a distillation-colorimetry method. [From Ref. 52, A. S. Black and S. A. Waring, Nitrate determination in an Oxisol using K_2SO_4 extraction and the nitrate-specific ion electrode, *Plant and Soil*, *49*(1):211 (1978), reproduced by permission of Martinus Nijhoff Publishers BV, The Hague, The Netherlands.]

a uniform ionic strength. They may also prevent dispersion of the clay particles that can cause errors in the measurement [50].

The most abundant interfering ion in these soil extracts is generally chloride. As already mentioned, Ag_2SO_4 has sometimes been used to precipitate the chloride. Raveh [53] added 0.1 *M* sodium citrate to aqueous soil extracts to overcome interferences (1 ml sodium citrate to 9 ml extract) and found it effective except where $Cl^-:NO_3^- > 100:1$.

Milham et al. [54] used a buffer of 0.01 *M* $Al_2(SO_4)_3$, 0.01 *M* Ag_2SO_4, 0.02 *M* boric acid, and 0.02 *M* sulfamic acid adjusted to pH 3 with sulfuric acid, to eliminate chloride, nitrite, and organic anion interferences in soil extracts. This was added 1:1 to the aqueous soil extracts prior to nitrate determination with the electrodes. Francis and Malone [55] added sulfanilamide with 0.01 *M* H_2SO_4 to soil extracts to complex nitrite as a diazonium salt. The

H_2SO_4 aids the reaction, destroys HCO_3^-, and maintains uniform ionic strength.

It is also possible to measure nitrate with an electrode directly in soil suspensions and pastes, and in moist soil in situ in the field. Bremner et al. [56] developed a technique for determining nitrate in 1:2 soil-water suspensions. The mixture was stirred and left for an hour, then the measurement was made while the suspension was magnetically stirred. They obtained almost perfect agreement between results from this method, from electrode measurements made in the supernatant solution, and from the steam distillation method. They also achieved 100% recovery of added nitrate on eight contrasting soils.

Hadjidemetriou [57] preferred to use a 0.01 *M* copper(II) sulfate extractant for the same determination because it prevented dinetrification, helped filtration, and eliminated interferences by bicarbonates in calcareous soils. Nitrites were removed from the extract with 1 *M* H_2SO_4 containing sulfamic acid. However, others have shown that nitrates can be measured in soil pastes and suspensions without the need for such chemical devices (see this section). Hadjidemetriou (private communication, 1986) found that the newer Orion modules are not effective for direct determination in the soil suspension of the nitrate ion because they are easily blocked.

Li and Smith [58] found that nitrate ISE measurements in soil extracts from dry and moist soils, using a $CaSO_4$-saturated solution, agreed quantitatively with those from continuous-flow analysis (CFA) in which the color developed by the reduction-diazotisation-amine coupling process was measured at 520 nm. In the ISE method, a nitrate interference suppressor, containing $Al_2(SO_4)_3$, H_3BO_3, Ag_2SO_4, and sulfamic acid, was added to the saturated $CaSO_4$ extract (10% v/v), and this was also put in the outer chamber of the Orion Model 90-02 double junction reference electrode. This method was much better in its precision and its correlation and quantitative comparison with CFA than ISE methods in which the outer chamber was filled with 0.04 *M* $(NH_4)_2SO_4$ and 0.4 *M* $(NH_4)_2SO_4$ in 1% and 2% v/v concentration was added to the saturated $CaSO_4$ extract.

Øien and Selmer-Olsen [47] recovered 98% of the added nitrate in 2:5 soil-water pastes but found that increasing the ratio decreased the recovery. Onken and Sunderman [59] obtained good recoveries in 513 soils in 1:5 soil-water suspensions with only 15 min shaking.

Bound [60] measured nitrate with a polymer membrane electrode in soil pastes made up in saturated $CaSO_4$. He found a better correlation between results obtained with an ISE and analyses done on a Technicon AutoAnalyzer when the ISE was coupled to an Ag-AgCl reference electrode with a salt bridge immobilized in PVC than to a saturated calomel reference electrode. With an immobilized salt bridge there is no liquid junction potential set up [61].

Measurements of soil nitrate in situ in the field were made on the Broadbalk Classical Experiment at Rothamsted by Nair and Talibudeen [62]. Permanent holes, lined with plastic tubes, were maintained in the soil at different depths, throughout the summer months, under winter wheat and in adjacent fallow soil. To make a measurement, the soil at the bottom of a hole was wetted with about 5 ml of deionized water then left to come to equilibrium for about 10 min. A Corning ISE, with a micro-calomel reference electrode, was inserted and the reading on an Orion Ionalyzer taken after 60 sec. Sometimes it was difficult to wet the soil when it was dry and cracked, so the method was later modified [63]; the soil was stirred to a paste immediately after the addition of the water and the measurement made 2 or 3 min later. The soil was then taken to be approximately at water holding capacity. Both methods proved to be reproducible. Typical results are shown in Fig. 19, showing the effect of the crop on nitrate levels on a plot given mineral nitrogen

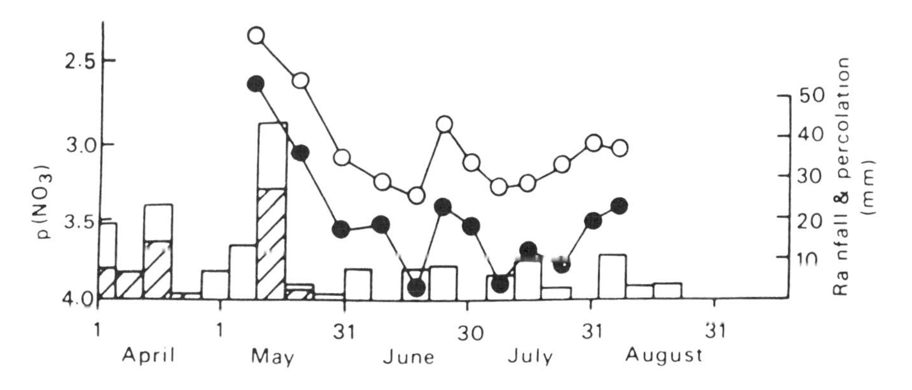

FIG. 19 Change of $p(NO_3) = -\log[NO_3]$ in fallow and cropped soil at 15 cm depth with time on the Broadbalk Classical Experiment, Rothamsted, Plot 07, and histograms of weekly rainfall and percolation through 0–50 cm soil (shaded), for April–August, 1175. o—o—o: fallow; •—•—•: cropped. [From Ref. 63, M. B. Page and O. Talibudeen, Nitrate concentrations under winter wheat and in fallow soil during summer at Rothamsted, *Plant and Soil*, *47*(3): 527 (1977), published by Martinus Nijhoff Publishers BV, The Hague, The Netherlands.]

fertilizer. By this method the weekly uptake of NO_3^--N by a winter wheat crop was estimated and compared with weekly uptakes as measured by regular sampling and analysis of the crop [64]. The correlation between the two methods was 0.77 ($p < 0.01$).

It seems quite likely that ion-selective electrodes will be used increasingly in monitoring nutrient ion concentrations in soil-free composts used in glasshouse crop production. In the relatively small volumes of composts used per plant, ion concentrations and salinity levels are critical. Preliminary investigations have been made on the use of electrodes in determining NO_3^-, K^+, and NH_4^+ concentrations in water extracts (with added ionic strength adjuster; see Sec. II.D) of sedge peat compost [65]. High recoveries of added nitrate have been achieved (90–100%) but only 50–60% of K^+ and NH_4^+, probably because of the cation-exchange properties of the peat.

E. Crystalline Solid-State Membranes

1. *Chloride and Bromide*

The halide electrode most important to soil science is that which is sensitive to chloride. In saline soils, the natural chloride concentration is important, because of its potential toxicity to plants; and in artificial systems created for cation exchange studies, chloride is often the only anion present. There is, therefore, much scope for the use of chloride-sensitive electrodes in soil science, although up to now reports of their use are few.

Hipp and Langdale [66] compared Cl^- determinations in water extracts of soil by titration, both with a silver/silver chloride electrode and a chloride ion-selective electrode as end point detectors, and by direct measurement with the chloride-selective electrode. The titrations gave identical results, but those from direct measurements were slightly higher. The regression equations were

$$y_1 = 1.055x - 0.085$$

$$y_2 = 1.046x + 2.58$$

where

x = results obtained by titration using the Ag-AgCl electrode
y_1 = results obtained by titration
y_2 = results by direct measurement, using the Cl^- electrode

Selmer-Olsen and Øien [67] recovered all the chloride added to a soil by shaking it with 0.5 *M* NH_4NO_3 (1:5, soil:solution) and measuring Cl^- in the filtered acidified extract with an Orion electrode.

They found no significant difference between measurements made with the electrode and by colorimetry.

The chloride electrode can be used in a rapid method for the determination of the cation exchange capacity (CEC) of soils. The soil is washed with ethanol to remove excess salts, then leached successively with 1 *M* NH_4Cl and 0.04 *M* NH_4Cl, and finally with 0.5 *M* Na_2SO_4. The Na_2SO_4 leachate is analyzed for NH_4^+ and Cl^- and the CEC calculated as the difference, in meq, between the two. Smart et al. [68] followed this procedure, analyzing the leachate for chloride by titration and with an Orion solid-state electrode. Calibrating the electrode with standard ammonium chloride solutions showed good agreement between the two methods, although with NaCl standards the comparison was poor. This was because NH_4Cl accounts for any complexing between NH_3 and the silver of the electrode. Greenhill and Peverill [69] used basically the same method, but with a mixed displacing solution of 1.5 *M* KNO_3 and 0.25 *M* $Ca(NO_3)_2$ and an ammonia probe to measure the ammonium.

Krupskiy et al. [70] made measurements of Cl^- in soil pastes with a solid-state electrode. The pastes were prepared with the Vasil'yev cone as described above. The reference electrode was a calomel with a KNO_3-agar salt bridge. The disadvantages of this bridge is the slow electrode response it creates, 10–15 min in this case. The same electrode was also used to make measurements in situ in the field, with soil moisture contents of 13–16%, but without the salt bridge.

In soil water, commonly occurring inorganic ions do not interfere with chloride analysis with the Ag-ISE, but in organic soils enough humic anions are released to complex and "poison" the Ag ion in the ISE, which prevents the full Nernstian potential from developing. Sikora and Stevenson [71] determined selectivity coefficients (Cl^- versus organic anion) for 11 humic and fulvic acids ranging from 0.59×10^{-4} to 1.3×10^{-4} mol g^{-1} at pH 6.5. They found that <0.1 g l^{-1} of humic substances did not interfere, but interference became serious when Cl was $<10^{-4.5}$ *M* and organic matter exceeded 0.1 g l^{-1}.

Sekarka and Lechner [72] employed an ISE and a calibration curve plus a "standard additions" technique (see Sec. II.E) to determine chloride concentrations in biological materials, food products, soils, and waste waters. The solids were digested in 0.1 *M* $HClO_4$ and $K_2S_2O_8$. The lowest concentrations that were determined in extracts of the organic materials were ca. 12 mg kg^{-1}, and of soils, 80 mg kg^{-1}. According to the authors, the extraction and measurement procedures "eliminated most interferences, including those caused by I^-, complexing and reducing compounds, macromolecules and surface-active agents."

Although the *bromide* ion is not of great importance in soil science, some workers have used electrodes to measure it. Abdalla and Lear [73] recovered an average of 97% of bromide added to soil by extraction with water (1:2, soil:water) and measurement in 0.01 *M* $NaNO_3$ with a solid-state bromide electrode. Saffigna et al. [74] determined bromide in lysimeter leachates and in 0.05 *M* K_2SO_4 soil extracts.

2. Fluoride

Fluoride is considered a pollutant on agricultural land, hence, increasing importance is attached to its measurement. It was determined in soils, rocks, and plants with an ISE after it had been rendered water-soluble by fusion of the sample with an alkali [75,76]. After fusion, McQuaker and Gurney [76] dissolved the sample in HCl, bringing the solution to pH 8. Aluminum, iron, calcium, and magnesium and also silicates were then filtered off as insoluble compounds to prevent complexation of the fluoride ion. The ISE measurement was made in a TISAB buffer (see Sec. II.D) at pH 5.2. A slightly acid pH was chosen because hydroxyl ions interfere. The citrate ion in TISAB also helps prevent Al-F complexes from forming; Crenshaw and Ward [75] increased the usual citrate concentration to enhance this effect, but without success. The EMF measurements were made in the buffered sample solutions by known addition techniques (see Sec. II.E). McQuaker and Gurney [76] recovered between 95 and 100% of fluoride added to grass, hemlock, and soil samples by this method.

Fluoride in plant materials has been determined with the Orion ISE, Model 94-09, in solutions from different digests to eliminate interference by multivalent metal ions. Thus, Takala [77] used microdiffusion with $HClO_4$ at 60°C for lichens, collecting the volatilized HF in 2 *M* NaOH. The fluoride was measured with the ISE in an acetate buffer, with the covalently bound F not extracted, and a recovery of ca. 90% of the total fluoride achieved. Villa [78], on the other hand, preferred to extract plant material with 0.1 *M* $HClO_4$ at 20°C for 20 min, coupling it with measurements of the F concentration with an ISE using the method of standard additions (see Sec. II.E). The author showed that recoveries of 98—102% were achieved and metal ions in the plant materials did not interfere.

3. Cyanide

The ISE electrode gives better results for CN^- analysis of aqueous solutions containing metal ions than titration and colorimetic methods [79]. Based on earlier work [80], Rands and Bain [81] developed a method to determine free and metal-ion-complexed CN^- in aqueous solutions, using an Orion ISE (Model 94-06A) whose surface was polished periodically with a commercial toothpaste "to prevent erratic response from surface coatings which developed after prolonged use... ." Free CN^- concentrations of ferricyanide solutions, in the presence of organic ligands, were determined with an ISE, before and after refluxing in a N_2 atmosphere. About 60–80% of complexed CN^- was released within 30 min of refluxing, and the authors concluded that their method was entirely satisfactory for determining the total CN^- content of aqueous solutions.

4. *Sulfide*

Sulfides form under anaerobic conditions in soil, and free hydrogen sulfide, toxic to plants even at low concentrations, may accumulate. Allam et al. [82] determined H_2S concentrations in submerged rice soils by measuring S^{2-} concentrations with a silver sulfide membrane electrode, and pH. Hydrogen sulfide concentrations were calculated using the equilibrium constant (K) for the reaction

$$H_2S \rightleftharpoons 2H^+ + S^{2-}$$

Thus,

$$K = \frac{[a_{(H^+)}]^2 \, a[S^{2-}]}{a_{(H_2S)}} = 10^{-20.9} \tag{18}$$

Electrode potential measurements were made in situ in the field and in samples taken back to the laboratory using a silver/silver chloride reference with a potassium nitrate salt bridge.

The sulfide ISE was calibrated in serially diluted sodium sulfide solutions (made with 1 *M* sodium hydroxide and purged with nitrogen) down to a sulfide concentration of 10^{-6} *M*. To obtain a calibration curve for sulfide concentrations below this value, the electrode potential was measured in a solution of constant *total* sulfide concentration at different pH values. In such a solution, the *free* sulfide ion and hydrogen ion concentrations are inversely proportional [Eq. (18)]. From this and an electrode potential/pH response curve, the sulfide ion concentrations for different pH values were calculated and a calibration curve for sulfide ion concentrations down to 10^{-20} *M* plotted.

Sanderson and Armstrong [83] investigated the concentrations of phytotoxic sulfides in water extracted from peaty forest soils subject to waterlogging, using an Orion ISE, Model 94-16. The electrode gave poor response times and poor reproducibility with a 34 mV rise per decade concentration increase. They found it useful for sulfide concentrations between 5×10^{-4} and 5×10^{-2} *M* (see Table 7, column 4, for the manufacturer's stated concentration range).

5. Copper and Other Divalent Metals

Copper is complexed by soil organic matter. Bloomfield and Sanders [84] measured Cu^{2+} concentrations with an Orion ISE, in solutions 10^{-4} M with respect to Cu^{2+} and 0.1 M with respect to KNO_3 (for constant ionic strength) in the presence of various amounts of organic matter. The latter consisted of the dialysate and colloidal fractions of extracts of composted lucerne, soil, and peat. The dialysate complexed more Cu and hence reduced the solution concentration of Cu^{2+} more than the colloidal fraction. The degree of complexing decreased with pH.

The Cu^{2+} ion-selective electrode (Orion), coupled with a double junction reference electrode, was used to determine stability constants and derive an affinity spectrum (Fig. 20) of Cu^{2+} with humic substances separated from the surface layers of river water [85]. The analysis of the data is based on the continuous function

$$\bar{\nu} = \frac{N_K KM}{1 + KM} \, d(\log K) \tag{19}$$

where

$\bar{\nu}$ = molar ratio of "bound metal" to "total ligand"
M = molar concentration of uncomplexed metal
K = site binding constant
n = number of equivalent binding sites

and the total number of sites is given by

$$n_0 = \sum_{i=1}^{m} n_i = \int_{-\infty}^{\infty} N_K \, d(\log K) \tag{20}$$

To solve for N_K, given experimental values of M and $\bar{\nu}$, a finite-difference numerical-approximation technique is used, resulting in an "affinity spectrum" of N_K versus log K, reflecting a distribution of K and an area under each peak proportional to ν. This approach views bonding as being represented by a distribution of sites where $[N_K d \log K]/n_0$ is the probability of finding a binding constant between log K and log K + d log K.

Sapek and Sapek [86] also investigated the binding of Cu^{2+} and Cd^{2+} by humic substances separated from peaty muck soils, using Orion ISEs (model numbers 94-29 and 94-48A, respectively), coupled with the single junction reference electrode 91-02. They found that the detection limit for both cations was about 10^{-6} M, the degree of

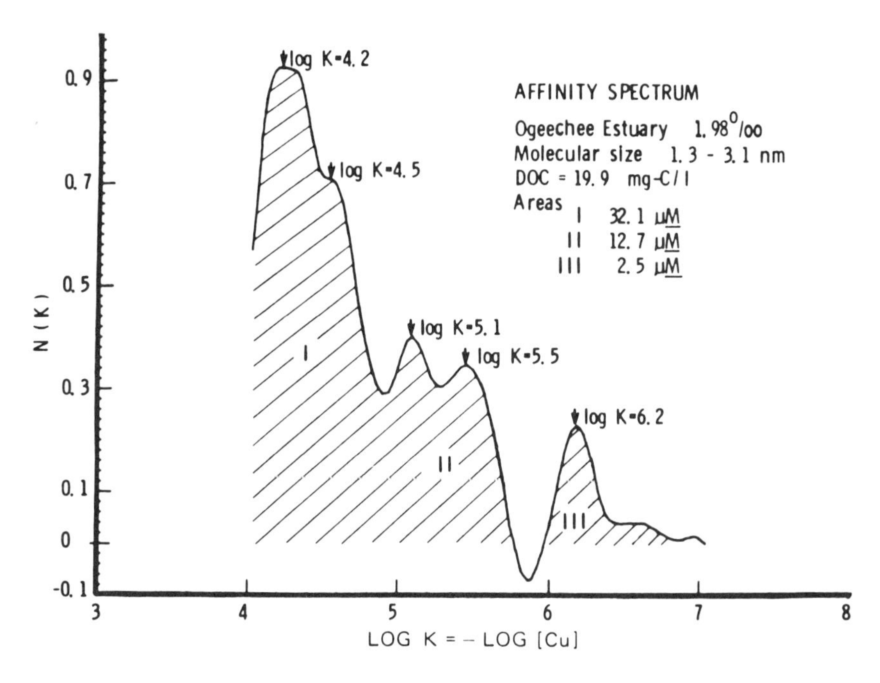

FIG. 20 Affinity spectrum of the titration of Ogeechee River estuary organics. Shading indicates areas under each region, I–III. (From Ref. 85.)

complexation (= total metal/uncomplexed ion in solution) increased with the concentration of the humic substance and with increasing humification of the muck soil. Under comparable conditions, Cu^{2+} was complexed more than Cd^{2+}, although such an analysis cannot be considered rigorous in a strictly thermodynamic sense. They suggested that when using ISEs, it was more advantageous to titrate a solution of the metallic ion with that of the complexing agent than the reverse procedure. Cavallero and McBride [87], using the same electrodes in an alternative "batch" procedure, obtained the expected result that organic matter in soil suspension delayed the formation of the hydroxides and hydroxycarbonates of both metals, although much more so with Cu^{2+}. They also demonstrated [88] that with 10^{-5} *M* concentrations of both ions, increasing the Al^{3+} concentration from 0 to 6.25×10^{-5} *M* caused the measured *apparent* concentration of Cu(II) to *decrease* to a third, whereas that of Cd(II) *increased* 1.8-fold when the Al^{3+} concentration was raised from 0 to 3.7×10^{-5} *M* (Fig. 21). Such opposing effects,

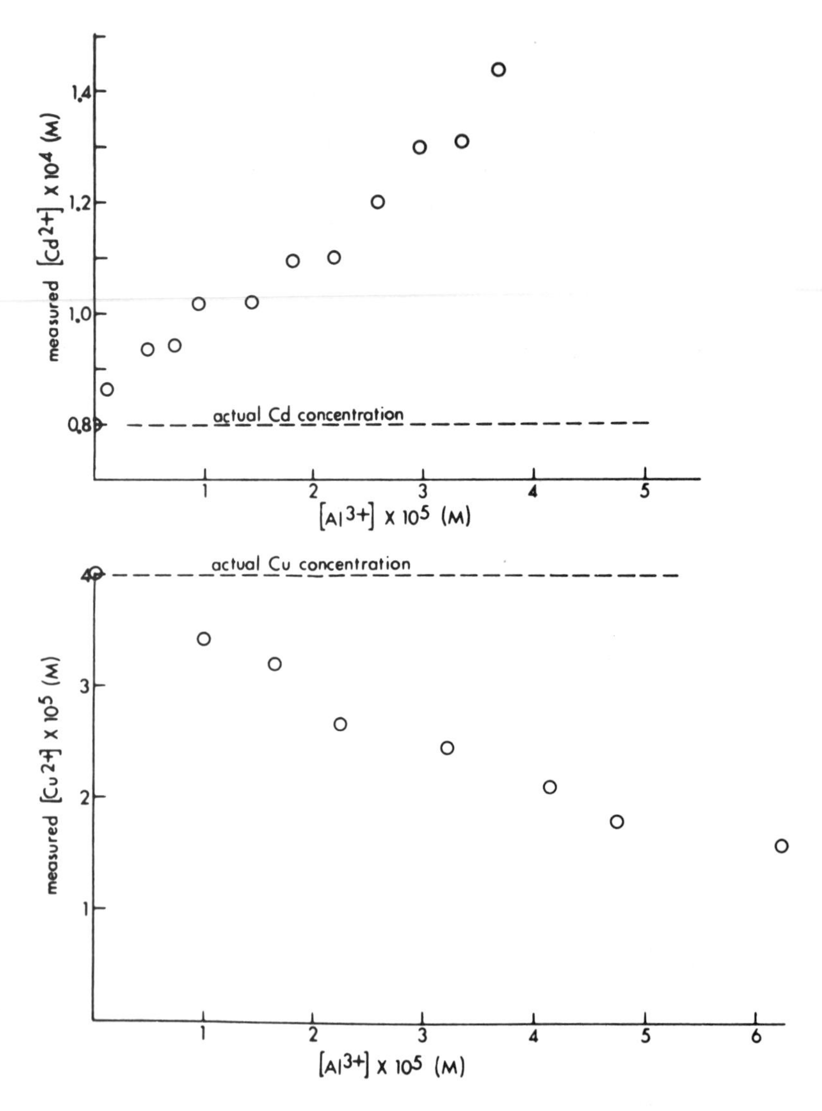

FIG. 21 Effect of dissolved Al on the apparent concentration measured by the Cd^{2+} and Cu^{2+} ion-selective electrodes. (From Ref. 88.)

antagonistic for Cu^{2+} but synergistic for Cd^{2+}, must result from different reactions of the Al^{3+} ion with the respective ion-selective membranes; this is an important point to appreciate by users of ISEs investigating mixtures of ions that are not mutually reactive.

Takamatsu and Yoshida [89] used ISEs to determine "successive" and "overall" stability constants for Cu^{2+}, Pb^{2+}, and Cd^{2+} complexes with 14 humic acids from diverse sources including humidified rice straw. At pH 5.0, the overall pK values for Cu, Pb, and Cd were 8.65, 8.35, and 6.25, respectively, increasing with pH, greater humification, and the concentration of weakly acidic functional groups in the humic acids. A more comprehensive examination of Cu^{2+} complexes at pH values 4.0, 5.0, and 6.0 with soil fulvic acids [90] combined the ISE method with electron paramagnetic resonance spectroscopy and differential pulse polarography. It confirmed the existence of two classes of binding sites [85] and that the number of sites increased with pH. In a more practically oriented context, Kerven et al. [91] used the Cu^{2+} ISE to determine Cu concentrations in extracts of peat soils and hence to elucidate Cu^{2+} complexation in the organic matter extract. Dalang et al. [92] extended the use of Cu^{2+} ISEs to study the effect of fulvates on the absorption of Cu^{2+} on kaolinite surfaces, whereas McBride and Boulden [93] were able to use this ISE for investigating the long-term reactions of Cu^{2+} in contaminated calcareous soils.

Copper activity determinations by the ISE and Donnan dialysis (DD) methods were compared for saturation extracts of soils treated with a factorial combination of Cu^{2+} salts and sludge [94]. The DD method was suitable for sludge treatments but gave Cu^{2+} activities for the Cu^{2+} salt treatments that were approximately a tenth of those obtained with ISE. However, ISE was only useful in aerobic conditions.

The Orion Cu^{2+} ISE was used by Corominas et al. [95], instead of atomic absorption analysis, to determine biuret impurities in urea and mixed fertilizers. The chemistry underlying both methods was identical, but the ISE method greatly simplified the procedure and equipment required. The Cu^{2+} activity in the solution in which the Cu-biuret complex was formed after adding excess cupric sulfate was measured directly by the Cu^{2+} ISE and the biuret concentration calculated by the difference. The effects of the starch solution and ethanol (the biuret solvent) had to be taken into account in the calibration curves for ISE. The authors claimed that the AA and ISE methods gave results that agreed at the 99% confidence level in the analysis of four fertilizers.

F. Gas-Sensing Electrodes

1. *Ammonia*

The ammonia gas probes have found two particular niches in soil science and agronomy: (1) in the determination of the cation exchange capacity of soils and clays and (2) ammonia determination in Kjeldahl digests for total N. An example of the use of an ammonia probe in soil CEC determinations has been mentioned already [69].

A simpler method that only involves one measurement has been devised for clays [96]. Salt-free ammonium-saturated clay (50—100 mg) is suspended in magnetically stirred ammonia-free water. 0.5 ml of 10 *M* NaOH solution is added to convert the adsorbed ammonium to ammonia, which is then determined directly with the probe (see Sec. III.D). This method gave satisfactory results for kaolinites, illites, and montmorillonites.

Ammonia formed in Kjeldahl digests is determined generally either by an automated colorimetric technique or by steam distillation and titration. The former method requires expensive equipment and the latter is slow and tedious. The probe offers a cheap and rapid alternative to these. Bremner and Tabatabai [97] described a method in which a sample containing approximately 1 mg of N was digested and the final digest made up to 100 ml. A 10 ml sample of this was mixed with 10 ml of 2.5 *M* NaOH and 80 ml of water. This solution was stirred magnetically and the NH_3 probe inserted to make the measurement.

Exchangeable and water-soluble ammonium in soil can also be determined with the ammonia probe [98]. Concentrations in filtered KCl and water extracts measured with the probe correlated almost perfectly with those measured by steam distillation, giving a regression coefficient of 1.

Viswanath et al. [99] extended the use of the ammonia electrode (Orion Model 95-10) to the determination of NH_3 in the presence of urea in urea ammonium phosphate-based NP and NPK fertilizers. Tartaric acid (10% v/v of a 10% solution) and NaOH (5% v/v of a 10 *M* solution) were added to test solutions of the fertilizers to suppress precipitation of heavy metals. Urea was not hydrolyzed under these conditions and the results compared very favorably and quantitatively with ammonia-N determined by the automated indophenol blue method.

2. *Nitrogen Oxides*

The nitrogen oxides gas-sensing probe can be used to determine nitrite in acidified solutions. Tabatabai [100] tested it in standard nitrite solutions made up with reagents commonly used as extractants for soil nitrate and nitrite (2 *M* KCl, 1 *M* KCl, saturated $CaSO_4$, 0.1 *M* LiCl, 0.15% $CaCl_2$, 500 $\mu g\ ml^{-1}$ P as $Ca(H_2PO_4)_2 \cdot H_2O$). One part of a H_2SO_4 + Na_2SO_4 solution (190 g anhydrous Na_2SO_4 + 53 ml concentrated H_2SO_4 per liter) was added to 10 parts of the test solution to give a final pH of 1.2 for nitrite measurement with the probe. Of the extractants, only 2 *M* KCl caused significant error, all the others gave results within 2% of the known nitrite concentration.

Nitrite measurements made in actual soil extracts and water samples with the probe and by the colorimetric method of Barnes and

Folkard [101] correlated perfectly ($r = 1$, $b = 1$). Also, more than 99% of nitrite added to soil extracts was measurable with the probe. Ammonium, Na^+, K^+, Ag^+, Ca^{2+}, Mg^{2+}, Cu^{2+}, Hg^{2+}, nitrate, chloride, sulfate, and phosphate did not cause any interference in the measurements when present in the test solutions at 0.1 *M* concentration. Of these, Cu^{2+} and Hg^{2+} are particularly important because they are often added to soil extracts to prevent microbial activity and they do cause errors in the colorimetric methods.

G. Continuous Flow Measurements

Because of the rapid response time of most ISEs, they can be used in continuous flow systems, either to measure ion concentrations in a series of solutions passed through sequentially or to monitor the concentration of a flowing solution. Greenhill and Peverill [69] used the flow cell shown in Fig. 22 for the measurement of NH_3 and Cl^- concentrations in their CEC determinations. They found that it was not possible to make both measurements concurrently because hydroxyl ions, from NaOH added to the samples for NH_3 determination, interfere

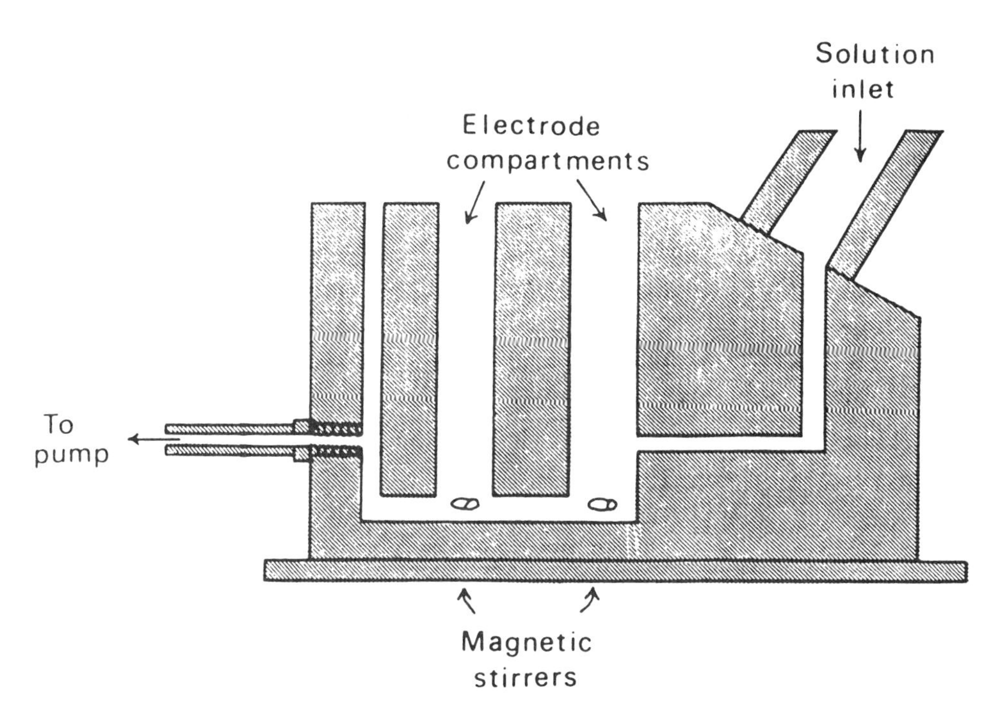

FIG. 22 Flow cell for measuring ammonia and chloride ions. (From Ref. 69.)

with the chloride determinations. If one ion at a time was determined, the flow cell gave results comparable to those made individually.

Mansell and Elzeftawy [102] designed a flow cell incorporating a combination chloride and reference electrode to monitor chloride percolating through a soil column (Fig. 23). Testing it with chloride solutions passed through a column of glass beads, they demonstrated that the response time of the electrode was sufficiently small to enable it to monitor solutions flowing at 35 ml hr^{-1}.

In a study of plant growth and nutrient uptake in flowing culture solutions, Clement et al. [103] used nitrate and potassium ISEs to monitor and control concentrations of these ions. In this system, output from the electrodes activates an automatic titrator that adds solutions to the system to bring up the ion concentration to a preset level. The electrodes operate in a flow cell with a reservoir that is filled at fixed intervals from a bypass pipe via solenoid valves (Fig. 24). One particularly interesting feature in the electrode system is the use of a servo-system incorporating a potentiometer to correct the potential between the ion-selective and reference electrodes in a standard solution to a preset value on the autotitrator

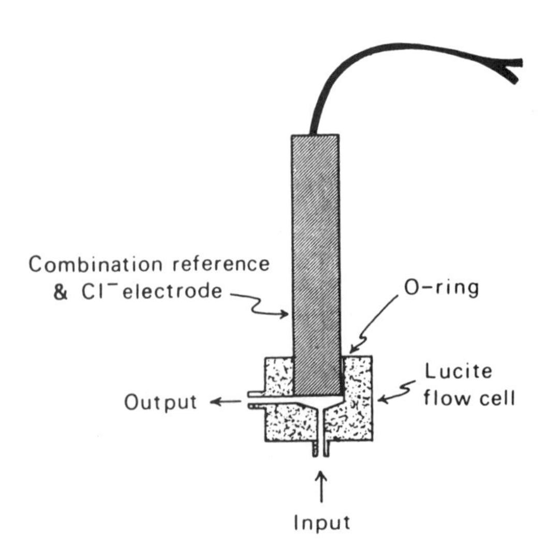

FIG. 23 Diagram showing a combination reference and chloride ion electrode in a lucite flow cell. [Reproduced from Ref. 102, *Soil Sci. Soc. Am. Proc.* *27*(2):378 (1972) by permission of the Soil Science Society of America.]

controller. This enables the output of an electrode pair at a standard ion concentration to be maintained for several weeks within ±0.75 mV.

H. Plant Analysis

This book is concerned with soil analysis, but because in many laboratories soil and plant analyses go on side by side, it is appropriate to mention briefly the use of ISEs in the analysis of plant material. The subject has been reviewed adequately by Moody and Thomas [104].

Plant materials are processed into three forms for such measurements:

1. Extracts from dried material, made directly or after ignition
2. Filtered extracts of fresh material
3. Macerates and juices of fresh material

Generally, problems concerned with electrode measurements in extracts prepared from dried plant material differ little from those

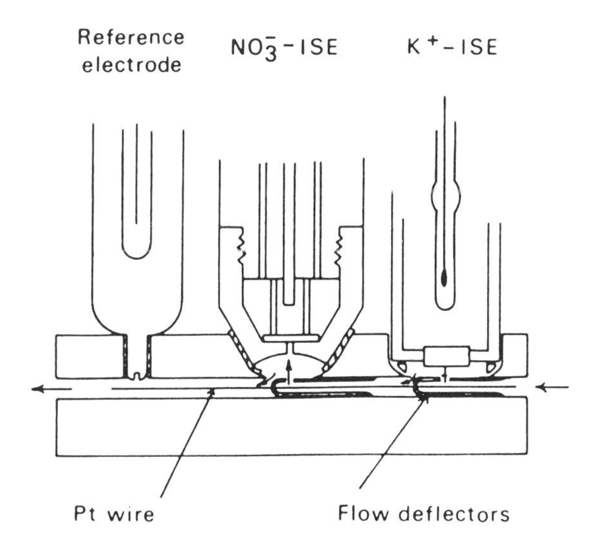

FIG. 24 Details of a flow cell with two ion-selective electrodes. (From Ref. 103.)

in soil extracts; indeed, the same extraction techniques are often used [53,54,73,76]. However, the greatest value of ISEs in plant analysis lies in the possibility of rapid measurements of nutrient or pollutant ions in fresh material. The technique has been used to determine levels of potentially toxic ions in foodstuffs (e.g., nitrate, cyanide, bromide [104]). No great success seems to have been achieved in using ISEs in the diagnosis of plant nutritional problems. In concentrated aqueous macerates of wheat leaves, a Corning nitrate electrode did not give reproducible results, apparently because of poisoning of the liquid ion exchanger [105]. However, Moody and Thomas [104] found it possible in preliminary experiments to measure chloride with a combination electrode directly on the freshly cut surfaces of oranges, apples, bananas, potatoes, carrots, and cucumbers.

For nitrate ion assay in plant material, principally fresh or dried leaves, the most common extracting agents are cold or boiling water and dilute aluminum sulfate solutions, the latter because the NO_3^-/SO_4^{2-} free ion ratio in plant material is usually small. Also, the sulfate ion does not interfere seriously. By contrast, the chloride ion frequently occurs in appreciable concentrations in plant material and interferes more than the sulfate ion with nitrate ISE measurements. In such conditions, the known additions method (Sec. II.E) has to be used to evaluate the correct nitrate concentration. This is doubly useful in decreasing Cl^- interference by increasing the NO_3^-/Cl^- ion ratio, and in giving a reliable measure of the extent of interference. Similar direct analysis of such aqueous extracts has been reported for Cl^- ions and should also be feasible for Na^+, K^+, Mg^{2+}, and Ca^{2+} in dilute acid extracts. For measurements of the fluoride, cyanide, and sulfur contents of plant material, however, plant material has to be digested such that the respective ionic forms are formed in the final extract by the hydrolysis, or oxidation, of the very stable covalent organic compounds of fluorine, cyanogen, or sulfur that can occur in plant material.

I. Microelectrodes in Plant Studies

In view of their rapid response times, ion selectivity, sensitivity, and portability, the logical extension of ISEs in plant studies is to the measurement of ion fluxes in various parts of the plant during growth. This is of special significance to research in the soil-plant environment if such measurements can be related to concomitant changes in ion concentrations in the root zone of the plant. The development in the last 20 years of microelectrodes with ion-sensitive tips, varying in length from 100 μm initially to a few micrometers at the present time, makes it feasible to measure ion concentrations in plant cells. No such measurements have been reported so far (although animal cells have been examined in this manner), neither have such electrodes been produced commercially, but their methods of manufacture are reported in detail in a book by Thomas [106], from which this brief account is abstracted.

Some examples of solid and liquid membrane microelectrodes are shown in Fig. 25, which are principally used for measuring H^+, Na^+, K^+, Ca^{2+}, and Cl^- ion concentrations. Figure 25a and 25b illustrate the design of the ion-sensitive glass electrode in which the membrane tip protrudes from a protective and insulating sheath of lead glass, the protrusion being overlong originally, as shown. In modern versions, this protrusion is about 10 μm long. The two glasses are joined together by a glass-to-glass seal (type a) or by dental wax (type b). For better protection of the ion-sensitive membrane, a recessed-tip electrode was later invented (type c) in which the lead glass sheath, drawn to an open jet point with an outer diameter of about 1 μm, encloses the ion-selective membrane. Such electrodes have been used for the H^+, Na^+, K^+, and Cl^- assay, the Cl^--sensitive element being an Ag/AgCl wire sealed by wax to the insulating glass.

The liquid ion-selective microelectrode is illustrated in Fig. 25d, which is self-explanatory. The tip diameter is of the order of 1 μm and is drawn by a mechanical device from borosilicate tubing. To date, calcium and chloride ion assays by such electrodes have been reported, but liquid membranes for Cl^- are not ion-selective enough for use in soil-plant research. For in vivo plant studies, the ion-selective liquid membranes, reported in Sec. III, presage in the near future the universal use of such microelectrodes for measuring also NH_4^+ and NO_3^- concentrations, especially if enterprising manufacturers take up the challenge of marketing them.

Samokhalov et al. [117] have reported a method for the rapid determination of nitrate ion in plant tissues with a needle-shaped ISE; their results agreed well with those obtained by the α-furyl-dioxime method. An account of the applications of ion-selective microelectrodes to biomedical research [118] contains some additional details of a technique that may interest and help potential users in the examination of plant cells.

J. Herbicide-Sensitive Liquid Membranes

Although herbicides are not yet directly and routinely analyzed in soils, plants and fertilizers, rapid methods for their analysis and of their residues are needed. For diquat and paraquat dications, Moody et al. [119,120] describe a PVC membrane impregnated with a cocktail of a macrocyclic polyether, dibenzo-30-crown-10 (DB10C10), and diquat-tetraphenylborate (DQT.2TPB), with *o*-nitrophenyl phenyl ether (NPPE) as a solvent mediator. (For paraquat, these electrodes

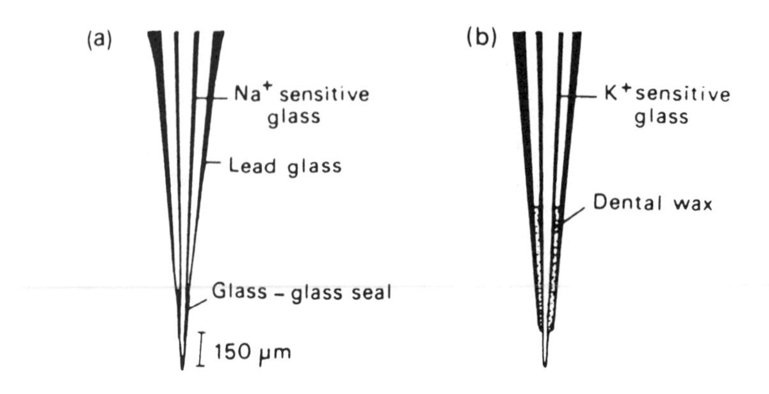

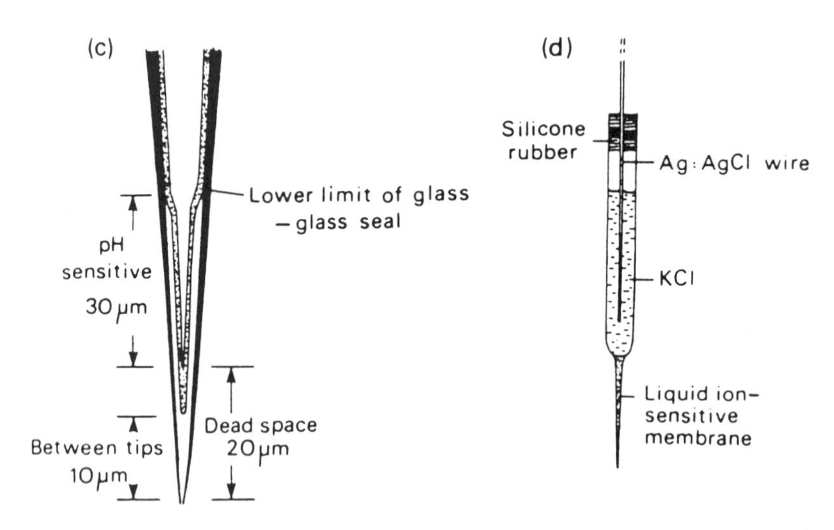

FIG. 25 Some ion-selective microelectrodes. (a) Na-sensitive glass electrode with glass seal; (b) K-sensitive glass electrode with wax seal; (c) recessed-tip type electrode; (d) liquid membrane electrode. [From Ref. 106, with permission from R. C. Thomas, *Ion-Sensitive Intracellular Microelectrodes*, 1978. Copyright © by Academic Press, Inc. (London) Ltd.]

were frequently poor in quality, but electrodes with only paraquat-tetraphenylborate, PQT.2TPB, on PVC, without a solvent mediator, were the best.)

For diquat analysis, a diquat:*bis*(tetra-4-chlorophenylborate)-based ISE had much better characteristics than a DQT-2TPB-based ISE [120]. Thus, its slope ranged from 30 ± 0.6–28 ± 1.7 mV/log*a*, and the corresponding minimum detectable DQT concentration from $4 \pm 3 \times 10^{-9}$ *M* to $3 \pm 0.2 \times 10^{-6}$ *M*. The usable pH range was 2–12 over a sample temperature range of 2–50°C. This very satisfactory performance led the authors to suggest that the potentiometric titration of diquat was now possible. They attributed the less satisfactory performance of the paraquat electrode to the PQT^{2+} ion being elongated and longer than the DQT^{2+} ion, causing a wider separation of the N^+ sites.

ACKNOWLEDGMENTS

The author acknowledges with thanks computer searches for up-to-date references on ion-selective electrodes and their applications by Dr. K. W. T. Goulding, Soils and Plant Nutrition Department, Rothamsted Experimental Station, Harpenden, Herts., United Kingdom.

REFERENCES

1. R. P. Buck, in *Ion Selective Electrodes* (E. Pungor and I. Buzas, eds.), Elsevier, Oxford, 1978, p. 21.
2. G. Eisenman, in *Ion Selective Electrodes* (R. A. Durst, ed.), Special Publ. 314, National Bureau of Standards, Washington, D.C., 1969, p. 1.
3. L. Ebdon, B. A. King, and G. C. Corfield, *Anal. Proc.*, *22*: 354 (1985).
4. J. P. Sandblom, G. Eisenman, and J. L. Walker, *J. Phys. Chem.*, *71*:3862 (1967).
5. J. Janata and R. J. Huber, in *Ion Selective Electrodes in Analytical Chemistry* (H. Freiser, ed.), Plenum, New York, 1980, p. 107.
6. P. N. Kember, *Water Sci. Technol. (Munich)*, *13*:255 (1981).
7. J. Janata, R. J. Huber, R. Cohen, and E. S. Kolesar, Jr., *Aviation, Space Environ. Med.* *52*:666 (1981).
8. W. E. Morf, E. Lindner, and W. Simon, *Anal. Chem.*, *47*:1596 (1975).
9. W. E. Morf, G. Kahr, and W. Simon, *Anal. Chem.*, *46*:1538 (1974).
10. R. P. Buck, *Anal. Chem.*, *40*:1432 (1968).
11. W. E. Morf and W. Simon, in *Ion Selective Electrodes* (E. Pungor and I. Buzas, eds.), Elsevier, Oxford, 1978, p. 149.

12. A. K. Covington, in *Ion Selective Electrodes* (R. A. Durst, ed.), Special Publ. 314, National Bureau of Standards, Washington, D.C., 1969, p. 107.
13. G. J. Moody and J. D. R. Thomas, *Lab. Practice, 28*:125 (1979).
14. M. S. Frant and J. W. Ross, *Anal. Chem., 40*:1169 (1968).
15. J. D. R. Thomas, in *Ion Selective Electrodes* (E. Pungor and I. Buzas, eds.), Elsevier, Oxford, 1978, p. 175.
16. R. A. Durst, in *Ion Selective Electrodes* (R. A. Durst, ed.), Special Publ. 314, National Bureau of Standards, Washington, D.C., 1969, p. 375.
17. P. Sorensen, *Kem. Maanedsbl., 32*:73 (1951).
18. G. Gran, *Analyst (Lond.), 77*:661 (1952).
19. T. M. Florence, *J. Electroanal. Chem., 31*:77 (1971).
20. J. Tacussel and J. J. Fombon, in *Ion Selective Electrodes* (E. Pungor and I. Buzas, eds.), Elsevier, Oxford, 1978, p. 567.
21. W. Misniakiewicz and K. Raszka, in *Ion Selective Electrodes* (E. Pungor and I. Buzas, eds.), Elsevier, Oxford, 1978, p. 467.
22. C. A. Bower, *Soil Sci. Soc. Am. Proc., 23*:29 (1959).
23. C. A. Bower, *Trans. 7th Int. Cong. Soil Sci., 2*:16 (1960).
24. J. B. Fehrenbacher, L. P. Wilding, and A. H. Beavers, *Soil Sci. Soc. Am. Proc., 27*:152 (1963).
25. S. A. El Swaify and M. N. Gazdar, *Soil Sci. Soc. Am. Proc., 33*:665 (1969).
26. C. A. Bower, *Soil Sci. Soc. Am. Proc., 25*:18 (1961).
27. J. Susini, M. Rouault, and A. Kerkeb, *Cah. ORSTROM Pedologie, 10*:309 (1972).
28. D. C. Orlov and A. Alzubaidi, *Agrokhimiya,* (2):135 (1965).
29. A. Alzubaidi, *Iraq J. Agric. Sci., 4*:13 (1969).
30. N. K. Krupskiy, L. A. Rappoport, and A. M. Aleksandrova, *Sov. Soil Sci., 6*:111 (1974).
31. N. K. Krupskiy, A. M. Aleksandrova, and L. A. Rappoport, *Sov. Soil Sci., 6*:241 (1974).
32. N. K. Krupskiy, A. M. Aleksandrova, L. A. Rappoport, and Yu. I. Lapkina, *Agrokhimiya* (7):152 (1974).
33. M. M. Mortland, *Quar. Bull. Michigan Agric. Exp. Sta., 43*:491 (1961).
34. R. E. Farrell and A. D. Scott, *Soil Sci. Soc. Am. J., 51*:594 (1987).
35. D. C. Orlov and A. Alzubaidi, *Agrokhimiya* (7):137 (1965).
36. V. C. Kennedy and T. E. Brown, *Clays and Clay Minerals, 13*:351 (1964).
37. R. L. Malcolm and V. C. Kennedy, *Soil Sci. Soc. Am. Proc., 33*:247 (1969).

38. A. Banin and D. Shaked, *Agrochimica*, *15*:238 (1971).
39. E. A. Woolson, J. H. Axley, and P. C. Kearney, *Soil Sci.*, *109*:279 (1970).
40. E. B. Doiron and E. Chamberland, *Commun. Soil Sci. Plant Anal.*, *4*:205 (1973).
41. E. O. McLean and G. H. Snyder, *Soil Sci. Soc. Am. Proc.*, *33*:388 (1969).
42. C. Wang and T. Yu, *Z. Pflanzen. Boden.*, *144*:514 (1981).
43. C. Wang and T. Yu, *Z. Pflanzen. Boden.*, *149*:598 (1986).
44. M. A. Parra and J. Torrent, *Soil Sci. Soc. Am. J.*, *47*:335 (1983).
45. D. Hornby and M. E. Brown, *Plant Soil*, *48*:455 (1977).
46. R. J. K. Myers and E. A. Paul, *Can. J. Soil Sci.*, *48*:369 (1968).
47. A. Øien and A. R. Selmer-Olsen, *Analyst (Lond.)*, *94*:888 (1969).
48. M. Willems, *Tidsskrift for Planteavl.*, *79*:495 (1975).
49. M. K. Mahendrappa, *Soil Sci.*, *108*:132 (1969).
50. A. R. Mack and R. B. Sanderson, *Can. J. Soil Sci.*, *51*:95 (1971).
51. A. Cottenie and G. Velghe, *Meded. Fakulteit Landbouwwetenschappen, Rijksuniversiteit Gent (Belgium)*, *38*:560 (1973).
52. A. S. Black and S. A. Waring, *Plant Soil*, *49*:207 (1978).
53. A. Raveh, *Soil Sci.*, *116*:388 (1973).
54. P. J. Milham, A. S. Award, R. E. Paul, and J. H. Bull, *Analyst (Lond.)*, *95*:751 (1970).
55. C. W. Francis and C. D. Malone, *Soil Sci. Soc. Am. Proc.*, *39*:150 (1975).
56. J. M. Bremner, L. G. Bundy, and A. S. Agarwal, *Anal. Lett.*, *1*:837 (1968).
57. D. G. Hadjidemetriou, *Analyst (Lond.)*, *107*:25 (1982).
58. S. Li and K. A. Smith, *Commun. Soil Sci. Plant Anal.*, *15*. 1437 (1984).
59. A. B. Onken and H. D. Sunderman, *Commun. Soil Sci. Plant Anal.*, *1*:155 (1970).
60. G. P. Bound, *J. Sci. Fd. Agric.*, *28*:501 (1977).
61. G. P. Bound and B. Fleet, *J. Sci. Fd. Agric.*, *28*:431 (1977).
62. P. K. R. Nair and O. Talibudeen, *J. Agric. Sci., Camb.*, *81*: 327 (1973).
63. M. B. Page and O. Talibudeen, *Plant Soil*, *47*:527 (1977).
64. M. B. Page, J. L. Smalley, and O. Talibudeen, *Plant Soil*, *49*: 149 (1978).
65. C. Turner and W. Carlisle, Private communication, 1979.
66. B. W. Hipp and G. W. Langdale, *Commun. Soil Sci. Plant Anal.*, *2*:237 (1971).
67. A. R. Selmer-Olsen and A. Øien, *Analyst (Lond.)*, *98*:412 (1973).

68. R. St. C. Smart, A. D. Thomas, and D. P. Drover, *Commun. Soil Sci. Plant Anal.*, *5*:1 (1974).
69. N. B. Greenhill and K. I. Peverill, *Commun. Soil Sci. Plant Anal.*, *8*:579 (1977).
70. N. K. Krupskiy, A. M. Aleksandrova, and Z. L. Boriskova, *Sov. Soil Sci.*, *5*:749 (1973).
71. F. J. Sikora and F. J. Stevenson, *Soil Sci. Soc. Am. J.*, *51*:924 (1987).
72. I. Sekarka and J. F. Lechner, *J. Assoc. Off. Anal. Chem.*, *61*:1490 (1978).
73. N. A. Abdalla and B. Lear, *Commun. Soil Sci. Plant Anal.*, *6*:489 (1975).
74. P. G. Saffigna, D. R. Keeney, and L. L. Hendrickson, *Commun. Soil Sci. Plant Anal.*, *7*:691 (1976).
75. G. L. Crenshaw and F. N. Ward, *Bull. U. S. Geol. Surv.* (1408):77 (1975).
76. N. R. McQuaker and M. Gurney, *Anal. Chem.*, *49*:53 (1977).
77. K. Takala, P. Kauranen, and H. Olkkonen, *Ann. Bot. Fennici*, *15*:158 (1978).
78. A. E. Villa, *Analyst (Lond.)*, *104*:545 (1979).
79. A. Schleuter, USEPA Rept. EPA-600/4-76-020, 1976.
80. J. L. Penland and G. Fischer, *Angew. Electrochem.*, *26*:391 (1972).
81. D. G. Rands and R. L. Bain, Univ. Illinois Rept. UILU-WRC-79-0143, 1979.
82. A. I. Allam, G. Pitts, and J. P. Hollis, *Soil Sci.*, *114*:456 (1972).
83. P. L. Sanderson and W. Armstrong, *J. Soil Sci.*, *31*:643 (1980).
84. C. Bloomfield and J. R. Sanders, *J. Soil Sci.*, *28*:435 (1977).
85. M. S. Shuman, B. J. Collins, P. J. Fitzgerald, and D. J. Olson, in *Aquatic and Terrestrial Humic Materials* (R. F. Christman and E. T. Gjessing, eds.), Ann Arbor Sci. Publ., Ann Arbor, Mich., 1983, p. 349.
86. B. Sapek and A. Sapek, *Polish J. Soil Sci.*, *13*:125 (1980).
87 N. Cavallero and M. B. McBride, *Soil Sci. Soc. Am. J.*, *44*:729 (1980a).
88. N. Cavallero and M. B. McBride, *Soil Sci. Soc. Am. J.*, *44*:881 (1980b).
89. T. Takamatsu and T. Yoshida, *Soil Sci.*, *125*:377 (1978).
90. W. T. Bresnahan, C. L. Grant, and J. H. Weber, *Anal. Chem.*, *50*:1675 (1978).
91. G. L. Kerven, D. G. Edwards, and C. J. Asher, *Soil Sci.*, *137*:91 (1984).
92. F. Dalang, J. Buffle, and W. Haerdi, *Environ. Sci. Technol.*, *18*:135 (1984).

93. M. B. McBride and A. Goulden, *Soil Sci. Soc. Am. J.*, *48*:56 (1984).
94. M. M. Minnich and M. B. McBride, *Soil Sci. Soc. Am. J.*, *51*:568 (1987).
95. L. Corominas, R. A. Navarro, P. Rojas, and J. L. Cruz, *J. Assoc. Off. Anal. Chem.*, *69*:119 (1986).
96. E. Busenberg and C. V. Clemency, *Clays and Clay Minerals*, *21*:213 (1973).
97. J. M. Bremner and M. A. Tabatabai, *Commun. Soil Sci. Plant Anal.*, *3*:159 (1972).
98. W. L. Banwart, M. A. Tabatabai, and J. M. Bremner, *Commun. Soil Sci. Plant Anal.*, *3*:449 (1972).
99. T. S. Viswanathan, M. Nagarajan, and B. Haribabu, *J. Assoc. Off. Anal. Chem.*, *63*:1248 (1980).
100. M. A. Tabatabai, *Commun. Soil Sci. Plant Anal.*, *5*:569 (1974).
101. H. Barnes and A. R. Folkard, *Analyst (Lond.)*, *76*:599 (1951).
102. R. S. Mansell and A. Elzeftawy, *Soil Sci. Soc. Am. Proc.*, *36*:378 (1972).
103. C. R. Clement, M. J. Hopper, R. J. Canaway, and L. H. P. Jones, *J. Exp. Bot.*, *84*:81 (1974).
104. G. J. Moody and J. D. R. Thomas, *J. Sci. Fd. Agric.*, *27*:43 (1976).
105. M. B. Page and O. Talibudeen, unpublished results.
106. R. C. Thomas, *Ion-Sensitive Intracellular Electrodes*, Academic Press, London, 1978.
107. J. H. Puffer and R. S. Cohen, *Chem. Geol.*, *15*:217 (1975).
108. D. T. Cooke, *Commun. Soil Sci. Plant Anal.*, *6*:501 (1975).
109. K. L. Chang, J. C. Hung, and D. H. Prager, *Microchem. J.*, *18*:256 (1973).
110. J. L. Paul and R. M. Carlson, *J. Agric. Fd. Chem.*, *93*:729 (1968).
111. A. S. Baker and R. Smith, *J. Agric. Fd. Chem.*, *17*:1284 (1969).
112. A. W. M. Sweetsur and A. G. Wilson, *Analyst (Lond.)*, *100*:485 (1975).
113. R. C. Baker, *Anal. Chem.*, *44*:1326 (1972).
114. P. M. McElfresh, *J. Agric. Fd. Chem.*, *26*:276 (1978).
115. J. E. Harwood, *Water Research*, *3*:273 (1969).
116. R. L. LaCroix, D. R. Kerrey, and L. M. Walsh, *Commun. Soil Sci. Plant Anal.*, *1*:1 (1970).
117. S. G. Samokhvalov, V. G. Prizhukova, L. I. Molkanova, A. M. Lapustin, A. I. Golubtsov, and T. N. Toroptseva, *Agrokhimiya* (4):106 (1983).
118. R. N. Khuri, in *Ion Selective Electrodes* (R. A. Durst, ed.), Special Publ. 314, National Bureau of Standards, Washington, D.C., 1969, p. 287.

119. G. J. Moody, R. K. Owusu, and J. D. R. Thomas, *Analyst (Lond.)*, *112*:121 (1987).
120. G. J. Moody, R. K. Owusu, and J. D. R. Thomas, *Anal. Lett.*, *21*(9):1653 (1988).

4

Continuous-Flow, Flow-Injection, and Discrete Analysis

KEITH A. SMITH *The Edinburgh School of Agriculture, Edinburgh, Scotland*

ALBERT SCOTT *Scottish Agricultural College, Edinburgh, Scotland*

I. INTRODUCTION

Routine chemical analysis involves a number of operations, such as pipetting, diluting, mixing, filtering, and transference of solutions to measuring instruments, that are repetitive and lend themselves to automation. When instrumental methods of measurement are involved, the recording of results may also be automated. The trend toward the elmination of manual operations, particularly when large numbers of samples are involved, has grown very rapidly in the last 20–25 years. A major contribution to this phenomenon was the development of *continuous-flow analysis* (CFA) by Skeggs [1], which was further developed by the Technicon Instruments Corporation and subsequently by several other manufacturers. A variant of the technique known as *flow-injection analysis* (FIA) has been developed [2,3], and another system known as *discrete* or *batch*, analysis, has provided an alternative approach to automation. Both continuous-flow and discrete analysis have their origins in clinical chemistry, but the former system has been applied extensively in soil and plant analysis and now discrete systems are also finding similar applications.

In this chapter, the principles underlying the systems of continuous-flow and discrete analysis are described, and examples of applications of these techniques to the analysis of soils and other environmental materials are described.

II. PRINCIPLES

A. Continuous-Flow Analysis

The basic system is that devised for colorimetric analysis, in which solutions of sample and reagents are made to flow through plastic tubes by the action of a peristaltic pump. Interaction between successive samples is minimized by segmentation of the liquid stream by air bubbles (hence, the alternative name of segmented-flow analysis, SFA). The relative quantities of reagents and sample are controlled by the diameters of the tubes. Streams are combined via T or Y pieces, mixing occurs during passage through coils of tubing, usually made of glass, and the color reaction takes place. The liquid then passes through the flow cell of a spectrophotometer, where the intensity of the color is measured and recorded on a chart. Comparable flow systems may be used for analyses involving flame photometry, atomic absorption spectrometry, or ion-selective electrodes as the method of measurement.

1. Dispersion and Interaction Between Samples

The flow of a segmented liquid stream through a tube that is wetted by the liquid is illustrated in Fig. 1. The thin film of liquid, thickness d_f, remaining on the tube wall from the passage of segment No. 0 will be overtaken by and mixed into segment No. 1, thus, solute present in the former segment will become transferred to the latter. The process is repeated between segments 1 and 2, 2 and 3, and so on over many following segments. Thus, the initially sharp transition in concentration (Fig. 2a) between a series of sample segments

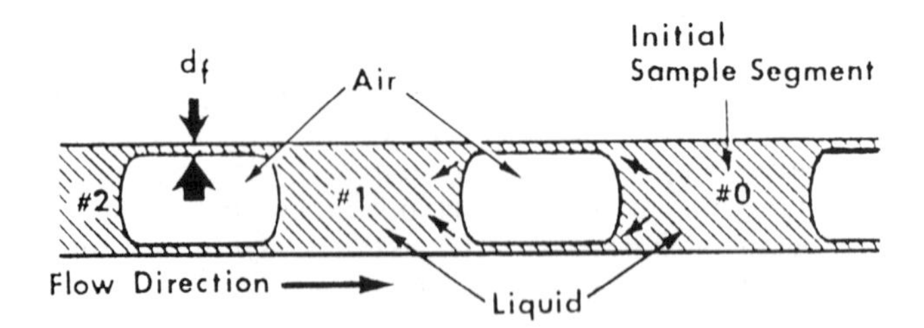

FIG. 1 Segmented flow through an open tube. (From Ref. 4, reprinted with permission from L. R. Snyder and H. J. Adler, Dispersion in segmented flow through glass tubing in continuous-flow analysis: The ideal model, *Anal. Chem.*, *48*:1017 (1976). Copyright © American Chemical Society, Washington, D.C.)

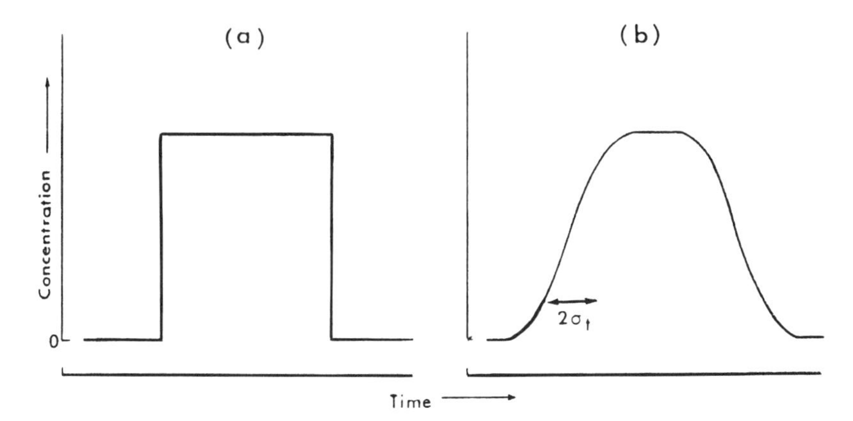

FIG. 2 Sample concentration profiles in segmented continuous-flow analysis. (a) At injection and (b) at detector. (After Snyder, Ref. 10.)

containing the analyte and preceding and following series containing a wash solution at zero concentration becomes transformed into a more gradual one (Fig. 2b). The relationship between concentrations in successive segments and sample-sample interactions has been investigated experimentally, mainly by Thiers and his coworkers. Mathematical models have been developed and used as a basis for computer programs to correct analytical data for interaction effects [4–9].

The dispersion process limits the spacing of samples and the resulting throughput rate. In Fig. 2b the increasing and decreasing parts of the curve are integrated Gaussian curves [4,10], and the extent of sample dispersion is given by σ_t, which is the standard deviation of the Gaussian curve, and is measured in seconds. The throughput rate f depends on the spacing of sample bands in units of σ_t. Snyder [10,11] has derived a value of $8\sigma_t$ for curves with flat plateaux as shown in Fig. 2b. Thus, for the latter, the throughput rate f (in samples per hour) is given by

$$f = \frac{3600}{8\sigma_t} = \frac{450}{\sigma_t} \tag{1}$$

According to the theory worked out by Snyder and Adler [4,12] and Snyder [13], the variance σ^2 of the dispersion occurring during segmented flow through a length of tubing is given by

$$\sigma^2 = \left[\frac{(\pi^2/72)d_t^4 u^{5/3}\eta^{2/3}}{\gamma^{2/3} V_s D_m'} + 1\right]\frac{0.5\pi L d_t^2 (u\eta/\gamma)^{2/3}}{V_s} \tag{2}$$

where d_t is the internal diameter and L is the length of the tube, u is the linear velocity of the liquid viscosity η and surface tension γ, V_s is the liquid-segment volume, and D_m' is a mass transfer coefficient that varies with the nature of the sample (solute) molecules and the viscosity of the liquid. For other liquids and/or temperatures, D_m' is related to the value of D_m for water at 25°C ($D_{w,25}$) and to liquid viscosity by the relationship

$$D_m' = D_{w,25}\left(\frac{\eta}{0.0089}\right)^{-5/3} \tag{3}$$

The dispersion σ in Eq. (2) can be expressed in terms of σ_t [Eq. (1)]

$$\sigma_t = \frac{\sigma}{n} \tag{4}$$

where n is the air-bubble rate. A number of other variables in Eq. (2) can be replaced by easily measured parameters such as the flow rate F_1. Equation (2) has been tested over a wide range of values of the various parameters, and general agreement within ±10% has been found [10]. However, continuous-flow systems are generally run at lower throughput rates than those indicated by Eqs. (1) and (2). This is because of the additional contribution to dispersion, resulting either from debubbling or, for the latest generation systems in which bubbles pass through the detector, from the requirement that each liquid segment be large enough to more than fill the detector flow cell [10].

2. *Mixing of Sample and Reagents*

Mixing is required whenever a reagent or a diluent is added and may be repeated several times in the course of one analysis. Complete mixing must occur in each liquid segment; otherwise, an irregular detector output will result. Mixing is brought about by the natural fluid motions that occur within a short, moving segment of liquid bounded at each end by a gas-liquid interface. These motions, known as *bolus* flow, result in rapid longitudinal mixing along the segment and in slow radial mixing [12]. In straight lengths of tubing, radial mass transfer occurs mainly by molecular diffusion, which is slow in liquids, but this can be overcome by inducing convective mixing simply by substituting a helical coil for the straight tube. Both the physical properties of the liquid and tube geometry are important in determining the rate of mixing. Viscosity, density, and flow rate all affect the process, with viscosity the most significant factor. Tube internal diameter, helix coil diameter, and segment length are also important [14].

3. *Flow Stability*

To achieve a stable detector output, the proportions of sample to reagents must be constant for all segments derived from the same sample, which requires constant flow rates and a very regular bubble pattern. Second-generation peristaltic pumps of the Technicon type run at a lower speed and with a higher roller frequency than in the original versions, and air injection is from a constant-pressure source under electronic control. Both of these developments contribute significantly to improved stability.

B. Flow-Injection Analysis

In the mid-1970s Růžička and Hansen [2] in Denmark and Stewart et al. [3] in the United States independently introduced a new concept of continuous-flow analysis called flow-injection analysis (FIA). In contrast with the conventional system involving a sample stream segmented with air bubbles, the flow-injection method is based, in its simplest form, on the rapid injection of the sample solution into an unsegmented carrier stream of reagent. The technique has many of the features of HPLC (see Chap. 13), and in their early developmental work Stewart et al. [3] did in fact use HPLC components for their system.

In FIA the sample is injected as a "plug" into the reagent stream. Very early in its progress along the tube, the shape of the sample distribution, known as the bolus shape, will be controlled predominantly by convection, but in the later stages diffusion will be dominant. Taylor [15,16] and Aris [17] derived the theory describing these two limiting situations over 20 years before the advent of flow-injection analysis. Taylor showed that when solute is introduced into a liquid stream moving by laminar flow with a mean velocity u and disperses by convection alone, the shape of the solute plug becomes distorted in time into a paraboloid (as shown in Fig. 3). He also showed that the effect of molecular diffusion gave rise eventually to dispersion with an apparent diffusion coefficient k given by

$$k = \frac{a^2u^2}{48D} \tag{5}$$

where D is the molecular diffusion coefficient and a the radius of the tube. Aris then found that the distribution tends to normality, and

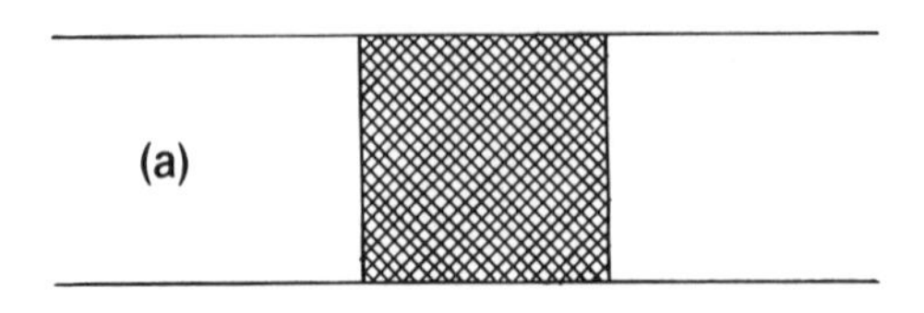

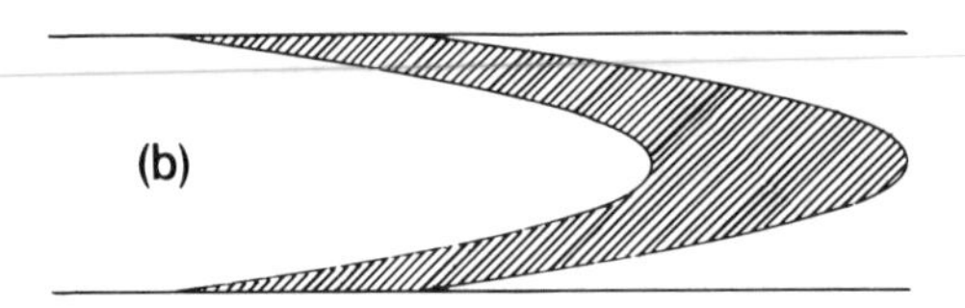

FIG. 3 Bolus shapes (a) at injection and (b) after injection. (From Ref. 18.)

that the effective distribution coefficient K is the sum of the molecular diffusion coefficient and Taylor's apparent diffusion coefficient

$$K = D + k$$

$$= D + \frac{a^2u^2}{48D} \quad (6)$$

In practice, the nature of the dispersion in FIA systems is the intermediate one in which both convection and diffusion play a significant part (Fig. 4c), and to which Taylor's solutions are therefore not applicable [19]. For example, studies with a typical FIA system showed that the volume dispersion increased rapidly with a flow rate up to 0.6 ml/min, but then leveled off [19]; such results cannot be explained by Taylor's equations, which predict a steadily increasing dispersion when flow rate is increased. Numerical methods for solution of the equations governing the diffusion-convection process in the regions of interest in FIA [20–22] have been applied by Vanderslice et al. [19] to obtain expressions for the initial appearance of a peak at the detector (travel time t_a) and the total time of observation of the peak (i.e., baseline to baseline, Δt_B)

$$t_a = \frac{109a^2D^{0.025}}{f}\left(\frac{L}{q}\right)^{1.025} \quad (7)$$

$$\Delta t_B = \frac{35.4a^2f}{D^{0.36}}\left(\frac{L}{q}\right)^{0.64} \quad (8)$$

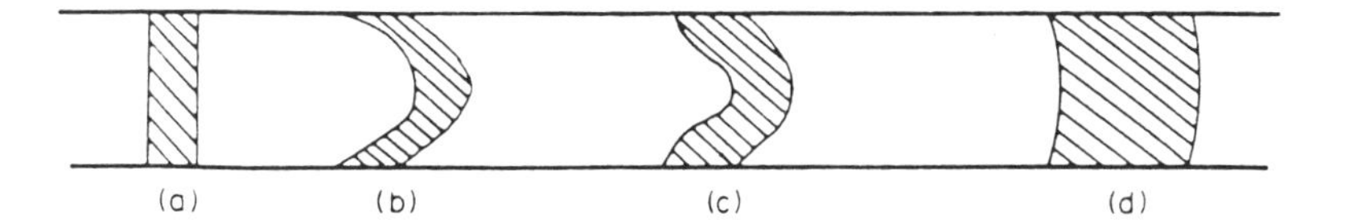

FIG. 4 Shapes of an injected bolus under different conditions. (a) Zero time, (b) convection-controlled region, (c) diffusion-convection region, (d) diffusion-controlled. (From Ref. 19.)

where q is the flow rate, L the tube length, D the diffusion coefficient, a the radius of the tube, and f an empirical factor that varies with experimental conditions between 0.5 and 1.0.

More recently, Tyson [23] derived equations for relating peak width to injection concentrations for single-line and merging stream manifolds. The equations are too detailed to be adequately summarized here, and the interested reader should consult this reference directly.

The magnitude of the dispersion in a practical system is dependent on the operating parameters, including sample volume, tubing bore size and length, and flow rate. By varying these parameters, the dispersion may be controlled, and this allows optimization of a flow-injection system for many diverse applications [24].

Limited dispersion is appropriate when FIA systems are used to feed such detectors as electrodes and atomic absorption spectrometers at high sample rates. Medium dispersion is often appropriate for color development in colorimetric applications. Finally, large dispersion can be utilized to give a substantial degree of mixing between sample and carrier stream to form a well-developed concentration gradient, e.g., for performing continuous-flow titrations [24].

Sample dispersion decreases with decreasing flow rate, and when the latter is zero, dispersion virtually ceases. This is exploited in so-called stopped flow measurements, which increase reaction times and thus increase the sensitivity of the determination.

C. Discrete Analysis

"Discrete" or "batch" analyzers are designed to simulate the operations commonly used in manual procedures. The advent of microprocessors has made possible very sophisticated systems, which can be programmed to suit the needs of each type of analysis required.

In contrast with continuous-flow analysis, in which each sample in turn traverses a common pathway of tubes to a detector, in discrete analysis each sample is contained in its own particular test tube or reaction vessel until it reaches the detector. In a typical integrated system, disposable plastic cups containing the samples are loaded, together with reaction tubes, into racks on a conveyor. A predetermined aliquot of sample is transferred by means of a pneumatic or motor-driven syringe into the tube. One or more reagents are dispersed in a similar manner, and the mixture is stirred mechanically. The time interval between additions may be programmed to allow the necessary reactions to take place. During this sequence of operations the tubes are immersed in a constant-temperature water bath to standardize reaction conditions, if necessary. The solution is finally pumped into the flow cell of a spectrophotometer and then to waste.

In many applications of the discrete analysis principle, only one or two of these operations may be employed. For example, automatic dispensing and dilution of samples may be all that is required prior to determination of metal ions by flame photometry or atomic absorption spectrometry. For pH determinations the automated steps could include the addition of water or calcium chloride solution to the soil, stirring, and insertion of the pH electrode into the suspension. In general, the large integrated analyzers are intended mainly for clinical chemical applications, although they can be adapted readily to a variety of other uses, whereas the simpler dispensers, diluters, and samplers are the instruments that have become more readily adopted for agricultural and environmental analysis.

III. PRACTICAL SYSTEMS: CONTINUOUS-FLOW ANALYSIS

A. General Aspects

For some years after the first publication of the continuous-flow method [1], the only commercially available equipment was the AutoAnalyzer produced by the Technicon Instruments Corporation, Tarrytown, New York. This equipment, in its newer versions, still occupies a significant part of the market, but similar systems are now available from manufacturers in several countries.

A continuous-flow system consists of a number of modules linked together in series (Fig. 5). The basic ones are an automatic sampler (Fig. 6a); a peristaltic pump (Fig. 6b); an analytical unit (otherwise known as a "cartridge" or "manifold"), in which reagents and sample are mixed in glass coils and reactions take place (Fig. 6c); a colorimeter or some other type of detector; and a data-recording system. Heating baths, units for Kjeldahl digestion, distillation, solvent extraction, dialysis, or filtration and chemical oxidation or reduction columns may also be included. Because of the modular nature of CFA apparatus, it is perfectly feasible to assemble

FIG. 5 Second-generation continuous-flow analysis system. From right to left: sampler, pump, analytical unit, colorimeter, chart recorder, data print-out unit. [Courtesy of Bran + Luebbe (G.B.) Ltd., Basingstoke, Hants, England.]

a system, or part of it, from apparatus already in the laboratory, rather than to purchase a complete assembly. For example, an existing conventional spectrophotometer can be used as the detector, instead of the purpose-built colorimeter.

B. Pumps and Tubing

1. Pumps

The flow of samples and reagents through a continuous-flow system is brought about by the action of a peristaltic pump. Constriction of a flexible tube, followed by displacement of the constriction along the tube, induces flow of the fluid within it. Some types of pump operate at a single speed, in which case flow rates are determined only by tube sizes, but a wide range of speed settings is available on other models. Reference has already been made to the need for a constant flow rate, to achieve a stable detector output. This requirement is generally more important than the absolute value of the flow rate achieved.

2. Pump Tubes

A number of different materials are available for use as pump tubes, depending on the chemicals passing through them. Those recommended

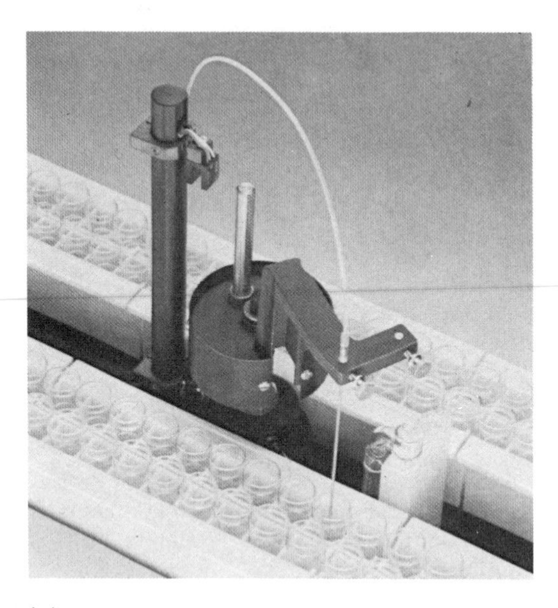

(a)

(b)

FIG. 6 Third-generation continuous-flow analysis modules. (a) Sampler, (b) pump module, (c) analytical unit. [Courtesy of Bran + Luebbe (G.B.) Ltd., Basingstoke, Hants, England.]

(c)

FIG. 6 (Continued)

for use in the Technicon AutoAnalyzer II system are shown in Table 1.

If the compatibility of tubing for a given solvent is not known, it may be tested as follows: Immerse a small piece of tubing in the solvent for about 30 min. If there is no sign of swelling or softening, proceed with a dynamic test, by fitting a pump tube of the same material and pumping the solvent through it for 20–30 min. Then, measure the flow rate by pumping into a graduated cylinder for 5–10 min. Pump with the tube for an hour or more and again check the flow rate. If the two values are similar, rapid degradation of the tube has not occurred [25].

Generally, if a pump tube is incompatible with a given liquid, there will be visible signs of deterioration during the above test. The tube can swell, crack, or snake, or can become cloudy or brittle, or turbidity may develop in the liquid as constituents leach from the tube. If a pump tube cannot be found to pump a given solvent, it may be necessary to change to another solvent or to adopt a different chemical reaction as the basis of the desired determination.

TABLE 1 Recommended Tubing for Specific Chemicals

Tubing	Chemical
Clear standard (polyvinyl chloride)	acetaldehyde (dilute) acids, mineral, dilute (to 50%) (not HF) aqueous solutions formaldehyde glycerine glycol sodium hydroxide water
Silicone	acetic acid 95%, water 5% acetone alcohols (low molecular weight) dioxane pyridine
Acid-resistant (fluoroelastomer)	acids, concentrated (>50%) benzene chloroform ethylene dichloride styrene toluene trichloroacetic acid xylene
Solvent-resistant	carbon tetrachloride hexane MIBK-ethanol (50:50) methanol, ethanol, 2-propanol

If neither alternative is practical, the displacement bottle technique can be used to introduce the solvent. A liquid that *can* be pumped satisfactorily, and that is immiscible with the solvent and either much heavier or much lighter than it, is pumped into a stoppered flask containing the solvent, thus displacing it through an outlet tube at a flow rate equal to the pumping rate of the other liquid.

Temperature affects chemical resistance, and the recommendations in Table 1 apply to room temperatures only. By diluting a solvent with a relatively inert diluent, it is possible to decrease the chemical effect on the tube [25]. Tubes may have a life of up to

500 hr, but this will be affected by many factors and cannot be specified very precisely.

At a given rate of rotation of the pump, flow rate is determined by tube internal diameter, and for accurate proportioning of samples and reagents the dimensions must be accurately known. Standard pump tubing is made from extrusions of polyvinyl chloride manufactured to very fine tolerance limits and also treated to relieve mechanical stress. Two grades of this tubing are currently available: the normal grade, with internal diameters varying from 0.13 mm (0.005 in.) to 2.79 mm (0.110 in.) (Table 2), which is usually quoted with approximate flow rates, and the more expensive flow-

TABLE 2 Internal Diameters and Color Codes of Tubes Used in Continuous-Flow Analysis

Tube number		Internal diameter		
Acid-resistant, solvent resistant	Silicone	in.	mm	Color
-01	—	0.005	0.13	orange & black
-02	—	0.0075	0.19	orange & red
-03	—	0.010	0.25	orange & blue
-04	—	0.015	0.38	orange & green
-05	—	0.20	0.51	orange & yellow
-06	-07	0.025	0.64	orange & white
-07	-08	0.030	0.76	black
-08	-09	0.035	0.89	orange
-09	-10	0.040	1.02	white
-10	-11	0.045	1.14	red
-11	-12	0.051	1.30	gray
-12	-13	0.056	1.42	yellow
—	-01	0.058	1.47	clear
-19	-14	0.060	1.52	yellow & blue
-13	-15	0.065	1.65	blue
-14	-16	0.073	1.85	green
-15	-17	0.081	2.06	purple
-16	-18	0.090	2.29	purple & black
-17	-19	0.100	2.54	purple & orange
-18	-20	0.110	2.79	purple & white

rated grade in which the tubes have precise flow rates varying from 0.015 to 3.90 ml min^{-1}.

All pump tubes have two colored identification collars exactly 6 in. (152 mm) apart to hold the tube in position when pumping. The color of the collars indicates the internal diameter, a standard color coding system having been adopted universally (Table 2).

In the Technicon Monitor 650 system, developed for industrial process control and monitoring of pollutants in effluents, pump tubes are replaced as a unit by means of a "Tech-Fit" quick-disconnect device. For each new determination, the entire pump harness is removed and replaced by a new preassembled one (Fig. 7).

3. *Transmission Tubing*

Tubing is required for interconnection of modules and for use as sleeving to connect glass tubing to glass tubing or plastic nipples. Similar considerations of solvent tolerance to those for pump tubes obviously apply, and when, for instance, acid-tolerant tubing is being used to pump a reagent, sleeving and transmission tubing downstream in the system must be made of the same material or of glass. It is recommended that glass transmission tubing be used when possible, to promote a smooth flow, and that the internal diameter of tubing, coils, and fittings should be the same, or at least within ±0.2 mm [25]. Plastic transmission tubing or sleeving should not be used for pump tubes because of the greater dimensional tolerances and lack of stress relief.

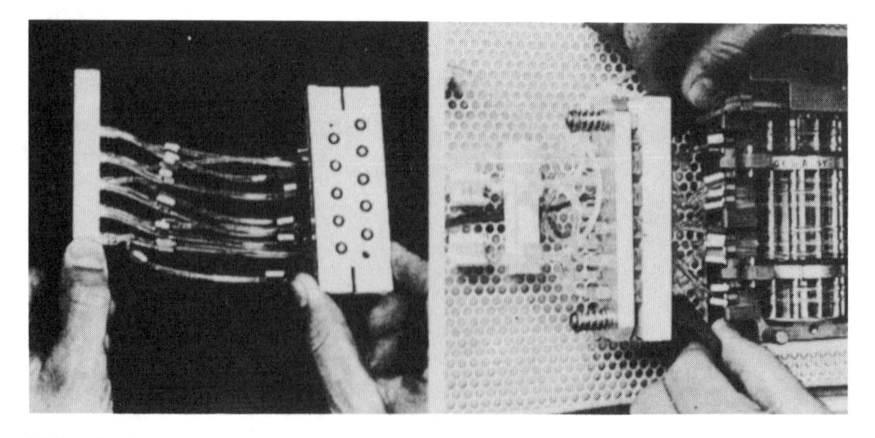

FIG. 7 "Tech-Fit" quick-disconnect system for pump tubes. (From Ref. 26.)

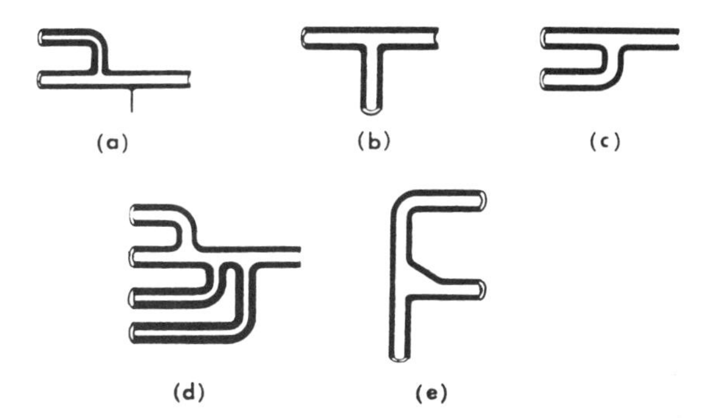

FIG. 8 Connectors and debubbler. (a) Connector with metal capillary side-arm; (b) and (c) standard-bore connectors; (d) multiple connector with two capillary side-arms; (e) debubbler.

C. Analytical Units

In the analytical unit, the sample is combined with one or more reagent streams, using connectors such as those shown in Fig. 8, and mixed by passage through a coil. Coils are usually made of glass, except when alkaline solutions are being pumped at elevated temperatures, for which application Teflon or Kel-F coils are available.

The analytical unit contains all the necessary coils, connectors, and other accessories for a specific analysis, including, if necessary, a miniaturized heating bath. Figure 6c shows a typical analytical unit with short mixing coils; a more elaborate unit including a dialyzer is shown in Fig. 9. In multichannel analyzer systems, it may be necessary to adjust the times of arrival of different streams at the detectors. This is accomplished by using so-called phasing coils, which are merely small delay coils of known internal volume.

Commercial instrument manufacturers routinely supply ready-made analytical units with all the necessary components for particular applications. However, the user can also make up assemblies in the laboratory to meet a particular need. When doing this, it is important to take account of the need to adjust tubing size to the rate of flow desired for each reagent. Attempting to pump large volumes through fine-bore tubing will create surging and a high back pressure, which may give rise to a leak at one of the connections.

D. Samplers

After air segmentation of the liquid stream, the sample solution is added by means of an automatic sampler. This device normally

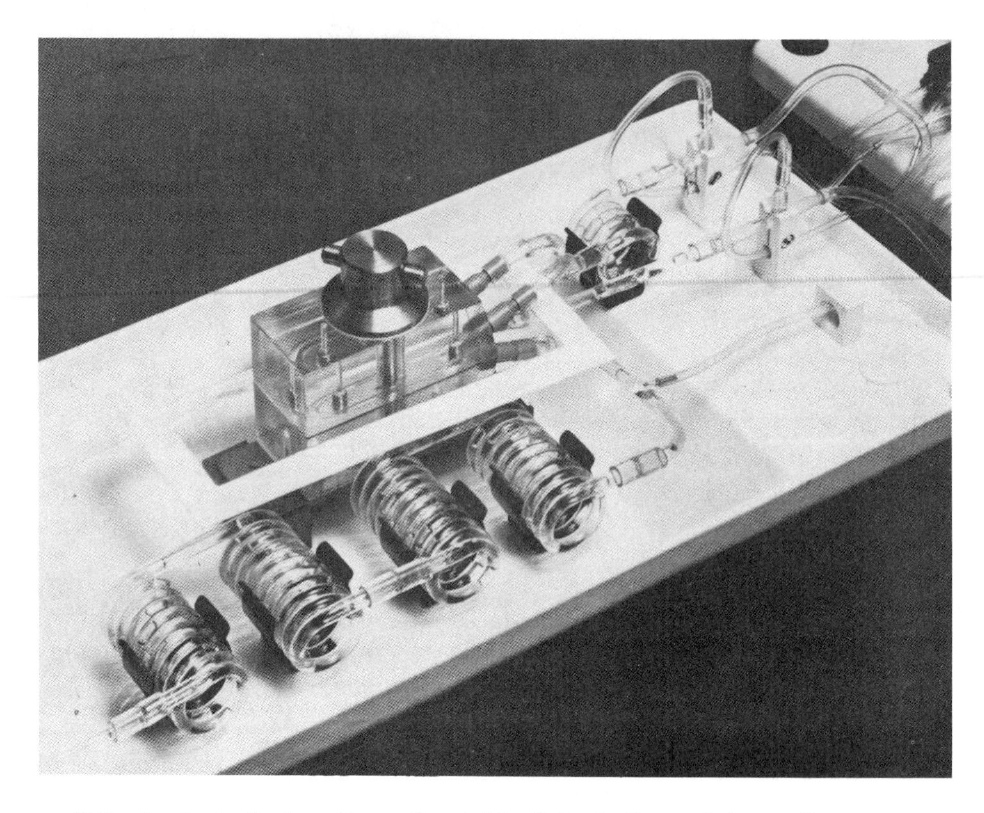

FIG. 9 Analytical unit or "cartridge" including mixing coils and dialyzer block. (Courtesy of Chemlab Ltd., Upminster, Essex, England.)

consists of a motor-driven turntable carrying a number of sample cups, typically of 2 or 4 ml capacity, into each of which in turn a sampling probe dips and aspirates solution into the continuous-flow system. The probe then aspirates a wash solution from a reservoir before the sample turntable moves the next cup into position, so that successive samples are separated in the flow system by the wash solution. In the modern Technicon Traacs-800 system, it is the sampling head rather than the sample cup that moves (Fig. 6a), and the order in which cups are sampled can be altered at will by preprogramming.

The sample probe, which is usually made from stainless steel, is connected to the sample tube with a short length of polyethylene tubing. This is used in preference to other materials because of its nonwetting properties and should be the same diameter as the sample tube or smaller.

Both sampling and wash times are controlled by an electronic timer and are adjusted in practice according to the requirements of the particular chemical reactions being employed. For optimum results, the minimum sampling time is that taken for the recorder trace of the most concentrated standard to reach a plateau and remain there for 5 sec. The time taken for the recorder to return to the baseline gives the maximum wash time. For most analyses, it is not necessary to have complete resolution of the individual peaks, and the wash time can be reduced progressively to the point where significant interference between a sample of high concentration and a following one of low concentration is just avoided. A typical recorder trace of adequately resolved peaks is illustrated in Fig. 10.

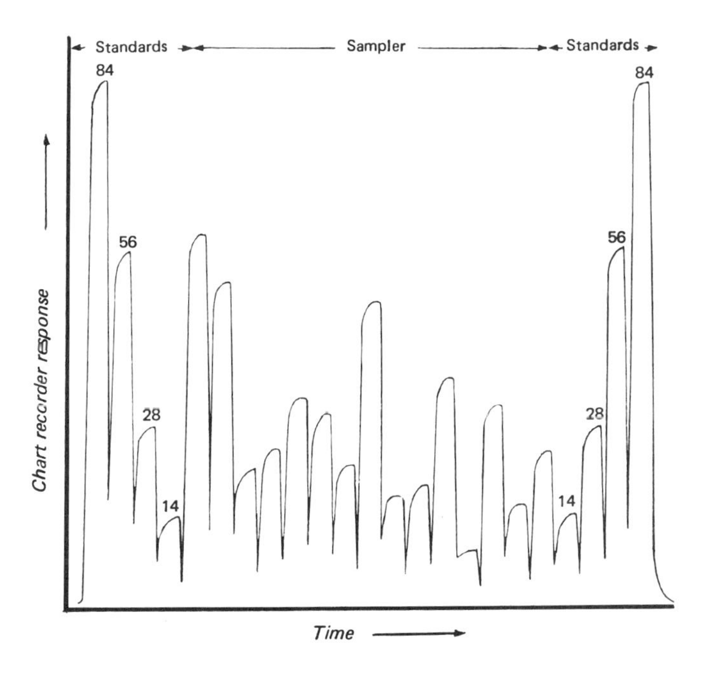

FIG. 10 Recorder trace of adequately resolved peaks in segmented continuous-flow analysis (of ammonium in soil extracts). Standards are 14, 28, 56, and 84 μg N ml^{-1}. Analysis rate 60 samples per hour; sample:wash ratio 5:1. (From Ref. 27.)

E. Other Modules

1. Dialyzer

In situations where soil extracts contain strongly colored humic substances or suspended matter that interfere in colorimetric or nephelometric determinations, dialysis provides a convenient method of separating the analyte from the interfering materials. It is also useful in reactions where colloids or precipitates are formed and can be used as an alternative to a dilution/recycling stage when sample concentrations are too high for satisfactory colorimetric determination.

A dialyzer is usually constructed from thick acrylic blocks (see Fig. 9) containing matching grooves up to 60 cm in length, with a membrane, usually made of cellulose acetate, sandwiched between the blocks. The segmented stream containing the sample is pumped through the groove above the membrane, while a segmented receiver stream (of similar composition, except for the analyte) flows at a similar rate below. The analyte passes through the membrane and reaches a concentration in the receiver stream proportional to the concentration in the original sample. The rate of dialysis depends on the nature of the membrane and the temperature (when required, elevated temperatures are achieved by preheating the donor solution in a coil). The final concentration in the receiver stream will depend also on the length of membrane exposed between the grooves.

2. Reductor

A common method for determining low concentrations of nitrate in water and soil extracts involves its reduction to nitrite, which is then determined colorimetrically. A method using copperized cadmium for this reduction [28] was modified for continuous-flow analysis by Armstrong et al. [29] and first applied to soil extracts by Henriksen and Selmer-Olsen [30]. The reductor is included in standard analytical cartridges for nitrate determinations. It can also be easily constructed in the laboratory by packing a small glass tube with 40–60 mesh (250–420 μm) cadmium that has been stirred briefly in copper sulfate solution. The solution enters the bottom after passing through a debubbler (Fig. 8e). Gas bubbles may form in the column, particularly after prolonged use, and when this occurs, they can be removed by locating the debubbler after the reductor column [30] (see Fig. 18, p. 218). Air is then reinjected into the stream in the normal controlled manner to create a segmented sample once again.

3. UV Digestor

Total nitrogen in soil extracts can be determined by passing the extract through a UV digestor unit. This consists of a long helical coil surrounding an axially mounted powerful UV lamp. The organically bound nitrogen in the extract is converted to ammonium, which may then be determined colorimetrically or by an ion-selective electrode.

F. Measuring Instruments

The liquid stream emerging from the manifold must enter some suitable measuring instrument to allow the concentration of the species of interest to be determined. A single instrument may be used when only one species is being determined, but when several measurements are required on a single sample, more than one may be involved. Multiple instruments are operated usually in parallel, following the splitting of the sample stream.

The most common instrument is the colorimeter, or spectrophotometer, used to measure the products of color-producing reactions. A variety of other instruments may also be used, e.g., flame photometers, atomic absorption spectrometers, ion-selective electrodes, nephelometers, and radioactivity detectors.

1. *Colorimeter or Spectrophotometer*

The colorimeter or spectrophotometer measures the transmission of light of a particular wavelength through a solution of colored species that absorbs strongly at that wavelength. The relationship between transmission and concentration is given by the Beer—Lambert law

$$I = I_0 e^{-kcd} \tag{9}$$

where I and I_0 are the intensities of the light transmitted and the incident light, respectively; c is the concentration of the species (moles per liter) and k is its *molar extinction coefficient* for the particular wavelength; and d is the path length of the light through the solution.

Rearranging Eq. (9) and taking logarithms, we obtain

$$-kcd = 2.303 \log \frac{I}{I_0} \tag{10}$$

Often, it is preferable to express the ratio in terms of the *absorbance A* (optical density), because this quantity is directly proportional to concentration

$$A = \log \frac{I_0}{I} = Kc \tag{11}$$

The Beer–Lambert law is an idealized relationship applying to monochromatic radiation. In actual practice, using filters, prisms, or gratings, it is impossible to obtain perfect monochromatic radiation. However, the narrower the band that can be obtained, the closer experimental data conform to the law [31]. Departures from a linear relationship may occur in systems involving ionization or dissociation equilibria that may be affected by such factors as pH and dilution, or when high concentrations cause changes in the refractive index of the solution [31,32].

Linearizing devices that convert the logarithmic concentration-transmission relationship to a linear one are commonly used to obtain output data in a more convenient form.

Continuous-flow analyzers are equipped with either a photometer with a monochromator adjustable to any wavelength in the 300–800 nm range, or a colorimeter containing a filter of the desired wavelength for a particular determination. Multichannel colorimeters make it possible to carry out up to six or eight determinations in parallel. Fiber optics are used to ensure efficient light transfer from the flow cells to the phototube detectors. Standard laboratory spectrophotometers equipped with flow cells may also be used as detector systems.

a. Flow Cells. These are usually about 1.5 mm in internal diameter, with a path length of 15–50 mm. These dimensions are considerably smaller than in the cells used with the original AutoAnalyzer systems, and consequently much smaller flow rates are needed to wash them out.

The segmented sample stream may be debubbled before entering the flow cell (Fig. 11). Alternatively, the bubbles may be allowed to pass through, using an electronic control to prevent the recorder from oscillating in response to the intermittent signal. This approach has been adopted in several modern commercial systems.

b. Turbidimetric Methods. Colorimeters or spectrophotometers may be used for turbidimetric determinations such as sulfate as barium sulfate [33–35]. Barium chloride solution containing a suspension-stabilizing agent such as gelatine or polyvinyl alcohol is added to the sample, and the absorbance caused by the barium sulfate precipitate is measured in the flow cell. The technique requires precautions to be taken to minimize adsorption of barium sulfate on the flow cell wall, which causes recorder baseline drift. The usual method is to wash with dilute HCl containing a wetting agent [35] or an EDTA buffer solution [33].

2. Flame Photometer and Atomic Absorption Spectrometer

Flame photometers are used in conjunction with continuous-flow systems principally for the determination of sodium and potassium. Atomic

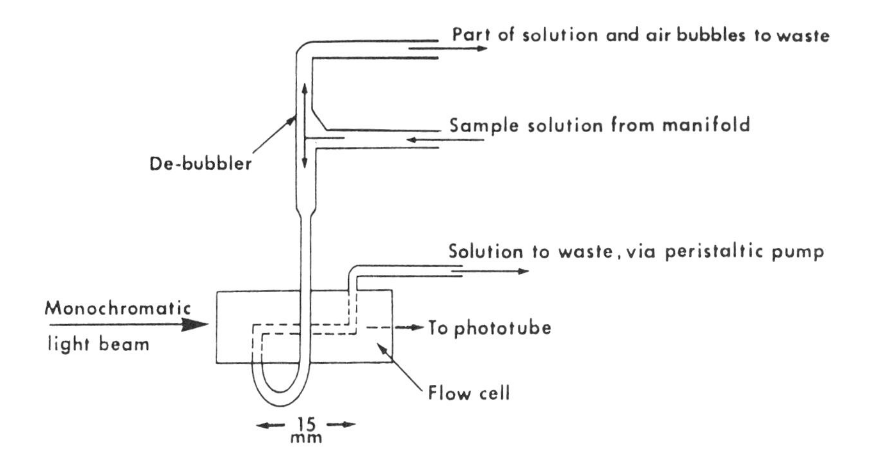

FIG. 11 Flow cell with external debubbler.

absorption spectrometers (see Chap. 1) provide the most satisfactory way of determining magnesium in soil extracts.

Both types of instrument measure simple elemental species, thus, the role of the continuous-flow system here is in automating not chemical reactions but sampling, dilution, and transfer to the aspirator-burner assembly. A major advantage of the continuous-flow system is that it is capable of making large-order dilutions with a high degree of accuracy and consistency.

The only modifications to standard procedures for flame photometric and atomic absorption measurements usually involved when methods are automated are in the introduction of the sample into the aspirator-burner assembly. The air-segmented stream must be debubbled before being aspirated to maintain a stable output.

3. *Nephelometer*

The determination of sulfate may be carried out by nephelometric (light scattering) methods instead of the turbidimetric method referred to above [36–38]. A nephelometer is essentially a modified fluorescence spectrophotometer (or *fluorimeter*) in which a square glass or quartz flow cell replaces the normal tubular one. A light beam that falls on the flowing stream is scattered by the suspended particles of barium sulfate, and some of the scattered light passes through the optical system to the detector. The special flow cell is necessary to reduce the background scattering to an acceptable level [39].

4. *Ion-Selective Electrodes*

Ion-selective electrodes or probes (see Chap. 3) are finding increasing application in continuous-flow systems. A higher precision is normally attainable with continuous-flow than with manual methods, because of the greater standardization of the conditions in which the sample is presented to the electrode [40].

To incorporate the ion-selective electrode into a continuous-flow system, it is normally fitted into a flow-through cap or cell. When a separate reference electrode is required, this can be located in a reservoir downstream from the indicator electrode, so that the cell dead volume can be kept as small as 5–10 μl [41]. It is sometimes necessary to debubble a segmented stream before entering the detector cell to avoid an erratic response. However, the ammonia probe has no external reference electrode, and the partial pressure of ammonia in the air bubbles is in equilibrium with the solution, so a steady response can be achieved without debubbling [40].

G. Optimization of Analytical Conditions

1. *Sensitivity*

To improve the sensitivity of an analytical system, a number of conditions may be altered [25]:

1. Sample size can be increased by increasing the diameter of the sample pump tube, and reagent concentration can be increased to counteract the resulting dilution of the final reaction mixture.
2. The path length of the flow cell may be increased, because according to the Beer–Lambert law the absorbance is directly proportional to path length.
3. When a reaction is dependent on elevated temperatures or takes a considerable time, the temperature of the heating bath can be raised, a larger volume heating coil can be used, or a time-delay coil added to allow the reaction to progress further.
4. When dialysis is involved, the path length can be increased, the temperature increased, or a more permeable membrane substituted for the one normally used.

2. *Accuracy and Precision*

a. Absorbance. For greatest accuracy, it is desirable that absorbance should be about 0.4 unit because this is near the point at which photometric error is minimized [25]. This is obviously much easier to accomplish for samples that are of similar concentration than for those that vary widely, and repeated analyses after dilution may be desirable for samples that give rise to very high absorbance.

b. *Drift and Noise.* Calibration drift may occur due to changes in baseline or method sensitivity, or both [42]. Causes may include changes in ambient, module or reagent temperatures, stray light and manifold stretching [43], coating of the flow cell, reagent degradation or degassing [25], or instrumental drift. Attempts should be made to eliminate the cause of drift, e.g., when coating occurs, wetting agents may be employed to overcome the problem. When drift cannot be eradicated, midrange standards and water blanks should be inserted at intervals to monitor it.

If recorder tracings are noisy, all fittings and connections should be checked to ensure that joints have been made properly. Uniformity of air-bubble injection should also be checked, transmission tubing made as short as practicable, and wetting agents used when appropriate [25].

c. *Sample and Wash Times.* Reference has already been made above (Sec. III.D) to the need for sampling times to be long enough to allow a response plateau to be sustained for at least 5 sec. If a digital printer is to be used, then a few more seconds should be added. Interaction between samples may be checked experimentally by analyzing several low- and high-concentration samples alternately, followed by several high-concentration ones in succession. If the second low sample produces a higher peak than the first, or if the second successive high sample is higher than the first, the wash time should be extended until all peaks from replicate samples agree within the limits required [25].

IV. PRACTICAL SYSTEMS: FLOW-INJECTION ANALYSIS

A. General Aspects

The basic components of an FIA system are an autosampler, a pump, a sample injection valve, an analytical manifold, and a detector/readout system. Autosamplers, several types of flow-through detectors (e.g., spectrophotometers, ion-selective electrodes, atomic absorption spectrometers), and units such as dialyzers are similar to those used in CFA systems and are not considered further here. The major differences from CFA apparatus are in the presence of an injection valve for the introduction of the sample into the liquid stream, and the use of much smaller bore tubing.

A typical FIA assembly is shown diagrammatically in Fig. 12, and a modern commercially manufactured system is shown in Fig. 13.

In comparison with air-segmented continuous-flow analysis, FIA has the following characteristics [44,46]:

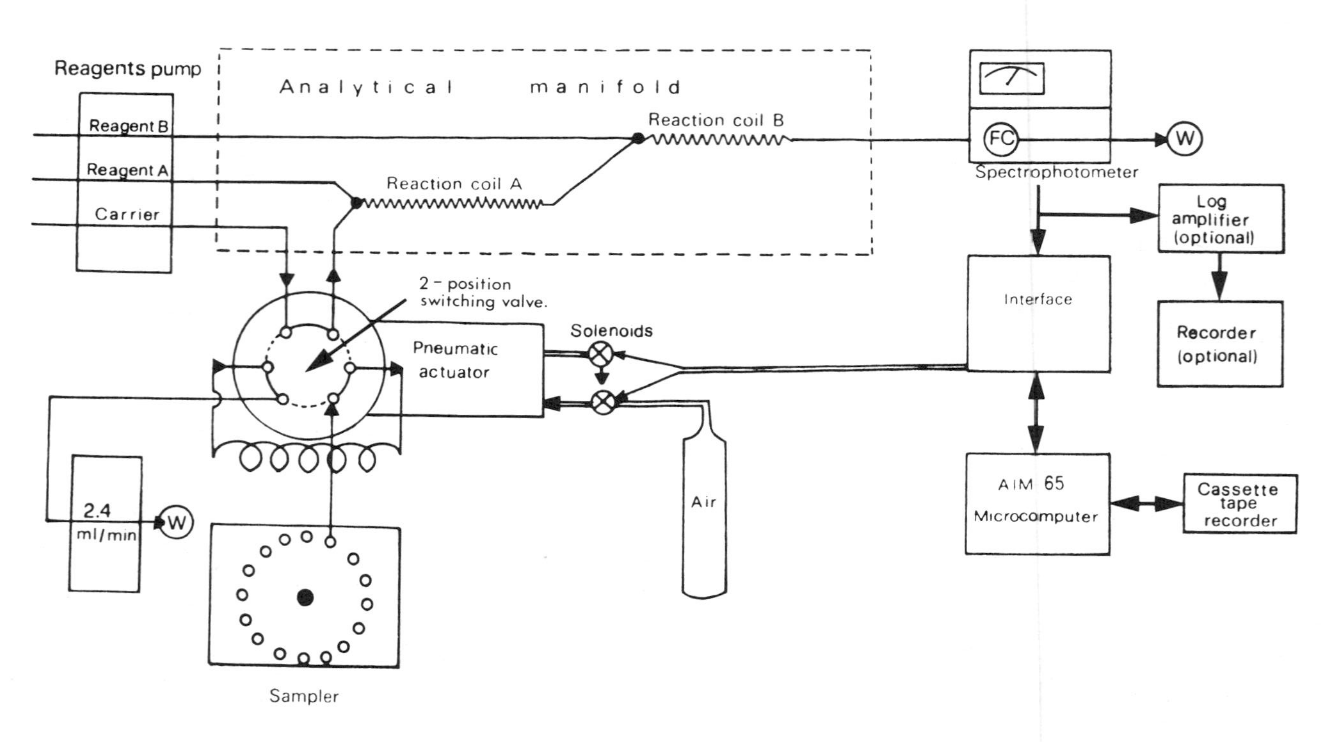

FIG. 12 Block diagram of flow-injection analysis system developed at University of Athens, Greece. (Redrawn from Ref. 45.)

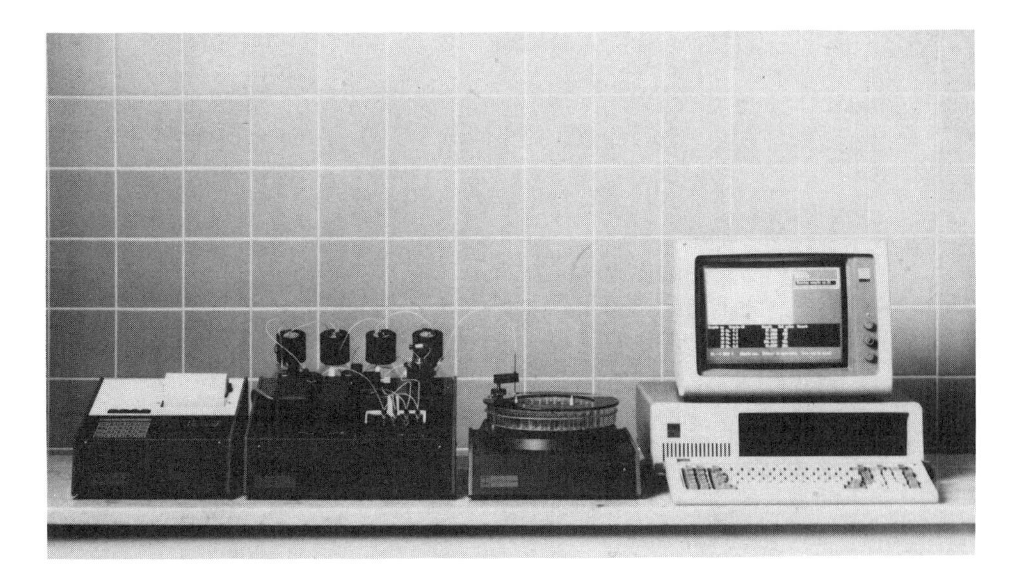

FIG. 13 Tecator FIA system. (Courtesy of Perstorp Analytical Ltd., Thornbury, Bristol, England.)

1. It is very considerably faster (Fig. 14). The output peaks are usually obtained within 30 sec of sample injection, as compared with some minutes for CFA. Provided that the time taken for each sample-reagent mixture to pass through the system to the detector remains constant, and provided that this time is long enough to produce an adequate amount of colored or other detectable product, there is no need for chemical equilibrium to be reached in order to obtain valid analytical results. According to Ranger [24], the most important parameter to control in FIA to ensure precision is the timing involved throughout the system. This demands, in particular, injection of precise volumes by the injection valve and reproducible slow rates of pumping.

2. It consumes less reagent and normally less sample, particularly when the merging zone technique is used. This technique was developed by Bergamin et al. [48] to minimize the consumption of expensive reagents. Both sample and reagent are injected into separate carrier streams that meet at a confluence, mix partially, and pass through a coil to promote further mixing and reaction. Thus, only enough reagent is introduced to meet the needs of the sample, whereas the washout is achieved by the inexpensive carrier-stream solution.

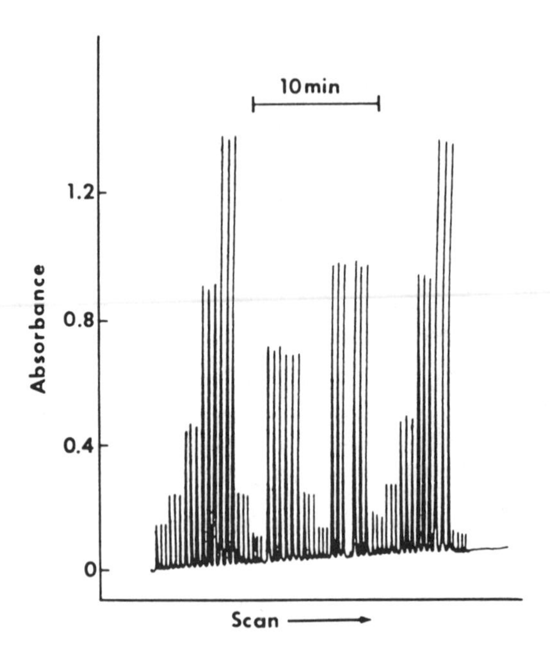

FIG. 14 Recorder trace from flow injection analysis of ammonium in soil extracts. From left to right are shown a series of ammonium standard solutions (0.5, 1.0, 2.0, 4.0, and 6.0 μg NH_4^+-N ml^{-1}), followed by eight sample solutions, a second set of standards, and finally a blank, all solutions being injected in triplicate. (From Ref. 47.)

3. It is ready for use almost immediately after start-up, whereas CFA systems require several minutes of operation for the baseline to stabilize.

Some of the most important practical aspects of FIA are outlined in the following sections. A more detailed discussion of FIA theory and practice may be found in the book by Růžička and Hansen [49] and the proceedings of an international conference held in 1985 [50], and in the numerous references cited therein.

B. Pumps, Tubes, and Coils

As with CFA, it is vital in FIA that a very reproducible flow rate be maintained. Various types of pump have been tried, such as syringe, peristaltic, progressive cavity, and single or dual piston pumps [24]. Peristaltic pumps are most commonly used, together with the same types of vinyl or silicone flow-rated tubing as is used for CFA, with the same color codes to indicate size (see Table 2).

The use of pumps originally intended for working against much higher back pressures (such as in HPLC) also appears to be satisfactory in some situations [51]. However, some of these pumps will not work satisfactorily with the low or even negative back pressures encountered when the FIA system is used to introduce samples into an atomic absorption spectrometer (i.e., when there are no mixing coils to create a resistance to flow). In such circumstances, back pressure may be created artificially by including a length of tubing between the pump and injector [52].

All connecting tubes, manifolds, and mixing coils are of narrow bore (commonly 0.4–1.0 mm). Mixing coils are usually made from rigid polyethylene, tygon (vinyl), or Teflon. Manifolds for dilution and reagent addition need nothing more complicated than Teflon T-pieces as confluence points. If elevated temperatures are required, the coils (and reagent reservoirs) may be immersed in a constant-temperature bath.

C. Injection Valves

In both commercial and laboatory-assembled systems, the sample is introduced into the moving stream of reagent by means of a rotary or sliding switching valve, which is actuated pneumatically or electrically. The sampling loop of a six-port, two-position valve (as illustrated in Fig. 12) is filled by aspirating sample from a cup on the auto-sampler while the preceding sample is passing through the manifold. The valve is then switched to the alternate position, diverting the carrier stream through the loop and pushing the sample as a "plug" into the manifold. The use of an eight-port valve with two loops makes it possible to fill one loop while the contents of the other are being injected, and vice versa, thus giving a higher rate of throughput [53]. Usually, the parts of valves coming into contact with the solutions are made from inert fluorocarbons and are free from corrosion problems [54].

Sherwood and co-workers [55,56] have described the use of a valveless injection technique to introduce very small samples into the FIA manifold. The technique, called "controlled dispersion analysis," uses a computer-controlled peristaltic pump and aspiration probe. To introduce a sample, the pump is stopped, the probe transferred from carrier to sample, the pump activated to draw up the desired volume, then stopped again, the probe returned to the carrier, and finally the pump restarted. This avoids the loss of sample that occurs with valves, as a certain volume is needed to fill the connecting lines and flush out the preceding sample [54].

D. "Heliflow" FIA System

A novel low-cost flow-injection analysis system has been introduced recently based on the helical flow analytical module developed by Brown and Smith [57]. This module is constructed from a cylinder of acrylic plastic (100 mm long by 50 mm in diameter) around which two connectable helical channels are cut (Fig. 15) [58]. The channels are sealed by a renewable sleeve of heat-shrink tubing, thus creating a reaction coil either 600 mm or 1200 mm in length.

Samples are introduced into the carrier stream by a simple two-way, variable-volume, manual valve that forms an integral part of the heliflow module. Reagents are introduced into the carrier stream via two entry ports mounted in each helical channel. All exterior tubing (0.4 mm ID) is connected to the module by a simple push-fit arrangement. With this system, it is possible to carry out a range of different analytical procedures with the single module, without the need for different manifold cartridges.

By incorporating a battery-powered peristaltic pump and colorimeter, the system may be used in the field as well as on the laboratory bench [58] and is manufactured in both versions by WPA of Linton, Cambridge, England. Originally developed for nitrate-nitrogen, the heliflow system has rapidly found new applications, e.g., in the determination of phosphorus, chloride, copper, zinc, and aluminum [59].

E. FIA: Atomic Absorption Spectrometer Systems

Flow-injection analysis equipment can be used for the supply of sample solutions directly to a detector such as an atomic absorption spectrometer. This topic has been reviewed by Tyson [54] and Fang et al. [60].

With a single-line manifold, the suitable dilution of samples can be achieved by choosing appropriate injection volumes and tube dimensions. Addition of an interference suppressant such as lanthanum is simply achieved by including it in the carrier stream. The alternative way of adding reagents is to merge a reagent stream with the carrier stream at a confluence point. A more economical approach, in terms of reagent consumption, is to use the merging zones method [54]. More complex manifolds have also been devised, including systems with phase separators to allow the determination of elements such as mercury and bismuth in gaseous form (mercury as the cold vapor [61], and bismuth as the hydride [62]).

Although the use of FIA as a system for feeding sample solutions directly to atomic absorption spectrometers (or other detectors such as ion-selective electrodes) undoubtedly can give satisfactory results, it is an expensive and somewhat inflexible approach compared with the use of programmable discrete analysis systems discussed in Sec. V.

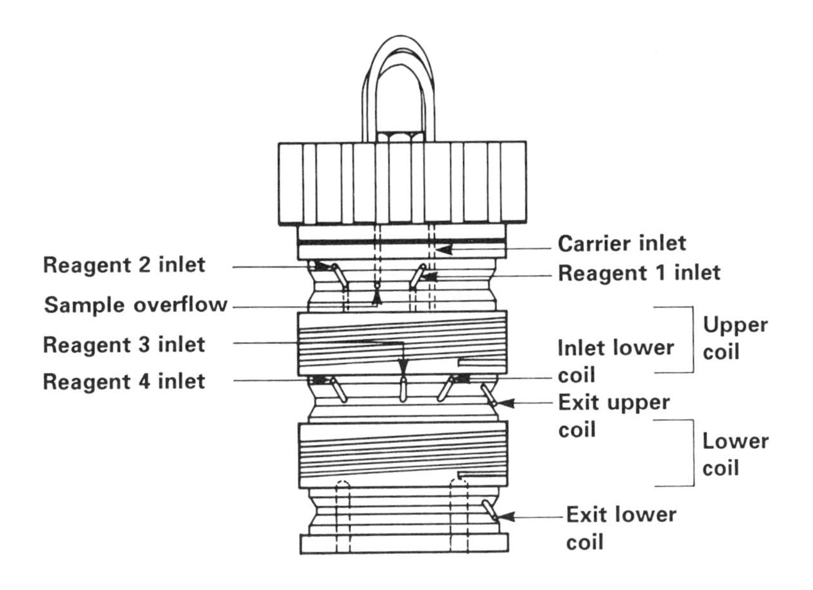

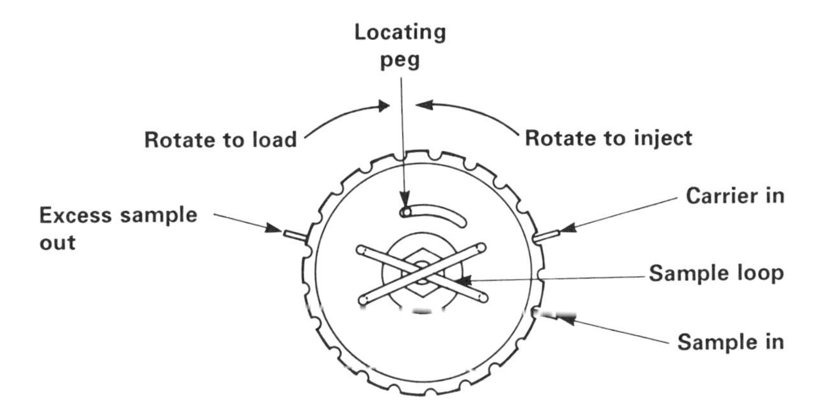

FIG. 15 Heliflow analytical module. (Courtesy of WPA Ltd., Linton, Cambridge, England.)

IV. PRACTICAL SYSTEMS: DISCRETE ANALYSIS

A. General Aspects

Major improvements have occurred in recent years in discrete or batch analyzers. Instruments designed to carry out a number of routine functions, such as pipetting, diluting, adding reagents, stirring, and transferring solutions to a flow cell or flame photometer, are available from a number of manufacturers in several different countries and have also been purpose-built in laboratory workshops and described in the scientific literature. The functions of single pieces of apparatus vary much more widely than in the case of continuous-flow systems, which are essentially modular in construction. It is convenient to consider separately here those multistage analyzers that carry out a complete sequence of operations from those that are intended to perform one or two functions only.

B. Automation of Individual Processes

1. *Sample Processors, Diluters, and Dispensers*

The recent introduction of robotic sampling systems, coupled to increasingly powerful microprocessors, and diluters/dispensers with high-resolution stepper-motors has made discrete analysis a very fast and accurate analytical tool.

The use of dual-syringe diluters, in which the relationship between sample and diluent syringe volumes can be optimized over a wide range, also enhances the precision and variety of tasks that can be undertaken with the modern discrete analyzer.

Two kinds of sampler are currently available, the robot arm type and the XYZ type with single or dual sampling heads. Of the two, the XYZ type is generally favored because of its simplicity and speed of operation, and because it is less prone to software problems or mechanical breakdown.

Systems utilizing integral microprocessors or stand-alone microcomputers demonstrate a high degree of flexibility in their programming and can have "random access" capabilities as well, i.e., the ability to repeat dubious samples or to dilute and rerun samples whose concentrations lie outside the calibration range. The introduction of a dual sampling head or even the use of multiple sampling probes (Fig. 16) can extend the analytical throughput to more than 1000 samples per hour, without any reduction in analytical performance.

A range of sophisticated accessories is also available from a number of manufacturers, e.g., conductive liquid sensors to minimize contamination of the sampling probes, or bar-code readers to assist with the large numbers of samples being processed.

As well as being used for sample preparation, these instruments can also be coupled directly to analytical instruments such as colorimeters, atomic absorption and emission spectrometers (Fig. 17), and inductively coupled plasma emission spectrometers. In this mode, one considerable advantage that discrete processes have, when compared with continuous-flow systems, is that results can be obtained

FIG. 16 Multiple sampling probe system at The Edinburgh School of Agriculture. In each operation of the four dual-syringe diluters, two soil extracts are simultaneously sampled and diluted for (a) phosphorus and (b) potassium and magnesium determinations.

virtually instantaneously at throughputs of up to 400 samples per hour.

Current techniques used for data capture in CFA involve recording electronically all the data relating to a particular analysis that have been recorded traditionally by a chart recorder (Fig. 10). From this electronic record, the sample peaks are identified and results calculated, with corrections being made for any acceptable baseline drift.

Discrete processors, by their very nature, produce near-square wave outputs (see Fig. 2a). When the system is interfaced to an external computer, two-way communication between the sample processor and the computer is possible. This simplifies both peak identification and the end of data collection, with results being calculated and verified for quality-control tests during the rinse cycle.

FIG. 17 Automatic sampling of diluted soil extracts for simultaneous determination of magnesium by atomic absorption and potassium by atomic emission spectrometry at The Edinburgh School of Agriculture.

2. *Systems for the Determination of pH*

The determination of soil pH is one of the most common tasks in soil analytical laboratories, and it is the type of repetitive operation that is admirably suited to automation.

Commercial instruments dedicated to the determination of soil pH are few, but modifications can easily be made to linear or rotary multisamplers designed for automatic titrators.

XYZ sample processors lend themselves very readily to the determination of pH; in the authors' laboratory one sampler has been modified to carry four pH electrodes and eight stirrers. The electrodes are wired through an interface to the analog/digital (A/D) converter of a microcomputer, bypassing the need for multiple pH meters. The system can accommodate 40 samples and determine their pHs in as little as 10 min.

Several other laboratory-made systems have been described. In Baker's system [63], 500 polystyrene containers are mounted in rows of 10 on aluminum plates, which are joined to form an endless conveyor belt. A fixed volume of each soil for analysis is introduced into each of the containers in the loading area, together with a small plastic-covered bar magnet. As each row of containers moves forward stepwise, a fixed volume of distilled water is added automatically to each, and the suspensions are mixed by a row of magnetic stirrers under the belt, stirring continually for 12 min. Ten pH electrodes mounted on a Perspex plate are lowered automatically into the suspensions. Each electrode is switched in sequence to a pH meter that is interfaced to a digital voltmeter and printer. After the readings have been made, the containers are inverted and washed with distilled water, and the bar magnets are collected in an aluminum mesh tray and recovered for reuse. The system has been used to analyze samples at the rate of 20,000 per year.

Goodman [64] developed a system to cope with batches of 6–60 samples that used motor-driven glass paddles for stirring and a single electrode. The apparatus could be fitted not only with a pH electrode but also an ion-selective electrode and has been applied to nitrate analysis. The paddles were necessary to break up the lumps of fresh soil that were used necessarily when nitrate was to be measured.

Grigg et al. [65] described a system in which an LKB Radirac fraction collector rotator is fitted with a 60-position aluminum turntable carrying PVC sample cups. To 12 ml soil samples, 30 ml of water is added from a dispenser, 10 cups at a time. The cups are stored at constant temperature overnight and then loaded onto the turntable, together with cups of buffer solutions. Stirrers and a combination glass-calomel electrode, attached to a single beam, are lowered into each cup in turn by means of a motor-driven cam. After 40 sec, a series of 12 measurements is made at 1 sec intervals. A programmable calculator rejects the highest and lowest value, works out the mean of the other 10, and adjusts the result for any drift in the pH meter during a run of samples. The result is then printed out. After the measurements, the electrode is withdrawn and automatically washed by a water jet.

V. APPLICATIONS

A. Nitrogen

1. *Total Nitrogen and Ammonium*

Wet chemical methods for the determination of total nitrogen in soils and plant material are all ultimately based on the conversion of the

TABLE 3 Regression Analysis of AutoAnalyzer and Distillation Values for Cation-Exchange Capacity (CEC) and Total Nitrogen

	CEC (me %)[a]	Total nitrogen (%)
Range	0–30	0–0.5
Regression[b]	$y = 0.9932x + 0.0947$	$y = 0.9907x + 0.0005$
r	0.9987***	0.9991***

[a]1 me % = 1 mmol ÷ (charge on free ion) per 100 g oven dry soil.
[b]x = AutoAnalyzer values, y = distillation values.
Source: Ref. 27.

nitrogen to ammonium ion, followed by measurement of the ammonium by colorimetry, ion-selective electrode, or some other convenient method. The conversion to ammonium is achieved by Kjeldahl digestion, either manually using a block digestor or in an automatic wet digestion device. The ammonium in the digests may then be analyzed by any of the alternative automated systems: CFA [27,66, 67], FIA [47], or discrete analysis [68].

A commonly used colorimetric method for ammonium in soil digests and extracts is the "indophenol" method, in which the ammonium present produces a blue color with alkaline phenol and sodium hypochlorite [27,66,67]. The quality of agreement possible between this continuous-flow method (modified by complexing interfering metals, e.g., iron and manganese, with citrate and tartrate) and the manual Kjeldahl distillation and titration method was explored by Searle [27]. Table 3 shows his results for 50 soils, both for total nitrogen and the measurement of cation-exchange capacity, in which the soils were first saturated with ammonium ions and then leached with sodium chloride. The values obtained by the two methods were very highly correlated ($p < 0.001$) and the regression coefficients were very close to unity, indicating no significant bias between the methods.

Other sources of interference in colorimetric determination of ammonium in soil extracts are colored humic substances or suspended matter, and the presence of labile organonitrogen compounds, e.g., amino acids [69,70]. The former may be overcome by incorporating a dialyzer into the manifold [69].

A flow-injection method for ammonium in soil extracts and natural waters has been described by Krug et al. [47]. This method uses Nessler's reagent [mercury (II) chloride and potassium iodide in alkaline solution], with the colored turbidity produced measured colorimetrically at 410 nm. An analysis rate of 120 samples per hour was achieved with the method, and a typical chart-recorder trace appears earlier in this chapter (Fig. 14, p. 208).

2. *Nitrate and Nitrite*

Continuous-flow methods for the determination of nitrate in soil extracts are based usually on its reduction to nitrite, with subsequent colorimetric determination of the nitrite formed plus any nitrite originally present in the sample. The latter may be determined separately by leaving out the reduction step, and the nitrate obtained by difference [30]. The reducing agent may be copperized cadmium [30,71], hydrazine and copper [72,73], or a suspension of *Escherichia coli* cells lacking nitrite reductase but containing nitrate reductase [74,75]. The nitrite formed is usually determined by diazotizing sulfanilamide and coupling with *N*-1-naphthylethylenediamine to form an azo-dye, which is measured at 520–540 nm. As in the case of ammonium determinations, when colored or turbid samples are being analyzed, a dialyzer can be incorporated into the manifold (Fig. 18). These azo-dye colorimetric procedures have also been applied to both conventional [76] and heliflow [58] FIA, as well as discrete analysis [77].

Hansen et al. [78] have used a different method of detection, a nitrate ion-selective electrode, with an FIA system to determine nitrate in soil extracts and fertilizer solutions. Such an electrode has also been applied in discrete analysis [64].

B. Phosphorus

A wide variety of extractants has been used in soil fertility studies for the determination of labile soil phosphorus. In the early days of automated analysis, methods were developed for use with the majority of these extractants (Table 4), based on standard colorimetric procedures such as the reduction of a molybdophosphate complex to "molybdenum blue." These methods have generally stood the test of time and form the basis of the standard automated procedures in use today. Continuous-flow methods are still widely employed after 20–30 years, but discrete methods have also become popular. Some discrete systems have achieved very high sample throughputs, e.g., 360 per hour was reported by Orr [79] using LKB equipment (LKB Instruments, Bromma, Sweden).

The molybdovanadatophosphate yellow colorimetric method for phosphorus has also been adapted for use in continuous-flow analysis. This method is less sensitive than the molybdenum blue method and has therefore mainly been applied to plant analysis [80–82], although Flannery and Markus [83] analyzed some soil extracts in this way.

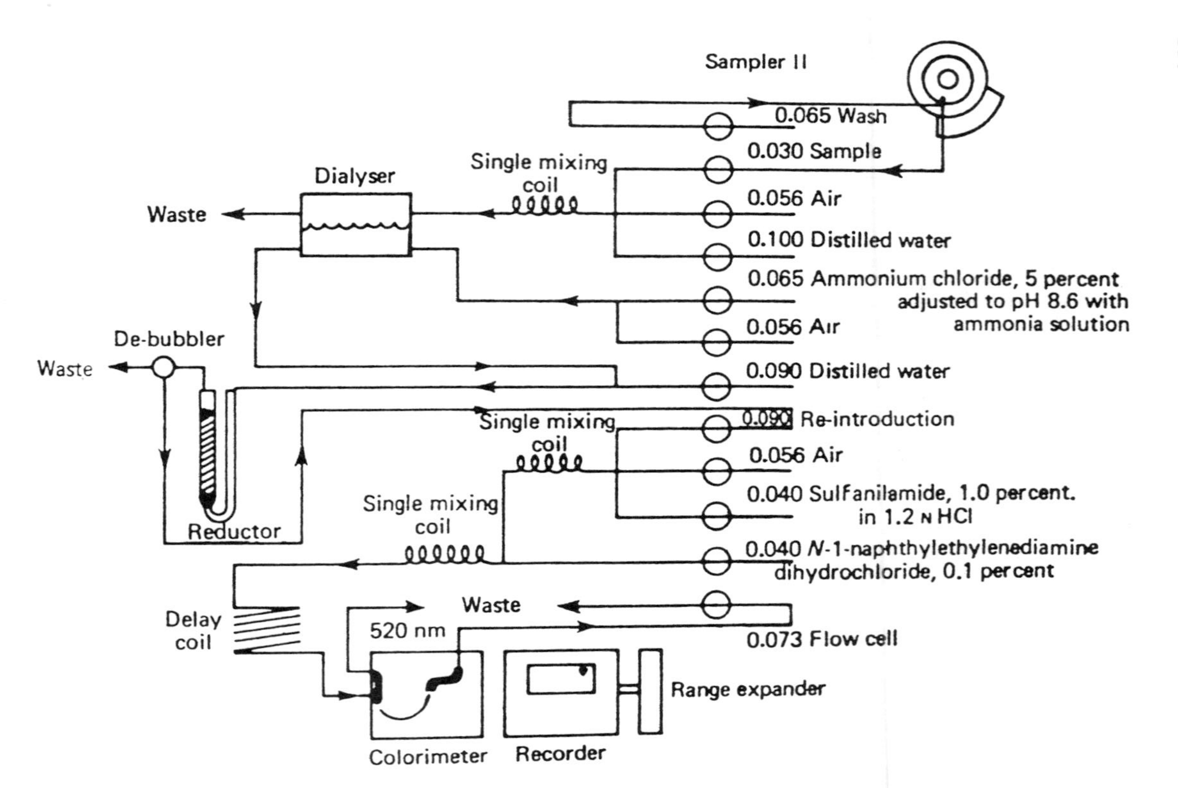

FIG. 18 Manifold for continuous-flow analysis of nitrate and nitrite in soil extracts. (From Ref. 30.)

TABLE 4 Applications of Automated Methods to the Determination of Phosphorus in Soil Extracts

Extractant	Refs.
0.001 *M* H_2SO_4/0.002 *M* $(NH_4)_2SO_4$, pH 3 (Truog)	[32,86]
0.5 *M* $NaHCO_3$/NaOH, pH 8.5 (Olsen)	[86, 88–90]
0.3 *M* NH_4F/0.025 *M* HCl (Bray and Kurtz No. 1)	79,83,86]
0.3 *M* NH_4F/0.1 *M* HCl (Bray and Kurtz No. 2)	[84,86]
0.2 *M* Ca-lactate, pH 3.8 (Egner–Riehm)	[90]
0.1 *M* NH_4-lactate/0.4 *M* acetic acid (Egner–Riehm–Domingo)	[90,91]
0.5 *M* NH_4-acetate/0.5 *M* acetic acid, pH 4.8 (modified Morgan)	[81]
0.4 *M* acetic acid, pH 2.6	[32]
0.01 *M* $CaCl_2$	[89]
0.01 *M* H_2SO_4	[92,93]
0.05 *M* HCl/0.012 *M* H_2SO_4	[83]
0.05 *M* Na_2SO_4, after sorption on an ion-exchange resin	[94]
Chang and Jackson's solutions (1 *M* NH_4Cl, 0.5 *M* NH_4F, 0.1 *M* NaOH, 0.3 *M* Na-citrate/$Na_2S_2O_4$)	[86]

In addition to being applied to phosphorus in the extractants listed in Table 4, automated methods are also suitable for solutions resulting from wet digestion procedures or fusions [32,84].

a. Interferences. Iron (III) is a common interferent, but its effect can be reduced by reduction to iron (II) by hydrazine [32] or hydroxylammonium chloride [85]. Organic matter can also interfere and as in the determination of ammonium or nitrate may be removed by dialysis [86]. Carbon dioxide liberation from alkaline bicarbonate (Olsen) soil extracts can cause problems in CFA [32,87], but has been overcome satisfactorily by passing the liquid stream through a debubbler to remove CO_2, after acidification [86,88].

C. Potassium, Calcium, Magnesium, and Sodium

In routine advisory soil analysis, some or all of the elements potassium, calcium, and magnesium, and occasionally sodium, have been determined by automated methods. Often, these determinations have been made, using continuous-flow methods, in parallel with the colorimetric determination of phosphorus (and sometimes ammonium or nitrate) and on the same soil extract. Zsoldos [90] has described such a system, fed by a 200-position sampler, in which potassium and sodium were determined by flame photometry. Interference from calcium in the determination of potassium was avoided, even in extracts of calcareous soils, by using a low-flame temperature (propane-butane/compressed air flame). Lithium was used as an internal standard.

Flannery and Markus [83] used independent flow systems when analyzing phosphorus, magnesium, and potassium plus calcium in soil extracts. These workers adopted a modified lake procedure for magnesium, in which $Mg(OH)_2$ was precipitated in alkaline solution and magnesium blue dye was adsorbed on the precipitate in the presence of a detergent (Brij 35) and suspending agent (polyvinyl alcohol). Potassium and calcium were measured simultaneously by flame photometry.

Clinton [95] developed a four-channel photometric system for soil leachates and plant digests in which calcium, sodium, and potassium were determined by flame emission and magnesium by atomic absorption, at a rate of 200 samples per hour. Instruments such as the Perkin-Elmer Model 5000 atomic absorption and flame emission spectrometer equipped with an automatic sampler may be programmed to determine Ca, Mg, K, and Na in 1 *M* ammonium acetate soil extracts without dilution [96]. Calcium and magnesium in waters have been determined sequentially in an FIA system: the calcium by a flow-through tubular electrode and the magnesium by atomic absorption spectrometry [97].

D. Aluminum, Silicon, and Iron

A continuous-flow method for aluminum in soils and plant digests, using the complexing dye sodium alizarin-3-sulfonate (alizarin red S) was reported as giving a very reproducible and linear response up to 15 μg Al ml^{-1} [98]. Allen et al. [32] recommend for aluminum a method using the complex formed with eriochrome cyanine R, and for silicon the reduction of molybdosilicic acid to a molybdenum blue complex, i.e., a reaction analogous to that used for phosphorus. A similar FIA method for silicon in soil extracts has been developed [99] (Fig. 19).

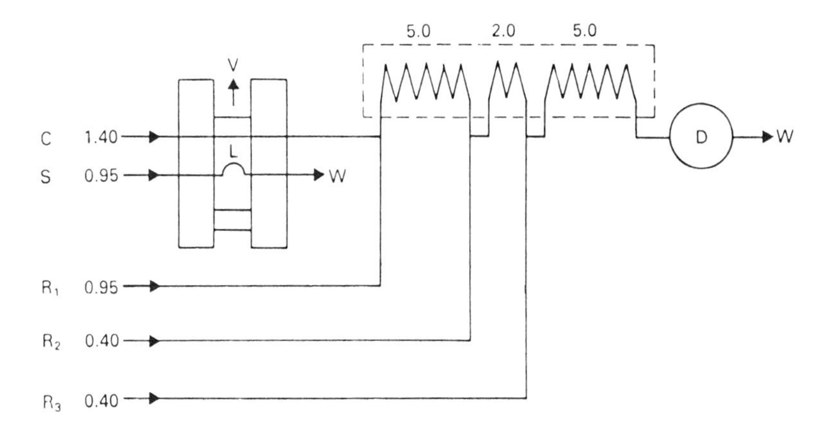

FIG. 19 Flow-injection manifold for determination of silicon in soil extracts. V, sliding injection valve (moves in direction of arrow); L, sampling loop (300 μl); D, detector; W, waste; C, carrier stream (water); S, sample; R_1, ammonium molybdate reagent, R_2, oxalic acid, 9%; R_3, ascorbic acid, 2%. Numbers are flow rates in ml min^{-1}. (From Ref. 99.)

Discrete analysis has been applied to extractable aluminum in soils, using a colorimetric method based on the complex with pyrocatechol violet [92].

Iron is a major component of most soils, and total iron or the fractions that are extractable with reagents such as oxalic acid or pyrophosphate are determined frequently in pedological investigations. For the determination of plant-available iron, 2.5% acetic acid is used commonly. Allen et al. [32] adapted the manual colorimetric method for iron using sulfonated 4,7-diphenyl-1,10-phenanthroline to CFA so that it could be used for both 3% oxalic acid and 2.5% acetic acid extracts.

E. Sulfur

Published methods for sulfur (as sulfate) in soil extracts and natural waters have involved turbidimetric [34,35], nephelometric [36, 38], and colorimetric [100,101] methods of measurement. A recent development for sulfur in water samples is the use of a lead-sensitive ion-selective electrode to measure the free lead(II) after mixing a reagent stream containing lead with the sample stream. Both FIA [102] and CFA [103] methods have been developed.

F. Trace Elements

Several automated methods for boron in plant material have been developed, including colorimetric methods based on reactions with azomethine-H [104,105] and quinalizarin [106]. The former was adapted with only slight modification to the analysis of hot water extracts of soils [32]. Also, a manual spectrofluorimetric method for boron using carminic acid was adapted to CFA by Ogner [107].

Automated procedures for chlorine, bromine, and iodine are all available. Chloride can be determined colorimetically by a reaction involving iron and thiocyanate [32]. Bromide can be determined by a CFA method involving oxidation of iodine to iodate [108], or by FIA using a tubular bromide-selective electrode [109]. Iodine in plant tissue and sodium hydroxide extracts of soil may be determined by a CFA procedure based on the catalytic action of iodine on the oxidation of arsenic(III) [110].

Trace metals in soil extracts and plant tissues have been determined by automated wet chemical procedures, although usually there are more suitable alternatives. Thus, copper [111] and molybdenum [112] in plants have been determined by CFA, and chromium(VI) in soil extracts and natural waters has been determined by FIA [113]. The measurement of arsenic in soil and plant digests [114], arsenic, antimony, and selenium in natural waters [115], and mercury in soils and sediments [116,117] are other examples.

G. pH

Reference has already been made (Sec. V.B) to several systems purpose-built for the automated measurement of soil pH. Discussion here is confined to the comparisons made with manual methods, which show the automated methods to give very satisfactory results.

Baker [63] compared results for 200 soil samples (136 temperate and 64 tropical) with pH ranging from 4.6 to 8.2, organic matter ranging from <1% to nearly 70%, and including the whole textural range. The following regression equation was obtained:

$$pH_{automated} = 1.0064\ (\pm 0.0191)\ pH_{manual} - 0.1460$$

and the correlation coefficient was 0.966. Goodman [64] found that the standard deviation for an individual measurement with his apparatus was 0.044 pH unit, compared with 0.039 for a manual measurement. The two sets of results were highly correlated (r = 0.999), and the regression gave a slope of 0.966.

Grigg et al. [65] found a significant but small bias between automated and manual methods of 0.038 pH unit, which they showed to be due to the slow attainment of equilibrium. They concluded that this difference was of little practical importance, that 95% of the pH values determined lay within 0.05 unit of the true value, and that the precision of the automated method was thus of the same order as that to which the results were reported (0.1 pH unit).

VI. CONCLUSIONS

Automated methods for wet chemical analysis have now been adopted widely in agricultural and environmental laboratories. The most widespread application has been in the routine analysis of soil extracts and plant materials in connection with advisory and extension services. The first phase of the switch to automated methods concentrated on the use of segmented continuous-flow systems, but more recently, the greater speed of flow-injection analysis and the improved speed and greater flexibility possible with discrete systems have made them popular alternatives. Whichever automated system has been adopted in soil and plant testing or soil research laboratories, the experience has invariably been that both sample throughput and the quality and reproducibility of the data obtained are generally superior to that which could be obtained normally by traditional manual methods.

Automation is not confined to chemical procedures; data capture by computer (whether a dedicated personal computer or a larger system, via a link) has both reduced the need for manual computation and increased the reliability of analytical data.

To the factors of speed and reliability must be added that of cost. The cost, in real terms, of microprocessor-based instrumentation has fallen considerably in the last few years relative to that of manual operations. This trend may be expected to continue and lead to the ever more widespread use of automated systems for chemical analysis.

REFERENCES

1. L. T. Skeggs, *Am. J. Clin. Pathol.*, *28*:311 (1957).
2. J. Růžička and E. H. Hansen, *Anal. Chim. Acta*, *78*:145 (1975).
3. K. K. Stewart, G. R. Beecher, and P. E. Hare, *Anal. Biochem.*, *70*:167 (1976).
4. L. R. Snyder and H. J. Adler, *Anal. Chem.*, *48*:1017 (1976).
5. R. E. Thiers and K. M. Oglesby, *Clin. Chem.*, *19*:246 (1964).
6. R. E. Thiers, R. R. Cole, and W. J. Kirsch, *Clin. Chem.*, *13*:451 (1967).
7. R. E. Thiers, J. Meyn, and R. F. Wilderman, *Clin. Chem.*, *16*:832 (1970).

8. M. A. Evenson, G. P. Hicks, and R. E. Thiers, *Clin. Chem.*, *16*:606 (1970).
9. R. E. Thiers, A. H. Reed, and K. Delander, *Clin. Chem.*, *17*:42 (1971).
10. L. R. Snyder, *Anal. Chim. Acta, 114*:3 (1980).
11. L. R. Snyder, in *Advances in Automated Analysis, 1976*, Technicon International Congress, Mediad Press, Tarrytown, N.Y., 1977, p. 76.
12. L. R. Snyder and H. J. Adler, *Anal. Chem.*, *48*:1022 (1976).
13. L. R. Snyder, *J. Chromatogr.*, *125*:287 (1976).
14. L. R. Snyder, J. Levine, R. Stoy, and A. Conetta, *Anal. Chem.*, *48*:942A (1976).
15. G. Taylor, *Proc. Roy. Soc. Lond. A, 219*:186 (1953).
16. G. Taylor, *Proc. Roy. Soc. Lond. A, 223*:446 (1954).
17. R. Aris, *Proc. Roy. Soc. Lond. A, 235*:67 (1956).
18. J. T. Vanderslice, A. G. Rosenfeld, and G. R. Beecher, *Anal. Chim. Acta, 179*:119 (1986).
19. J. T. Vanderslice, K. K. Stewart, A. G. Rosenfeld, and D. G. Higgs, *Talanta, 28*:11 (1981).
20. V. Ananthakrishnan, W. N. Gill, and A. J. Barduhn, *J. Am. Inst. Chem. Eng.*, *11*:1063 (1965).
21. H. Bate, S. Rowlands, J. A. Sirs, and H. W. Thomas, *Brit. J. Appl. Phys.*, *2*:1447 (1969).
22. H. Bate, S. Rowlands, and J. A. Sirs, *J. Appl. Physiol.*, *34*: 866 (1973).
23. J. F. Tyson, *Anal. Chim. Acta, 179*:131 (1986).
24. C. B. Ranger, *Anal. Chem.*, *53*:20A (1981).
25. Technicon Instruments Corporation, Tarrytown, N.Y., Manual TN1 0170-01, 1972.
26. J. DiLiddo, A. Conetta, and W. Dorsheimer, Int. Lab., Oct. 1984, p. 63.
27. P. L. Searle, *N.Z. J. Agric. Res.*, *18*:183 (1975).
28. E. D. Wood, F. A. J. Armstrong, and F. A. Richards, *J. Mar. Biol. Assoc. U.K.*, *47*:23 (1967).
29. F. A. J. Armstrong, C. R. Stearns, and J. D. H. Strickland, *Deep Sea Res.*, *14*:381 (1967).
30. A. Henriksen and A. R. Selmer-Olsen, *Analyst (Lond.)*, *95*: 514 (1970).
31. A. B. Prince, in *Methods of Soil Analysis* (C. A. Black, ed.), Am. Soc. Agron., Madison, Wis., 1965, p. 866.
32. S. E. Allen, H. M. Grimshaw, J. A. Parkinson, and C. Quarmby, *Chemical Analysis of Ecological Materials*, Blackwells, Oxford, 1974.
33. W. D. Basson and R. G. Böhmer, *Analyst (Lond.)*, *97*:266 (1972).
34. A. G. Sinclair, *N.Z. J. Agric. Res.*, *16*:287 (1973).
35. O. R. Mulcahy, *Lab. Practice*, *26*:679 (1977).

36. G. M. Ogner and A. Haugen, *Analyst (Lond.)*, *102*:453 (1977).
37. G. Toennies and B. Bakay, *Anal. Chem.*, *25*:160 (1961).
38. J. R. Bettany and E. H. Halstead, *Can. J. Soil Sci.*, *52*:127 (1972).
39. W. B. Furman, *Continuous-Flow Analysis*, Marcel Dekker, New York, 1976.
40. P. L. Bailey, *Analysis with Ion-Selective Electrodes*, Heyden, London, 1978.
41. E. H. Hansen, A. K. Ghose, and J. Ruzicka, *Analyst (Lond.)*, *102*:705 (1977).
42. A. Bennet, D. Gartelmann, J. I. Mason, and J. A. Owen, *Clin. Chim. Acta*, *29*:161 (1970).
43. J. A. Nisbet and E. Simpson, *Clin. Chim. Acta*, *39*:339 (1972).
44. C. Riley and B. F. Rocks, *J. Automat. Chem.*, 5:1 (1983).
45. M. Koupparis and P. Anagnostopoulou, *J. Automat. Chem.*, *6*:186 (1984).
46. T. Greatorex and P. B. Smith, *J. Inst. Water Eng. Sci.*, *39*:81 (1985).
47. F. J. Krug, J. Růžička, and E. H. Hansen, *Analyst (Lond.)*, *104*:47 (1979).
48. H. Bergamin, E. A. G. Zagatto, F. J. Krug, and B. F. Reis, *Anal. Chim. Acta*, *101*:17 (1978).
49. J. Růžička and E. H. Hansen, *Flow Injection Analysis*, Wiley, New York, 1981.
50. A. M. G. Macdonald, H. L. Pardue, A. Townshend, and J. T. Clerc (eds.), "*Flow Analysis III*," Proc. Int. Conf., Birmingham, U.K., 1985, *Anal. Chim. Acta*, *179* (special issue) (1986).
51. W. R. Wolf and K. K. Stewart, *Anal. Chem.*, *51*:120 (1979).
52. N. Yoza, Y. Aoyogi, and S. Ohashi, *Anal. Chim. Acta*, *111*:163 (1979).
53. W. D. Basson, *Lab. Practice*, *26*:541 (1977).
54. J. F. Tyson, *Analyst (Lond.)*, *110*:419 (1985).
55. R. A. Sherwood, B. F. Rocks, and C. Riley, *Analyst (Lond.)*, *110*:493 (1985).
56. C. Riley, L. H. Aslett, B. F. Rocks, R. A. Sherwood, J. D. McK. Watson, and J. Morgan, *Clin. Chem.*, *29*:332 (1983).
57. R. H. Brown and S. R. Smith, U.K. Patent No. 2189597, 1988.
58. S. R. Smith, R. H. Brown, L. W. Chubb, and P. Hadley, *Lab. Practice*, *37*:99 (1988).
59. Heli-flow Information Sheet, WPA Ltd., Linton, Cambridge, England, 1988.
60. Z. Fang, S. Xu, X. Wang, and S. Zhang, *Anal. Chim. Acta*, *179*:325 (1986).
61. J. C. De Andrade, C. Pasquini, N. Ballan, and J. C. Van Loon, *Spectrochim. Acta, Part B*, *38*:1329 (1983).
62. O. Aström, *Anal. Chem.*, *54*:190 (1982).

63. K. F. Baker, *Analyst (Lond.)*, *95*:885 (1970).
64. D. Goodman, *Analyst (Lond.)*, *101*:943 (1976).
65. J. L. Grigg, H. J. Flewitt, G. A. Baird, R. B. Jordan, and K. V. Vo, *Analyst (Lond.)*, *105*:1 (1980).
66. G. E. Schuman, M. A. Stanley, and D. Knudsen, *Soil Sci. Soc. Am. Proc.*, *37*:480 (1973).
67. D. K. Markus, J. P. McKinnon, and A. F. Buccafuri, *Soil Sci. Soc. Am. J.*, *49*:1208 (1985).
68. A. Scott, unpublished results.
69. A. R. Selmer-Olsen, *Analyst (Lond.)*, *96*:565 (1971).
70. F. J. Adamsen, D. S. Bigelow, and G. R. Scott, *Commun. Soil Sci. Plant Anal.*, *16*:883 (1985).
71. S. Li and K. A. Smith, *Commun. Soil Sci. Plant Anal.*, *15*:1437 (1984).
72. A. P. Rowland, H. M. Grimshaw, and O. M. H. Rigaba, *Commun. Soil Sci. Plant Anal.*, *15*:337 (1984).
73. S. Ananth and J. T. Moraghan, *Soil Sci. Soc. Am. J.*, *51*:664 (1987).
74. R. H. Lowe and M. C. Gillespie, *J. Agric. Fd. Chem.*, *23*:783 (1975).
75. C. W. Rice, M. S. Smith, and J. M. Crutchfield, *Commun. Soil Sci. Plant Anal.*, *15*:663 (1984).
76. M. F. Gine, H. Bergamin, E. A. G. Zagatto, and B. F. Reis, *Anal. Chim. Acta*, *114*:191 (1980).
77. C. T. Cottrell, "*Automatic Chemistry Systems for Water Analysis*," Tech. Rept., Pye Unicam Ltd., Cambridge, England, 1978.
78. E. H. Hansen, A. K. Ghose, and J. Růžička, *Analyst (Lond.)*, *102*:705 (1977).
79. C. H. Orr, *Commun. Soil Sci. Plant Anal.*, *2*:85 (1971).
80. W. D. Basson, D. A. Stanton, and R. G. Böhmer, *Analyst (Lond.)*, *93*:166 (1968).
81. P. Crooks and A. Scott, unpublished results.
82. J. A. Varley, *Analyst (Lond.)*, *91*:116 (1966).
83. R. L. Flannery and D. K. Markus, *J. Assoc. Off. Anal. Chem.*, *63*:779 (1980).
84. S. K. Ng, *J. Sci. Fd. Agric.*, *21*:275 (1970).
85. S. McLeod and A. R. P. Clarke, *Analyst (Lond.)*, *103*:238 (1978).
86. J. L. Grigg, *Commun. Soil Sci. Plant Anal.*, *6*:95 (1975).
87. H. G. Zandstra, *Can. J. Soil Sci.*, *48*:219 (1968).
88. P. D. Salt, *Chem. Ind.*, *18*:584 (1968).
89. M. W. Brown, *Analyst (Lond.)*, *109*:469 (1984).
90. L. Zsoldos, *Hung. Sci. Instr.*, *39*:19 (1977).
91. K. Darab and E. Akos, *Agrokem Talajtan (Hungary)*, *21*:115 (1972); *Chem. Abs.*, *77*:150847 (1972).

92. J. R. Burrows and J. H. Meyer, *Proc. S. Afr. Sugar Technol. Assoc.*, *50*:114 (1976).
93. J. R. Burrows, *Proc. S. Afr. Sugar Technol. Assoc.*, *52*:195 (1978).
94. J. Hislop and I. G. Cooke, *Soil Sci.*, *105*:8 (1968).
95. O. E. Clinton, *Chem. N.Z.*, *43*:143 (1979).
96. M. Cooksey and W. B. Barnett, *At. Absorp. Newsletter*, *18*:1 (1979).
97. J. Alonso, J. Bartroli, J. L. F. C. Lima, and A. A. S. C. Machado, *Anal. Chim. Acta*, *179*:503 (1986).
98. L. A. Lancaster and R. Balasubramaniam, *J. Sci. Fd. Agric.*, *25*:381 (1974).
99. O. K. Borggaard and S. S. Jorgensen, *Analyst (Lond.)*, *110*: 177 (1985).
100. J. Keay, P. M. A. Menagé, and G. A. Dean, *Analyst (Lond.)*, *97*:897 (1972).
101. C. S. Cronan, *Anal. Chem.*, *51*:1333 (1979).
102. J. F. Coetzee and C. W. Gardner, *Anal. Chem.*, *58*:608 (1986).
103. H. Hara, G. Horvai, and E. Pungor, *Analyst (Lond.)*, *113*: 1817 (1988).
104. W. D. Basson, R. G. Böhmer, and D. A. Stanton, *Analyst (Lond.)*, *94*:1135 (1969).
105. S. R. Porter, S. C. Spindler, and A. E. Widdowson, *Commun. Soil Sci. Plant Anal.*, *12*:461 (1981).
106. A. L. Willis, *Commun. Soil Sci. Plant Anal.*, *1*:205 (1970).
107. G. Ogner, *Analyst (Lond.)*, *105*:916 (1980).
108. G. S. Pyen, M. J. Fishman, and A. G. Hedley, *Analyst (Lond.)*, *105*:657 (1980).
109. J. F. van Staden, *Analyst (Lond.)*, *112*:595 (1987).
110. H. van Vliet, W. D. Basson, and R. G. Böhmer, *Analyst (Lond.)*, *100*:405 (1975).
111. J. V. Purves, *Commun. Soil Sci. Plant Anal.*, *5*:261 (1974).
112. E. G. Bradfield and J. F. Stickland, *Analyst (Lond.)*, *100*:1 (1975).
113. S. S. Jorgensen and M. A. B. Regitano, *Analyst (Lond.)*, *105*: 292 (1980).
114. P. N. Vijan, A. C. Rayner, D. Sturgis, and G. R. Wood, *Anal. Chim. Acta*, *82*:329 (1976).
115. P. D. Goulden and P. Brooksbank, *Anal. Chem.*, *38*:1431 (1974).
116. H. Agemian and A. S. Y. Chau, *Analyst (Lond.)*, *101*:91 (1976).
117. A. M. Jirka and M. J. Carter, *Anal. Chem.*, *50*:91 (1978).

5

Ion Chromatography

M. ALI TABATABAI and NICHOLAS T. BASTA *Iowa State University, Ames, Iowa*

I. INTRODUCTION

Ion chromatography (IC) is a term that describes recent major advances in the determination of ions. Since its introduction in 1975 by Small et al. [1], research in the area of IC has made significant advances in separation and determination of ionic species. IC has become a rapid and sensitive technique for analyzing complex mixtures of ions. Now, ion chromatographs are available that feature high-speed separation and continuous monitoring by detector-analyzer systems, which yield the instantaneous readout of analytical data. Several books have been published on IC, including the development and use of its components, and the potential of this technique as an analytical tool [2–7]. The purpose of this chapter is to describe the IC instruments and methods developed for the determination of anions and cations in soils and soil solutions. The application of these methods to plant and water analysis will also be described. Several IC methods are available for the determination of ions other than those discussed in this chapter, but these methods have not been evaluated for soil analysis [5,6,8,9].

II. BASIC PRINCIPLES

Ion chromatography has its roots in pioneering work in the area of ion exchange, including the development of synthetic ion-exchange resins. A review of the work published on these topics is beyond the scope of this chapter, but information on the basic principles involved in the operation of ion chromatographs is presented.

Ion chromatographs can be divided into two major groups: those that operate on the principle of eluent suppression (dual-column system) and those with no suppressor column (single-column system). The electrical conductivity of a solution varies with the concentration of ions present, and this principle is the basis of a universal IC detector system used for the determination of all ionic species in solution.

A. Eluent-Suppressed Ion Chromatography

The only suppressor-type IC is marketed by Dionex Corporation (Sunnyvale, Calif.). Figure 1 shows the basic components of a suppressed-type ion chromatograph. For simplicity, the reservoirs of the eluent and water used for regeneration of the suppressor column and the valving system involved in the IC are not shown. The instrument employs the following components: (1) Eluent pump and reservoir, (2) sample injection valve (the sample loop can be adjusted from about 50 μl to several hundred μl), (3) ion-exchange separation column, (4) suppressor column coupled to conductivity detector, meter, and output device, and (5) regenerating pump with

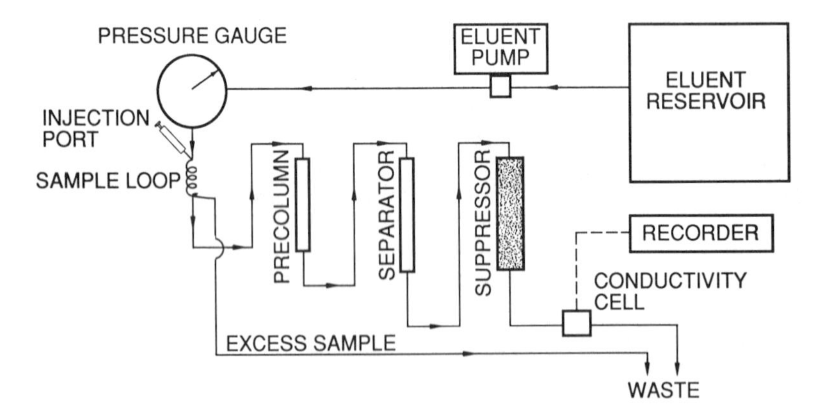

FIG. 1 Simplified schematic diagram of suppressed-type IC (Dionex Model 10). [Reprinted from Ref. 10, *Soil Sci. Soc. Am. J.*, *43*:899 (1979) by permission of the publisher, Soil Science Society of America.]

electronic timer and controls. Several types of columns are commercially available for the ion-exchange separation of the common inorganic and organic anions via eluent-suppressed IC. Information on the type of resin used and columns available from Dionex Corp. is summarized by Johnson [9].

In the eluent-suppressed IC, the ion species are resolved by conventional elution chromatography followed by passage through an eluent stripper, or "suppressor," column, wherein the eluent coming from the separating column is stripped or neutralized. Thus, only the ion species of interest leave the bottom of the suppressor column in a background of H_2CO_3, which exhibits a low conductivity, in the case of anions, or water in the case of cations, whereupon they are monitored subsequently in the conductivity cell/meter/recorder (integrator) combination. The eluent flow rate can be varied by adjusting the pump pressure, but normally it is about 2–3 ml min^{-1}. An aliquot (~2 ml) of a suspension-free soil extract is injected by a plastic syringe into the IC. The sample loop on the injection valve can be adjusted, but normally a volume of 100 μl is used. The 2-ml volume is convenient to insure proper flushing of the injection valve loop and lines.

The resin of the suppressor column must be regenerated after about 50 analyses (8–10 hr) by flushing the suppressor column with 0.5 *M* H_2SO_4 (15 min) in the case of anions or 1 *M* NaOH (15 min) in the case of cations, followed by deionized water (25 min). In some Dionex models (e.g., Model 10 IC) this can be accomplished without attending the instrument after each working day.

Table 1 shows the reactions in the separator and suppressor columns in the determination of anions, alkali metals, and alkaline earth metals. In the determination of anions, the IC is equipped with a separator column packed with a Dionex low-capacity, anion-exchange agglomerated resin in the HCO_3^- form, and the suppressor column contains a strong acid, high-capacity cation-exchange resin in the H^+ form. The eluent used normally is a mixture of dilute $NaHCO_3$ and Na_2CO_3, although other dilute mixtures (e.g., $NaHCO_3$ + NaOH, Na_2CO_3 + NaOH) are also used [9]. The anions are separated and converted to their strong acids in a background of H_2CO_3, which has low conductivity. The presence of strong acids in H_2CO_3 is measured by a conductivity cell and reported as peaks on a stripchart recorder or integrator. The peak height is directly proportional to the concentration of ions in solution. From calibration graphs prepared for peak height versus concentration of ions in standard solutions containing the ions of interest, the concentrations of the ionic species in the sample are calculated. The mixture of the standards can be prepared from reagent-grade chemicals. Figure 2 shows a tpical chromatogram of a standard solution containing 2 mg l^{-1} each of F^-, Cl^-, PO_4^{3-}–P, NO_3^-–N, and SO_4^{2-}–S. The retention times of these and other anions are listed in Table 2.

TABLE 1 Reactions of Separator and Suppressor Columns Used in Determination of Anion, Alkali Metals, and Alkaline Earth Metals by Ion Chromatography

Component	Reactions in separation of ion specified		
	Anion[a]	Alkali metal[b]	Alkaline earth metal[c]
Eluent	3 m*M* $NaHCO_3$ + 1.8 m*M* Na_2CO_3	5 m*M* HCl	2.5 m*M* HCl + 2.5 m*M* *m*-PDA·2HCl[d]
Displacing ion	HCO_3^-	H^+	*m*-$PDAH_2^{2+}$
Separator column			
Eluent	R–HCO_3 + $NaHCO_3 \rightleftharpoons$	R–H + HCl $\rightleftharpoons$	R–$PDAH_2$ + $PDAH_2^{2+}$ + $2Cl^- \rightleftharpoons$
	R–HCO_3 + $NaHCO_3$	R–H + HCl	R–$PDAH_2$ + $PDAH_2^{2+}$ + $2Cl^-$
Sample	R–HCO_3 + MA $\longrightarrow$	R–H + MA $\longrightarrow$	R–$PDAH_2$ + MA + $2Cl^- \longrightarrow$
	R–A + $MHCO_3$	R–M + HA	R–M + $PDAH_2^{2+}$ + $2Cl^-$ + A
	R–A + $NaHCO_3 \longrightarrow$	R–M + HCl $\longrightarrow$	R–M + $PDAH_2^{2+}$ + $2Cl^- \longrightarrow$
	R–HCO_3 + NaA	R–H + MCl	R–$PDAH_2$ + MCl_2
Suppressor column			
Eluent	R–H + $MHCO_3 \longrightarrow$	R–OH + HCl $\longrightarrow$	R–OH + $PDAH_2^{2+}$ + $2Cl^- \longrightarrow$
	R–M + H_2CO_3	R–Cl + H_2O	R–Cl + PDA + H_2O
Sample	R–H + NaA $\longrightarrow$	R–OH + MCl $\longrightarrow$	R–OH + $MCl_2 \longrightarrow$
	R–Na + HA	R–Cl + MOH	R–Cl + $M(OH)_2$

[a]M = Na, A = anion.
[b]M = alkali metal, A = associated anion.
[c]M = alkaline earth metal, A = associated anion.
[d]*m*-PDA·2HCl = *m*-phenylenediamine dihydrochloride.
Source: Ref. 11.

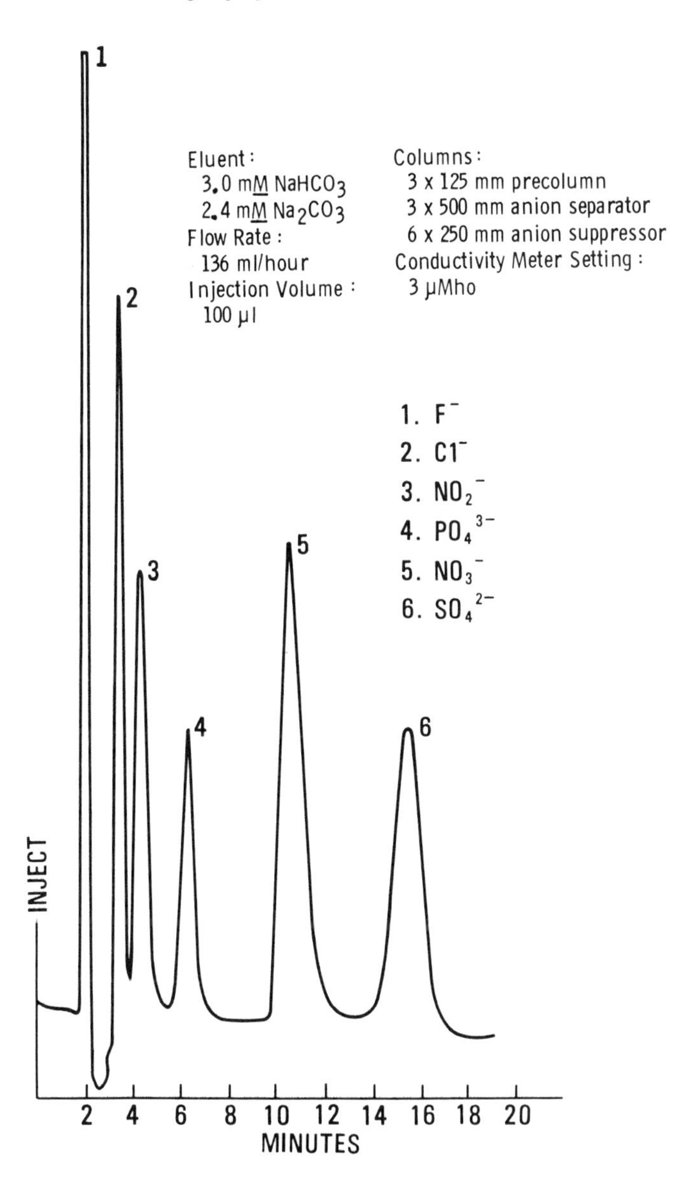

FIG. 2 Typical anion chromatogram of a solution containing 2 mg l^{-1} each of F^-, Cl^-, NO_2^-–N, PO_4^{3-}–P, NO_3^-–N, and SO_4^{2-}–S obtained by using a suppressed-type IC. [Reprinted from Ref. 10, *Soil Sci. Soc. Am. J.*, *43*:899 (1979) by permission of the publisher, Soil Science Society of America].

TABLE 2 Retention Times for Selected Anions

Anion or compound	Retention time (min): Precolumn and[a] 250-mm separator column	Retention time (min): Precolumn and[b] 500-mm separator column
$S_2O_3^{2-}$	1.2	ND
S^{2-}	1.3	ND
F^-	1.4	2.3
Cl^-	1.9	3.7
NO_2^-	2.3	4.5
PO_4^{3-}	3.4	6.5
NO_3^-	4.8	10.9
SO_3^{2-}	5.2	ND
$S_2O_5^{2-}$	5.2	ND
SO_4^{2-}	8.2	15.7
$S_4O_6^{2-}$	R	R
SCN^-	R	R
Thiourea	R	R
Glucose-6-sulfate	R	R
p-Nitrophenyl sulfate	R	R

[a]IC conditions: 125-mm precolumn connected to 250-mm separator column; eluent = 3.0 m*M* $NaHCO_3$ + 1.8 m*M* Na_2CO_3; flow rate = 3 ml min^{-1}; recorder = 3 μS full scale. R, retained by column.
[b]IC conditions: 125-mm precolumn connected to 500-mm separator column; eluent = 3.0 m*M* $NaHCO_3$ + 2.4 m*M* Na_2CO_3; flow rate = 2.25 ml min^{-1}; recorder = 3 μS full scale. ND, not determined. R, retained by column.
Source: Ref. 10.

In the determination of the alkali metals (Li, Na, K, Rb, and Cs), the separator column is a Dionex low-capacity, cation-exchange agglomerated polystyrene divinylbenzene copolymer cation resin in the H^+ form, and the suppressor column contains a strong base, high-capacity, anion-exchange resin in the OH^- form. The eluent is 5 m*M* HCl. The alkali metals are separated and converted to their hydroxides in a background of H_2O, which has very low conductivity. The conductivity of the metal hydroxides is measured by a conductivity cell and reported as peaks on a stripchart recorder or integrator. The reactions involved in the separator and suppressor columns are shown in Table 1. The separation and detection of the alkaline earth metals (Mg, Ca, Sr, and Ba) are similar to the procedure for the alkali metals, except that a mixture of 2.5 m*M* HCl + 2.5 m*M* *m*-phenylenediamine dihydrochloride is used as the eluent (Table 1). Typical chromatograms of standard solutions containing mixtures of Li, Na, K, Rb, and Cs, and Mg, Ca, Sr, and Ba, respectively, are shown in Fig. 3.

B. Single-Column Ion Chromatography

Two alternative methods to that described in Sec. II.A are now available for ion separation and determination. In both methods no suppressor column is needed (single-column system). Instead, moderately conducting eluents are used to elute a variety of ions. One technique is a variation of conventional HPLC, in which silica-based column packings provide ion separations. In a second similar approach, specially synthesized macroporous polystyrene-divinylbenzene resins with low capacities are coupled with moderate-conductivity mono- or polyvalent eluting ions [4]. Dedicated systems for single-column IC have been introduced recently by Wescan Instruments, Inc. (Santa Clara, Calif.), Hewlett-Packard Instruments, Inc. (Palo Alto, Calif.), and Brinkman Instruments, Inc. (Westbury, N.Y.). The instrument distributed by Brinkman is manufactured by Metrohm in Switzerland.

Figure 4 shows the basic components of a nonsuppressed-type (single-column system) ion chromatograph (SCIC). The technique employs the following components: (1) eluent pump and eluent reservoir, (2) sample injection valve (a sample loop of ≥500 μl is normally used), (3) ion-exchange separator column and conductivity detector coupled to an output device. In this system, a low-capacity exchange column and low-conductivity eluent are used without the need for a suppressor column [12–15]. Eliminating the suppressor column reduces the post-column dead volume, resulting in faster analyses, but the SCIC system is about 20 times less sensitive than the eluent-suppressed system. Appropriate low-capacity exchange columns used in the SCIC systems include a macroporous polystyrene divinylbenzene resin [13,14], or surface-quaternized silica [16]. Organic acids (phthalate, benzoate, or citrate) are often used in the mobile phase of SCIC [13,17], with phthalic acid being the most common because of its wide range of retention control (via pH adjustment)

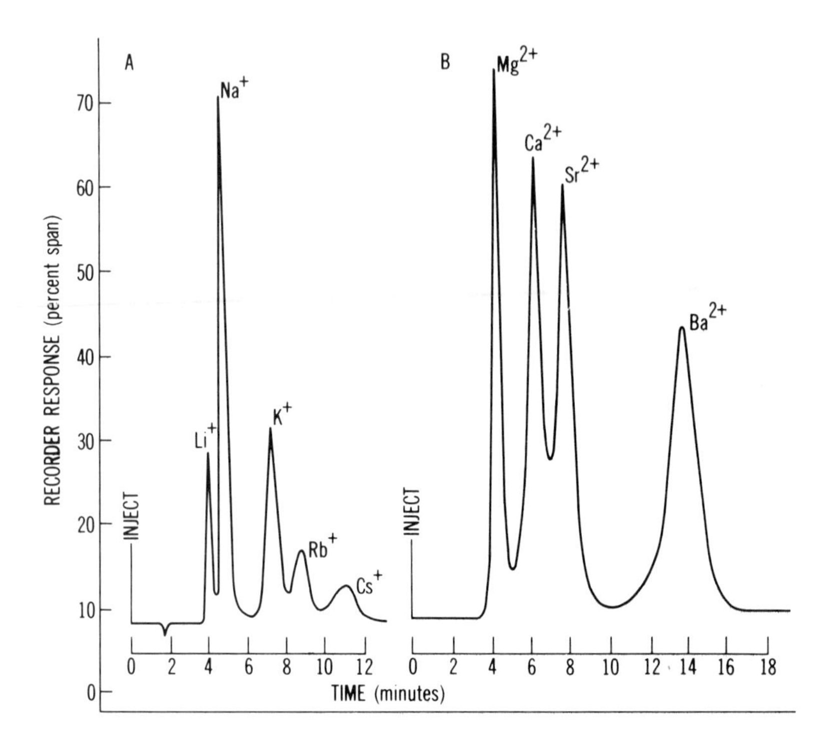

FIG. 3 Typical chromatograms of alkali and alkaline earth metals. (A) Alkali metals (2 mg l^{-1} each of Li^+, Na^+, K^+, Rb^+, and Cs^+). IC operation conditions: eluent 5 m*M* HCl; columns, 3 mm × 150 mm precolumns, 6 mm × 250 mm separator, 9 mm × 100 mm suppressor; flow rate, 2 ml min^{-1}; conductivity meter setting, 30 μS; injection volume, 100 μl. (B) Alkaline earth metals (1 mg l^{-1} Mg^{2+}, 2 mg l^{-1} Ca^{2+}, 5 mg l^{-1} Sr^{2+}, and 10 mg l^{-1} Ba^{2+}). IC operational conditions: eluent 2.5 m*M* HCl + 2.5 m*M* *m*-phenylenediamine dihydrochloride; columns, 4 mm × 50 mm precolumn, 3 mm × 250 mm separator, 9 mm × 100 mm suppressor; flow rate 2 ml min^{-1}; conductivity meter setting, 30 μS; injection volume, 100 μl. (From Ref. 11.)

and equivalent conductance [18]. Anions of interest elute in the hydrogen form (e.g., HCl, HNO_3, H_2SO_4) against a background of ionized phthalate ions. A number of equilibria affect SCIC. Buffer ions (usually weak acid ions) equilibrate with the free acid in solution. Both of these species, in turn, equilibrate with their bound forms at the surface of the stationary phase [17]. Details of the

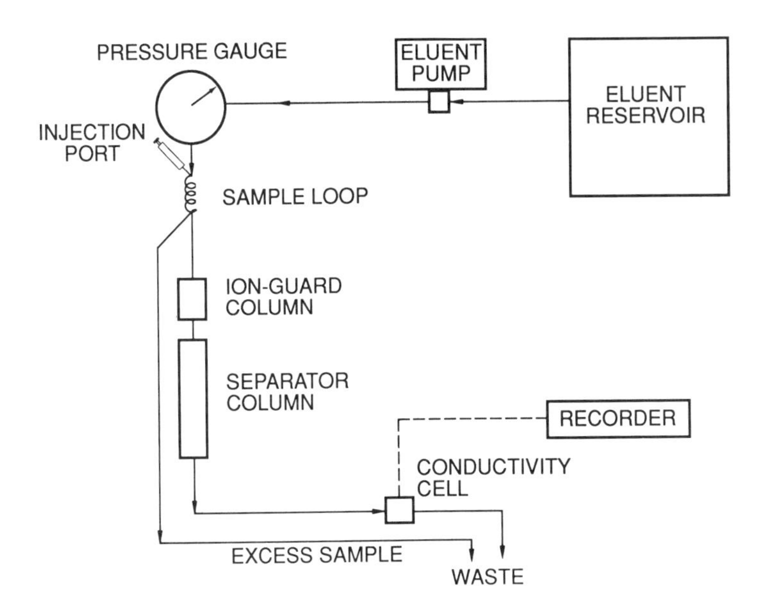

FIG. 4 Simplified schematic diagram of single-column-type IC.

reactions involved and factors affecting ion-exchange separations in the SCIC system, information on other types of separations, and column technology are presented by Jupille [17]. Most of these systems, however, have not been used for soil analysis.

For the determination of NH_4^+ and alkali metals, the mobile phase used in the SCIC system must have a strong affinity for the ion-exchange resin in order to displace separated ions from the analytical column. Maximum sensitivity is achieved when the equivalent conductance of the ionic species gives a detection signal well above the eluent background [19]. Dilute HNO_3 is used for the determination of NH_4^+ and the alkali metals [20]. Use of 10 m*M* HNO_3 (pH 2.1) has been shown to give excellent resolution of monovalent cations with elution complete in <6 min when a Vydac 401 TP cation-exchange column (Separation Group, Hesperia, Calif.) is used [21]. Ethylenediammonium dinitrate (5 m*M*, pH 6.1) competes more strongly with divalent cations in solution than does HNO_3, thus providing better resolution and peak symmetry for the divalent cations. Fritz et al. [20] recommended a solution pH of 6.1 so that all carbonic acid species will elute as bicarbonate and cause no interference with the analysis.

Background conductance and minimum detection limits of both alkali and alkaline earth metals increase with increasing concentration of the electrolyte mobile phase [22], whereas retention times decrease with increasing eluent concentration and decreasing resin capacity (k') [19]. The commercially available columns (e.g., Vydac 401 TP cation-exchange column) have relatively low k' [0.10 mol(-) kg^{-1}], although resins of even lower k' have been synthesized for chromatographical separation of ions [12,19,23–25].

C. Design and Operational Features

In a variant of the suppressor column system, the resin in the suppressor column is replaced by an ion-exchange membrane in tubular form to condition the eluent continuously [26]. This membrane (sulfonated polyethylene hollow fiber) acts exactly like the suppressor resin in that ions are exchanged from the membrane for ions in the eluent stream. The innovation is that the membrane is regenerated continuously by a gravity-fed flow of low-concentration H_2SO_4 (in analysis of anions) that continuously replaces the ions that are exchanged onto the fiber with ions from the regenerant. Thus, separate regeneration steps are eliminated. The replacement of the conventional ion-exchange resin bed suppressor column with the hollow fiber suppressor allows continuous operation of an IC without varying interference from baseline dips, ion-exclusion effects, or chemical reaction. Stevens et al. [26] concluded that conventional suppressor column systems have less band spreading than those using hollow fiber suppressors, and this results in slightly poorer resolution of early eluting ions with the latter type of eluent suppression technique. However, our experience is to the contrary. Furthermore, recent work by Weiss [2] showed that because of the low dead volume of a membrane suppressor, mixing and band broadening effects are minimized and the sensitivity is generally enhanced compared with the more traditional packed bed suppressor. Details of the theory of operation of the hollow fiber suppressor are discussed by Weiss [2], Stevens et al. [26], Hanaoka et al. [27], and Small [28]. This eluent-suppression system, however, has not been used with the IC instruments that have thus far been evaluated for soil analysis.

The columns used in eluent-suppressed-type IC were initially made of glass. Present columns are made of plastic. The performance of these columns is equivalent to or better than that of glass columns and there is no breakage. Typical plastic column internal diameters are from 4–9 mm; their lengths vary from 50–250 mm. Dionex Corporation is the sole distributor of these columns.

Columns in single-column-type (nonsuppressed) IC can be glass, plastic, or stainless steel. Most columns now in use are made of stainless steel. The phthalate and benzoate eluents used in SCIC have pH values ranging from 3–7; therefore, little corrosion is expected. The injection of samples with pH values higher than 7 is not advisable because the silica packing will degrade severely. Samples with pH values >7 are normally treated with eluent until the proper pH balance is achieved.

The commonly used detector is a high-sensitivity flow cell conductivity meter. The cell body is constructed of Kel-F plastic with an internal volume ranging from 2–6 μl. The electrodes are made of 316 stainless steel. The conductivity meter setting ranges from 0.1–1000 μS (Dionex Model 10) or up to 10,000 μS (Dionex Model 2002i). The conductivity meter setting commonly used, however, is 3, 10, or 30 μS. The meter signal is either displayed on a meter or digital readout. Conductivity measurements are quite sensitive to temperature fluctuations, and adequate temperature control is desirable. Some of the early eluent-suppressed IC instruments manufactured by Dionex Corp. (e.g., Model 10) did not have this temperature control, but this has been remedied in later models. To facilitate temperature compensation a thermistor is placed in the liquid line just after the electrode of the conductivity detector module (e.g., Dionex Model 2002i). The cell is driven by a high-frequency oscillator from the main circuit board. The cell output drives an amplifier, and changes in the ionic composition in the cell result in signal changes to the amplifier. The signal caused by the presence of conductive ions in the cell, after temperature compensation, results in meter and recorder-pen deflection.

The SCIC instruments have many of the components of eluent-suppressed-type instruments. The main difference between these two types of instruments lies in the column packing, and the lack of a regeneration pump and timer in the SCIC systems.

Many inorganic ions display strong absorbance in the lower range of UV. In the past, these wavelengths were not readily accessible to IC photometers. But now with the availability of UV detectors that reach down to 200 nm and less, inorganic anions (e.g., NO_3^-, NO_2^-, Br^-, I^-, BrO_3^-, IO_3^-, and $S_2O_3^{2-}$) can be determined [28]. Although NO_3^-, NO_2^-, and Br^- in such diverse environments as river and waste treatment waters, rain, eutectic salt mixtures, and saliva have been determined [29], little information is available on the use of UV detectors for the determination of these or other inorganic ions in soils. Also, ions such as SCN^-, $S_2O_3^{2-}$, and several polythionate species have been measured successfully by using low-capacity resins and $NaClO_4$ as an eluent. Cortes [30] used silica-based columns with amino functional groups for the effective separation of both organic and inorganic anions that are UV-absorbing.

One of the most important requirements of the IC technique is that the sample injected for analysis should be free of particulates. Loss of resolution can result from a contaminated precolumn or analytical column. Reproducibility may be affected by a contaminated column, an insufficiently conditioned column, or microbial growth in the eluents when stored at room temperature for several days, especially those used with single-column systems. Therefore, the eluent used with the single-column system should be prepared freshly on a daily basis. The precolumn, analytical column, and suppressor column can be used for several months. Degradation of the columns' resins can be detected easily from inconsistent peak heights and lack of peak resolution. The relative retention time for the eluent-suppressed IC system is affected by the eluent composition, and for the single-column system is affected by the pH and ionic strength of the mobile phase. An increase in the ionic strength and pH of the eluent causes the solute retention time to decrease. In general, retention of ionic species is directly proportional to column length and inversely related to eluent flow rate. Increasing the column length generally results in greater resolution of the solute; however, the time required for analysis is increased [10,31]. Analysis time is decreased at high flow rates, but this can lead to poor resolution with overlapping peaks. Analyte retention time is also indirectly proportional to the concentration of the sample injected, but this effect is minor at low analyte concentrations [1,22]. Another factor that significantly affects peak height, peak resolution, and reproducibility of results is temperature (approximately 2% $°C^{-1}$ for peak height). Temperature variations may also cause changes in retention time and produce baseline drift. Waterbaths, jackets, or column heaters may be used to eliminate fluctuations in laboratory temperature, but these are costly and difficult to operate at room temperature with small columns. In most situations, this fluctuation in laboratory temperature can be overcome easily by running standard samples more often during the work day or placing the instrument in an air-conditioned room to eliminate severe fluctuations in daily temperature. The sensitivity of the IC system can be adjusted by changing the range of the conductivity detector and/or sample size (sample loop).

In addition to the instruments described in this section, other IC instruments are available from Dionex Corp. that involve postcolumn reaction systems for the determination of polyphosphates and the transition metals in aqueous solutions. No information is available, however, on the use of these instruments for soil and plant analysis.

Several advanced eluent-suppressed IC models are manufactured by Dionex Corp., but these instruments have not been evaluated for soil analysis. In using any IC instrument, the operator must be familiar with the principle of operation and the reactions involved. In addition to those reactions, knowledge of the sample components is very useful. Clean-up procedures for most IC instruments are provided by the manufacturers. Although most of these procedures are not difficult to perform, experience with the IC system and familiarity with functions of its components are very helpful.

III. APPLICATIONS

Application of IC to soil analysis was pioneered by the senior author and his associates at Iowa State University in the late 1970s [10]. Since then several papers have appeared in the soil science literature on the use of suppressed and nonsuppressed IC systems for the determination of anions and cations in soil solutions and exchangeable bases in soils. Some of these methods have been applied successfully to plant and water analysis. The IC system should be useful for a variety of methods used in soil and plant analysis, provided that the reagents used in the procedures are compatible with the basic principles of operation of the IC. As such, many of the current methods used in soil and plant analysis produce ionic species in a background of either highly acidic media or high salt concentrations. Consequently, new approaches or modifications of current methods are essential before using the IC system for the determination of the ionic species produced. Therefore, in this section, the IC systems that have been evaluated for the analysis of soils will be discussed and the application of the method to plant and water analysis will be integrated into this discussion.

A. Anions

The first report on the application of IC for the determination of anions (NO_3^- and SO_4^{2-}) in soils is that by Dick and Tabatabai [10]. In this work, these anions were extracted with water or salt solutions and determined by using a Dionex Model 10 IC (Dionex Corp., Sunnyvale, Calif.), which basically is a low-pressure dual-column (suppressed-type IC system) ion chromatograph, and the results obtained by IC were compared with those obtained by the steam distillation method for NO_3^- and methylene blue method for SO_4^{2-}. The system involved a separator column (3 mm by 250 mm) packed with a Dionex low-capacity anion-agglomerated resin (in addition to this column, a precolumn, 3 mm by 125 mm, containing the same resin was used to protect the separator column by removing particulates and other potentially poisonous substances from the eluent stream) and a suppressor column (6 mm by 250 mm). The eluent was 3.0 m*M* $NaHCO_3$ + 1.8 m*M* Na_2CO_3 at a flow rate of 3 ml min^{-1} and pump pressure of 3.1 MPa (450 psi). The sample loop on the injection valve contained a volume of 100 μl. Typical chromatograms obtained for soil extracts are shown in Fig. 5.

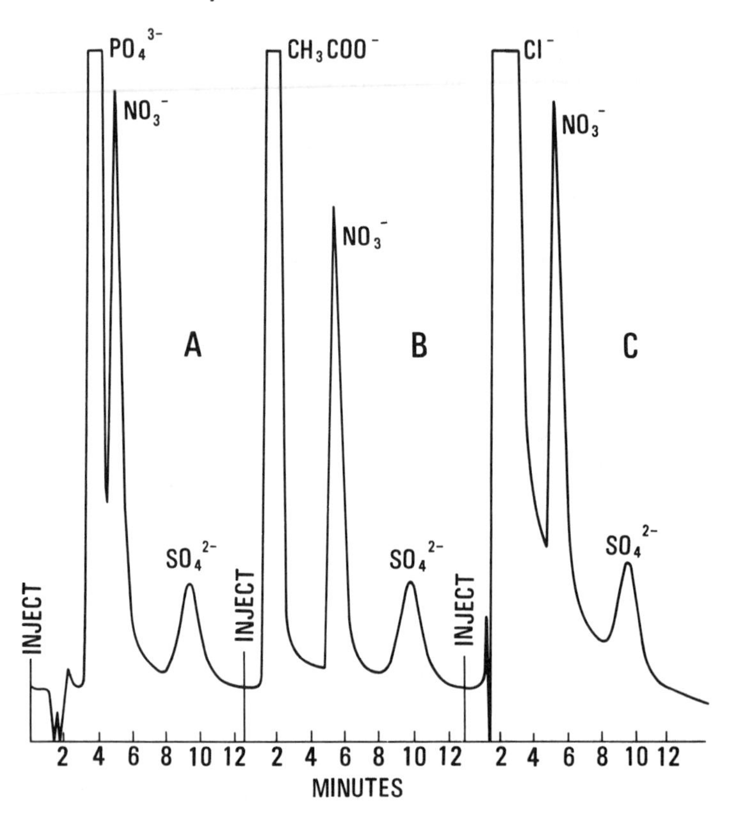

FIG. 5 Typical anion chromatogram of soil extractants containing 4 mg l^{-1} NO_3^-—N and 1.5 mg l^{-1} SO_4^{2-}—S. The extractants were (A) 100 mg l^{-1} P as $Ca(H_2PO_4)_2$, (B) 10 m*M* $Ca(C_2H_3O_2)_2$, (C) 0.15% $CaCl_2$. [Reprinted from Ref. 10, *Soil Sci. Soc. Am. J.*, *43*:899 (1979) by permission of the publisher, Soil Science Society of America.]

Tests of the effect of extractants on the NO_3^-—N values obtained for soils showed that the results by the IC method were in close agreement with those obtained by the steam distillation method

TABLE 3 Comparison of Nitrate N in Soils Obtained by Ion Chromatographic and Steam Distillation Methods

Soil	Method[a]	Nitrate N extracted by reagent specified ($mg\ kg^{-1}$ soil)[b]						
		A	B	C	D	E	F	G
Ackmore	IC	61.8	59.9	60.5	60.1	59.8	59.7	67.6
	SD	60.2	60.2	59.8	59.1	60.6	60.3	68.6
Downs	IC	158	157	159	160	159	160	160
	SD	157	156	157	158	158	157	168
Webster	IC	52.3	53.5	52.9	53.1	52.8	53.2	53.3
	SD	51.9	52.5	51.7	53.6	52.5	52.8	52.6
Canisteo	IC	34.4	35.1	35.2	34.6	34.5	34.4	35.1
	SD	34.0	34.5	34.8	35.1	33.9	33.8	34.5
Okoboji	IC	45.1	45.5	45.6	45.5	45.1	44.9	45.2
	SD	45.6	44.9	45.4	44.8	45.2	45.3	44.7
Mean	ID	70.3	70.4	70.6	70.7	70.2	70.4	73.4
	SD	69.7	69.6	69.7	70.1	70.0	69.8	73.7

[a]SD = steam distillation.
[b]A, water; B, 10 m*M* LiCl; C, 10 m*M* KCl; D, 0.15% $CaCl_2$; E, 10 m*M* $Ca(C_2H_3O_2)_2$; F, 100 mg l^{-1} P as $Ca(H_2PO_4)_2$; G, 3.0 m*M* $NaHCO_3$ + 1.8 m*M* Na_2CO_3.
Source: Ref. 10.

(Table 3). Further evidence that the IC method is satisfactory for the determination of NO_3^-—N in soil extracts was obtained by showing that the IC method gave quantitative recovery of NO_3^-—N added to soils [10]. The results obtained by this method are very precise (Table 4).

Studies of the effect of extractants on the SO_4^{2-}—S values obtained for soils showed that the results by the IC method are in close agreement with those obtained by the methylene blue method (Table 5) and that H_2O, 10 m*M* LiCl, 10 m*M* KCl, 0.15% $CaCl_2$, 10 m*M* $Ca(C_2H_3O_2)_2$, and a solution containing 100 mg l^{-1} P as $Ca(H_2PO_4)_2$ extract almost identical amounts of SO_4^{2-}—S from soils. Further evidence that the IC method is satisfactory for the determination of SO_4^{2-}—S in soil extracts was obtained by showing that the recovery of SO_4^{2-} added to soils is quantatitive [10]. The precision of the results obtained by this method is shown in Table 6.

TABLE 4 Precision of IC Method for Analysis of NO_3^-–N in Soils

Soil	Extractant[a]	Nitrate N (mg kg^{-1} soil)[b]			
		Range	Mean	SD	CV (%)[c]
Muscatine	A	29.7–30.8	30.1	0.16	0.5
	B	29.9–31.2	30.8	0.17	0.6
	C	29.5–31.0	30.2	0.15	0.5
Tama	A	169–173	171.7	1.95	1.1
	B	170–175	173.0	1.96	1.1
	C	170–174	172.1	1.95	1.1
Okoboji	A	44.6–47.3	45.5	0.20	0.4
	B	44.2–46.2	44.5	0.25	0.6
	C	44.9–46.5	45.7	0.26	0.6

[a]A, 0.15% $CaCl_2$; B, 10 m*M* $Ca(C_2H_3O_2)_2$; C, 100 mg l^{-1} P as $Ca(H_2PO_4)_2$.
[b]Results of seven analyses of each soil extract; SD = standard deviation.
[c]CV = coefficient of variation.
Source: Ref. 10.

Busman et al. [32] used a Dionex model 10 IC for the simultaneous determination of total S and Cl in plant samples after combustion of the sample in a Schöniger-type oxygen flask containing deionized water (the S and Cl gases produced were absorbed in water, producing SO_4^{2-} and Cl^-). The IC was equipped with a precolumn and separator column (each 4 mm by 50 mm, Dionex concentrator columns), and a suppressor column (9 mm by 100 mm). The eluent was 3.0 m*M* $NaHCO_3$ + 1.8 m*M* Na_2CO_3 pumped at a flow rate of 2.5 ml min^{-1} and a pump pressure of 3.1 MPa (450 psi).

The average total S values of 15 plant samples by the IC method and by the methylene blue method after digestion with NaOBr were 0.255 and 0.259%, respectively. Comparison of the results obtained for Cl by the IC method and by a colorimetric method involving the use of mercuric thiocyanate and ferric ammonium sulfate showed that the average Cl values for the plant samples were 0.380 and 0.377%, respectively [32]. Busman et al. [32] reported, however, that the recovery of total N and P by the IC method after combustion in an oxygen flask was not quantitative. This is expected, because combustion of organic N compounds results in the production of dinitrogen (N_2) and nitrogen oxides (NO_X), including nitrous oxide (N_2O). The N_2 and N_2O gases produced are inert and are not soluble in

TABLE 5 Comparison of Sulfate S in Soils Obtained by Ion Chromatographic and Methylene Blue Methods

		Sulfate S extracted by reagent specified ($mg\ kg^{-1}$ soil)[b]						
Soil	Method[a]	A	B	C	D	E	F	G
Ackmore	IC	5.6	5.5	5.7	5.6	5.5	5.6	5.6
	MB	5.5	5.7	5.5	5.5	5.7	5.7	5.5
Downs	IC	18	18	17	18	17	18	24
	MB	17	18	18	17	18	17	23
Webster	IC	5.2	5.1	5.0	5.2	5.2	5.1	6.8
	MB	5.1	5.2	5.2	5.0	4.9	5.0	6.7
Canisteo	IC	8.6	8.8	8.6	8.7	8.8	8.6	8.8
	MB	8.7	8.7	8.8	8.6	8.7	8.8	8.6
Okoboji	IC	8.9	9.0	9.0	8.9	9.0	8.9	9.0
	MB	8.8	8.9	9.0	8.8	8.9	9.0	8.9
Mean	IC	9.3	9.3	9.1	9.3	9.1	9.2	10.8
	MB	9.0	9.3	9.3	9.0	9.2	9.1	10.5

[a]MB = methylene blue.
[b]A, water; B, 10 m*M* LiCl; C, 10 m*M* KCl; D, 0.15% $CaCl_2$; E, 10 m*M* $Ca(C_2H_3O_2)_2$; F, 100 mg l^{-1} P as $Ca(H_2PO_4)_2$; G, 3.0 m*M* $NaHCO_3$ + 1.8 m*M* Na_2CO_3.
Source: Ref. 10.

water. Therefore, quantitative recovery is not expected for total N by the ignition and IC procedure. In addition to N_2 and N_2O, however, the ignition of plant material would produce NO and NO_2, which are highly soluble in water. When dissolved in water, these gases produce HNO_2 and HNO_3, respectively. The anions of these acids can be determined by the IC instrument described.

Busman et al. [32] reported that the total P values obtained by the ignition of plant samples and subsequent analysis by the IC procedure were very low as compared with those obtained by the molybdenum blue method after oxidation with NaOBr. They examined the fate of the unrecovered P by this IC procedure and concluded that it seems either that the plant P was oxidized to polyphosphate(s), which are not detectable by the IC method (polyphosphates are retained by the IC analytical column), or that insoluble metal phosphates (in the ash) are produced on the ignition of plant materials.

TABLE 6 Precision of IC Method for Analysis of Sulfate S in Soils

Soil	Extractant[a]	Sulfate S ($mg\ kg^{-1}$ soil)[b]			CV (%)[c]
		Range	Mean	SD	
Muscatine	A	4.3–4.6	4.5	0.10	2.2
	B	4.3–4.5	4.5	0.08	1.8
	C	4.1–4.3	4.2	0.10	2.5
Tama	A	10.9–11.8	11.7	0.37	3.1
	B	10.9–11.8	11.4	0.49	4.3
	C	10.9–11.8	11.5	0.37	3.2
Okoboji	A	8.7–8.9	8.6	0.15	1.7
	B	8.5–9.0	8.7	0.20	2.3
	C	8.7–9.1	8.5	0.16	1.9

[a]A, 0.15% $CaCl_2$; B, 10 m*M* $Ca(C_2H_3O_2)_2$; C, 100 mg l^{-1} P as $Ca(H_2PO_4)_2$.
[b]Results of seven analyses of each soil extract; SD = standard deviation.
[c]CV = coefficient of variation.
Source: Ref. 10.

In more recent work, Tabatabai et al. [33] developed a method for the determination of total S in soils and plant materials by using a Dionex Model 2002i IC. In this method, the sample was ignited with $NaHCO_3$ and Ag_2O at 550°C for 3 hr. The ignition mixture was dissolved in acetic acid, and an aliquot was injected into the IC instrument after filtration through a 0.45 μm Metrical GA-8 membrane filter. The Dionex 2002i employed is a high-pressure liquid chromatograph. This instrument was equipped with a precolumn (AG-3, 3 mm by 50 mm) and a separator column (AS-3, 3 mm by 250 mm), both packed with a low-capacity anion-exchange resin, and a suppressor column (9 mm by 100 mm) packed with a high-capacity cation-exchange resin in the H^+ form. The injection valve was attached to a 50-μl sample loop, and the eluent was 3.0 m*M* $NaHCO_3$ + 2.4 m*M* Na_2CO_3 at a flow rate of 2.3 ml min^{-1} and pump pressure of 4.8 MPa (700 psi). The mode of detection was conductimetric by using a Dionex conductivity detector module. This detector consists of a conductivity cell fitted with a thermistor and microprocessor-based detection system that allows temperature compensation. The conductivity cell output was measured with a Perkin-Elmer Laboratory Computing Integrator (LCI-100). Tabatabai et al. [33] reported that the total S values in soils and plant materials agreed closely with those obtained by a methylene blue method after alkaline oxidation with NaOBr.

Kalbasi and Tabatabai [34] developed a method for the determination of NO_3^-, Cl^-, SO_4^{2-}, and PO_4^{3-} in water extracts of plant material by eluent-suppressed IC. They used a Dionex Model 10 IC instrument equipped with a precolumn (3 mm by 40 mm, No. 022596), separator column (3 mm by 240 mm, HPLC-AS-3, P/N 030985), suppressor column (9 mm by 80 mm, No. 025670), and dual pen recorder (Honeywell) for the independent recording of peaks at a conductivity setting of 30 μS full scale. The eluent was 3.0 m*M* $NaHCO_3$ + 1.8 m*M* Na_2CO_3, with a flow rate of 2.1 ml min^{-1} and a pump pressure of 3.4 MPa (500 psi). The averages of 10 plant samples by this IC method agreed closely with those by the steam-distillation method for NO_3^-—N (0.158 versus 0.164%), with the titrimetric method using $AgNO_3$ and K_2CrO_4 for Cl^- (1.22 versus 1.19%), and with the colorimetric heteropoly blue method for PO_4^{3-}—P (0.335 versus 0.330%). The average SO_4^{2-}—S value of the 10 plant samples by the IC method (0.067%) was somewhat lower than the corresponding value (0.075%) obtained by the reduction to H_2S and colorimetric determination as methylene blue. This disagreement between the average values of SO_4^{2-}—S is expected because the reagent used for the reduction of SO_4^{2-}—S to H_2S in the methylene blue method is not specified for SO_4^{2-}; it reduces a variety of inorganic and organic S compounds. In addition to these anions, Kalbasi and Tabatabai [34] reported the separation and determination of malic acid by the same IC system. They also showed that a peak for F^- was eluted with an unidentified organic acid. Therefore, the amount of water-soluble F^- in plant materials could not be determined accurately.

Work by Tabatabai and Dick [31] and Krupa and Tabatabai [35] showed that the suppressed IC (Dionex Model 10) gives accurate and precise results in the determination of NO_3^-, Cl^-, SO_4^{2-}, and PO_4^{3-} in natural waters.

Nieto and Frankenberger [36] evaluated a single-column ion chromatograph for the determination of Cl^-, NO_2^-, NO_3^-, and SO_4^{2-} in soil extracts obtained with water or salt solutions containing LiCl, KCl, $Ca(C_2H_3O_2)_2$, or $CaCl_2$. They showed that, in addition to these anions, SO_3^{2-}, Br^-. I^-, and ClO_4^- can be separated within 10 min. These anions, however, were not detected in soils.

The SCIC system that they used was based on a commercially available HPLC instrument, by using a 500-μl injection loop and an eluent containing 4 m*M* phthalic acid (pH 4.5). The anion elution sequence corresponds to the following order:

$$Cl^- > NO_2^- > Br^- > NO_3^- > I^- > ClO_4^- > SO_4^{2-} > SO_3^{2-}$$

These SCIC conditions, however, do not detect PO_4^{3-}. The inorganic anions were quantified by using a conductivity detector. The results by this IC system were close to those obtained by conventional methods [36]. When a 500-μl injection loop is used, this single-column technique has detection limits, expressed in mg l^{-1} of soil extract, as follows: Cl^-, Br^-, NO_3^-–N, and SO_4^{2-}–S (0.025), ClO_4^-, and I^- (0.5), and NO_2^-–N and SO_3^{2-}–S (1.0). This technique proved reproducible for determination of anions in soil extracts as is evident from the low coefficient of variation values reported, which ranged from 3.6–8.4%.

Because the SCIC conditions described above do not detect PO_4^{3-}, Karlson and Frankenberger [37] modified these conditions for the determination of this form of P in aqueous extracts of soils. The SCIC system that they used consisted of a Beckman model 332 HPLC equipped with a Model 110A pump and Model 210 sample injector. This system was connected to a Vydac 3021C4.6 anion-exchange column (4.6 mm by 250 mm), a Wescan 269-003 ion-guard column (4.6 mm by 40 mm), an Eldex Model III thermostat column heater, a Wescan Model 213 conductivity detector, and a Hewlett-Packard Model 3390A integrator with variable input voltage. The column was maintained at 27°C throughout the analysis. The eluent stream consisted of 1.5 m*M* phthalic acid adjusted to pH 2.7 with formic acid. The method allows precise (relative standard deviation = 1.1 to 3.3%) measurements of trace amounts of orthophosphate in the presence of high background levels of Cl^- and NO_3^-. Karlson and Frankenberger showed that orthophosphate values obtained by the SCIC system agreed closely with those by an Autoanalyzer based on the Mo blue chromophore reaction. Under these SCIC conditions, the elution of orthophosphate occurred after 6 min, whereas Cl^- and NO_3^- had respective retention times of 11 and 20 min.

Recently, Barak and Chen [38] used a SCIC system for the determination of Cl^-, NO_3^-, and SO_4^{2-} in 3 min by using 15 m*M* phthalic acid (pH = 2.5) as an eluent and a guard column as an analytical column. The equipment configuration that they used consisted of a Perkin-Elmer Series 10 Liquid Chromatograph pump, 50-μl sample injection loop, Wescan ion-guard anion cartridge (269-003, 4.6 mm by 30 mm), Wescan ion-guard holder (269-002), Jasca Uvidec-100-V UV spectrophotometer with a flowthrough cell (10-mm optical path), and LDC/Milton Roy CI-10 computing integrator. The eluent flow rate was 5 ml min^{-1}, producing a back pressure of 7×10^5 kg m^{-2} or less. The detection was based on UV absorbance at 300 nm.

Although no results for soils were presented, Barak and Chen [38] claimed that they used the above system for the analysis of soil extracts prepared from saturated pastes. Concentration ranges that can be determined by this IC system are to a large extent a function of sample injection loop size, which thereby determines the amount of analyte injected. In addition to the 50-μl loop, Barak and Chen [38] used loop sizes ranging from 6 μl for an extremely saline soil extract to 200 μl for rainwater analysis. They recommended the routine measurement of electrical conductivity in order to estimate total anion concentration and thereby ensure that the sample falls within the range of the calibration graph. They reported that hundreds of analyses were performed on a single column over periods of up to 3 months, and they attributed the relative longevity of the guard column used to the fact that, at pH 2.5, the organic contaminants, including humic substances, are almost uncharged, it resulting in much less adsorption in the resin.

B. Cations

The IC technique can be used for the simultaneous determination of NH_4^+ and alkali and alkaline earth metals in soil and plant extracts and water samples provided that the aliquot analyzed is free from interfering substances such as organic materials, high concentrations of soluble salts, and extreme pH values. Both the eluent-suppressed and single-column IC systems have been evaluated for the determination of these metals in soils, plant materials, and natural waters.

Basta and Tabatabai [39] developed a method of determination of exchangeable bases in soils by an eluent-suppressed IC system equipped with a precolumn, separator column, and suppressor column. Separate precolumns and separator columns were employed for the determination of alkali (monovalent) and alkaline earth (divalent) metals. A 3 mm by 150 mm precolumn and 6 mm by 250 mm separator column were used for the determination of the alkali metals, and a 4 mm by 50 mm precolumn together with 3 mm by 250 mm separator column for the alkaline earth metals. A separate suppressor column (9 mm by 100 mm) was used for the determination of each of the monovalent and divalent metals. The injection valve contained a 100 μl sample loop. The eluent for the determination of the monovalent bases was 5 m*M* HCl at a flow rate of 3 ml min^{-1} and pump pressure of 2.8 MPa (400 psi). The eluent for the determination of the divalent bases was 2.5 m*M* HCl + 2.5 m*M* *m*-phenylenediamine dihydrochloride [$C_6H_4(NH_2)_2 \cdot 2HCl$] at a flow rate of 2 ml min^{-1} and pump pressure of 3.4 MPa (500 psi).

This method [39] involves extraction of the exchangeable bases with neutral 1 *M* NH_4OAc, followed by ignition of the soil extract at 400°C for 30 min, dissolution of the residue in 5 m*M* HCl, and determination of the exchangeable bases by using a Dionex Model 10 IC equipped as described above. Typical chromatograms obtained for such extracts are shown in Figs. 6 and 7. Results obtained for Na and K by this IC method were in close agreement with those

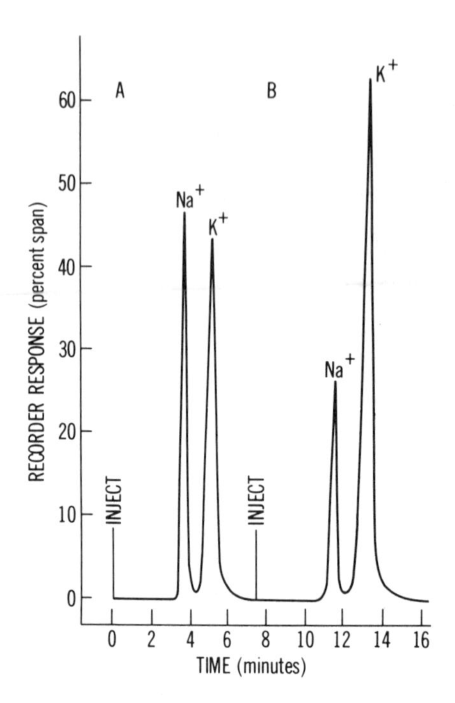

FIG. 6 Typical chromatogram of monovalent exchangeable bases in soils. (A) Nicollet soil, (B) Webster soil. For IC operational conditions, see Fig. 3's caption. [Reprinted from Ref. 39, *Soil Sci. Soc. Am. J.*, *49*:84 (1985) by permission of the publisher, Soil Science Society of America.]

obtained by both atomic emission spectrometry (AES) and atomic absorption spectrometry (AAS) (Table 7). Equally good agreement was obtained for Ca and Mg with results by AAS (Table 8). The eluent-suppressed IC method has the same degree of precision as those of AAS and AES (Tables 9 and 10). This IC method has an excellent sensitivity with detection limits of 0.1, 0.05, 0.1, and 0.03 mg l^{-1} for K, Na, Ca, and Mg, respectively. With this method, K and Na can be determined simultaneously in 6 min, and Ca and Mg can be determined in 7 min.

Basta and Tabatabai [40] applied the eluent-suppressed IC system described above to the determination of total K, Na, Ca, and Mg in plant materials. This method involves dry ashing approximately 0.1 g of plant material, followed by the dissolution of the residue in 5 m*M* HCl and the determination of K, Na, Ca, and Mg

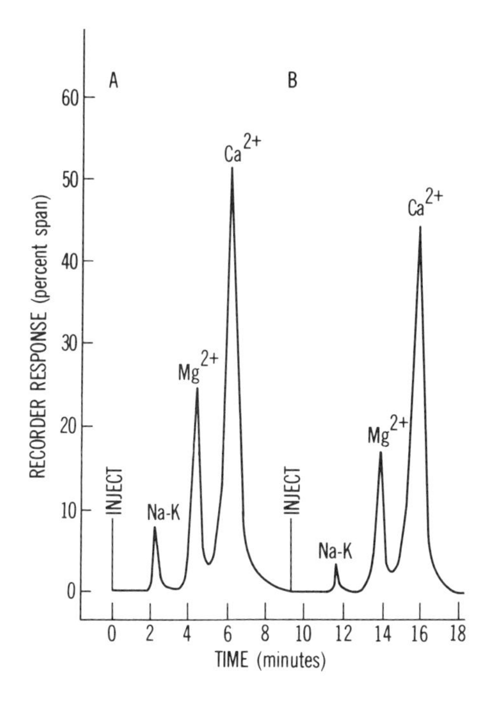

FIG. 7 Typical chromatograms of divalent exchangeable bases in soils. (A) Nicollet soil, (B) Webster soil. For IC operational conditions, see Fig. 3's caption. [Reprinted from Ref. 39, *Soil Sci. Soc. Am. J.*, *49*:84 (1985) by permission of the publisher, Soil Science Society of America.]

in the digest by using a Dionex Model 10 IC under the same conditions described for the determination of exchangeable bases in soils [39]. They showed that the average total K values in 25 plant materials by this IC method, atomic absorption spectrometry (AAS), and atomic emission spectrometry (AES) were 3.08, 3.08, and 3.07%, respectively. The corresponding average total Na values were 0.258, 0.264, and 0.256%. The average total Ca values in the same plant samples by the IC and AAS methods were 1.20 and 1.19%, respectively. The corresponding average total Mg values were 0.327 and 0.331% [40].

Recent work by Basta and Tabatabai [41] showed that the eluent-suppressed IC system described above is also readily applicable to the determination of K, Na, Ca, and Mg in natural waters.

TABLE 7 Comparison of Exchangeable K and Na Values Obtained by IC Method and by AAS and AES Methods

Soil	Base	Exchangeable K or Na obtained by the method specified [cmol (i) kg^{-1} soil][a]		
		IC	AAS	AES
Hagener	K	0.37	0.37(0.35)	0.37(0.36)
	Na	0.04	0.04(0.03)	0.04(0.03)
Hayden	K	0.28	0.32(0.29)	0.27(0.26)
	Na	0.10	0.12	0.09
Clarion	K	0.26	0.23(0.23)	0.24(0.24)
	Na	0.07	0.07	0.07
Marshall	K	0.71	0.64(0.63)	0.66(0.65)
	Na	0.04	0.03(0.03)	0.04(0.04)
Nicollet	K	0.24	0.23(0.25)	0.24(0.25)
	Na	0.21	0.22(0.19)	0.22(0.19)
Fayette	K	0.66	0.61(0.59)	0.61(0.60)
	Na	0.07	0.10	0.08
Muscatine	K	0.59	0.52(0.53)	0.53(0.54)
	Na	0.03	0.03	0.02
Tama	K	0.63	0.62(0.61)	0.61(0.62)
	Na	0.13	0.09	0.11
Webster	K	0.48	0.47(0.50)	0.49(0.51)
	Na	0.08	0.10	0.08
Lester	K	0.64	0.66(0.68)	0.65(0.68)
	Na	0.08	0.09	0.08
Average	K	0.49	0.47	0.47
	Na	0.09	0.09	0.08

[a]Figures in parentheses are exchangeable K or Na values obtained by analyzing directly the NH_4OAc extract of soil by the method indicated.
Source: Ref. 39.

TABLE 8 Comparison of Exchangeable Ca and Mg Values Obtained by IC method and by AAS Method

		Exchangeable Ca or Mg obtained by the method specified [cmol (i) kg^{-1} soil][a]	
Soil	Base	IC	AAS
Hagener	Ca	6.86	6.71(6.44)
	Mg	1.44	1.45(1.49)
Heyden	Ca	11.0	11.3(11.3)
	Mg	2.08	2.28(2.24)
Clarion	Ca	12.1	11.7(12.5)
	Mg	3.41	3.36(3.31)
Marshall	Ca	12.8	13.6(13.5)
	Mg	2.14	2.28(2.25)
Nicollet	Ca	14.6	14.6(14.9)
	Mg	4.46	4.35 (4.49)
Fayette	Ca	13.2	13.4(11.0)
	Mg	2.68	2.62(2.65)
Muscatine	Ca	13.4	13.9(13.7)
	Mg	3.88	3.95(4.10)
Tama	Ca	15.3	14.7(14.9)
	Mg	3.95	4.04(4.03)
Webster	Ca	17.3	16.5(15.7)
	Mg	5.38	5.57(5.50)
Lester	Ca	18.0	18.5(18.4)
	Mg	5.56	5.53(5.54)
Average	Ca	13.5	13.5(13.2)
	Mg	3.50	3.54(3.56)

[a]Figures in parentheses are exchangeable Ca or Mg values obtained by analyzing directly the NH_4OAc extract of soil by the method indicated.
Source: Ref. 39.

TABLE 9 Comparison of Precision of IC, AAS, and AES Methods for Determination of Exchangeable K and Na in Soils

Exchange-able base	Soil	IC ($cmol\ kg^{-1}$ soil)[a]			CV (%)[b]	AAS ($cmol\ kg^{-1}$ soil)			CV (%)	AES ($cmol\ kg^{-1}$ soil)			CV (%)
		Range	Mean	SD		Range	Mean	SD		Range	Mean	SD	
K	Hagener	0.35–0.38	0.36	0.01	2.8	0.37–0.40	0.39	0.01	2.6	0.34–0.42	0.38	0.03	7.9
	Nicollet	0.25–0.32	0.29	0.02	6.9	0.26–0.31	0.29	0.02	6.9	0.27–0.30	0.28	0.02	7.1
	Fayette	0.60–0.64	0.62	0.02	3.2	0.57–0.62	0.60	0.02	3.3	0.59–0.65	0.61	0.02	3.3
	Tama	0.58–0.62	0.61	0.03	4.9	0.60–0.62	0.60	0.01	1.7	0.59–0.63	0.61	0.02	3.3
	Webster	0.41–0.45	0.43	0.01	2.3	0.45–0.48	0.45	0.02	4.4	0.42–0.46	0.44	0.02	4.5
Na	Hagener	0.04–0.06	0.05	0.01	20.0	0.03–0.06	0.04	0.01	25.0	0.03–0.06	0.04	0.01	25.0
	Nicollet	0.13–0.17	0.15	0.02	13.3	0.14–0.17	0.16	0.01	6.3	0.15–0.19	0.18	0.02	11.1
	Fayette	0.06–0.12	0.09	0.02	22.2	0.05–0.09	0.08	0.02	25.0	0.07–0.11	0.08	0.02	25.0
	Tama	0.09–0.13	0.10	0.02	20.0	0.11–0.12	0.12	0.004	3.3	0.12–0.14	0.13	0.01	7.7
	Webster	0.07–0.11	0.09	0.02	22.2	0.07–0.10	0.08	0.01	12.5	0.09–0.12	0.10	0.01	10.0

[a]Range of six replicate extractions and analyses; SD, standard deviation.
[b]CV, coefficient of variation.
Source: Ref. 39.

TABLE 10 Comparison of Precision of IC and AAS Methods for Determination of Exchangeable Ca and Mg in Soils

Exchange-able base	Soil	IC ($cmol\ kg^{-1}$ soil)[a]				AAS ($cmol\ kg^{-1}$ soil)			
		Range	Mean	SD	CV (%)[b]	Range	Mean	SD	CV (%)
Ca	Hagener	6.83–7.34	7.21	0.19	2.6	6.84–7.25	7.01	0.17	2.4
	Nicollet	13.3–14.1	13.7	0.4	2.9	11.3–14.8	14.6	0.2	1.4
	Fayette	11.8–13.3	12.6	0.8	6.3	12.0–13.8	12.5	0.9	7.2
	Tama	14.5–16.3	15.3	0.6	3.9	15.2–16.7	15.7	0.6	3.8
	Webster	16.2–18.3	17.2	0.7	4.1	16.0–17.1	16.5	0.4	2.4
Mg	Hagener	1.46–1.53	1.52	0.03	2.0	1.46–1.54	1.51	0.04	2.6
	Nicollet	3.93–4.04	3.98	0.05	1.3	3.75–4.14	4.09	0.06	1.5
	Fayette	2.62–2.73	2.68	0.04	1.5	2.64–2.85	2.75	0.08	2.9
	Tama	4.03–4.35	4.22	0.12	2.8	4.06–4.33	4.21	0.09	2.1
	Webster	5.18–5.36	5.27	0.08	1.5	5.05–5.27	5.14	0.10	1.9

[a]Range of six replicate extractions and analyses; SD, standard deviation.
[b]CV, coefficient of variation.
Source: Ref. 39.

Although Basta and Tabatabai [39] showed that all alkali and alkaline earth metals could be separated and determined by the suppressed-type IC system, no Li, Rb, Cs, Sr, or Ba could be detected in the soil, plant materials, or water samples that they analyzed [39–41]. However, the use of cation concentration columns that are commercially available makes possible the detection of cations at the part-per-billion level, which may be useful for the determination of these metals when present in very low concentrations.

Nieto and Frankenberger [36] evaluated a single-column IC for the determination of NH_4^+, alkali metals, and alkaline earth metals in soil extracts by using 10 m*M* HNO_3 (pH 2.1) and 5 m*M* ethylenediammonium dinitrate (pH 6.1) for the mono and divalent cations, respectively. The metal ions were separated on a commercially available, low-capacity, cation-exchange column, with conductimetric detection. Simultaneous determination of Li^+, Na^+, NH_4^+, and K^+ or Mg^{2+}, Ca^{2+}, Sr^{2+}, and Ba^{2+} was performed on standards in <7 min with precision ranging from 3–8%. The monovalent cation elution sequence was

$$Li^+ > Na^+ > NH_4^+ > K^+$$

whereas that for the divalent cations was

$$Mg^{2+} > Ca^{2+} > Sr^{2+} > Ba^{2+}$$

Minimum detectable concentrations of the selected cations by using a 500-μl loop were Na^+, Ca^{2+}, and Mg^{2+}: 0.05 mg l^{-1}; Li^+ and Ba^{2+}: 0.1 mg l^{-1}; NH_4^+: 0.5 mg l^{-1}; K^+ and Sr^{2+}: 1 mg l^{-1}. Results obtained by this single-column IC system agreed closely with those obtained by other methods: steam distillation for NH_4^+ ($r = 0.996$***), AES for Na^+ ($r = 0.992$***), AES for K^+ ($r = 0.999$***), AAS for Ca^{2+} (0.996**), and AAS for Mg^{2+} (0.995***) [36].

IV. SUMMARY AND CONCLUSIONS

Several IC instruments permitting the simultaneous determination of anions, alkali metals, and alkaline earth metals have been evaluated for the analysis of soils, plant materials, and natural waters. The principles of these methods and studies to evaluate these instruments for soil, plant, and water analysis were reviewed.

Several IC instruments and separator columns and other accessories are commercially available; a few of these have been evaluated for soil analysis, and it has been shown that they give very satisfactory results for nutrients in both anionic and cationic forms in soil extracts, plant materials, and natural waters. The IC is well suited to the analysis of complex mixtures of ions, but care must be taken to obtain the samples in a suitable solution. This is because many of the procedures employed in soil analysis result in matrices that are not compatible with the chemistry of the IC systems. Therefore, such procedures have to be modified before using them in conjunction with any IC system.

This field of analytical chemistry is in its infancy, and it will be undoubtedly several years before IC enjoys the use and wide distribution it deserves as an analytical technique for soil and plant analysis. However, because of the potential it offers, particularly for the simultaneous analysis of several anions (for which there is no alternative instrumental method, unlike the situation for metal cations), widespread future use may be predicted with confidence.

REFERENCES

1. H. Small, T. S. Stevens, and W. C. Baumann, *Anal. Chem.*, *47*:1801 (1975).
2. J. Weiss, *Handbook of Ion Chromatography*, Dionex Corp., Sunnyvale, Calif., 1986.
3. J. S. Fritz, D. T. Gjerde, and C. Pohlandt, *Ion Chromatography*, Hüthig, New York, 1982.
4. F. C. Smith, Jr. and R. C. Chang, *The Practice of Ion Chromatography*, Wiley, New York, 1983.
5. E. Sawicki, J. D. Mulik, and E. Wittgenstein (eds.), *Ion Chromatographic Analysis of Environmental Pollutants*, Ann Arbor Science, Ann Arbor, Mich., 1978.
6. J. D. Mulik and E. Sawicki, *Ion Chromatographic Analysis of Environmental Pollutants*, Vol. 2, Ann Arbor Science, Ann Arbor, Mich., 1979.
7. J. G. Tarter (ed.), *Ion Chromatography*, Marcel Dekker, New York, 1987.
8. J. D. Mulik and E. Sawicki, *Environ. Sci. Technol.*, *13*:804 (1979).
9. E. L. Johnson, in *Ion Chromatography* (J. G. Tarter, ed.), Marcel Dekker, New York, 1987.
10. W. A. Dick and M. A. Tabatabai, *Soil Sci. Soc. Am. J.*, *43*:899 (1979).
11. M. A. Tabatabai, in *Proceedings of Sulphur-84*, Sulphur Development Institute of Canada, Calgary, Alberta, Canada, 1985, p. 661.
12. D. T. Gjerde and J. S. Fritz, *J. Chromatogr.*, *176*:199 (1979).
13. D. T. Gjerde and J. S. Fritz, *Anal. Chem.*, *53*:2324 (1981).

14. D. T. Gjerde and J. S. Fritz, *J. Chromatogr., 186*:509 (1979).
15. D. T. Gjerde, G. Schmuckler, and J. S. Fritz, *J. Chromatogr., 187*:35 (1980).
16. J. E. Girard and J. A. Glatz, *Am. Lab., 13*:26 (1981).
17. T. Jupille, in *Ion Chromatography* (J. G. Tarter, ed.), Marcel Dekker, New York, 1987.
18. T. H. Jupille, D. W. Togami, and D. E. Burge, *A Single-Column Ion Chromatography Aids Rapid Analysis*, Wescan Instruments Inc., Santa Clara, Calif., 1983, p. 1.
19. D. J. Gjerde, J. S. Fritz, and G. Schmuckler, *J. Chromatogr., 186*:509 (1979).
20. J. S. Fritz, D. J. Gjerde, and R. M. Becker, *Anal. Chem., 52*:1519 (1980).
21. K. F. Nieto and W. T. Frankenberger, Jr., *Soil Sci. Soc. Am. J., 49*:592 (1985).
22. Z. Iskandaranl and D. J. Pletrzyk, *Anal. Chem., 54*:2427 (1982).
23. G. E. Boyd, B. A. Soldano, and O. D. Bonner, *J. Phys. Chem., 58*:456 (1954).
24. J. S. Fritz and J. N. Story, *J. Chromatogr., 90*:267 (1974).
25. J. S. Fritz and J. N. Story, *Anal. Chem., 46*:825 (1974).
26. T. S. Stevens, J. C. Davis, and H. Small, *Anal. Chem., 53*:1488 (1981).
27. Y. Hanaoka, T. Murayama, S. Muramoto, T. Matsuura, and A. Nanba, *J. Chromatogr., 239*:537 (1982).
28. H. Small, *Anal. Chem., 55*:235A (1983).
29. V. T. Turkelson and R. P. Himes, 24th Rocky Mountain Conference, Denver, Colorado, 1982.
30. H. J. Cortes, *J. Chromatogr., 254*:517 (1982).
31. M. A. Tabatabai and W. A. Dick, *J. Environ. Qual., 12*:209 (1983).
32. L. M. Busman, R. P. Dick, and M. A. Tabatabai, *Soil Sci. Soc. Am. J., 47*:1167 (1983).
33. M. A. Tabatabai, N. T. Basta, and H. J. Pirela, *Commun. Soil Sci. Plant Anal., 19*:1701 (1988).
34. M. Kalbasi and M. A. Tabatabai, *Commun. Soil Sci. Plant Anal., 16*:787 (1985).
35. S. V. Krupa and M. A. Tabatabai, in *Sulfur in Agriculture*, (M. A. Tabatabai, ed.). Am. Soc. Agron., Madison, Wis., 1986, p. 491.
36. K. F. Nieto and W. T. Frankenberger, Jr., *Soil Sci. Soc. Am. J., 49*:587 (1985).
37. W. Karlson and W. T. Frankenberger, Jr., *Soil Sci. Soc. Am. J., 51*:72 (1987).
38. P. Barak and Y. Chen, *Soil Sci. Soc. Am. J., 51*:257 (1987).
39. N. T. Basta and M. A. Tabatabai, *Soil Sci. Soc. Am. J., 49*:84 (1985).

40. N. T. Basta and M. A. Tabatabai, *Soil Sci. Soc. Am. J.*, *49*: 76 (1985).
41. N. T. Basta and M. A. Tabatabai, *J. Environ. Qual.*, *14*:450 (1985).

6

Automated Instruments for Determination of Total Carbon, Nitrogen, and Sulfur in Soils by Combustion Techniques

M. ALI TABATABAI and JOHN M. BREMNER *Iowa State University, Ames, Iowa*

I. INTRODUCTION

Dry combustion techniques have not been used extensively for the determination of total carbon, total nitrogen, or total sulfur in soils because until recently the techniques available were complicated and time-consuming compared with wet oxidation methods. These defects of dry combustion methods of analysis have been eliminated by the development of automated techniques that are very simple and rapid compared with their predecessors. The purpose of this chapter is to describe the automated combustion instruments that have been evaluated for the determination of total C, total N, and total S in soils and to review current information concerning the use of these instruments for soil analysis. Several automated instruments besides those discussed have been used to analyze soils, but no work to evaluate these instruments for soil analysis has been reported. Most of these instruments were designed for the analysis of pure compounds or homogeneous materials having high C, N, or S contents, and it seems unlikely that they will prove satisfactory for the analysis of soils or other heterogeneous materials with low contents of these elements.

II. CARBON ANALYZERS

The only automated combustion instrument thus far evaluated for total C analysis of soils is the Leco 70-Second Carbon Analyzer

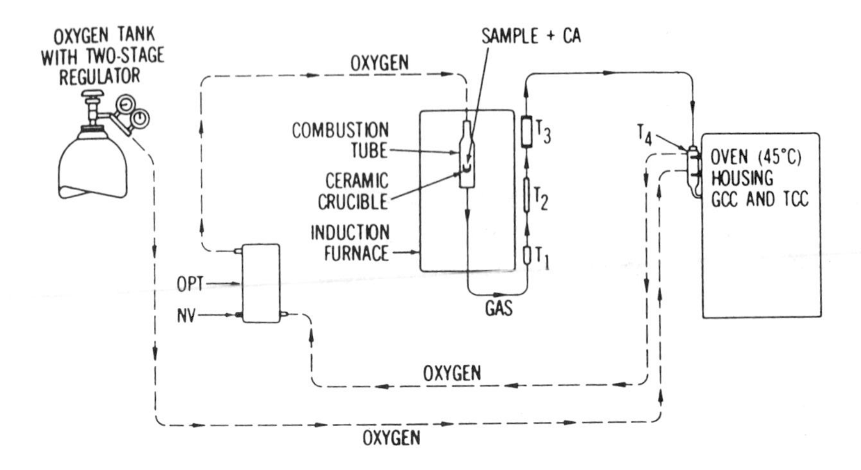

FIG. 1 Schematic diagram of Leco 70-Second Carbon Analyzer. CA, combustion accelerators; T_1, dust trap; T_2, trap containing activated MnO_2; T_3, trap containing heated catalyst (hopcalite); T_4, trap containing anhydrous $Mg(ClO_4)_2$; OPT, O_2-purification train; NV, needle valve, GCC, gas collection cylinder; TCC, thermal conductivity cell.

developed by the Laboratory Equipment Corporation, St. Joseph, Mich. This instrument was designed originally for total C analysis of iron and steel, but its speed and simplicity have led to its use for analysis of nonferrous as well as ferrous materials, and Tabatabai and Bremner [1] have shown that it is very satisfactory for the determination of total carbon in soils.

A. Principles

The Leco 70-Second Carbon Analyzer (Fig. 1) consists of an O_2-purification train, a high-frequency induction furnace, and a carbon dioxide analyzer that utilizes the difference in thermal conductivity between CO_2 and O_2. In total C analysis by this instrument, the sample is treated in a ceramic crucible with combustion accelerators (usually iron plus tin and tin-coated copper) and heated to a high temperature (>1650°C) in a stream of purified O_2, the high temperature being generated by using an induction furnace to induce an electrical field in the accelerator-treated sample. The gases evolved are passed through a dust trap (to remove metal oxides), a trap containing activated MnO_2 (to remove S oxides, N oxides, and halogen gases), a heated catalyst (hopcalite) tube (to convert

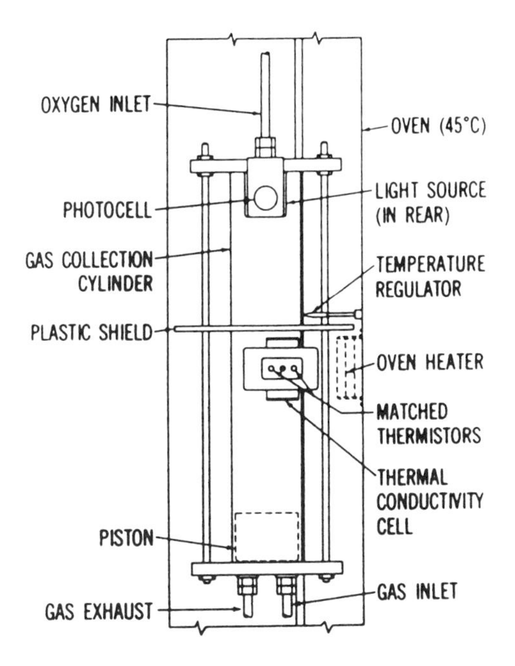

FIG. 2 Details of gas collection cylinder in Leco 70-Second Carbon Analyzer.

CO to CO_2), and a trap containing anhydrous $Mg(ClO_4)_2$ (to remove water vapor) (see Fig. 1). The purified CO_2–O_2 mixture thus obtained is collected in a cylinder maintained at a constant temperature (45°C) and is analyzed for CO_2 by measuring its thermal conductivity with a thermistor-type thermal conductivity cell. The gas collection cylinder (Fig. 2) is maintained at 45°C to eliminate temperature variations that would cause an imbalance in the thermal conductivity cell. This cell consists of a pair of matched thermistors in two arms of a Wheatstone bridge. The reference thermistor is kept in an environment of constant gas pressure, gas flow, and temperature, and the measuring thermistor is maintained in the same environment, except that the gas composition varies according to the CO_2 content of the sample. The sensitivity of the analysis for CO_2 is governed by the difference in thermal conductivity between O_2 and CO_2. The thermal conductivities of O_2 and CO_2 are 24 and 14 mJ m^{-1} sec^{-1} $°C^{-1}$ (57 and 33 μcal cm^{-1} sec^{-1} $°C^{-1}$), respectively.

The Leco 70-Second Carbon Analyzer is designed so that it functions automatically once the accelerator-treated sample is placed in the induction furnace and the cycle switch is pressed to start the analysis. The sequence of events initiated when this switch is pressed is as follows: (1) The O_2 flow is turned on and solid-state timers are activated in sequence through the contacts of a "holding" relay tube. (2) The piston in the gas collection cylinder drops, thereby purging gas from the cylinder. (3) O_2 is swept through the system and the gas collection cylinder for a period of 20 sec (the induction furnace is off during this time). (4) The induction furnace is turned on (this is indicated by a red light) to initiate a 10 sec preburn cycle (O_2 is supplied to the combustion tube during this time, but the flow of O_2 to the cylinder is stopped). (5) At the end of the 10 sec preburn, a solenoid valve is opened to admit the gas from the combustion tube to the bottom of the cylinder. As the gas flows into the cylinder, the piston is forced upward until it interrupts the light source of the photocell (Fig. 2). This causes a "holding" relay tube to turn off the induction furnace and close the gas inlet at the bottom of the cylinder. (6) After a 20 sec delay, a clamp controlling a voltmeter is released to allow it to respond to the measuring thermistor in the conductivity cell. (7) At the end of a 10 sec "read" period, the output of the thermal conductivity cell is amplified by a DC amplifier, and the output of the amplifier is recorded automatically as milligrams of carbon on a digital meter. The response of the digital meter is linear between 0 and 8 mg C.

B. Calibration

Before use, the thermal conductivity cell is balanced so that, with pure O_2 in the gas collection cylinder, it gives zero output on the digital meter. With the cell thus balanced, its output as measured on the digital meter is proportional to the amount of CO_2 in the cylinder. Before use, the analyzer is calibrated with carbon standards. The metallic-carbon standards supplied by the Laboratory Equipment Corporation are convenient for this purpose.

C. Commercial Instruments

The Laboratory Equipment Corporation supplies five automated carbon analyzers besides the Leco 70-Second instrument, namely, the Leco IR-12 Carbon Determinator, the Leco DC-12 Duo-Carb Analyzer, the Leco CS-46 Simultaneous Carbon/Sulfur Determinator, and the Leco CHN-600 and CHN-800 Analyzers. The DC-12 Duo-Carb instrument permits the determination of both total C and organic C, the CS-46 instrument allows the determination of both total C and total S, and the CHN-600 and CHN-800 instruments allow the simultaneous determination of total C, H, and S.

In total C analysis by the Leco IR-12 instrument, the sample is mixed with a metal combustion accelerator (usually Cu or Fe) in a ceramic crucible and is heated in an induction furnace in an atmosphere of O_2 to oxidize C to CO and CO_2. The gas mixture thus obtained is dried, passed through a heated Pt catalyst to convert CO to CO_2, and collected in a cell for CO_2 analysis by an infrared detector. This instrument has been used successfully for the determination of total C in complex materials such as automotive catalyst support formulations [2], which contain about 97% of inorganic material, such as clay, talc, and alumina, and about 3% of organic extrusion aids, such as methyl cellulose and diglycol stearate.

In total C analysis by the Leco DC-12 Duo-Carb Analyzer, the sample is mixed with vanadium pentoxide and heated to about 1000°C in a stream of O_2 in an induction furnace. The CO_2 thus produced is collected in a cylinder and measured by a thermal conductivity detector. To determine organic C, a second sample is combusted at about 600°C, and the CO_2 thus produced is determined as in the procedure for the determination of total C. Inorganic C in the sample is calculated as the difference between the results of the analyses for total C and organic C. This instrument was designed for the rapid determination of total C and organic C in shales, soils, and limestones, but it has not been evaluated for soil analysis.

In total C analysis by the Leco CS-46 instrument, the sample is mixed with a combustion accelerator and combusted in O_2 in an induction furnace. The CO and CO_2 thus produced are measured individually by separate infrared detectors, and the outputs of these detectors are added electronically to obtain total C. The SO_2 produced by combustion is measured by a third infrared detector. This instrument differs from previous carbon analyzers in that a recirculating O_2 system replaces the conventional flow-through system, solid-state infrared detectors replace the conventional Luft cell for CO_2 analysis, and direct measurement of CO replaces catalytic conversion of CO_2. It permits the simultaneous determination of total C and total S within 30 sec and clearly deserves evaluation for the routine determination of total C and S in soils.

Another automated combustion instrument now available for the simultaneous determination of total C and total S is the I.R.-Matic "C-S" VK-111 AS Analyzer supplied by Kukusai Electric Co., Japan. In the determination of total C and total S by this instrument, the sample is mixed in a ceramic crucible with combustion accelerators (iron powder and tungsten chips) and combusted in a high-frequency induction furnace using O_2 as the combustion and carrier gas and N_2 as purge gas. The CO_2 and SO_2 thus produced are measured by Luft-type nondispersive infrared detectors. The instrument is supplied with an automatic balance and accessories such as linearizers, integrators, blank adjusters, single-point calibrating adjusters, and weight compensators, and it permits the determination of both total C and total S in less than 60 sec. It has given satisfactory results when used for the analysis of standard rocks, NBS minerals, and geological exploration reference samples and it seems promising for the determination of C and S in soil samples [3].

Stewart et al. [4] demonstrated that the Coleman Model 29 Nitrogen Analyzer (an automated Dumas combustion instrument described in Sec. III) can be adapted for the determination of total C in soils.

Dugan [5] has described an automatic combustion instrument developed by Hercules, Inc., Wilmington, Del. for the simultaneous determination of C, H, N, and S. The procedure in analysis by this instrument involves an uncatalyzed, flash-combustion of the sample in a quartz tube in an O_2-He atmosphere. The CO_2, H_2O, N_2, and SO_2 formed by this combustion are separated and determined by a gas chromatographic procedure involving the use of a thermal conductivity detector. This instrument has not been evaluated for the determination of C, N, or S in soils, but it seems unlikely to prove useful for soil analysis.

Another automated combustion instrument that involves flash combustion in a He-O_2 atmosphere is the Model 1106 Elemental Analyzer developed by Carlo Erba Strumentazione, Milan, Italy, for the determination of C, H, N, O, and S. For C, H, and N analysis, the gases produced by flash combustion are passed over Cr_2O_3 to ensure quantitative combustion, and the excess O_2 is removed by Cu at 650°C before the separation and determination of N_2, CO_2, and H_2O by gas chromatography. For S analysis, the gases produced by flash combustion are passed over WO_3 to ensure the quantitative conversion of S to SO_2, which is determined by gas chromatography.

III. NITROGEN ANALYZERS

The only automated combustion instruments thus far evaluated for the total N analysis of soils are the Model 29 and Model 29A Nitrogen Analyzers developed by Coleman Instruments, Inc., Maywood, Ill. (now Coleman Instruments Division of the Perkin-Elmer Corporation, Oak Brook, Ill.). These analyzers are automated Dumas combustion instruments (see Table 1 for specifications).

A. Principles

A schematic diagram of the Coleman Model 29 Nitrogen Analyzer is shown in Fig. 3. In total N analysis with this instrument, the sample is mixed with combustion catalyst (usually CuO) and heated to above 900°C in a quartz combustion tube filled with a catalyst. This heating to pyrolyze the sample is carried out in an atmosphere of high-purity CO_2, and the gases evolved are swept by purified CO_2

TABLE 1 Coleman Nitrogen Analyzers (Manufacturer's Specifications)

Specification	Analyzer model 29	Analyzer model 29A
Sample size, mg	2–50	25–500
Maximal amount of N in sample, mg	4	40
Number of analyses per hour	4–5	3–4
Capacity of nitrometer syringe, ml	5	50
Reproducibility of nitrometer syringe, μl	±2	±10
Operating cycle, min	8	12
Readout	Digital counter	Motor-driven digital counter

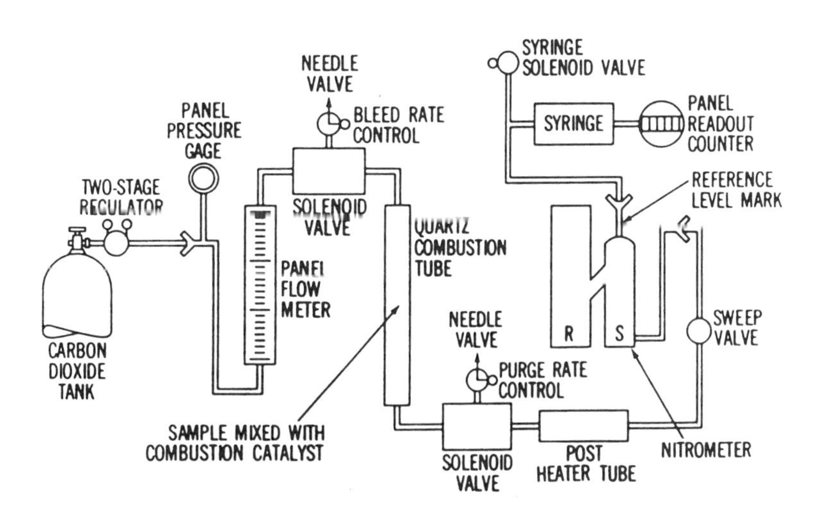

FIG. 3 Schematic diagram of Coleman Model 29 Nitrogen Analyzer. Diagram does not show furnace used to heat combustion tube. Nitrometer has reservoir (R) and magnetic stirrer (S) and contains concentrated KOH solution.

through a postheater tube containing Cu in the first half of its length and CuO in its terminal section. This tube is heated by an auxiliary furnace (usually to 600–700°C). Its function is to reduce N oxides to N_2, oxidize any H_2 to H_2O, and convert CO to CO_2. The gas stream from the postheater tube passes into a glass nitrometer containing concentrated alkali (KOH solution), where it is released from a nozzle under mercury as a stream of small bubbles. These bubbles are brought into contact with the alkali by a magnetic stirrer which ensures complete absorption of CO_2 and the N_2 collects in the body of the nitrometer, displacing alkali into a reservoir. The N_2 in the nitrometer is then measured volumetrically by drawing it into a precisely calibrated stainless steel syringe. The plunger of this syringe is driven by a micrometer screw linked to a precision digital counter. Manual adjustment of this counter provides the readout of N_2 volume directly in microliters. The Model 29 Analyzer has a syringe capacity of 5 ml, which permits the consecutive analysis of microsamples without resetting the N_2 measurement system. The Model 29A Analyzer has a syringe capacity of 50 ml. Its syringe plunger is driven to approximate adjustment by a reversible electric motor, the final adjustment being made manually.

B. Calculation of Results

The percentage of N in the sample analyzed is calculated from R_1 (initial counter reading), R_2 (final counter reading), P (barometric pressure), t_1 (initial syringe temperature), and t_2 (final syringe temperature). The steps in the calculation are as follows:

1. Record the observed volume of N_2, V_o

$$V_o = R_2 - R_1$$

where V_o is the observed N_2 volume (in microliters) and R_1 and R_2 are as defined above.

2. Determine the corrected N_2 volume, V_c

$$V_c = V_o - (V_b + V_t)$$

where

V_b = volume of blank (in μl)
V_t = volume correction for temperature (in μl)
$= C_f\ (t_2 - t_1)$

where C_f is a correction factor based on the final counter reading (this factor is readily available from tables supplied by the manufacturer) and t_2 and t_1 are in degrees Kelvin.

3. Determine the corrected barometric pressure P_c

$$P_c = P_o - (P_b + P_v)$$

where P_o is the observed barometric pressure (in torr), P_b the barometric temperature correction, and P_v a correction for the vapor pressure of the KOH solution in the nitrometer (P_b and P_v are readily available from tables supplied by the manufacturer).

An empirical approximation of $P_b + P_v = 11.0$ is sufficiently accurate for P_o between 740 and 780 torr (99–104 kPa) and syringe temperatures between 298 and 305 K.

4. Calculate the percentage of N in the sample from the following equation:

$$\%N = \frac{P_c}{T} \times \frac{V_c}{W} \times 0.04493$$

where T is the final syringe temperature in degrees Kelvin, W is the sample weight in milligrams, and P_c and V_c are as defined above.

C. Commercial Instruments

The Laboratory Equipment Corporation has developed three automated nitrogen analyzers (the Leco UO-14SP Nitrogen Determinator, the Leco TN-15 Nitrogen Determinator, and the Leco TC-36 Nitrogen/Oxygen Determinator) in which total N is converted to N_2 by fusing the sample in a graphite crucible at a high temperature in an inert (He) atmosphere. These instruments were developed for the rapid determination of total N in ferrous materials, but it seems likely that they could be adapted for the routine determination of total N in soils.

In total N analysis by the Leco UO-14SP instrument, the sample is mixed with CuO-Pt catalyst, encased in tinfoil, and fused (ca. 2600°C) in a graphite crucible in an impulse furnace. The gases evolved are passed over CuO at 450°C (to convert N oxides to N_2), through ascarite (to remove CO_2), and through $Mg(ClO_4)_2$ and P_2O_5 (to remove H_2O). The remaining gases are passed through a molecular sieve column maintained at 10–11°C. This column separates N_2 from other gases and permits the quantitative determination of N_2 by a thermal conductivity detector. The analysis time is approximately 7 min. There seems very little doubt that this instrument will prove satisfactory for the determination of total N in soils, because Wong and Kemp [6] found that it gave satisfactory results when used for the total N analysis of 18 sediments containing from 0.02–2.3% N, and the problems in sediment analysis are similar to those encountered in soil analysis.

The procedure for total N analysis by the Leco TN-15 instrument is similar to that employed in the UO-14SP instrument, but the sample is fused in a He atmosphere in an electric furnace at 3000°C, and the analysis time is only 80 sec.

The Leco TC-36 instrument permits the simultaneous determination of nitrogen and oxygen. In analysis by this instrument, the sample is fused in a He atmosphere in an electric furnace at 2500°C, and a stream of He is used to carry the gases evolved over CuO (to oxidize CO to CO_2 and H_2 to H_2O) and through a column containing $Mg(ClO_4)_2$ (to remove H_2O). The N_2 and CO_2 in the stream of He emerging from this column are then separated and determined by a gas chromatographic procedure involving the use of a thermal conductivity detector (oxygen in the sample is calculated from the CO_2 determined by this procedure). The analysis time is about 3.5 min.

The Laboratory Equipment Corporation, St. Joseph, Mich., recently introduced two automated elemental analyzers (the Leco CHN-600 and CHN-800 Analyzers) that permit the simultaneous determination of total C, H, and N in solid or liquid organic materials. The CHN-600 Analyzer is a macrosample instrument capable of analyzing samples ranging in weight from 100–200 mg, whereas the CHN-800 Analyzer is a microsample instrument capable of analyzing samples ranging in weight from 3–15 mg. In total N analysis by these instruments, a weighed sample is encapsulated in tin or copper and dropped into a reusable ceramic crucible centered in the primary hot zone of a U-shaped combustion tube located in a resistance furnace, and the sample is burned in O_2 at 950°C. The potential combustion products are CO_2, water vapor, oxides of N, N_2, and oxides of S. Oxides of S are removed with a reagent in the secondary hot zone to prevent the formation of H_2SO_4. The secondary hot zone also ensures the complete combustion of all volatile gases. The remaining products of combustion (CO_2, H_2O, O_2, N_2, and NO_X) are collected and mixed thoroughly in a glass tube under a sliding PVC piston. The CO_2 and H_2O levels are constantly monitored during combustion by two independent selective IR detectors, and when they drop to predetermined levels, combustion is terminated. At this stage, an aliquot of the combustion products is removed automatically and carried by H_2 gas through a reagent train containing hot Cu for the reduction of NO_X to N_2, ascarite for the removal of CO_2, and anhydrone for the removal of H_2O. The N_2 thus obtained is then collected and measured by a thermal conductivity detector. The measurements are weight-compensated and displayed digitally as percent C, H, and N. Total analysis time for all three elements is 4–5 min with the CHN-600 Analyzer and less than 2.5 min with the CHN-800 Analyzer. Recent work by Sheldrick [7] indicates that the Leco CHN-600 Analyzer may be useful for the routine estimation of total N in soils.

Carlo Erba Strumentazione, Milan, Italy, has developed automated combustion instruments, the Model 1400 Nitrogen Analyzer and the later version, the Model 1500, that merit evaluation for the determination of total N in soils. In total N analysis by these instruments, the sample is oxidized in a furnace containing NiO and O_2 maintained at 1000°C. The gases thus produced are swept from the oxidation furnace by a stream of He and passed through a reduction furnace containing Cu at 750°C to reduce oxides of N to N_2. The N_2 thus obtained is then passed through traps to remove moisture, CO_2, and any inorganic acids in the gas stream and is determined by a gas chromatographic procedure using a thermal conductivity detector.

Antek Instruments, Inc., Houston, Texas, recently introduced automated N analyzers in which the sample is pyrolyzed in an O_2-Ar atmosphere to oxidize N to NO, which is determined by using a chemiluminescent detector to measure the light emitted by metastable NO_2 produced by mixing NO with O_3. These instruments have been used to determine N in petroleum fractions [8] and to measure gaseous loss of N from plants during transpiration [9], but they have not been evaluated for the total N analysis of soils.

IV. SULFUR ANALYZERS

The only automated combustion instrument thus far evaluated for the total sulfur analysis of soils is the Leco Sulfur Analyzer supplied by the Laboratory Equipment Corporation, St. Joseph, Mich. Like the Leco Carbon Analyzer discussed in Sec. II, this instrument was designed originally for the analysis of iron and steel.

A. Principles

The Leco Sulfur Analyzer consists of an O_2-purification train, a high-frequency induction furnace fitted with a timer to select the time of induction, and an automatic sulfur dioxide (SO_2) titrator (Fig. 4). In total S analysis by this instrument, the sample is treated in a ceramic crucible with combustion accelerators (usually iron plus tin, copper, or tin-coated copper), and the crucible is covered with a porous ceramic cover and heated to a high temperature (ca. 1600°C) in a stream of purified O_2, the high temperature being generated by using an induction furnace to induce an electrical field in the accelerator-treated sample. The SO_2 liberated by combustion of the sample is collected in a dilute solution of hydrochloric acid containing KI, starch, and a trace amount of KIO_3 and is determined by the titration of this solution with a standard KIO_3 solution. This titration is performed automatically through

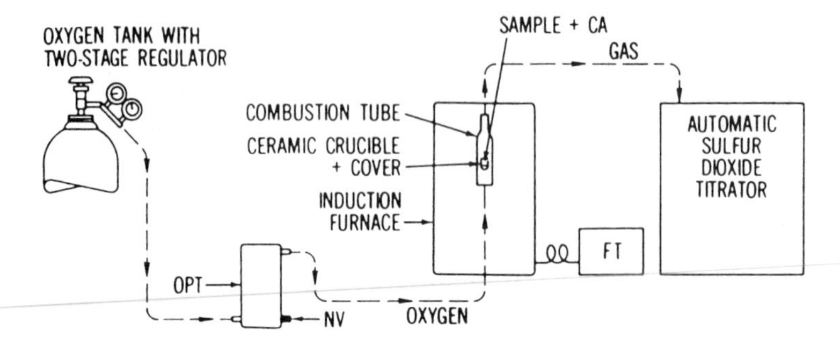

FIG. 4 Schematic diagram of Leco Sulfur Analyzer. CA, combustion accelerators; OPT, O_2-purification train; NV, needle valve; FT, furnace timer.

the use of a photocell to maintain the initial optical density of the blue color formed by the reaction of starch with the iodine formed by the reaction of KI and KIO_3 in the presence of HCl. The reactions that occur during this titration are indicated by the following equations:

$$5KI + KIO_3 + 6HCl = 3I_2 + 6KCl + 3H_2O$$

$$SO_2 + I_2 + 2H_2O = H_2SO_4 + 2HI$$

The Leco Sulfur Analyzer functions automatically once the accelerator-treated sample is placed in the induction furnace and the timer switch is pressed to initiate the analysis.

B. Calibration

Before use, the automatic sulfur dioxide titrator is calibrated by analyzing sulfur standards. The metallic-sulfur standards supplied by the Laboratory Equipment Corporation are convenient for this purpose.

C. Commercial Instruments

The Laboratory Equipment Corporation has developed three automated sulfur analyzers that have not thus far been evaluated for soil analysis, but clearly deserve evaluation. One is the Leco CS-46 Simultaneous Carbon/Sulfur Determinator described in Sec. II.C.

The others are the Leco SC-32 and SC-132 Automatic Sulfur Determinators. These two instruments differ only in the ranges of total S that can be determined, the ranges being from 0.005–99.99% and 0.001–99.99%, respectively. In total S analysis by these instruments, the sample is combusted at about 1350°C in a stream of O_2, and the SO_2 thus produced is measured by a solid-state infrared detector. The results can be displayed in any of three forms: percent sulfur, percent SO_4, or percent SO_3. The analysis time is only 2 min.

V. APPLICATIONS

A. Carbon Analyzers

Table 2 and Fig. 5 show results obtained by Tabatabai and Bremner [1] in work to evaluate the Leco 70-Second Carbon Analyzer for the total C analysis of soils. The 46 soils used in this work were selected so that they differed markedly in organic-matter content (0.3–34% organic C), texture (2–94% sand, 1–43% clay), and carbonate content (0–26.5% $CaCO_3$), and the total C values obtained in the analysis of these soils by the Leco 70-Second Carbon Analyzer were compared with those obtained by the wet combustion method of Allison [10], which has gained acceptance as a reliable method for determining total C in soils. This comparison showed that the total C values obtained by the two methods agreed closely (Fig. 5) and that the Leco method of analysis was at least as precise as Allison's method (Table 2). Other studies reported by Tabatabai and Bremner [1] showed that the Leco 70-Second Carbon Analyzer gave the quantitative recovery of carbonate-C and that the results of soil analysis by this instrument were not affected by the presence of substantial amounts of chloride. A study of the effect of sample mesh-size on the total C analysis of mineral soils by the Leco Carbon Analyzer showed that the precision of the results obtained with this instrument usually increased with a decrease in soil mesh-size but that, with most soils studied, the results with <40 mesh (420 μm) soil were almost as precise as those obtained with <100 mesh (150 μm) or <300 mesh (50 μm) soil (Table 3). Soil samples containing 2–10 mg of C (usually 0.2–0.3 g of mineral soil) were found satisfactory for total C analysis by the Leco Analyzer. With several soils, incomplete combustion was observed when the amount of soil taken for analysis exceeded 0.5 g or contained more than 12 mg of C.

Total C analysis of soils with the Leco 70-Second Carbon Analyzer is very simple and rapid, and a single operator can easily perform more than 150 analyses in a normal working day (including

TABLE 2 Precision of Results Obtained in Total Carbon Analysis of Soils by Leco 70-Second Carbon Analyzer and by Wet Combustion Method of Allison

Soil	Method of analysis[a]	Number of analyses	Total C content (%)		
			Range	Mean	SD[b]
Buckner	L	8	0.29–0.33	0.32	0.02
	A	5	0.28–0.34	0.31	0.02
Ida	L	8	1.06–1.13	1.10	0.02
	A	6	1.10–1.18	1.15	0.03
Hayden	L	8	3.02–3.11	3.06	0.02
	A	6	2.97–3.11	3.07	0.02
Oxbow[c]	L	5	3.30–3.38	3.36	0.03
	A	5	3.34–3.46	3.38	0.03
Glencoe	L	8	5.76–5.84	5.80	0.13
	A	6	5.65–5.82	5.67	0.13
Peat	L	5	33.9–34.5	34.2	0.25
	A	5	32.5–34.5	33.5	0.90

[a]L, Leco C Analyzer; A, Allison method [10].
[b]SD, standard deviation.
[c]26.5% $CaCO_3$.
Source: Ref. 1.

weighing of soil samples). Well over 250 analyses of preweighed samples can be performed within 8 hr.

B. Nitrogen Analyzers

Stewart et al. [11] analyzed 47 soils containing 0.02–0.47% N by the Coleman Model 29 Nitrogen Analyzer and compared the results with those obtained by a Kjeldahl method commonly used for the total N analysis of soils. They found that the results of these two methods agreed closely for soils containing less than about 0.3% N, but that the Model 29 Analyzer gave 5–20% higher values than the Kjeldahl method for soils containing substantial amounts of organic matter. This confirmed previous work [12,13] showing that the Dumas and Kjeldahl procedures gave almost identical results with mineral soils, but that the Dumas method gave significantly higher values than the

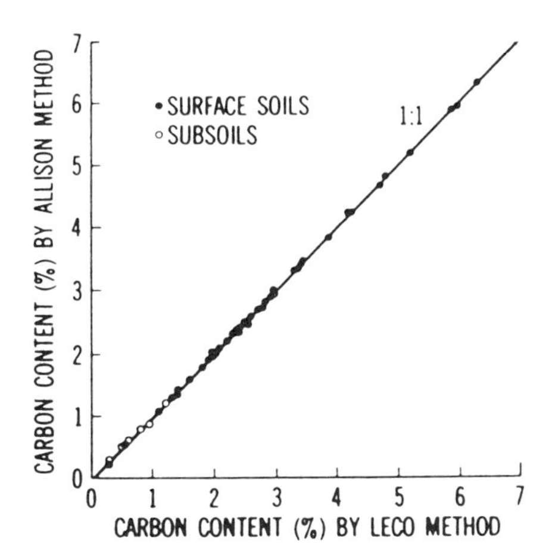

FIG. 5 Results of total carbon analysis of 46 soils by Leco 70-Second Carbon Analyzer versus results by method of Allison [10]. [Reprinted from Ref. 1, *Soil Sci. Soc. Am. Proc.*, *34*:607–610 (1970) by permission of the publisher, Soil Science Society of America.]

Kjeldahl method with organic soils. Stewart et al. [11] showed that the higher values obtained by the Dumas procedure with organic soils were caused by incomplete combustion, which resulted in methane (CH_4) being formed and measured with the N_2 produced by combustion. They later found [4] that this problem could be eliminated by the use of a CuO-Pt catalyst instead of CuO in the postheater tube of the Model 29 Analyzer and showed that when the Model 29 Analyzer was modified accordingly, the total N values obtained with both mineral and organic soils by this instrument agreed closely with those obtained by Kjeldahl analysis (Table 4).

The Coleman Model 29 Analyzer was not designed for the analysis of soils or other heterogeneous materials having low N contents, and its use for the total N analysis of such materials is complicated by sampling problems imposed by sample-size limitations [4]. To overcome this difficulty and eliminate the need for the fine grinding of samples and use of a microbalance in the analysis of heterogeneous materials having low N contents, the Coleman Company developed the Model 29A Nitrogen Analyzer, which permits the analysis of relatively large samples of heterogeneous materials (see Table 1).

Keeney and Bremner [14] evaluated the Coleman Model 29A Nitrogen Analyzer for total N analysis of soils. They found that a technique involving the combustion of soil with a CuO-Pt catalyst at 940°C for 12.6 min in a tube filled with this catalyst and the use of

TABLE 3 Effect of Sample Mesh-Size on Total Carbon Analysis of Soils by Leco 70-Second Carbon Analyzer

	Max. particle size		Total C content by Leco Analyzer (%)[a]		
Soil	Mesh	μm	Range	Mean	SD
Lindley	10	2000	1.16–1.41	1.29	0.11
	40	420	1.31–1.38	1.33	0.03
	100	150	1.35–1.38	1.36	0.01
Sharpsburg	10	2000	2.30–2.45	2.36	0.06
	40	420	2.28–2.34	2.31	0.02
	100	150	2.28–2.32	2.29	0.02
	300	50	2.26–2.29	2.28	0.01
Grundy	10	2000	2.60–2.69	2.63	0.04
	40	420	2.63–2.75	2.68	0.04
	100	150	2.62–2.69	2.66	0.03
	300	50	2.60–2.68	2.64	0.03
Glencoe	10	2000	5.65–6.05	5.81	0.15
	40	420	5.68–5.86	5.78	0.07
	100	150	5.77–5.84	5.80	0.03

[a]Five analyses. SD, standard deviation.
Source: Ref. 1.

a postheater tube containing Cu and a CuO-Pt catalyst maintained at 700°C gave higher results than other techniques studied and that the total N values obtained by this technique with surface samples of mineral and organic soils agreed closely with those obtained by a Kjeldahl method previously evaluated for soil analysis (Table 5). This combustion technique differed from that used by Stewart et al. [4] for soil analysis by the Model 29 Analyzer mainly in that the time of combustion was 12.6 instead of 4.5 min and the soil sample was mixed with a CuO-Pt catalyst instead of CuO and was pyrolyzed in a tube filled with this catalyst.

Stewart et al. [11] found that the coefficient of variation of the results obtained by the technique they originally adopted for the total N analysis of soils by the Coleman Model 29 Analyzer was about 2% for soils containing less than 0.1% N and about 1% for soils containing more than 0.1% N. They subsequently found [4] that when this combustion technique was modified to eliminate interference by CH_4, the average relative standard deviation for the mean of the

TABLE 4 Comparison of Results Obtained in Total Nitrogen Analysis of 15 Soils by Coleman Model 29 Nitrogen Analyzer and by Kjeldahl Method

	Total N content (%)	
Soil number	Coleman analyzer	Kjeldahl method
1	0.060	0.061
2	0.102	0.103
3	0.103	0.110
4	0.154	0.155
5	0.141	0.145
6	0.168	0.167
7	0.189	0.191
8	0.194	0.196
9	0.215	0.217
10	0.299	0.300
11	0.330	0.339
12	0.349	0.358
13	0.481	0.481
14	1.178	1.202
15	1.313	1.316

Source: Ref. 4.

results obtained with mineral and organic soils was 1.61%, which was slightly higher than the corresponding deviation for the mean of the results obtained by Kjeldahl analysis (1.42%). Keeney and Bremner [14] found that the precision of the technique they adopted for the total N analysis of soils by the Coleman Model 29A Analyzer was slightly lower than that of the Kjeldahl method. For example, the coefficient of variation of the results they obtained in six analyses of a Marshall soil (0.212% N) was 1.5% for the Model 29A Analyzer and 0.9% for the Kjeldahl method.

Stewart et al. [4] used 30–500 mg of soil (0.2–0.3 mg of N) for the determination of total soil N by the Coleman Model 29 Analyzer and emphasized the need for the fine grinding of the soil sample to eliminate sampling error. Keeney and Bremner [14] used 0.2–1.0 g of soil (2.8–4.7 mg of N) for total N analysis by the Coleman Model 29A Analyzer, but found that this instrument permitted analysis of up to 3 g of mineral soil and 1 g of organic soil and that the sample analyzed need not be finely ground [satisfactory results were obtained with <40-mesh (420 μm) soil]. They also found that the

TABLE 5 Comparison of Results Obtained in Total Nitrogen Analysis of 12 Soils by Coleman Model 29A Nitrogen Analyzer and by Kjeldahl Methods

		Total N content (%)		
Soil	Sample depth (cm)	Coleman analyzer	Kjeldahl method	HF-Kjeldahl method
Buckner	0–15	0.028	0.028	0.028
Yolo	38–58	0.086	0.083	0.092
Promise	75–112	0.065	0.055	0.068
Clarence	38–58	0.092	0.092	0.093
Ida	0–15	0.108	0.108	0.109
Monona	0–15	0.169	0.169	0.170
Marshall	0–15	0.212	0.211	0.213
Galva	0–15	0.281	0.280	0.281
Clyde	0–15	0.411	0.410	0.412
Glencoe	0–15	0.887	0.890	0.889
Howard peat	0–15	1.64	1.64	1.63
Colo peat	0–15	2.34	2.34	2.33

Source: Ref. 14.

total N analysis of soil by the Model 29A Analyzer was not affected if the sample analyzed contained a substantial amount of water. In contrast, Stewart et al. [11] obtained evidence that a sweeping action caused by the vaporization of water during combustion could vitiate the total N analysis of undried soil samples by the Model 29 Analyzer.

Stewart and Porter [15] showed that indigenous fixed ammonium N in profile samples of two clay soils was fully recovered by the Kjeldahl method only when these samples were treated with HF before analysis. Stewart et al. [11] found that their procedure for the total N analysis of soils using the Coleman Model 29 Analyzer did not recover any more or less of the fixed ammonium N in these profile samples than did the Kjeldahl method. Keeney and Bremner [14] found that the technique they adopted for the total N analysis of soils by the Model 29A Analyzer gave higher values than the Kjeldahl method with two subsoils showing a pronounced HF effect in Kjeldahl analysis, but did not give the quantitative recovery of the fixed ammonium N in these subsoils (Table 5).

Stewart et al. [4] obtained 94.5–100.0% recovery of nitrate N in the analysis of soil by the Coleman Model 29 Analyzer, and Keeney and Bremner [14] obtained 96.0–97.6% recovery of both nitrate N and nitrite N in the analysis of soil by the Coleman Model 29A Analyzer. This suggests that soils contain materials that promote the recovery of nitrate N and nitrite N in Dumas analysis because difficulties have been encountered in the total N analysis of nitrates and nitrites by Dumas procedures [16–18].

The possibility that Coleman Nitrogen Analyzers or other automated Dumas combustion instruments can be adapted successfully for ^{15}N analysis in tracer studies of N transformations in soils deserves attention because the Dumas combustion technique converts labeled N directly to N_2 and is simple and rapid compared with the three-stage technique currently used to convert labeled N to N_2 for the mass spectrometer assay of ^{15}N [19]. Attempts to adapt the Coleman Model 29 Nitrogen Analyzer for ^{15}N analysis have been reported by Barsdate and Dugdale [20] and by Desaty et al. [21], but the only automated Dumas combustion instrument that has thus far shown promise for ^{15}N analysis is the RoboPrep-CN Analyzer marketed by Europa Scientific Ltd., Crewe, England. This is essentially a Carlo–Erba Model 1500 Nitrogen Analyzer modified so that it can be linked via a capillary interface to a mass spectrometer for the ^{15}N analysis of the N_2 produced by Dumas combustion. No studies to evaluate this instrument for total N and ^{15}N analyses of soils or plant materials have been published, but unpublished work in several laboratories indicates that it will prove valuable for such analyses.

Several studies have indicated that Coleman Nitrogen Analyzers can be adapted successfully for the total N analysis of plant materials [4,22–26].

In summary, work to evaluate Coleman Nitrogen Analyzers for the determination of total N in soils indicates that these instruments can be used satisfactorily for the total N analysis of most soils but do not give the quantitative recovery of fixed ammonium N in some soils. These analyzers have the advantage that they require very little bench space compared with the equipment needed for total N analysis by the Kjeldahl method. However, they have gained little acceptance because besides requiring considerable skill and experience for satisfactory use, they permit only 30–40 analyses per day and do not compare favorably with semiautomated Kjeldahl methods for the routine total N analysis of soils.

C. Sulfur Analyzers

Tabatabai and Bremner [27] attempted to adapt the Leco Sulfur Analyzer for determination of total S in soils but were unable to obtain accurate and precise results with this instrument. The procedure giving results in closest agreement with those obtained by three procedures used as reference methods involved the use of methionine to calibrate the burette of the automatic SO_2 titrator and combustion of < 100 mesh (150 μm) soil (usually about 0.25 g) for 7 min using iron and tin-coated copper as combustion accelerators. However, the results obtained by this procedure were less precise and usually higher than those obtained by the methods used for comparison (Tables 6 and 7). The combustion technique evaluated by Tabatabai and Bremner was adopted after the systematic study of factors affecting the analysis of soils by the Leco Sulfur Analyzer. The factors studied included the amount and mesh-size of soil sample, amount and type of combustion accelerator (iron, copper, tin, tin-coated copper), method of treating soil sample with accelerator, type of combustion crucible, time of combustion, and method of calibrating the burette of the automatic SO_2 titrator. The results showed that <100 mesh (150 μm) soil gave more reproducible

TABLE 6 Comparison of Results Obtained in Total Sulfur Analysis of Soils by Leco Sulfur Analyzer and by Other Methods

	Total S content (mg S kg^{-1} soil)			
	Leco S	Other methods[a]		
Soil	analyzer	A	T	S
Buckner	60	56	56	55
Tama	257	246	248	238
Marshall	330	310	315	294
Hayden	354	334	333	324
Webster	405	410	411	392
Primghar	405	418	418	392
Marcus	472	438	441	440
Mean	326	316	317	305

[a]A, acid oxidation method of Arkley [28]; T, alkaline oxidation method of Tabatabai and Bremner [29]; S, dry ashing method of Steinbergs et al. [30].
Source: Ref. 27.

results than < 60 mesh (250 μm) soil and that a combustion time of 7 min gave higher sulfur values than a combustion time of 4 min. They also showed that none of the ceramic crucibles tested gave satisfactorily low and precise blank values and that the results of soil analysis by the Leco Sulfur Analyzer depended on the type of S standard used to calibrate the burette of the automatic SO_2 titrator. The total S values obtained when inorganic S standards were used to calibrate the burette of the titrator were consistently lower than those obtained when calibration was performed with organic S standards.

The Leco method of S analysis is based on the assumption that the combustion technique in this method converts S quantitatively to SO_2. Since the procedure used to determine SO_2 in this method does not recover S evolved as sulfur trioxide (SO_3), Tabatabai and Bremner [27] studied the possibility that their failure to obtain accurate and precise results in the total S analysis of soils by the

TABLE 7 Precision of Results Obtained in Total Sulfur Analysis of Soils by Leco Sulfur Analyzer and by Other Methods

		Total S content (mg S kg^{-1} soil)[b]		
Soil	Method[a]	Range	Mean	SE
Tama	L	224–272	252	5.0
	A	236–254	244	1.8
	T	241–252	247	1.3
	S	227–247	237	3.0
Webster	L	368–445	398	8.7
	A	406–424	411	1.6
	T	406–420	410	1.3
	S	369–406	391	3.5
Marcus	L	394–506	470	10.0
	A	430–447	438	1.6
	T	434–446	441	1.1
	S	399–446	432	7.0

[a]L, Leco S Analyzer; A, acid oxidation method of Arkley [28]; T, alkaline oxidation method of Tabatabai and Bremner [29]; S, dry ashing method of Steinbergs et al. [30].
[b]Ten analyses. SE, standard error.
Source: Ref. 27.

Leco Sulfur Analyzer might be due to the formation of SO_3 as well as SO_2 during the combustion of soils by the Leco technique. To check this possibility, they used 1.0 *M* NaOH instead of dilute HCl to absorb the S-containing gases formed by the combustion of soil samples in the Leco instrument and analyzed this solution for S by the Johnson–Nishita procedure [31] (SO_2 and SO_3 are rapidly absorbed by 1.0 *M* NaOH with the formation of sodium sulfite and sodium sulfate, and both sulfite-S and sulfate-S are determined quantitatively by the Johnson–Nishita procedure). They found, however, that this modification did not improve the accuracy or precision of the Leco method.

Searle [32] described a method for the total S analysis of soils in which the soil sample is heated with molybdenum trioxide, chromium trioxide, and iron powder in a Leco high-frequency induction furnace, and the S-containing gases evolved are absorbed in 1 *M* NaOH, which is analyzed subsequently for S by the Johnson–Nishita procedure [31]. He evidently found that it was necessary to use catalytic oxidants, such as molybdenum and chromium trioxides, to obtain the complete combustion of soil samples with the Leco induction furnace. However, his results showed that when catalytic oxidants were used, the recovery of soil S was far from quantitative if it was determined by the absorption and titration of the SO_2 evolved as in the normal Leco technique. Searle concluded that catalytic oxidants promote the formation of SO_3 during the combustion of soils by the Leco technique and cited work by Green [33] indicating that catalytic oxidants promote the formation of SO_3 during the S analysis of cast iron by an induction-furnace technique. But he did not report the results of S analysis of soils by the normal Leco procedure or give any basis for his conclusion that the use of catalytic oxidants is essential for the complete combustion of soil samples with the Leco induction furnace. Moreover, examination of his data shows that the modified Leco method he proposed for the total S analysis of soils had low precision and gave results that did not agree closely with those obtained by an unpublished acid oxidation procedure used for comparison.

Tiedemann and Anderson [34] attempted to adapt the Leco Sulfur Analyzer for the total S analysis of both soil and plant material. Initially they employed a Leco method recommended for the total S analysis of hydrocarbons, but they found that this method was unsatisfactory for soil or plant analysis and recommended a procedure involving four modifications of this method, including modifications to eliminate interference by "N and Cl from the sample." However, the procedure recommended gave only 80–90% recovery of inorganic S (potassium sulfate) or organic S [*p*-(methylamino) phenol sulfate] added to soil or plant material, and it was not evaluated by comparing its results with those obtained by accepted methods of determining the total S in soils or plants.

Lowe [35] used the Leco Sulfur Analyzer in work to characterize the S in some Canadian soil profiles but did not describe fully the procedure he used for the total S analysis of soils with this instrument. He noted, however, that the procedure did not have good precision, and his results showed that it gave very poor recovery of total S in the soil samples analyzed. This is evident from Table 8, which shows that the total S values he obtained by the Leco Analyzer with several profile samples were lower than the values obtained for HI-reducible S and that, with many samples, they were much lower than the sum of the values obtained by analysis for HI-reducible S and carbon-bonded S.

Kaplan et al. [36] found that the Leco Sulfur Analyzer did not give precise results when used for the total S analysis of marine sediments and that the total S values obtained by this instrument with surface samples of these sediments were much lower than those obtained by a wet combustion method of determining total S. They attributed the low results with surface samples to the incomplete combustion of organic matter and reported that the combustion of these samples produced smoke that interfered with the operation of the automatic photoelectric titrator used for the determination of SO_2

TABLE 8 Sulfur Analyses of Some Alberta Soil Profile Samples

Soil	Depth of sample (cm)	Sulfur content (mg S kg^{-1} soil)[a]			
		T	H	C	H + C
1	0–10	343	338	231	569
	10–18	139	143	73	216
	10–28	87	70	21	91
	28–43	96	89	32	121
2	0–12	869	465	484	949
	12–25	148	253	20	273
	25–37	78	145	14	159
	37–51	183	310	—	—
3	0–3	1040	590	643	1233
	3–18	396	189	144	333
	18–38	204	129	68	207
	38–88	72	76	21	97

[a]T, total S (Leco S Analyzer): H, HI-reducible S; C, carbon-bonded S.
Source: Ref. 35.

in the Leco method of analysis. Tabatabai and Bremner [27] did not experience this smoke problem in the analysis of soils.

Smith [37] recently described a method for the total S analysis of soils and plant material using the Leco Sulfur Analyzer, but he reported no evidence that this method was satisfactory for soil analysis.

VI. SUMMARY

Several instruments permitting the rapid determination of total C, total N, or total S by automated combustion methods have been evaluated for soil analysis. The principles of these methods are discussed, and studies to evaluate these instruments for soil analysis are reviewed.

Work to evaluate the Leco 70-Second Carbon Analyzer has shown that this instrument permits the rapid, accurate, and precise determination of total C in soils and is very satisfactory for routine soil analysis. The Coleman Model 29 and Model 29A Nitrogen Analyzers have been adapted successfully for the total N analysis of soils, but they are not simple and rapid enough for routine soil analysis. Attempts to adapt the Leco Sulfur Analyzer for the accurate and precise determination of total S in soils have not thus far been successful, but this instrument may prove useful for S analysis in some types of soil investigations. Currently available automated combustion instruments that merit evaluation for soil analysis include the Leco range of instruments (the DC-12 Duo-Carb Analyzer, the IR-12 Carbon Determinator, the CS-46 Simultaneous Carbon/Sulfur Determinator, the SC-32 Sulfur Analyzer, the UO-14SP Nitrogen Determinator, the TC-36 Nitrogen Analyzer, the TN-15 Nitrogen Determinator, and the CHN-600 and CHN-800 Analyzers) and the Carlo-Erba Model 1500 Automatic Nitrogen Analyzer.

REFERENCES

1. M. A. Tabatabai and J. M. Bremner, *Soil Sci. Soc. Am. Proc.*, *34*:608 (1970).
2. B. A. Swinehart, *Anal. Chim. Acta*, *91*:417 (1977).
3. S. Terashima, *Anal. Chim. Acta*, *101*:25 (1978).
4. B. A. Stewart, L. K. Porter, and W. E. Beard, *Soil Sci. Soc. Am. Proc.*, *28*:366 (1964).
5. G. Dugan, *Anal. Lett.*, *10*:639 (1977).
6. H. K. T. Wong and A. L. W. Kemp, *Soil Sci.*, *124*:1 (1977).
7. B. H. Sheldrick, *Can. J. Soil Sci.*, *66*:543 (1986).
8. H. V. Drushel, *Anal. Chem.*, *49*:932 (1977).

9. C. A. Stutte and R. T. Weiland, *Crop. Sci.*, *18*:887 (1978).
10. L. E. Allison, *Soil Sci. Soc. Am. Proc.*, *24*:36 (1960).
11. B. A. Stewart, L. K. Porter, and F. E. Clark, *Soil Sci. Soc. Am. Proc.*, *27*:377 (1963).
12. A. W. J. Dyck and R. R. McKibbin, *Can. J. Res.*, *13B*:264 (1935).
13. J. M. Bremner and K. Shaw, *J. Agric. Sci. Camb.*, *51*:22 (1958).
14. D. R. Keeney and J. M. Bremner, *Soil Sci.*, *104*:358 (1967).
15. B. A. Stewart and L. K. Porter, *Soil Sci. Soc. Am. Proc.*, *27*:41 (1963).
16. G. M. Gustin and C. L. Ogg, in *Treatise on Analytical Chemistry*, Part II, Vol. 2 (I. M. Kolthoff and P. J. Elving, eds.), Wiley-Interscience, New York, 1965, pp. 405–498.
17. D. F. Ketchum, *Anal. Chem.*, *36*:957 (1964).
18. A. Steyermark, *Quantitative Organic Microanalysis*, Academic Press, New York, 1961, p. 152.
19. J. M. Bremner, in *Methods of Soil Analysis*, Part 2 (C. A. Black, ed.), Am. Soc. Agron., Madison, Wis., 1965, pp. 1256–1286.
20. R. J. Barsdate and R. C. Dugdale, *Anal. Biochem.*, *13*:1 (1965).
21. D. Desaty, R. McGrath, and L. C. Vining, *Anal. Biochem.*, *29*:22 (1969).
22. M. E. Ebeling, *J. Assoc. Off. Agric. Chem.*, *50*:38 (1967).
23. M. E. Ebeling, *J. Assoc. Off. Agric. Chem.*, *51*:766 (1968).
24. S. N. Farmer, C. J. Howarth, and L. B. Hughes, *Chem. Ind. (Lond.)* (4):154 (1967).
25. F. G. Hamlyn and J. K. R. Gasser, *Chem. Ind. (Lond.)* (35):1142 (1970).
26. G. F. Morris, R. B. Carson, D. A. Shearer, and W. T. Jopkiewicz, *J. Assoc. Off. Agric. Chem.*, *51*:216 (1968).
27. M. A. Tabatabai and J. M. Bremner, *Soil Sci. Soc. Am. Proc.*, *34*:417 (1970).
28. T. A. Arkley, Ph.D dissertation, Univ. Calif., Berkeley, 1961.
29. M. A. Tabatabai and J. M. Bremner, *Soil Sci. Soc. Am. Proc.*, *34*:62 (1970).
30. A. Steinbergs, O. Iismaa, J. R. Freney, and N. J. Barrow, *Anal. Chim. Acta*, *27*:158 (1962).
31. C. M. Johnson and H. Nishita, *Anal. Chem.*, *24*:736 (1952).
32. P. L. Searle, *Analyst (Lond.)*, *93*:540 (1968).
33. H. Green, *Metallurgia*, *60*:229 (1959).
34. A. R. Tiedemann and T. D. Anderson, *Plant Soil*, *35*:197 (1971).
35. L. E. Lowe, *Can. J. Soil Sci.*, *49*:375 (1969).

36. I. R. Kaplan, K. O. Emery, and S. C. Rittenberg, *Geochim. Cosmochim. Acta, 27*:297 (1963).
37. G. R. Smith, *Anal. Lett., 13*:465 (1980).

7

X-Ray Fluorescence Analysis

ANGELA A. JONES *The University of Reading, Reading, Berkshire, England*

I. INTRODUCTION

X-ray fluorescence spectrometry (XRFS) is a method of elemental analysis that assesses the presence and concentration of various elements by measurement of secondary X-radiation from the sample that has been excited by an X-ray source. The method is rapid, does not destroy the sample, and, with automatic instruments, is suitable for routine operation. Elements from the heaviest down to atomic number 9, F, can be determined at levels of a few mg kg^{-1} or less, often with only simple methods of preparation.

In recent years, greatly increased interest of problems of environmental pollution and trace element deficiencies and toxicities in agricultural soils has meant that analytical data are required on some elements that have not previously been subject to routine analysis. XRFS offers a rapid and accurate technique for routinely analyzing these elements, particularly in work of a survey nature. For example, Br and S can be determined readily by XRFS while presenting problems with other techniques. Furthermore, new developments may allow determination of O, N and C in soils.

Revision of Chapter 5 by Carolyn Wilkins in first edition (1983).

In the past few years, much progress has been made in the field of energy-dispersive X-ray spectrometry, which has several advantages, as well as some drawbacks, over the previously more widely used wavelength-dispersive spectrometry. This chapter will, however, deal mainly with wavelength-dispersive spectrometry, which has been applied to many problems in soil analysis, while discussing the advantages and disadvantages of the energy-dispersive systems that may be used increasingly for soil analysis in the next few years. Various technological improvements will lead to these systems becoming more compact and to the analysis of smaller samples. Other forms of X-ray spectrometry, such as electron spectroscopic methods, are not yet widely used and will not be discussed here but have been reviewed briefly by Jenkins [1]. Further gradual developments in instrumentation are expected, particularly toward more reliable equipment and automation.

In this chapter, as in another similar one [2], a brief description is given of the theory of X-ray spectrometry and the component parts of the spectrometer, to enable a new operator to understand the spectrometer, to be aware of the pitfalls, and to be able to assess the possibilities of the various techniques. Methods of sample preparation and other practical considerations are discussed, and applications to the quantitative analysis of different types of samples are reviewed.

II. X-RAY PRODUCTION AND CHARACTERISTICS

Before describing the design and working of an X-ray spectrometer, it is necessary to discuss briefly the behavior and nature of X-rays. Accounts of the physics of X-rays and information related to XRFS can be found in several books, including Liebhafsky et al. [3], Bertin [4], Jenkins and de Vries [5], and Jenkins [1], the last mentioned being a useful general book on many aspects of XRFS. Up-to-date reviews of all aspects of XRFS can be found biennially in *Analytical Chemistry* (e.g., [6]).

A. Production of X-Rays

Characteristic X radiation is produced when excited electrons of an atom return from outer orbitals to inner levels and the atom returns to the ground state. When an atom is bombarded by particles of sufficient energy, the electrons filling the subshells are excited and may fill outer orbitals. On returning to the original atomic configuration, the atom will emit energy, in the form of X radiation, with wavelengths depending on the excited and final orbital in which the electron rests. Many such transitions take place in one atom and a number of characteristic wavelengths can be detected when any element has been excited. Fortunately, the X-ray spectrum of a particular element is simple and the main analytical lines used in XRFS rarely overlap. The commonly used analytical lines have wavelengths in the region 0.02–2 nm (0.2–20 Å) although X-rays can range in wavelength from 0.01–20 nm, bounded on one side by the vacuum ultraviolet region and on the other side by γ rays.

The wavelengths of radiation emitted from any one transition can be determined from the following equation:

$$E = h\nu = \frac{hC}{\lambda} \qquad (1)$$

where E is the difference in energy of the electron in its final and excited states, h is Planck's constant, ν is the frequency of the radiation, C is the velocity of light, and λ is the wavelength of the radiation.

From (1), relationship (2) can be derived

$$\lambda = \frac{1.24}{E} \qquad (2)$$

where λ is expressed in nanometers and E in kilo-electron volts (keV).

Continuous radiation is produced when high-speed electrons enter matter and decelerate, and in X-ray fluorescence spectrometry will be produced in the X-ray tube. Secondary excitation, e.g., by X-rays, will not produce continuous radiation.

B. Types of Radiation

The two types of radiation, continuous and characteristic, are illustrated in Fig. 1. It can be seen that the former has a continuous range of wavelengths. The intensity of this radiation emitted by an X-ray tube varies with the tube current, voltage, and atomic number of the material of the anode. Continuous radiation usually provides the main part of the radiation for specimen excitation.

Characteristic radiation is unique for every element and the spectrum is governed by the electron configuration within the atom. If the reader is not familiar with the structure of the atom and the nomenclature for the shells around the nucleus (K, L, M, etc.), Bertin [4] provides an easily followed summary of the subject. The characteristic X-ray spectrum of an element is formed by X-ray photons emitted as the atom returns from an excited state to the ground state. An electron falling into a vacancy in the K shell will give K

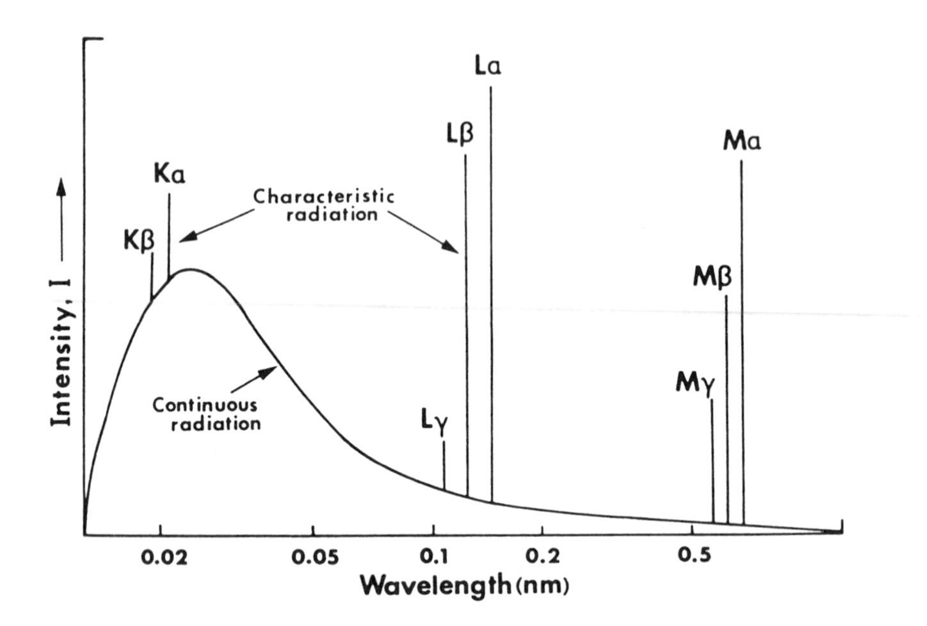

FIG. 1 X-ray spectrum from cellulose samples irradiated by tungsten tube, showing continuous and characteristic radiation.

series lines, i.e., $K\alpha_1$, $K\alpha_2$, $K\beta$, etc., as well as a series of transitions in outer shells. All lines in one series result from electron transitions from different levels to one shell and the number of transitions is limited by various rules applying to quantum numbers (see Bertin [4]). For general analytical work $K\alpha$, $K\beta$, $L\alpha$, and $L\beta$ spectral lines are most important. For the same spectral line, the wavelength decreases as the atomic number increases (Moseley's law)

$$\frac{1}{\lambda} = K_1(Z - K_2)^2 \tag{3}$$

where λ is the wavelength, K_1 and K_2 are constants, and Z is the atomic number.

To eject an electron from the inner shells of an atom, the incident electron must have sufficient energy. The energy needed is called the excitation potential and is measured in electron volts.

C. X-Ray Characteristics of Importance in XRFS

When X-rays interact with matter, they are both absorbed and scattered. Relative to absorption, scatter is small or very small except for elements of low atomic number at short wavelengths.

1. Absorption

As X-rays pass through matter, radiation is absorbed and the intensity of the emergent beam is given by

$$I = I_0 e^{-\mu x} \tag{4}$$

where I, I_0 are the radiation intensities of a monochromatic beam of X-rays as they enter (I_0) and leave (I) an absorber of thickness x with a linear absorption coefficient μ. I is always less than I_0 and the difference is mainly due to photoelectric absorption caused by bombardment of the sample and ejection of electrons by photons. The mass absorption coefficient of a sample is the most useful measure of absorption for practical purposes and the term is derived from Eq. (4) as follows:

$$I = I_0 e^{-(\mu/\rho)\rho x} \tag{5}$$

where μ/ρ is the mass absorption coefficient in square centimeters per gram, x the thickness, and ρ the density of the sample.

μ/ρ is a function of the wavelength of the incident radiation and the atomic number of the absorber. If the absorber contains more than one element, μ/ρ is equal to the sum of the weight fractions multiplied by their individual mass absorption coefficients, i.e.,

$$\frac{\mu}{\rho} = W_1\left(\frac{\mu_1}{\rho_1}\right) + W_2\left(\frac{\mu_2}{\rho_2}\right) + \cdots \tag{6}$$

where W_1, W_2 are weight fractions.

Tables of mass absorption values are available [3,5] and examination of these will show a number of sudden drops in μ/ρ with increasing wavelengths (Fig. 2). These coincide with the K, three L, five M, etc. absorption edges that equate to the binding energies of electrons at these levels. Incident radiation of shorter wavelength than an absorption edge can excite electrons from the same level. In practical terms, if the mass absorption coefficient of a sample is determined on either side of the absorption edge due to a major component, very different values will be obtained and, conversely, two samples containing high and low concentrations of a particular element will have very different mass absorption coefficients at some wavelengths.

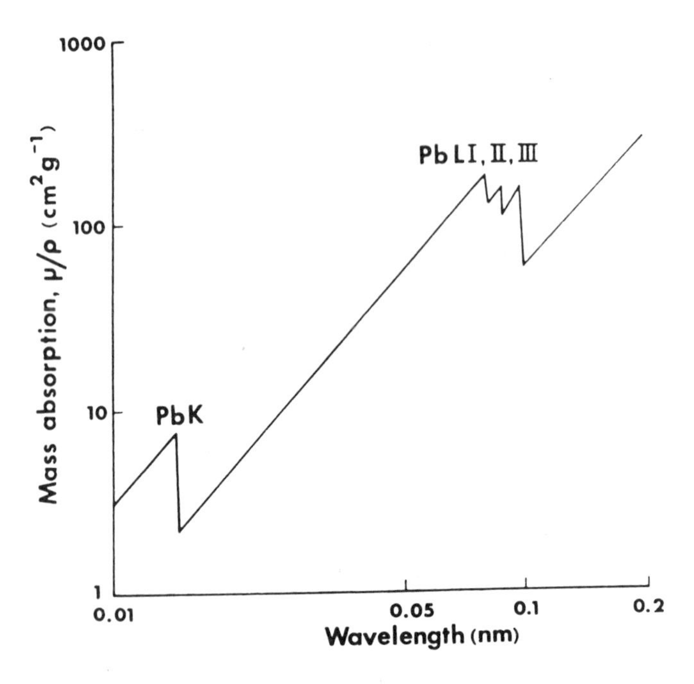

FIG. 2 Variation of mass absorption with wavelength for lead.

2. *Scattered Radiation*

There are two forms of scattered radiation, known as coherent and incoherent scatter. Coherent, or Rayleigh, scatter occurs when a photon hits an electron and is deflected without losing energy. Incoherent, or Compton, scatter is caused by a photon hitting an electron with loss of energy (see also Chaps. 8 and 9). In an X-ray spectrum, this Compton scatter is shown as a diffuse peak of wavelength slightly greater than lines caused by the tube target (coherently scattered radiation). The difference in wavelength (λ) varies only with the angle between scattered and unscattered X-rays, and for most commercial spectrometers of $\Delta\lambda \cong 0.0024$ nm [4]. The ratio of intensity between the coherent and Compton scatter is related to the atomic number of samples and the wavelength of the radiation. Compton scatter can be seen very clearly when looking at samples of low Z and also when using an X-ray tube with high-energy radiation. For example, more Compton scatter would be apparent when irradiating an organic soil than a mineral soil and would also be greater for Mo Kα radiation (0.071

nm) than Cr Kα radiation (0.229 nm). Care must be taken to account for any Compton scattered radiation when interpreting X-ray spectra, especially if the wavelength coincides with that of some element present in the sample.

3. *Diffraction*

When X-rays come into contact with a crystal lattice, they are diffracted according to Bragg's law

$$n\lambda = 2d \sin \theta \quad (7)$$

where n is the line order (e.g., 1, 2, 3), d the distance between diffracting planes of the crystal, and 2θ the angle through which the X-rays are diffracted.

This equation is used in XRFS to identify the elements present in analytical samples; usually, tables [7,8] are available to facilitate identification.

D. Instrumentation

Figure 3 shows generalized diagrams of the components of the two main types of X-ray spectrometer. The bulk of the discussion on instrumentation, however, will refer to the wavelength-dispersive spectrometer. Most other sections of this chapter are equally applicable to both types of spectrometer.

1. *Excitation of Sample*

The excitation source of an X-ray fluorescence spectrometer must be stable and have sufficient output to excite the sample. An X-ray tube (generally the Coolidge type) is most commonly used in XRFS [1,5], especially in wavelength-dispersive spectrometers, and has a high stability.

X-ray tubes can be made with targets composed of various elements, commonly Sc, Cr, Ag, Au, Mo, W, and Rh. The choice of tube is governed by the wavelength of the element of interest, with regard to both the excitation of this wavelength and the likelihood of interference by characteristic tube radiation. The Cr tube or, a recent product, the Sc tube [9], is most suitable for light elements, as a large proportion of long wavelength X-rays is produced. A Rh tube is a good all-purpose one. Tubes have also been developed, the anode of which has a thin layer of a light metal on a heavy metal; with these, different voltages on the tube can excite either light or heavy elements.

Radionuclides are often used as sources of excitation in energy-dispersive X-ray spectrometers, but the low photon yield of these

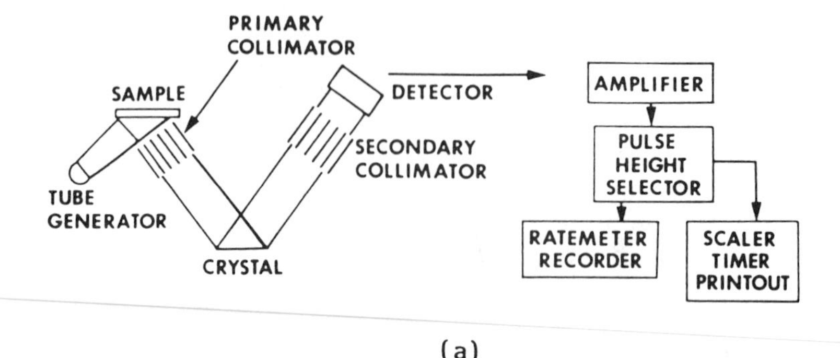

(a)

SAMPLE
EXCITATION
SOURCE
DETECTOR
PREAMPLIFIER
AMPLIFIER
MULTI-
CHANNEL
ANALYZER
DISPLAY
PRINT
OUT

(b)

FIG. 3 Diagrammatic representations of (a) wavelength-dispersive spectrometer and (b) energy-dispersive spectrometer.

sources makes them of no use in a wavelength-dispersive spectrometer.

2. *Crystal Dispersion*

Radiation that is emitted by an excited sample is collimated, usually by a set of Soller slits, and is then diffracted by a crystal of known $2d$ spacing, d being the interplanar spacing between diffracting planes of the crystal. The choice of crystals is large (Table 1) and some crystals can be cut in several ways to give more than one $2d$ spacing.

The crystal sledge inside the goniometer chamber of the spectrometer is linked to the detectors, and as the goniometer is turned,

TABLE 1 Common[a] Diffraction Crystals Most Efficient for Soil Analysis

Substance	2*d* (nm)	Useful range (nm)	Reflection	Comments
Lithium fluoride				
LiF 220	0.2848	0.025–0.272	high	Similar range to LiF 200 used when greater resolution is needed.
LiF 200	0.4028	0.035–0.384	very high	Useful for Ca and heavier elements.
Germanium[b]				
Ge 111	0.6532	0.057–0.623	high	Mainly used for S, P, and Cl.
Pentaerythritol[b]				
PET 002	0.8742	0.076–0.834	high	Temperature-affected 2*d* spacing. Mainly used for K, Al, and Si.
Acid phthalates[b]				
KAP 001	2.664		medium	All unstable in acid environments. TlAP
RbAP 001	2.612	0.22–2.5	high	is most efficient, mainly used for F,
TlAP 001	2.575		high	Na, and Mg.

[a]Less common but potentially very useful crystals include: Graphite 002 ($2d = 0.6715$ nm), for determination of P, S, Cl, and K, very expensive; lead stearate ($2d = 10.04$ nm), for determination of very light elements (Mg-B); indium antimonide ($2d = 0.8406$ nm), without second-order reflections for P, S, and Cl, and great intensity for Si.

[b]Crystal is sometimes mounted 30° advanced in crystal sledge; therefore 2θ angle set on goniometer is 30° less than actual 2θ.

the crystal travels half the angular distance of the detector (see Fig. 3). The goniometer can rarely be used outside a range for 2θ of 0–147° and, accordingly, several crystals must be used to determine all the elements. Tables of wavelengths are available [7,8] and include information on energies, wavelengths, and 2θ for various elements and crystals.

Crystals can be curved or flat, depending on the geometry of the spectrometer. A curved crystal can give a strongly focused beam of radiation and, therefore, greater intensity than a flat crystal. As the optimum curvature of the crystal alters with wavelength, curved crystals are used often in fixed-channel spectrometers where channels are set up for many analyses at one particular wavelength, whereas flat crystals are used in spectrometers for more varied purposes. Multilayer monochromators that span the gap between crystals and gratings increase sensitivity in measuring Na, Mg, and Al and allow the measurement of even lighter elements [10].

3. *Detection of X-Rays*

After the radiation has been diffracted, it is again collimated and detected, usually by a scintillation or gas flow proportional counter. The photons that register in the counter are turned into electrical pulses that can then be measured.

a. Gas Flow Proportional Counter. Photons enter the counter via a thin window (1–6 μm thick) generally made of Mylar, polypropylene, or polycarbonate and react with the inert gas that flows through the counter (generally argon plus 10% methane) to form ion pairs of one positive ion and one electron. As the electrons move toward the wire anode that runs through the center of the counter, they ionize argon atoms, causing a large gain in the number of electrons that reach the anode. The mean number of ion pairs produced, and the size of the electronic output pulse, is proportional to the energy of the radiation. However, if a flow proportional counter wire is allowed to become very pitted or dirty, resolution will deteriorate. The "dead time" is the recovery time after the counter has received a pulse when it cannot function until the effect has dispersed (1–2 μsec).

A peak resulting from the characteristic radiation of the flow counter gas will occur if the energy of photons entering the detector is greater than the absorption edge energy of the counter gas. This is called an escape peak and has energy slightly less than that of the main peak being analyzed. This peak can usually be removed by pulse height selection (see below) but causes interference in some analyses.

b. Scintillation Counter. Photons are converted into light pulses when they fall on a phosphor (generally a sodium iodide crystal). These light pulses are converted into electrical energy by a photomultiplier. The dead time is slightly shorter in the scintillation counter than the gas flow proportional counter, but the resolving power with regard to photon energy is less.

The most useful wavelength range for the scintillation counter is 0.02–0.2 nm, whereas the flow proportional counter becomes sensitive at about 0.15 nm and is useful until about 2 nm if a very thin (1 μm) window is fitted.

c. Semiconductor Detectors. These are used in energy-dispersive spectrometers in which the detector has to resolve all the pulses of different energies. The technological development of these detectors has not advanced in recent years, and they still show significant spectral distortion when recording X-ray photons with energies <2 keV. Descriptions of semiconductor detectors can be found in Bertin [4] and Jenkins [1]; for a more detailed discussion of radiation detector systems, see Chap. 8.

d. Pulse-Height Selection. In the wavelength-dispersive spectrometer, pulses arising from radiations that are diffracted at the same angles but have different voltages can be sorted by pulse-height selection. This can arise when a first-order line for one element has the same diffraction angle as a higher-order line for a different element or, as mentioned above, if any escape peak of different energy overlaps the peak for the element to be analyzed. A lower-level voltage barrier is set and a "window" of an acceptable voltage allows some pulses to pass while blocking those of very high or very low voltage. The multichannel analyzer of an energy-dispersive spectrometer can be thought of as a large number of individual pulse-height selectors.

After pulse-height selection, the pulses are counted by a scaler and timer and can be recorded on a chart for visual appraisal or can be printed. In a wavelength-dispersive spectrometer, the scaler and timer readouts near and at the peak of sample and standard will give information with which to assess the concentration of a particular element. In some cases, dead time correction and correction for interference by other elements must be applied. Information from the multichannel analyzer of an energy-dispersive system must be subjected to peak stripping, live time corrections, escape peak interference corrections, and others before interpreting. This is usually done by computer. Bertin [4] discusses procedures for energy-dispersive analysis.

III. CHOICE OF OPERATING CONDITIONS FOR WAVELENGTH-DISPERSIVE SPECTROMETER

When choosing optimum operating conditions for various parts of the wavelength-dispersive spectrometer, many factors must be considered, such as type of sample, element to be determined, elements likely to interfere, concentration of the element, and availability of equipment.

It is assumed, when discussing operating conditions, that the sample is solid, of a size suitable to be held in a standard sample holder, and infinitely thick (i.e., a sample of sufficient thickness that an increase does not result in an increase in intensity).

A. Generator Setting and X-Ray Tube

Tube voltage should be chosen first to ensure excitation of the element of interest. The minimum voltage (in kilovolts) may be calculated from $V_{min} = 1.235/\lambda_{min}$ if the absorption edge is substituted for λ_{min}. In addition, the voltage chosen should maximize the difference between peak and background relative count rates, and the current (which affects equally both peak and background) should be chosen so that the spectrometer runs at slightly less than the maximum quoted power rating. In the analysis of soils, voltages of 40–60 kV are recommended to excite Kα radiation from lighter elements and Lα radiation from heavier elements while at the same time avoiding high background counts and prolonging tube life.

Although some new spectrometers are not designed to enable easy changing of the X-ray tube for different analyses, if selection is possible, it is governed by the following three main factors:

1. Whether tube lines may interfere with analyte lines, although it may be possible to remove such interference by placing a filter over the tube window to absorb this radiation.
2. The wavelength of the analyte line and the thickness of the tube window. (A tube with a thin window, e.g., a Cr tube, transmits greater amounts of long wavelength radiation that more efficiently excites elements of low atomic number.)
3. The energy of the characteristic tube radiation with regard to optimum excitation of the sample.

The higher the atomic number of the X-ray tube target, the more energy is produced by the continuous radiation.

B. Primary Collimator

A fine collimator will resolve adjoining peaks better than a coarse collimator, but intensity will be lower. If high resolution is needed, the fine collimator should be used, but for the analysis of elements with low atomic numbers, where peaks of neighboring elements are widely separated, the coarse collimator is often adequate. However, when analyzing for Mg with a TlAP crystal, the fine collimator is recommended, due to the overlap of the Mg Kα line by the Ca Kα (111) line in calcareous samples [11].

C. Analyzing Crystal

The angular range of the goniometer is limited and so more than one analyzing crystal must be used to diffract the entire range of wavelengths required in XRFS (Table 1). For the determination of any element, the crystal with the most appropriate $2d$ spacing, greatest reflectivity, and greatest angular resolution should usually be chosen. However, the crystal that gives the greatest resolution rarely gives the greatest intensity and choice depends on individual analyses. For example, when analyzing soil samples for Zr, Y, Sr, Rb, and Br, interfering Kβ lines (i.e., Sr Kβ with Zr Kα, etc.) are minimized by using a LiF 220 crystal. If greater intensity is needed, the LiF 200 crystal is used and correction factors are applied to reduce interference. (Two lines that cannot be resolved are Pb Lα and As Kα, so a correction for Pb must be made if determining As using As Kα, and for Pb determinations, the Pb Lβ line should be used.)

Care should be taken when handling crystals and when using them in the spectrometer. Although many crystals are fairly stable and robust, the following need special attention:

1. *Pentaerythritol (PET)*. This crystal is commonly used for K, Si, and Al analysis but has a large coefficient of thermal expansion. Therefore, the $2d$ spacing, and hence the goniometer reading for the peak of a particular element, varies with temperature. Jenkins and de Vries [5] have plotted temperature sensitivities of various analyzing crystals. It is recommended that the peak position angle be checked frequently during analysis.
2. *Thallium acid phthalate (TlAP)*. This crystal is used for Mg and Na analysis. It is attacked by airborne acid vapors and should, therefore, be stored in a dessicator when not in use.

Reflectivity can decrease as crystals age, so old crystals are often less efficient.

D. Detector

Most wavelength-dispersive spectrometers have a choice of two detectors, the scintillation counter and the flow proportional counter, which can be used separately or, preferably, in tandem. Other sealed gas detectors are also in fixed positions in modern instruments.

The scintillation counter is most efficient for the detection of short wavelengths and can be used up to about 0.3 nm, whereas the flow proportional counter is more efficient for longer wavelengths (above 0.15 nm). A thin window, usually aluminum-coated Mylar (the aluminum coating on the inside of the counter enables a symmetrical field to be maintained around the counter wire) is fitted to the entry port of the flow proportional counter; the 6 μm window is popular and robust, but much greater intensity for the analysis of light elements can be obtained by using a thinner window of 1 or 2 μm that absorbs less long wavelength radiation.

E. Path

The path of the X-rays in the spectrometer is usually through air or vacuum. X-rays are absorbed by air, the extent of absorption increasing with wavelength. At wavelengths of greater than 0.15 nm, a vacuum path is more efficient and becomes increasingly necessary for longer wavelengths. It is, however, not possible to analyze liquids in a conventional sample cell if the system is under vacuum; to overcome this problem, either a special sample cell is used that will withstand a vacuum, or the system can be flushed with helium that absorbs less long wavelength radiation than air [1].

F. Electronic Parameters

Optimum counter voltage (EHT) will vary with attenuation. The attenuation level should be chosen to be only just high enough to eliminate most of the amplifier noise. Any higher setting (lower amplifier gain) necessitates an increase in EHT, which in turn leads to drift in pulse height and broadening of the pulse amplitude distribution. Pulses that are of too large or too small voltage are rejected by upper and lower level discriminators, respectively. Only those falling in the window between these levels are recorded.

Figure 4 shows a plot of count rate against EHT when no discrimination is in operation. The voltage range of the plateau varies with wavelength and attenuation, and an EHT setting should be chosen that is fairly near the middle of the plateau. Figure 5 shows a plot of count rate against increasing lower discriminator level using a very narrow window and constant EHT. For analytical applications, the lower discriminator level should be set at A and the upper at B, giving a window of B-A volts and allowing all the counts falling in the peak to be recorded.

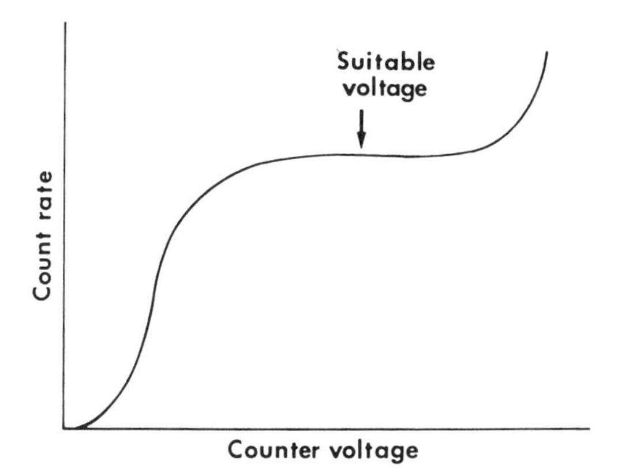

FIG. 4 Plot of X-ray count rate against counter voltage, showing suitable voltage for analysis.

Several checks may be made regularly on the performance of XRFS instruments. These generally include a determination at intervals of the resolution, i.e., the peak width at half height expressed as a percentage of the pulse amplitude distribution, of the counters and especially of the flow counter. In addition, a daily

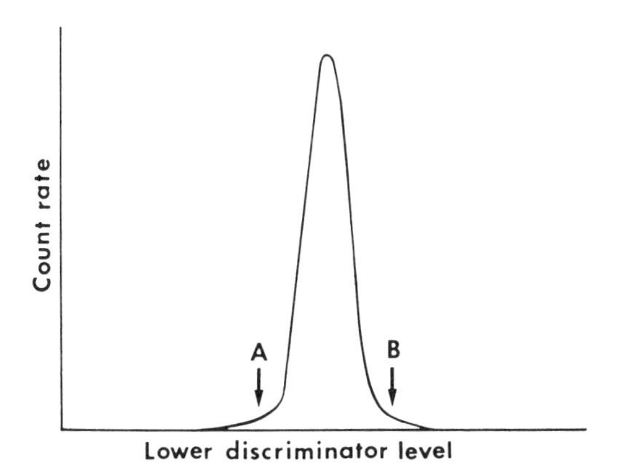

FIG. 5 Variation in count rate with lower discriminator level, using a narrow window (0.05 V). A is a suitable lower level, and the interval A-B a suitable window, for analysis.

check may be made of the counts from a standard (monitor) in standard conditions (see Sec. VI.A).

Choice of count rates and counting times is discussed in Secs. V and VI.

IV. SAMPLE PREPARATION

There are many different ways of preparing a sample for XRFS analysis and the choice depends on several factors, including desired accuracy and precision of results, amount and nature of the sample, and concentration of the element of interest.

For qualitative analysis, very little sample preparation is usually necessary. For example, a rock sample can be placed in a sample holder, dry soil can be poured into a sample cup that has Mylar or polycarbonate at the base, or a powder can be spread on Scotch tape and fixed into a sample holder. Samples for quantitative analysis, however, must be much more carefully prepared to ensure the homogeneity of the sample, sufficiently fine particle size, flat surface of the sample, and other factors to be discussed below.

A. General Procedures

When choosing methods of sample preparation for XRFS, it is necessary to consider sampling in general. It is difficult to obtain a sample of soil representative of a particular area, and intrafield variation of various soil properties is often as great as interfield variation. However, when a sample of soil enters the laboratory, care should be taken to ensure that the specimen taken for analysis is as representative and homogeneous as possible. First, if the sample is to be dried, it must be realized that some elements may volatilize at fairly low temperatures, e.g., S, Se, Hg [12,13]. Also, elements will be more easily lost from peaty soils than from mineral samples, in which they have more stable forms. The following scheme of sample preparation should ensure that a fairly representative specimen is analyzed:

1. Air-dry soil.
2. Select a random subsample of the whole, if necessary by quartering or riffling.
3. Grind the subsample in a dry state.

For most chemical analyses, dry grinding of the sample is necessary and this should be done by a method that will not contaminate the sample with the elements of interest. To check for contamination, duplicate samples can be ground by separate methods, e.g., using apparatus made of different materials, and analytical results compared. Information on the materials used in grinding equipment can be obtained from the manufacturers. There is also a possibility of contamination by solder from brass sieves, and with all trace element determinations, if sieving is necessary, nylon sieves should be used.

B. Sample Preparation for XRFS

1. Undiluted Samples

In soils, for the analysis of elements of atomic number greater than 23, or for the analysis of some of the very light elements (e.g., F, Na, Cl, S), a convenient method of sample preparation is to grind the dry soil finely (until it feels like flour when rubbed between the fingers; approximately 50 μm in diameter) and then to press the soil into an aluminum cap or a boric acid backing of an appropriate size to fit in the sample holder. A detailed account of sample grinding for XRFS has been written by Jenkins and de Vries [5], and dies for pressing disks can be obtained commercially or constructed from the drawings of Norrish and Chappell [14]. Samples and standards for comparison must be pressed at the same pressure; 100–150 MPa (1–1.5 t cm^{-2}) is appropriate [14], but some samples, e.g., quartzose soils, are still friable at this pressure and it may be necessary to increase the pressure on the sample or add a binding agent. Useful binding agents for soils are polyvinyl alcohol, urea, and cellulose, added in the smallest quantities possible (thus reducing dilution of the sample and matrix changes). (It should be remembered that all such materials may, of course, contain a proportion of the element of interest, e.g., Cu in cellulose, P in Mylar.) A sample of about 2 g will make an adequately thick disk for analysis of most soils using K spectra (up to $Z = 40$) but for very peaty soils, more sample may be needed to make an infinitely thick disk.

Samples to be analyzed for Na and Cl should be handled particularly carefully to avoid transferring these elements from the hands to the analytical surfaces.

2. Fused Samples

Fused samples are used in many laboratories when analyzing soils and other silicate-rich materials for some major elements (Mg, Al, Si, P, K, Ca, Ti, and Fe). Several methods of fusion are discussed in detail by Hutchinson [15], and the choice of fusion mixture will depend on the chemical composition of the sample. Probably the most useful established method for major element determinations in soils is that of Norrish and Hutton [16]. In this method, the soil sample, usually ignited, is converted into a glass by fusing at 950° with a borate flux and oxidizer in the ratio 0.28 g soil:1.5 g flux. The borate flux is composed of lithium tetraborate (38.0 g), lithium carbonate (29.6 g), and lanthanum oxide (13.2 g), which are fused at 1000°C, quenched, and homogenized. Full details of the preparation of the fusion mixture and the glass disk are to be found in Norrish and Hutton [16], but the following points should be noted:

1. The fusion mixture is available commercially as Spectroflux 105 (Johnson Matthey Chemicals Ltd., Hatton Garden, London). The La content of this flux has, unfortunately, been found to vary by as much as 6%, which can alter the mass absorption characteristics significantly, e.g., a 6% increase in La content depresses counts for Si Kα by 2% [17].
2. Glass disk preparation is made considerably easier if a mechanical plunger is used to press the disk.
3. The weight loss on ignition should be determined for each batch of fusion mixture by igniting a weighed amount, cooling and reweighing, using the same conditions as those used when fusing samples to make glass disks. Alternatively, the flux should be weighed immediately after being dried at 400°C.
4. The recommended temperatures can be obtained easily with a compressed air/natural or coal gas burner, a gas/oxygen burner, or a propane/oxygen burner.
5. The asbestos parts of the apparatus can be replaced by silica or alumina material, e.g., "Supalux."
6. The graphite disks can be replaced by duralumin disks.

The pressing of the disks by this method takes a certain amount of skill but it is worth persevering. For a full explanation of experimental and theoretical considerations, the reader should refer to Norrish and Hutton [16] and also to Harvey et al. [18], who suggested the modifications listed above to make sample preparation easier. Hutchinson [15] has given a very detailed and simple account of sample preparation for this method.

When performing major element analysis in some laboratories, it may be convenient to determine Na and Mg by other chemical methods, as Na determinations done by XRFS are not very sensitive and Mg determinations may be affected by errors caused, particularly by background variations in samples and standards. The Na and Mg determinations can be done by other methods on the redissolved disk. The determination of Mg is explained in detail by Norrish and Hutton [16] but it is likely to be the least accurate and precise of the elements determined.

Other fusion methods that have potential for soil investigations have been described by Haukka and Thomas [19], Thomas and Haukka [20], and Hutton and Elliott [21]. A glass is made with a 2:1 mixture of sample and lithium metaborate that can be used to determine major and minor elements. Glasses are made in a similar way to that described by Norrish and Hutton [16] and matrix corrections are given. The fact that major and minor elements can be determined on the same sample, which has been diluted very little without the addition of a heavy absorber, is very useful, especially on automatic spectrometers and equipment with multichannel analyzers. However, matrix coefficients and standardization constants must be determined by the user.

The preparation of standard samples needs particular care. Standards made with Al powder as a binder are very stable even for light elements, whereas those made with epoxy as a binder will degrade [22]. If mixtures are to be used, these must be mixed very carefully, and often it is only after using a combination of grinding and ball milling that homogeneous pellets can be produced.

3. *Liquids*

It is often useful to determine concentrations of elements in solution by XRFS when the solution can be poured into a liquid sample holder to be analyzed directly. Solutions containing concentrations below the detection limit can often be concentrated (see Sec. VII).

The size of any sample may be made smaller than the usual maximum if smaller sample holders are used or liners and masks are inserted into the holders. As the size of the area of the sample irradiated is defined by the window from the X-ray tube, this should be reduced at the same time by a suitable mask. Lower counts inevitably result but can often be tolerated, or compensated for by longer counting times.

V. QUALITATIVE AND SEMIQUANTITATIVE ANALYSIS

X-ray fluorescence spectrometry, especially energy-dispersive spectrometry, is extremely useful for identifying the elements contained in a sample. The wavelength-dispersive spectrometer, although spanning the same wavelength range as the energy-dispersive system but with greater sensitivity, has to be run using several changes of crystal, counter, and electronics, whereas the energy-dispersive spectrometer can display the whole spectrum at the same time, making a rapid assessment of major and moderate constituents very easy.

For this purpose, an energy-dispersive system is often fitted to a scanning electron microscope and can then be used qualitatively to determine the constituents of individual grains in soils or soil fractions. Figure 6a shows the part of the spectrum from a soil sample obtained with a lithium fluoride crystal in a wavelength-dispersive spectrometer, and Fig. 6b shows the corresponding spectrum with an energy-dispersive system.

A. Operation of the Wavelength-Dispersive Spectrometer for Qualitative Analysis

The goniometer arm of the spectrometer, to which the detectors are coupled, is driven twice as fast as the diffraction crystal inside the crystal chamber. The goniometer can be driven automatically at various speeds, and the detector is maintained in the correct position to receive the diffracted X rays.

To identify the elements in a sample by XRFS, the procedure is as follows:

1. Present sample to spectrometer. For qualitative purposes, the sample can be stuck onto Scotch tape and placed in a liquid sample holder, absorbed onto a filter paper, or irradiated in solution.
2. Select generator current and voltage. Count rates of below 50,000 and above 100 counts per second should be aimed for. (N.B. Do not operate the generator above the approved safe power rating.)
3. Select primary collimator and crystal (Table 3).
4. Select counter (Table 3).
5. Select counter EHT. This can be done as described in Sec. III.F for an element in the range of the scan.
6. Select window. Use a very wide window or no window.
7. Record trace. A trace of the form of Fig. 6a will be recorded and the peaks can then be identified. The alignment of the chart with the 2θ angle marker pen makes identification of the 2θ angle straightforward.
8. Identify elements. Tables of X-ray wavelengths are available that give the 2θ angles for particular crystals and the main analytical lines can be picked out [7,8]. It is easy to mistake a peak for one element when it is, in fact, a line resulting from the X-ray tube, including Compton scatter, contamination from the sample holder, or a less intense peak from a major constituent of the sample. When identifying peaks, make sure that the $K\beta$ peak is identified, as well as the $K\alpha$ peak (intensity is generally 5:1, $K\alpha$:$K\beta$) or the $L\alpha$, β, and γ peaks. It is useful

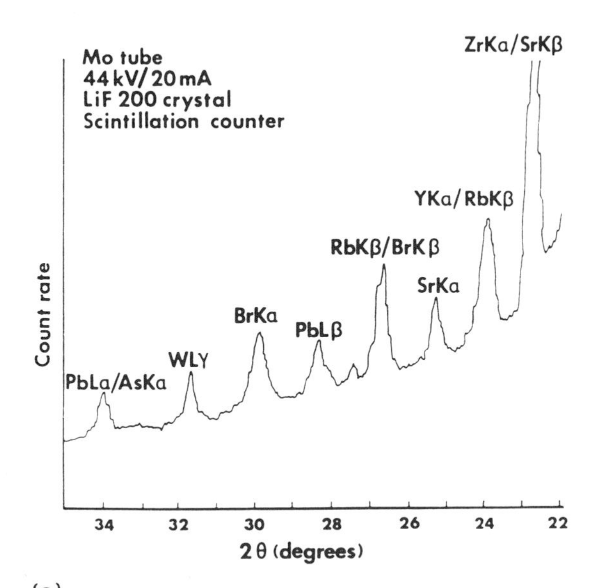

(a)

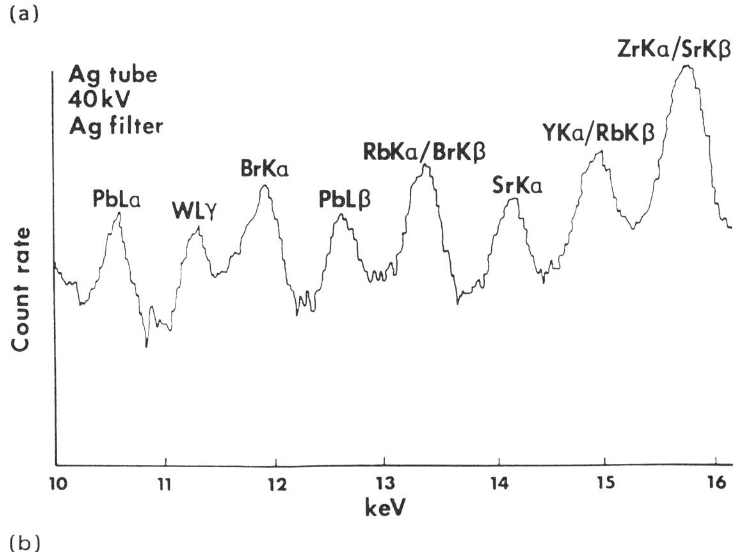

(b)

FIG. 6 (a) Spectrum from soil sample using wavelength-dispersive spectrometer. Soil concentrations (mg kg^{-1}): Zr:403; Y:91; Sr:35; Rb:91; Pb:100; Br:89; As:4. The presence of W is due to contamination from the grinding barrel. (b) Spectrum from soil sample using energy-dispersive spectrometer. Soil concentrations as shown for Fig. 6a.

to make a transparent overlay of the tube and common elemental lines for easy identification in qualitative analysis.

B. Semiquantitative Analysis

If reasonably flat and homogenous samples are used, the X-ray spectrum described above can be used to measure intensities of different elements and, by comparison with standards of similar composition, a semiquantitative estimate of elemental concentrations can be made. For soil determinations, standards made up of Al_2O_3 or Al_2O_3 + SiO_2 (more friable) plus 5–10% Fe_2O_3 contain some of the main constituents of a natural soil and have a similar matrix to a mineral soil for the purpose of XRFS. Standards for plant analysis should be made from cellulose. Known additions of elements of interest enable the standards to be compared with samples.

This method of semiquantitative analysis was used by Williams [23] to obtain information on 12 elements (Mn, Fe, Ni, Cu, Zn, As, Br, Rb, Sr, Y, Zr, Pb) in a large number of soils. The peak height above background for a particular element was compared with that of standards, any interference from other peaks was subtracted, and a simple matrix correction was applied to allow for differences in composition between sample and standard. For the last 10 of the elements, the matrix correction was obtained very simply from the measurement of the background between peaks, but iron and manganese needed a more complicated treatment depending on measured iron content.

When energy-dispersive XRFS is being used, energy tables can be used to identify peaks. However, care must be taken when interpreting the spectra (Fig. 6b), as the resolution is poorer than that of the wavelength-dispersive spectrometer and misleading peaks can arise, including escape peaks and $K\beta$ peaks.

VI. QUANTITATIVE ANALYSIS

In XRFS, the count rate obtained from a particular element when a specimen is irradiated is proportional to the concentration of that element. Thus, if the count rate for the element in a sample is compared with the count rate of a standard of known composition, an estimate of the amount of the element can be made. There are, however, factors caused by differences in composition in different specimens that will influence count rate and must be corrected for.

If there are large amounts of the element of interest in the sample and the background near the peak is low, the background can be ignored. However, in most minor element determinations, background count rates should be measured on either side of the peak and as near to the peak as possible, without being influenced by radiation due to the element being determined or by any other peaks in the vicinity. The background count rate at the peak position can be estimated also by measuring background and peak count rates on a specimen prepared from a pure chemical, e.g., silica, that does not contain the element to be determined. A peak-to-background ratio can be determined and used to correct peak count rates for the background in samples.

Results obtained from comparing the count rates of samples with those of standards must be corrected for variations in the mass absorption coefficient between samples and standards. Methods of determining mass absorption coefficients are described in Sec. VI.B.

A. Sensitivity, Accuracy, and Precision

Limits of detection for elements being determined by XRFS vary with the matrix of the sample and with the equipment being used. Figure 7 shows a rough guide to limits of detection readily obtained on a wavelength-dispersive spectrometer. However, sensitivity, accuracy, and precision are based on careful measurement and correction procedures. The main effects that will influence analyses are absorption, particle size, and enhancement. The first two have a large effect in soil analysis but enhancement, which will occur when the photoelectric absorption of a characteristic line by a matrix element takes place, may often be ignored. It is especially noticeable where radiation from a characteristic line is of slightly higher energy than the element being measured and has the effect of increasing the radiation emitted by this element. In soils, the analysis of major elements may be affected by this phenomenon, but if fused samples are analyzed by the method of Norrish and Hutton [16], enhancement effects are accounted for in the computation of results. The determination of minor elements in soil Fe/Mn concretions can also be affected strongly by enhancement and absorption.

Bertin [4] lists sources of errors that are likely to mar the accuracy and precision of quantitative XRFS determination. They are dealt with, in turn, below.

1. *Counting Errors*

Counting statistics are dealt with in great detail by Jenkins and de Vries [5]. Many X-ray analysts find it convenient to set a fixed time for counting, although this is not always as accurate as setting fixed counts. Counting errors become large at very high count rates due to dead time and pulse drift. At count rates of over 5000 sec^{-1}, a dead time correction must be applied (see Chap. 6). Dead time can be calculated by several methods comparing ratios of high

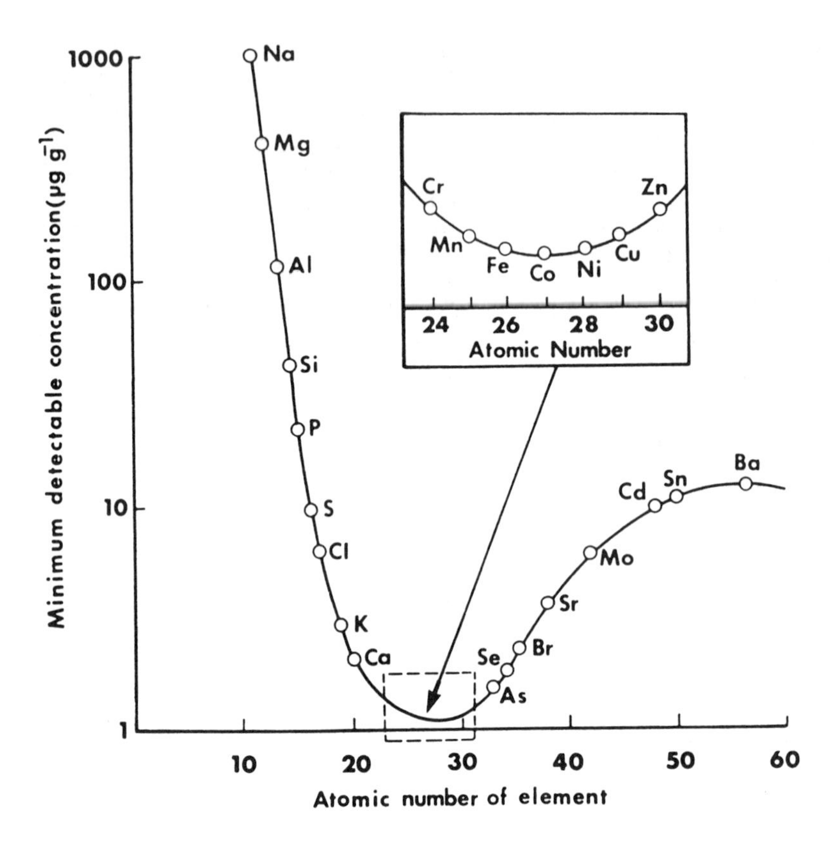

FIG. 7 Approximate guide to minimum concentrations detectable by XRFS. (After G. Brown, unpublished results.)

and low X-ray intensity [14,24]. The optimum distribution of counting time between peak and background varies with the difference in count rates of peak and background in the following way:

$$\frac{t_p}{t_b} = \left(\frac{C_p}{C_b}\right)^{1/2} \tag{8}$$

where t_p and t_b are the counting times for peak and background, respectively, and C_p and C_b are the corresponding count rates.

Therefore, although the time should be divided equally between peak and background at very low concentrations, as concentration increases, so should the ratio $t_p:t_b$. The lower limit of detection Q for a particular analysis is given by

$$Q = \frac{3}{m}\left(\frac{C_b}{t_b}\right)^{1/2} \tag{9}$$

where m is counts per second per unit of concentration (e.g., micrograms per gram or percent). However, the practical limit for a determination is higher than that evaluated statistically and is discussed by Jenkins [1].

2. *Instrumental Errors*

These are generally small errors only, caused by generator instability, pulse drift in counters, and crystal instability with temperature changes (mainly PET). Generators are usually stable in the short term provided that adequate time is given (usually 1 hr) to warm up and the main supply is reasonably quite stable. Changes in the $2d$ spacing of temperature-sensitive crystals make it advisable to use a standard to every three or four samples and to check frequently the 2θ angle. Energy-dispersive systems have very stable electronics and the temperature of the detector is closely controlled.

3. *Variations in Operating Conditions*

Errors due to this cause have been reduced drastically by automatic and semiautomatic spectrometers. With manual spectrometers, care must be taken to reproduce operating conditions exactly when changing such parameters as 2θ angle, crystal, and EHT. Careful sample loading and labeling are, of course, essential.

4. *Variations in Samples*

Errors due to variations in samples are probably among the most important in quantitative analysis. The most important factors are homogeneity, particle size, flatness, and thickness.

Norrish and Chappell [14] outlined the theory involved in the choice of particle size for XRFS and recommended grinding until there is no change in X-ray intensity. Care must be taken, however, that some crystals, e.g., mica, do not orientate themselves to form a film over the top of a pressed pellet of a micaceous sample; also, saline samples may form a layer of salt on the surface. Berry et al. [25], Hunter and Rhodes [26], and Rhodes and Hunter [27] have all discussed particle size effects in detail.

If a sample is not infinitely thick with regard to X-rays, there will be loss of intensity. The amount of sample necessary varies with wavelengths, and samples for light element determinations can be thinner than samples for the heavier elements. The thickness necessary for a particular wavelength and type of sample can often be calculated readily or established by preparing samples of varying thickness until no increase in the count rate is detected with increased thickness.

5. *Line Interferences*

These cause errors that can increase the intensity of the line being analyzed or of background determinations on either side of the peak. Choice of crystals can reduce interferences, but some peaks cannot be resolved, e.g., Pb Lα and As Kα, and with others a loss of intensity has to be accepted if some lines are to be separated, e.g., using the LiF 220 crystal to distinguish Co Kα from Fe Kβ. First-order interferences, i.e., those where $n = 1$ in the Bragg equation (7), can cause large errors (e.g., Sr Kβ with Zr Kα; Rb Kβ with Y Kα); tube lines, and their Compton scatter, can interfere (e.g., Cr Kβ with Mn Kα, Mo Kα with Zn Kα); and high-order interferences ($n > 1$) can cause error although they can be reduced by pulse height discrimination. Escape peaks can also cause difficulties in some analyses, and as well, the spectral purity of the tube should be monitored from time to time.

6. *Calibration Errors*

Calibration errors can be kept to a minimum by the choice of carefully chosen and prepared standards. For soils, artificial standards can be made up in a "soil base" mixture of iron, aluminum, and silicon oxides or iron and aluminum oxides [24]. Adequately analyzed standards and soils are uncommon but Abbey [28] offers some; other soils and sediments with certified contents are also available [29,30] and working values for a wide range of rocks and minerals have been compiled [31]. It is common practice to use U.S. Geological Survey standard rocks [32]. However, it must be noted that when using these rocks as standards, the matrices vary greatly, with resulting big differences in mass absorption coefficients.

B. Analytical Procedures

Quantitative analysis almost always involves more than a straightforward comparison between samples and standards. For samples all of similar composition, with similar mass absorption coefficients, this is possible but unlikely in soil analysis. One of the few exceptions is the analysis of some of the very light elements in soils, e.g., F, Na, and S, where large differences are needed in major element concentrations to cause large differences in mass absorption [33]. The Na peak, therefore, can be measured and intensity versus concentration can be read straight from a calibration. Sulfur, another fairly minor light element, can also be measured this way but Brown and Kanaris-Sotiriou [34] recommend a correction for mass absorption in highly organic soils.

Methods of correcting for different matrices are described below.

1. *Addition of Internal Standard*

The sample can be "spiked" with a known amount of the element to be analyzed and the initial concentration of the element in the sample is derived by extrapolation. This can be a very accurate and precise way of determining the amount of a particular element in a sample, as matrix effects are avoided [35].

2. *Dilution*

Dilution of the sample is a very efficient way of decreasing matrix effects and also, for major element analysis, bringing down the concentration to a reasonable level so that dead time corrections, pulse drift, and other phenomena found at high count rates are reduced.

3. *Determination of Mass Absorption Coefficient*

The mass absorption coefficient (MAC) can be measured directly for wavelengths shorter than around 0.3 nm by using a small, thin pellet of the sample to attenuate the X-ray beam.

MAC is calculated from the equation

$$A = \frac{1}{\rho x} \ln \frac{I_0}{I} \tag{10}$$

where A is the MAC, ρx is in units of grams per square centimeter, and I_0 and I are the intensities of the unattenuated and attenuated X-ray beam, respectively.

The method and apparatus has been described in detail by Norrish and Chappell [14] and Hutchinson [15]. The pellet must be placed in a holder between the scintillation counter and the diffraction crystal, but in the newer spectrometers, it is difficult to position the pellet, and this method of MAC measurement is therefore not suitable for modern, and especially automatic, spectrometers.

Indirect measurements of mass absorption can be achieved by several methods. A correction based on the intensity of background radiation near the peak is a convenient estimate of scattered radiation as this usually has to be measured if determining minor elements (e.g., [23]). Table 2 shows the relationship between background intensities and the mass absorption coefficient, both expressed as ratios to an artificial soil base. This method is only reliable with samples and standards of fairly similar compositions and will not be accurate if extrapolated past an absorption edge of an element of high and variable concentration (e.g., iron in soils).

TABLE 2 Comparison of Mass Absorption Ratios Obtained by Quantitative and Semiquantitative Methods

	Mass absorption ratio[a] for Sr		Mass absorption ratio for Zn	
Sample	Quantitative[b]	Semi-quantitative[c]	Quantitative	Semi-quantitative
1	0.83	0.88	0.86	0.88
2	0.97	1.03	0.98	0.94
3	1.16	1.18	1.15	1.03
4	0.82	0.92	0.84	0.84
5	1.00	1.01	1.01	0.97

[a]Mass absorption coefficient expressed as ratio to that of artificial soil blank.
[b]Measured by attenuating X-ray beam [14].
[c]Calculated from background count rate.

The intensity of scattered tube radiation can also be measured to estimate matrix variations and is a good approximation if a coherently or incoherently scattered line can be found near the line being analyzed [36]. Norrish and Chappell [14] point out that there are limitations to these methods, including the following points: the background can be affected by the tail of the analyte line (this can be corrected for), by interference from other lines including satellite lines [1], and by crystal fluorescence. Measurements of Compton scattered radiation are less misleading in these respects. Methods using a modified background ratio [37] or the linear relationship between the reciprocal of analyte-line intensity and the reciprocal of concentration [38] show that a variety of approaches can be successful in dealing with the problems associated with variations in MAC. Rousseau [39] has proposed a new method that accurately corrects for matrix effects.

If the major element composition of the sample is known, the mass absorption of the sample can be calculated from mass absorption tables using Eq. (6). A method using Rayleigh scattering to determine MAC has also been described [40].

VII. EXAMPLES IN SOIL ANALYSIS

X-ray fluorescence spectrometry has been applied to many aspects of soil science and compares favorably with other methods of analysis [41,42]. Table 3 summarizes analytical conditions for common soil elements, gives examples of XRFS determination of these elements in soils, and also quotes the mean concentrations expected in soils [43] and the minimum detection limits for some elements [14]. The ease of sample preparation and the fact that the same sample can be used to determine many elements often simplify the analysis of "difficult" elements or samples [44] and make XRFS a useful tool for multielement surveys for both major elements (e.g., [45–47]) and minor elements (e.g., [48–54]).

The determination of sulfur by XRFS is rapid and sensitive and is of great value in soil science and plant analysis [55,56]. For example, XRFS has been used to determine S to locate acid sulfate soils [57], and S-deficient sites and S patterns in soils and herbage in a general survey [58]. The sulfur content can be determined on a pressed powder pellet and peak count rates of samples and standards can be compared directly. When analyzing highly organic samples, however, a matrix correction for organic carbon is necessary [34,59]. If the vacuum pump oil of the spectrometer contains S, the absorbent surface of the sample pellet may become contaminated, and Kanaris–Sotiriou and Brown [60] recommended coating the sample with a polycarbonate film to cut down S absorption from contaminants. Sulfur has been determined by XRFS in soil extracts [61,62] and various soil fractions [63], by adding a variety of diluents to the sample to reduce matrix effects [64,65]. However, Bolton et al. [55] found that S could be determined rapidly on pressed powders and McLachan and Crawford [66] found the same with loose powders. When making standards for plant analysis, it must be realized that during irradiation S can be vaporized from samples containing organic or elemental S [67]. Chlorine can also be determined by XRFS, but there are many forms of contamination when preparing samples and standards that may interfere with this analysis (Sec. IV.B).

Many investigations of Mn, Co, Cu, Zn, and Mo have been done on soils, both to determine the total amounts of these plant constituents and to assess the extractable percentages. The determination of total Mn, Cu, and Zn in the range found in soils is straightforward by XRFS [68], although any contamination in the tube anode by Ni, Cu, and Zn must be corrected for [69] and Au L Compton scatter overlaps the Zn Kα peak. The other two important nutrients mentioned here, Co and Mo, cannot be determined satisfactorily by XRFS at deficiency levels without preconcentration.

TABLE 3 Conditions for XRF Analysis of Soils

Element	Mean soil concentration [43] ($mg\ kg^{-1}$)[a]	Lower limit of detection ($mg\ kg^{-1}$)[b]			
			Optimal radiation[c]	Crystal	Line
F	200		Sc	TlAP	F Kα
Na	0.6%	15	Sc	TlAP	Na Kα
Mg	0.8% MgO	0.04% MgO	Sc	TlAP	Mg Kα
Al	13.4% Al_2O_3	0.016% Al_2O_3	Sc	PET	Al Kα
Si	70.5% SiO_2	0.025% SiO_2	Sc	PET	Si Kα
P	0.14% P_2O_5	0.008% P_2O_5	Sc	Ge 111	P Kα
S	700		Sc	Ge 111	S Kα
Cl	100		Sc	Ge 111	Cl Kα
K	1.7% K_2O	0.004% K_2O	Cr	PET	K Kα
Ca	1.92% CaO	0.002% CaO	Cr	LiF 200	Ca Kα
Sc	7		Cr	LiF 200	Sc Kα
Ti	0.83% TiO_2	0.003% TiO_2	Cr	LiF 200	Ti Kα
V	100	1	Au/W	LiF 220	V Kα
Cr	100	<1	Au/W	LiF 200	Cr Kα
Mn	850	<1	Au/W	LiF 200	Mn Kα
Fe	5.4% Fe_2O_3	<1	Au/W	LiF 200	Fe Kα
Co	8	<1	Au/W	LiF 220	Co Kα
Ni	40	<1	Au	LiF 200	Ni Kα
Cu	20	<1	Au	LiF 200	Cu Kα
Zn	50	<1	Au	LiF 200	Zn Kα
Ga	<1–30	<1	Mo/Au/W	LiF 200	Ga Kα
Ge	1		Mo/Au/W	LiF 200	Ge Kα
As	6		Mo/Au/W	LiF 200	As Kα
Se	<1				
Br	5		Mo/Au/W	LiF 200	Br Kα
Rb	100	<1	Mo/Au/W	LiF 200	Rb Kα
Sr	300	<1	Mo/Au/W	LiF 200	Sr Kα
Y	50	<1	Mo/Au/W	LiF 200	Y Kα
Zr	300	<1	Mo/Au/W	LiF 200	Zr Kα
Mo	2		Mo/Au/W	LiF 200	Mo Kα
Cd	0.06		Mo/Au/W	LiF 200	Cd Kα
			Mo/Au/W		Cd Lα
Sn	10		Cr/W	LiF 200	Sn Kα
			Cr/W		Sn Lα
I	5		Cr	LiF 200	I Lα
Cs			Cr	LiF 200	Cs Lα
Ba	500	4	Au/W/Mo	LiF 200	B Lβ

Operating conditions				
2θ	Counter[d]	Sample preparation	Interferences in soils	References
90.62	FPC	pressed powder		
55.08	FPC	pressed powder		
45.18	FPC	glass disk	Ca	[11]
145.07	FPC	glass disk		
109.21	FPC	glass disk		
140.90	FPC	glass disk		
110.68	FPC	pressed powder		[33,55–67]
92.75	FPC	pressed powder		[33,91,92]
50.69	FPC	glass disk		
113.06	FPC/SA	glass disk		
97.65	FPC/SA	concentration req.		[93]
86.12	FPC/SA	glass disk		[70,72]
123.16	FPC/SA	pressed powder	Ti	[94]
69.34	FPC/SA	powder/glass	V, Ba	
62.96	FPC/SA	powder/glass	Cr	[68,95,96]
57.51	FPC/SA	powder/glass	Mn	[94,99,100]
52.79	FPC/SA/SC		Fe	[24,68,94,97]
48.66	FPC/SA/SC	pressed powder		[69]
45.02	FPC/SA/SC	pressed powder	Ni	[68,96]
41.79	FPC/SA/SC	pressed powder	Au(tube), Cu	[50,68,98]
38.91	SC/SX	pressed powder		[35]
	SC/SX	concentration req.		
33.99	SC/SX	pressed powder	Pb	
	SC/SX	concentration req.		
29.96	SC/SX	pressed powder		[99–101]
26.91	SC/SX	pressed powder	Br	
25.14	SC/SX	pressed powder		[49]
23.79	SC	pressed powder	Rb	[73]
22.55	SC	pressed powder	Sr	[48,71,73]
20.84	SC	concentration req.		
15.32	SC			
158.3	FPC	concentration req.		
14.04	SC	concentration req.		
126.7	FPC			
102.80	FPC	concentration req.		
91.72	FPC	pressed powder		
79.18	FPC	pressed powder	Ti	

(continued)

TABLE 3 (Continued)

Element	Mean soil concentration [43] (mg kg^{-1})[a]	Lower limit of detection (mg kg^{-1})[b]	Optimal radiation[c]	Crystal	Line
La	30		Au/W/Mo	LiF 220	La Lα
Ce	50		Au/W/Mo	LiF 220	Ce Lα
Hg	0.03 0.8		Au/Mo	LiF 200	Hg Lα
Pb	10	5	Au/Mo	LiF 200	Pb Lβ
U	1	1	Au/Mo	LiF 220	U Lα

[a]Except where shown.
[b]Detection limits for minor elements in quartz matrix [14].
[c]Rh can provide suitable radiation over wide range of elements, but sensitivity improved greatly by using specific energizing radiation.
[d]FPC: flow proportional counter; SA: sealed argon counter; SC: scintillation counter; SX: sealed xenon counter.

Elements contained in resistate minerals, e.g., Zr and Ti, may often give information about the pedogenic history of a soil, including erosion. Both these elements can often be determined more easily by XRFS [70–73] than by other methods of analysis. Another approach to the study of erosion has involved the comparison of profiles defined by the MAC at various depths [74].

The study of soil pollution has aroused much interest in the past few years and such elements as Cr, Ni, Zn, As, and Pb emitted during smelting are simply determined by XRFS. Unfortunately, three elements of importance, Cd, Sn, and Hg, are usually below the detection limit of XRFS in soils and preconcentration would be necessary before analysis. However, the quantities of polluting elements that often occur in flyash and sewage sludge (except B) are often within the range of detection of XRFS. Lead pollution of the atmosphere [75] and of urban and rural soil and vegetation has been detected by XRFS determination [76,77], which compared favorably with other methods [42].

Naturally large occurrences of many elements in soils are detected easily and rapidly by semiquantitative XRFS techniques, and in this way, unusually large amounts of uncommon elements have been discovered, e.g., [24,78]. A special application of XRFS allows the determination of the cation-exchange capacity of soil materials [79].

Operating conditions				
2θ	Counter[d]	Sample preparation	Interferences in soils	References
138.58	FPC	concentration req.		
128.11	FPC	concentration req.		[88]
35.89	FPC	concentration req.		[88]
28.23	SC	pressed powder	Sn	[75–77]
37.30	SC	pressed powder (concentration req.)	Rb Kα	[78]

When the highly linear excitation from synchroton radiation can be applied, XRFS can detect concentrations as low as 0.02 mg kg^{-1} of most elements. In addition, this offers the possibility of using much smaller samples [80].

Although, as has been indicated above, XRFS can be used advantageously in many branches of soil science and often with very little sample preparation, preconcentration is sometimes necessary. Procedures to concentrate those elements for which XRFS is less sensitive [81–84] or those present in very low concentrations are being developed rapidly and improved to the extent that detection at the nanogram level of some elements has been reported [81,83, 85]. The methods may involve precipitation and filtration [86] or depend on exchange with ion-exchange resins, sometimes followed by filtration [85–88]. Special filter papers have been developed for these methods [85], but it has been shown that, as with other "thin film" methods, X-ray intensities are affected greatly by matrix effects and these become particularly important with light elements [88,89]. Nevertheless, the detection of metals after preconcentration is improved greatly compared with pressed pellets [86], and some of the problems with filter papers can be overcome by batch equilibration with a resin that is then dried and presented to the XRF spectrometer on an adhesive surface [87].

A special advantage of XRFS is its ability to scan a sample and make a qualitative or semiquantitative estimate of any unusual element, or any usual element that is present in particularly large or small concentration. In the nondispersive spectrometer, all the pulses are processed simultaneously, making multielement determinations very rapid.

Finally, it should be stressed that X rays are dangerous and great care should always be taken when dealing with X-ray equipment. The reader is recommended to read Jenkins and Haas [90] that outlines the hazards involved. However, provided precautions are taken, XRFS is a safe and reliable method for the determination of many elements in soils.

ACKNOWLEDGMENTS

George Brown of Rothamsted Experimental Station made many helpful comments on the original manuscript and Link Systems Ltd., High Wycombe, Buckinghamshire, England, supplied information used in Fig. 6b.

REFERENCES

1. R. Jenkins, *An Introduction to X-Ray Spectrometry*, Heyden, London, 1974.
2. A. A. Jones, in *Methods of Soil Analysis, Part 2, Chemical and Microbiological Properties* (A. L. Page, ed.), Am. Soc. Agron., Madison, Wis., 1982.
3. H. A. Liebhafsky, H. G. Pfeiffer, E. H. Winslow, and P. D. Zemany, *X-Rays, Electrons and Analytical Chemistry*, Wiley-Interscience, New York, 1978.
4. E. P. Bertin, *Introduction to X-ray Spectrometric Analysis*, Plenum Press, New York, 1978.
5. R. Jenkins and J. L. de Vries, *Practical X-Ray Spectrometry*, Macmillan, London, 1970.
6. A. A. Markowicz and R. E. Van Grieken, *Anal. Chem.*, *58*: 279R (1986).
7. E. W. White and G. G. Johnson, *X-Ray Emission and Absorption Wavelengths and Two-theta Tables*, 2nd ed., ASTM Data Series 37A, American Society for Testing and Materials, Philadelphia, Pa., 1970.
8. M. C. Powers, *X-Ray Fluorescent Spectrometer Conversion Tables for Topaz, LiF, NaCl, EDDT and ADP Crystals*, Philips Electronic Instruments, New York, 1957.
9. J. Kikkert and G. Hendry, *Advan. X-Ray Anal.*, *27*:423 (1984).

10. J. V. Gilfrich, *Anal. Chim. Acta, 188*:51 (1986).
11. W. P. van der Burg, *The Analysis of Mg, Choosing the Crystal*, Philips Analytical Equipment Department Bulletin, Eindhoven, The Netherlands, 1974.
12. J. M. McNeal and A. W. Rose, *Geochim. Cosmochim. Acta, 38*: 1759 (1974).
13. R. E. Collier and J. Parker-Sutton, *J. Sci. Fd. Agric., 27*: 743 (1976).
14. K. Norrish and B. W. Chappell, in *Physical Methods in Determinative Mineralogy* (J. Zussman, ed.), Academic Press, London, 1977.
15. C. S. Hutchinson, *Laboratory Handbook of Petrographic Techniques*, Wiley, New York, 1974.
16. K. Norrish and J. T. Hutton, *Geochim. Cosmochim. Acta, 33*: 431 (1969).
17. K. Norrish, private communication, 1986.
18. P. K. Harvey, D. M. Taylor, R. D. Hendry, and F. Bancroft, *X-Ray Spectr., 2*:33 (1973).
19. M. T. Haukka and I. L. Thomas, *X-Ray Spectr., 6*:204 (1977).
20. I. L. Thomas and M. T. Haukka, *Chem. Geol., 21*:39 (1978).
21. J. T. Hutton and S. M. Elliott, *Chem. Geol., 29*:1 (1980).
22. M. Forte, *X-Ray Spectr., 12*:115 (1983).
23. C. Williams, *J. Sci. Fd. Agric., 27*:561 (1976).
24. S. E. Calvert, B. L. Cousens, and M. Y. S. Soon, *Chem. Geol., 51*:9 (1985).
25. P. F. Berry, T. Furuta, and J. R. Rhodes, *Advan. X-Ray Anal., 12*:612 (1969).
26. C. B. Hunter and J. R. Rhodes, *X-Ray Spectr., 1*:107 (1972).
27. J. R. Rhodes and C. B. Hunter, *X-Ray Spectr., 1*:113 (1972).
28. S. Abbey, Geol. Surv. Can. Paper 80-14, Washington, D.C., 1980.
29. NBS Standard Reference Materials Catalog 1984-85, NBS Special Pub. 260, 1984.
30. B. Griepink, H. Muntau, H. Gonska, and E. Colinet, *Fresenhus Z. Anal. Chem., 318*:588 (1984).
31. K. Govindaraju, *Geostand. Newsletter, 8* (1984).
32. F. J. Flanagan, U.S. Geol. Surv. Prof. Paper 840, 1976.
33. L. Leoni, M. Menichini, and M. Saitta, *X-Ray Spectr., 11*:156 (1982).
34. G. Brown and R. Kanaris-Sotiriou, *Analyst (Lond.), 94*:782 (1969).
35. I. M. Zsolnay, J. M. Brauer, and S. A. Sojka, *Anal. Chim. Acta, 162*:423 (1984).
36. C. W. Childs and R. J. Furkert, *Geoderma, 11*:67 (1974).
37. L. G. Livingstone, *X-Ray Spectr., 11*:89 (1982).
38. C.-H. Huang and T. E. Smith, *X-Ray Spectr., 12*:87 (1983).

39. R. M. Rousseau, *X-Ray Spectr.*, *13*:115 and 121 (1984).
40. A. H. Beavers and K. R. Olson, *Soil Sci. Soc. Am. J.*, *50*: 1088 (1986).
41. D. C. Bartenfelder and A. D. Karathanasis, *Comm. Soil Sci. Plant Anal.*, *19*:471 (1988).
42. C. F. Pavely, B. E. Davies, and K. Jones, *Comm. Soil Sci. Plant Anal.*, *19*:107 (1988).
43. H. J. M. Bowen, *Environmental Chemistry of the Elements*, Academic Press, London, 1979.
44. J. Wankava, B. Knob, F. Moudry, and M. Kuba, *X-Ray Spectr.*, *11*:109 (1982).
45. S. H. Watts, *Geochim. Cosmochim. Acta*, *41*:1164 (1977).
46. T. Wakatsuki, H. Furukama, and K. Kyama, *Geochim. Cosmochim. Acta*, *41*:891 (1977).
47. C. Williams and J. H. Rayner, *J. Soil Sci.*, *28*:180 (1977).
48. L. R. Drees and L. P. Wilding, *Soil Sci. Soc. Am. Proc.*, *37*: 82 (1973).
49. L. J. Evans and W. A. Adams, *J. Soil Sci.*, *26*:319 (1975).
50. A. C. Oertel, *J. Soil Sci.*, *12*:119 (1961).
51. R. J. Gilkes, G. Scholz, and G. M. Dimmock, *J. Soil Sci.*, *24*:523 (1973).
52. C. W. Childs, *Geoderma*, *13*:141 (1975).
53. U. M. Cowgill, *Dev. Appl. Spectrosc.*, *5*:3 (1966).
54. I. Bradley, C. C. Rudeforth, and C. Wilkins, *J. Soil Sci.*, *29*:258 (1978).
55. J. Bolton, C. Brown, G. Pruden, and C. Williams, *J. Sci. Fd. Agric.*, *24*:557 (1973).
56. M. A. Tabatabai and J. M. Bremner, *Soil Sci. Soc. Am. Proc.*, *34*:417 (1970).
57. C. Bloomfield, J. K. Coulter, and R. Kanaris-Sotiriou, *Trop. Agric.*, *45*:289 (1968).
58. C. Williams, *J. Agric. Sci. Camb.*, *82*:189 (1974).
59. R. G. Darmody, D. S. Fanning, W. J. Drummond, and J. E. Foss, *Soil Sci. Soc. Am. Proc.*, *41*:761 (1977).
60. R. Kanaris-Sotiriou and G. Brown, *Analyst (Lond.)*, *94*:780 (1969).
61. S. Roberts and F. E. Kochler, *Soil Sci.*, *106*:164 (1968).
62. A. R. Gibson and D. J. Giltrap, *N.Z. J. Agric. Res.*, *22*: 439 (1979).
63. R. G. McLaren and R. S. Swift, *J. Soil Sci.*, *28*:445 (1977).
64. C. C. Evans, *Analyst (Lond.)*, *95*:919 (1970).
65. N. Souty and R. Guennelon, *Ann. Agron.*, *18*:653 (1967).
66. K. D. McLachan and M. J. Crawford, *J. Sci. Fd. Agric.*, *21*: 408 (1970).
67. R. T. King, *X-Ray Spectr.*, *8*:9 (1979).

68. C. Wilkins, *J. Agric. Sci., 92*:61 (1979).
69. V. S. Keramidas and D. S. Fanning, *Soil Sci. Soc. Am. Proc., 40*:857 (1974).
70. D. C. Bain, *J. Soil Sci., 27*:68 (1976).
71. D. S. Fanning and M. L. Jackson, *Soil Sci., 103*:253 (1967).
72. A. E. Hubert and T. T. Chao, *Anal. Chim. Acta, 92*:197 (1977).
73. E. Murad, *J. Soil Sci., 29*:219 (1978).
74. K. R. Olson and A. H. Beavers, *Soil Sci. Soc. Am. J., 51*: 441 (1987).
75. P. A. Cawse, *A Survey of Atmospheric Trace Elements in the U.K. (1972–1973)*, U.K. Atomic Energy Authority, Harwell, 1974.
76. C. Williams, *J. Agric. Sci. Camb., 82*:189 (1974).
77. C. Wilkins, *Environ. Poll. 15*:23 (1978).
78. C. Williams and G. Brown, *Geoderma, 6*:223 (1971).
79. D. J. Greenland, *Trans. 10th Int. Congr. Soil Sci. (Moscow), 2*:278 (1974).
80. R. D. Giauque, J. M. Jaklevic, and A. C. Thompson, *Anal. Chem., 58*:940 (1986).
81. M. Murata, M. Onatsu, and S. Mushimoto, *X-Ray Spectr., 13*:83 (1984).
82. W. J. Campbell, E. F. Spano, and T. E. Green, *Anal. Chem., 38*:987 (1966).
83. R. E. Van Grieken, C. M. Bresseleers, and B. M. Vanderborght, *Anal. Chem., 49*:1326 (1977).
84. H. H. LeRiche, *Geochim. Cosmochim. Acta, 32*:791 (1968).
85. M. Murata and K. Murokado, *X-Ray Spectr., 11*:159 (1982).
86. C. L. Luke, *Anal. Chim. Acta, 41*:237 (1968).
87. L. M. Bernardo, R. B. Clark, V. Knudsen, and J. W. Maranville, *Comm. Soil Sci. Plant Anal., 16*:823 (1985).
88. I. Roelandts, *Anal. Chem., 53*:676 (1981).
89. G. K. H. Tamm and G. Lacroix, *Anal. Lett., 15*:1373 (1982).
90. R. Jenkins and D. J. Haas, *X-Ray Spectr., 2*:135 (1973).
91. B. P. Fabbi and L. F. Espos, *Appl. Spectr., 26*:293 (1972).
92. N. B. Price, *Geochim. Cosmochim. Acta, 41*:1769 (1977).
93. G. Nelson Eby, *Anal. Chem., 44*:2137 (1972).
94. R. M. Taylor and J. B. Giles, *J. Soil Sci., 21*:203 (1970).
95. L. A. Wolfe and H. Zeitlen, *Anal. Chim. Acta, 51*:349 (1970).
96. J. H. Ellis, R. I. Barnhisel, and R. E. Phillips, *Soil Sci. Soc. Am. Proc., 34*:866 (1970).
97. J. H. Ellis, R. E. Phillips, and R. I. Barnhisel, *Soil Sci. Soc. Am. Proc., 34*:591 (1970).
98. H. Bergseth, *Analyst (Lond.), 100*:96 (1975).

99. G. Brown and D. A. Jenkinson, *Comm. Soil Sci. Plant Anal.*, *2*:45 (1971).
100. C. Wilkins, *J. Agric. Sci. Camb.*, *90*:109 (1978).
101. P. van Cauwenberge and L. A. Gordts, *J. Agric. Fd. Chem.*, *25*:1000 (1977).

8
Nuclear and Radiochemical Analysis

KEITH A. SMITH *The Edinburgh School of Agriculture, Edinburgh, Scotland*

I. INTRODUCTION

Most substances may be measured either by instrumental methods, such as those described elsewhere in this book, or by conventional methods of analysis. Radionuclides, however, are measured universally by instrumental methods. Their essential characteristic, the emission of penetrating ionizing radiation, has been the basis of methods of measurement since the early part of this century, and a wide variety of instruments capable of detecting and recording this radiation has been developed.

The relative ease with which radionuclides may be detected, and quantitatively measured, has led to their widespread use as tracers in physical and biological systems. In soil science the major application in this category has been to investigate the fate of labeled nutrient ions added to the soil, e.g., ^{32}P-labeled phosphate. The phenomenon of isotopic exchange, whereby ions present in the soil in exchangeable form can exchange with chemically identical but radioactively labeled ions in solution, has been exploited widely to demonstrate the dynamic nature of the exchange equilibrium and to determine the size of the labile pools of various nutrient ions. Measurements of this type can be made without destruction of the system, making in vivo applications possible. Another important application of radionuclides has been the elemental analysis of soil and plant materials, either by radio-activation techniques in which the activity is induced in the sample as a result of irradiation by neutrons or other nuclear particles, or by isotope dilution methods.

For more than 30 years, much effort has been devoted to the determination, and the assessment of the biological significance, of radionuclides distributed in the environment as a result of atmospheric explosions of nuclear weapons or releases from nuclear reactor installations and processing plants. Many of the techniques involved in tracer studies have also been employed in the investigation of these pollutant radionuclides, and vice versa.

The consideration of all aspects of the use and measurement of radionuclides in soil science and related fields of study would occupy a whole volume. Accordingly, the scope of this chapter is restricted to an examination of available methods for determining those radionuclides used as tracers for biologically important elements in soils and soil-plant systems, nuclear and radiochemical methods in elemental analysis, and the determination of the labile pools of nutrient ions in the soil and their interaction with soil materials.

An outline is given of the theory of radioactive decay, the properties of radionuclides, and the principles of radiation detection, as a necessary basis for understanding and solving the practical problems involved in nuclear and radiochemical methods. For more detailed treatments, and a discussion of essential safety precautions associated with the handling of radionuclides, a number of excellent books are available [1–9]. Another very useful source of information is the series of biennial reviews in *Analytical Chemistry* [10–13].

II. THEORY OF RADIOACTIVITY

A. Nature of Radionuclides

Radionuclides are characterized by unstable nuclei that spontaneously disintegrate or *decay*, with the emission of charged particles or the capture of orbital electrons. This decay results in a change in the charge on the nucleus, giving a product (a *daughter nuclide*) of different atomic number, i.e., one that is a different chemical element. The latter may be stable, or it, in turn, may be radioactive, in which case it will also decay. Ultimately, a stable form will be achieved.

Some radionuclides with an atomic number ≥ 82, and a very small number of others, decay by the emission of an alpha (α) particle (a helium nucleus). This decreases atomic mass by 4 and atomic number by 2 (Fig. 1). Most other radionuclides emit beta (β) radiation, i.e., electrons (β^-) or positrons (β^+), which increase and decrease atomic number, respectively, by 1, leaving the mass unchanged (Fig. 2). If the daughter nuclide produced by the decay is in an excited state, transition to the ground state occurs with

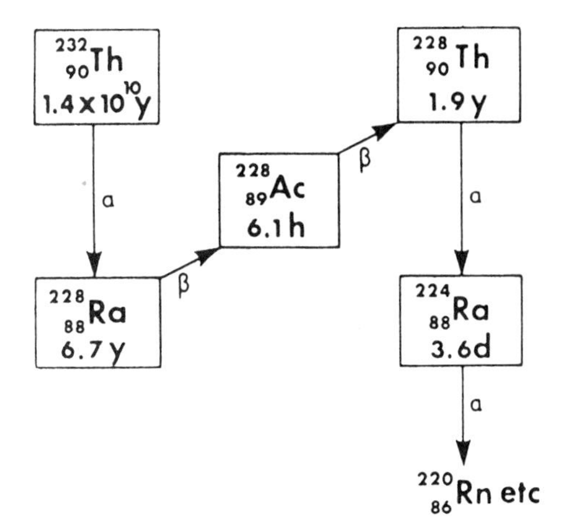

FIG. 1 Changes in atomic mass and atomic number as a result of α and β decay.

the emission of the excess energy in the form of electromagnetic radiation, or gamma (γ) rays (Fig. 2c). Gamma emission leaves both atomic mass and atomic number unchanged. Another mode of decay involves the capture of an external electron by the nucleus. This results in the emission of X rays known as *Bremsstrahlung* of "braking radiation" with properties identical to those of γ radiation. Bremsstrahlung production also occurs when β particles lose kinetic energy by interaction with matter.

Decay schemes, illustrating pure β^-, pure β^+, and β^- decay accompanied by γ radiation, respectively, are shown in Fig. 2. Many radionuclides have much more complex decay schemes than that of ^{60}Co shown in Fig. 2c and decay frequently occurs by more than one process, e.g., by electron capture and positron emission. The relative proportions in which the two processes occur remain constant for any particular radionuclide. When a positron loses its energy by interaction with electrons, it finally combines with an electron and the particles are annihilated, resulting in the liberation of two γ rays, each of which has an energy of 0.51 MeV. Thus, a positron-emitting nuclide always has this characteristic "annihilation radiation" associated with it, in addition to its β^+ activity.

B. Laws of Radioactive Decay

Radioactive decay processes are random, and the number of decay events that occur within a given period of time, i.e., the *activity*,

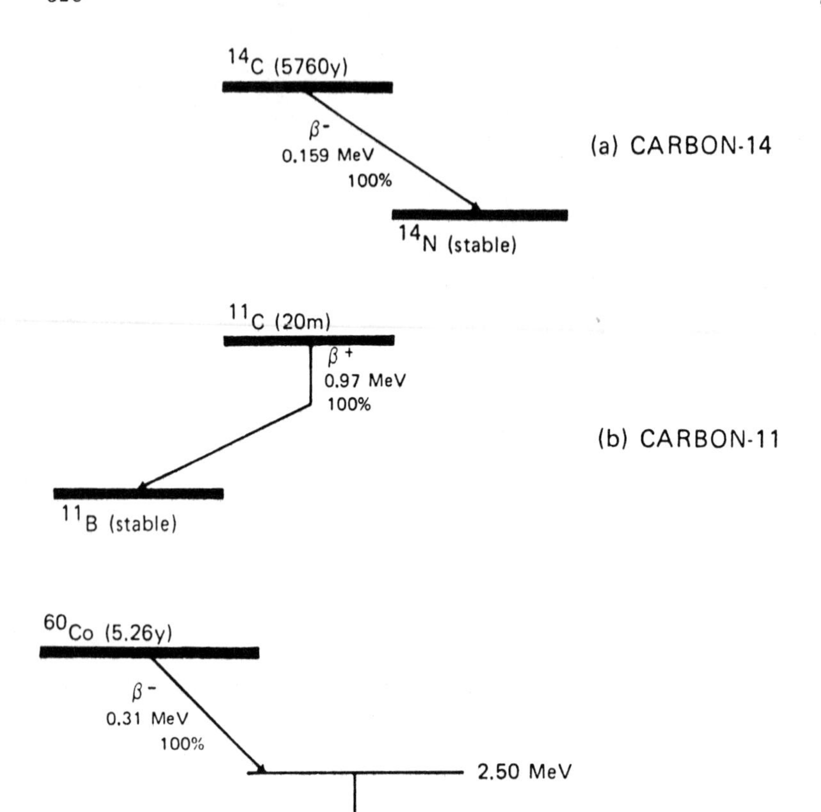

FIG. 2 β-decay schemes (a) β^-, (b) β^+, (c) β^- accompanied by emission of γ radiation.

A, is directly related to the number, N, of unstable atoms of the radionuclide present

$$A = -\frac{dN}{dt} = \lambda N \tag{1}$$

where λ is the decay constant for the particular radionuclide in question.

Integration of Eq. (1) gives

$$N = N_0 e^{-\lambda t} \tag{2}$$

where N_0 is the number of radioactive atoms existing at time $t = 0$.

As $A = \lambda N$,

$$A = A_0 e^{-\lambda t} \tag{3}$$

1. *Half-Life*

This is the time taken for the activity to decrease to half the original value. Thus,

$$A = \frac{A_0}{2} = A_0 e^{-\lambda t_{1/2}} \tag{4}$$

where $t_{1/2}$ is the half-life. Taking logarithms, we obtain

$$t_{1/2} = \frac{\ln 2}{\lambda} = \frac{0.693}{\lambda} \tag{5}$$

The observed activity of a radioactive source will be related to its absolute disintegration rate by a proportionality factor that will depend on the measuring instrument used and the particular conditions under which the measurement is made. This factor will remain constant if the conditions also remain constant; thus, the observed activity of a source containing a single radionuclide will decrease exponentially with time, and a plot of the logarithm of the activity against time will be linear. The half-life of a radionuclide can be determined from the slope of this plot; the determination of half-life is the best criterion of the radiochemical purity of a material.

If two unrelated radionuclides are present in the same sample and each decays to a stable daughter nuclide, the overall rate of decay is a composite of the individual exponential decay rates. Provided the half-lives of the nuclides are sufficiently different, the decay curve may be resolved into its individual components, so that the half-lives may be determined and also the initial activities by extrapolating back to $t = 0$ (Fig. 3).

Where a radionuclide decays to a radioactive daughter nuclide, the change in activity with time (starting with the pure parent nuclide at $t = 0$) depends on the relative half-lives of parent and daughter. Equations for "secular equilibrium" (the steady-state condition of equal activities of a long-lived parent radionuclide and short-lived daugher, e.g., ^{90}Sr-^{90}Y), for "transient equilibrium"

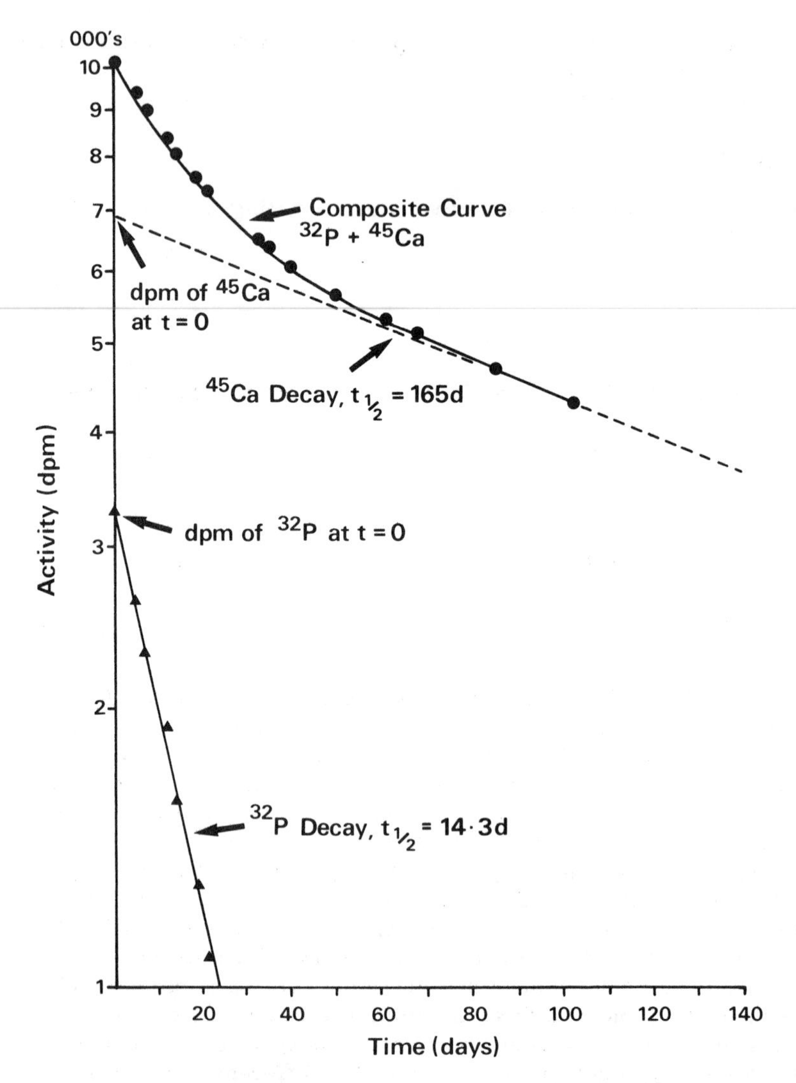

FIG. 3 Semilogarithmic decay curves of ^{32}P and ^{45}Ca resolved from composite decay curve. (From Ref. 9, with permission from M. F. L'Annunziata, *Radionuclide Tracers*, 1987, p. 46. Copyright © by Academic Press, London.)

[where $t_{1/2}$ (parent)/$t_{1/2}$ (daughter) lies between 10^4 and 1], and where no equilibrium exists [$t_{1/2}$ (parent) < $t_{1/2}$ (daughter)] are given by L'Annunziata [9].

2. *Units*

The basic S.I. unit of radioactivity is the becquerel (Bq); however, it is likely to take a considerable time for the familiar curie (Ci) to cease to be used.

$$1 \text{ Bq} = 1 \text{ disintegration per second} = 2.7 \times 10^{-11} \text{ Ci}$$

For most practical purposes, radionuclides are likely to be used in megabecquerel quantities or even higher, but sources that are actually measured ("counted") may have activities of the order of 0.1–100 Bq (i.e., approximately the picocurie to nanocurie range).

C. Production of Radionuclides

Some radionuclides occur in nature, e.g., ^{40}K, ^{14}C, and nuclides with $Z \geq 82$ resulting from the decay of uranium and thorium. Others, including some now widely used as tracers, are produced by the nuclear fission of uranium or plutonium. However, the great majority of useful tracers are produced artificially by the bombardment of target elements with α particles, protons, neutrons, or other particles to bring about the desired nuclear transformation [4]. For example, irradiation of phosphorus (which has only the single stable isotope ^{31}P) with thermal neutrons results in the capture of a neutron and thus the formation of ^{32}P; excess energy is lost by the emission of a γ ray. This reaction is normally written in the form ^{31}P$(n,\gamma)^{32}$P. The same product can also be produced from sulfur by the reaction ^{32}S$(n,p)^{32}$P if the neutrons are sufficiently energetic. The irradiation of elemental phosphorus transforms only a small fraction of the ^{31}P atoms to ^{32}P and thus the target material, after irradiation, contains both nuclides. The inactive ^{31}P is the isotopic carrier for the ^{32}P, so-called because it is normally present in sufficient quantity to be precipitated, and thus carry down the trace quantity of the radionuclide. The amount of radioactivity per unit mass of stable element (i.e., the *specific activity*) will depend on the irradiation conditions. In contrast, the formation of ^{32}P from sulfur is an important example of the formation of a radionuclide in a *carrier-free* state, i.e., in a state in which the only atoms present, after chemical separation from the parent element, are radioactive ones of the new element.

Some of the radionuclides that have been used in soil and plant studies are listed in Table 1. It will be seen that the half-lives vary from a few hours to 3×10^5 years (for ^{36}Cl). This greatly affects the usefulness of the various nuclides. Normally, experiments can only be carried out over a period of a few half-lives because of inadequate activity remaining for measurement, or because

TABLE 1 Some Useful Tracer Radionuclides

Radionuclide	Half-life[a]	Principal radiations (MeV)[a]		
		β^-	β^+	γ
^{3}H	12.3 yr	0.02		
^{14}C	5700 yr	0.2		
^{22}Na	2.6 yr		0.5	1.275, 0.51[b]
^{24}Na	15.0 hr	1.4		2.75, 1.37
^{32}P	14.3 days	1.7		
^{33}P	25.3 days	0.2		
^{35}S	87.5 days	0.2		
^{36}Cl	3×10^5 yr	0.7		
^{42}K	12.4 hr	3.5		1.525
^{45}Ca	163 days	0.3		
^{54}Mn	312 days			0.835
^{59}Fe	44.6 days	0.5, 1.6		1.10, 1.29
^{58}Co	70.8 days		0.5, 1.3	0.81, 0.51[b]
^{60}Co	5.3 yr	0.3		1.33, 1.17
^{64}Cu	12.7 hr	0.6	0.7	0.51[b]
^{65}Zn	244 days		0.3	1.11, 0.51
^{75}Se	120 days			0.136, 0.265
^{86}Rb	18.7 days	1.8		1.08
^{85}Sr	64.9 days			0.51
^{89}Sr	50.5 days	1.5		
^{99}Mo	66 hr	1.2		0.74
^{131}I	8.0 days	0.6		0.36
^{134}Cs	2.1 yr	0.7		0.605, 0.80
^{137}Cs	30.1 yr	0.5, 1.7		0.66[c]

(continued)

TABLE 1 (Continued)

Radio-nuclide	Half-life[a]	Principal radiations (MeV)[a]		
		β^-	β^+	γ
^{133}Ba	10.5 yr			0.36
^{140}Ba	12.8 days	0.5, 1.0 (1.4[d])		0.54 (0.50, 1.6[d])

[a]Principal source of nuclear data: W. Seelman-Eggebert, G. Pfennig, and H. Münzel, *Chart of the Nuclides*, 4th ed., Gersbach, Munich, 1974.
[b]Annihilation radiation.
[c]From ^{137m}Ba daughter ($t_{1/2}$ = 2.6 min).
[d]From ^{140}La daughter ($t_{1/2}$ = 40 hr).

longer-lived activities present as impurities become increasingly significant. Even shorter-lived nuclides than those listed in Table 1 have been used in some studies, e.g., ^{13}N($t_{1/2}$ = 10 min) [14] and ^{15}O($t_{1/2}$ = 2 min) [15], but their usefulness is rather limited.

III. DETECTION AND COUNTING

A. General Aspects

The radiations emitted by radionuclides when they decay cause ionization or molecular excitation in the substances through which they pass. It is these phenomena that are utilized in all the systems that have been devised to detect and count radioactive emissions. The ionization process can be detected electrically, as in ionization chambers, semiconductor detectors, and gas counters such as Geiger—Muller tubes, by the emission of light pulses, as in inorganic or plastic scintillator detectors, or by interaction with photographic emulsions.

In some laboratories a very narrow range of radionuclides is used, leading to the use of a single specialist item of detection and counting equipment. For instance, many biological and medical laboratories use only the low-energy β emitters ^{3}H (tritium) and ^{14}C, and the counting is carried out almost exclusively by liquid scintillation counters. However, as work with soils involves the use of a wide range of nuclides (Table 1), emitting radiation of different types and energies, many forms of detector have been employed.

An outline of the principal types is given in this section. More detailed information is available in a number of books [2,4–7,9] and the extensive literature cited therein.

Some of the detectors described here have existed in one form or another for many years. Although they are being superseded in some areas of research by newer devices, modern versions of traditional types of detector, e.g., Geiger–Muller tubes, are still commercially available. They are now normally combined with modern solid-state electronic circuitry, usually in modular form, e.g., the NIM series, and frequently form part of automated counting systems. Low-cost, manually operated instruments, however, continue to serve a useful function in soil science and related disciplines. Whatever the level of sophistication of the sample input or data output systems, satisfactory counting procedures depend on an appreciation of the characteristics of the particular radiation to be determined and selection of an appropriate detector.

1. *Interactions of Radiation with Matter*

Alpha particles are intensely ionizing and lose their energy so rapidly by multiple collisions with bound electrons, that their tracks are only a few centimeters long in air, and they will not penetrate a thin sheet of plastic or paper. This has important implications for the quantitative measurement of α activity. Unless the matrix material that contains the α-emitting radionuclide is spread in an extremely thin layer, self-absorption of the particles within the source may be severe. The ranges of the α particles from the nuclides of the Uranium and Thorium Series have been calculated to be 13–45 μm in soil and 20–70 μm in plant ash [16], and the ranges in precipitates of insoluble salts, following radiochemical separation, will be of the same order.

The path length of β particles will vary greatly, depending on their emission energy and the density of the medium through which they are traveling. A β particle with an energy of 1 MeV will have a range of approximately 3 m in air, 4 mm in water or biological tissue, and somewhat less than this in solids. Absorption by air and by the "windows" of conventional gas ionization detectors (see Sec. III.B) always occurs, but it can be particularly important for α and low-energy β particles, such as those emitted by ^{14}C. With such a nuclide it is necessary to place the source very close to the detector and to make the window as thin as possible within practical limits. Self-absorption of β particles within the source is discussed in Sec. III.B.

It should be noted that the figures for β-particle energies listed in Table 1 are maximum values (E_{max}), and that particles are actually emitted with a range of energies, from very low values up to E_{max}.

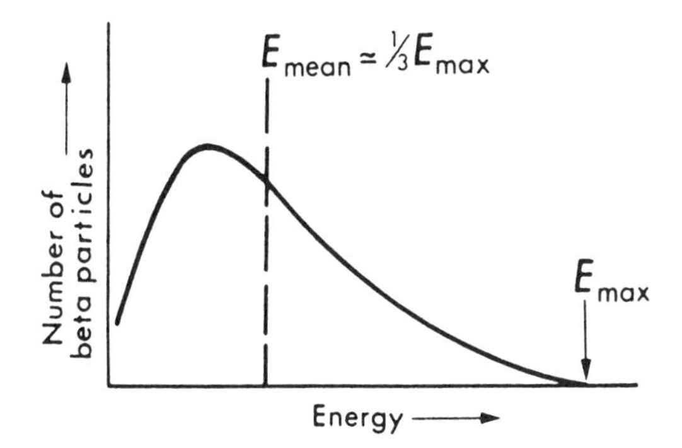

FIG. 4 β-spectrum. (From Ref. 2.)

Thus, β particles emitted from a radionuclide form a continuous spectrum of the type shown in Fig. 4. This can be important when considering the merits and disadvantages of β-counting systems. In some systems many of the lower-energy particles are lost by absorption and scattering before they even reach the detector. The different β-energy profiles exhibited by different radionuclides can be utilized to count a mixture of two together in a single sample, thus allowing double labeling experiments to be carried out.

Unlike α and β radiation, γ rays consist of electromagnetic radiation rather than charged particles. Consequently, their interaction with matter is much reduced and their penetrating power is greatly increased. This greatly affects the methods used for satisfactory counting. While there is relatively little self-absorption in the source (which can therefore consist of up to several kilograms of soil or other material [17]), the detector must be large enough to absorb the energy of the γ ray, wholly or partially, in order to initiate some process that culminates in a detectable pulse.

2. *Background and Shielding*

The inevitable presence of cosmic radiation, and also natural radioactivity and fallout in the materials from which laboratories, and even counting equipment are constructed, results in a background counting rate in the absence of a source. This background (particularly the cosmic ray element) is minimized normally by housing the detector in a 50 mm thick lead shield or "castle."

In many tracer applications, source count-rates are sufficiently high for the background contribution to be insignificant, but in some circumstances it may be required to count very low activities.

This requires special techniques to reduce the background, such as the use of anticoincidence circuitry. A typical anticoincidence system for β counting consists of a central detector surrounded by a guard detector or detectors. When radiation passes through the guard and the central detector (triggering both almost simultaneously), the pulse is not recorded. However, a particle emitted from a source into the central detector will not have sufficient energy to pass through it and into the guard detector, so only the former is triggered. The resulting pulse is recorded. Modern anticoincidence systems have backgrounds of less than 1 count per minute. Other precautions include the use of old lead [in which the naturally occurring ^{210}Pb ($t_{1/2}$ = 22 yr) has decayed away] for shielding [1,18], and materials of low potassium content in counter construction [18]. The performance of a counting system in detecting any particular source of radiation is assessed normally on the basis of a *figure of merit*, equal to S^2/B, where S and B are the sample and background counts, respectively.

3. Counting Efficiency and Standardization

Radiation is emitted from an active source in all directions; to maximize detection efficiency the sample source is usually placed as close to the detector as possible, so that almost all the radiation emitted within a solid angle of 2π steradians is intercepted. If the source is moved away from the detector, then the solid angle it subtends decreases and the count rate falls. Self-absorption of radiation by the source and scattering also have major effects on counting efficiency (see Sec. III.B).

For absolute standardization of β-emitting radionuclide preparations, 4π counters are used. As the name applies, the radiation emitted in all directions is counted, and β-γ coincidence techniques are also employed. Most users are not equipped to undertake absolute standardization of the radionuclide preparations they wish to use, but standardized solutions are available from the major suppliers and can be used to prepare sources for counter calibration. For many applications, measurements need only be made on a relative basis, and the absolute calibration of counters is therefore not so important. It is usually necessary, however, to ensure that counter performance does not fluctuate over the period when experimental studies are being carried out, and it is desirable to check this by the regular counting of standard sources.

4. Statistical Accuracy of Counting

Radioactive decay is a random process; thus, the counting rate observed from a given source will vary from measurement to measurement. It is not possible to define a true counting rate, but only a mean rate about which the observed rate will vary according to a Poisson distribution. For a Poisson distribution, the standard deviation σ is given by

$$\sigma = (N)^{1/2} \tag{6}$$

where N is the number of events recorded in a given observation.

When low count rates have to be measured, fluctuations in the background rate become significant. It can be shown [20] that if the net count rate for a sample is R_S, with a standard deviation σ_S,

$$\sigma_S = \left[\frac{R_{(s+b)}}{t_{(s+b)}} + \frac{R_b}{t_b}\right]^{1/2} \tag{7}$$

where

$R_{(s+b)}$ = observed sample-plus-background count rate
$t_{(s+b)}$ = counting period for the sample
R_b = background count rate
t_b = background counting period

Thus for measurements of sample counting rates similar to the background, the error for the sample is much higher than that for the sample plus background. This factor must be taken into account when intending to count low-activity samples to a given statistical accuracy. Fuller treatments of the statistical theory are given in Ref. 20 and other standard radiochemistry texts.

B. Gas Ionization Detectors

In its simplest form, a gas ionization detector is a chamber containing a gas and two electrodes, across which an electric potential is maintained. When radiation passes through this chamber, it causes the gas molecules within it to ionize. These ions then move toward the appropriate electrode and a pulse of current flows between the electrodes. Measurement of the total current or the number of pulses can be used to determine the amount of radiation reaching the chamber. This system is very flexible, and by varying the dimensions and materials of the gas chamber and its operating conditions, it can be used to detect α and β particles, γ rays, and neutrons. However, the counting efficiency for γ rays is very low and depends on the emission of secondary electrons from a heavy metal absorber (see Sec. III.E).

The pulse height from the ionization chamber varies with the applied voltage as shown in Fig. 5, and this affects the way in which it can be used. At voltages less than V_1, any ions present within the chamber move relatively slowly toward the electrodes. Some of the ions recombine before reaching the electrodes, so that this region of the curve is unsuitable for counting purposes. Between V_1 and V_2, the ions travel sufficiently quickly to reach the electrode without recombination occurring, giving rise to a *plateau* (the saturation current). Under these conditions the chamber can be used as a counter and this region is known as the *ionization chamber region*.

Above V_2 (Fig. 5), the pulse size increases again with voltage. The ions are subject to greater acceleration, sufficient to cause the production of secondary ions by collision with further gas molecules. This behavior results in an avalanche of ions and a corresponding

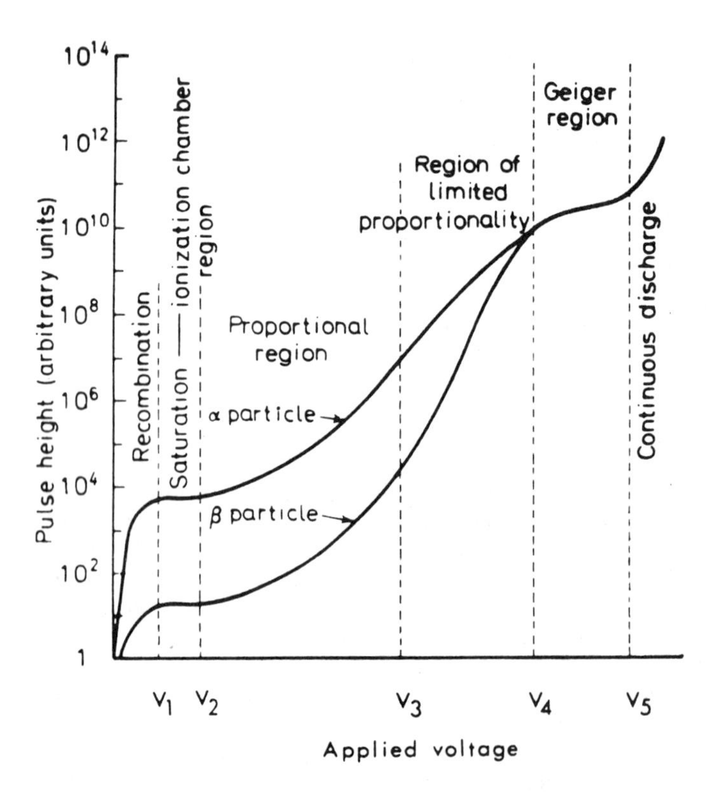

FIG. 5 Relationship between pulse height and applied voltage for a gas ionization detector. (From Ref. 21.)

increase in pulse size; this process is known as *gas amplification*. Furthermore, the size of the pulse obtained is proportional to the energy of the ionizing radiation, and the region between V_2 and V_3 is known as the *proportional region*. A detector operating under these conditions is known as a proportional counter and can be used to differentiate between radiations of different energies.

Between V_4 and V_5, secondary ionization processes are so great that the pulse size always reaches its maximum value and is independent of the energy of the incident radiation, although it does increase somewhat with voltage. This is known as the Geiger–Muller (G-M) region and is the basis of the *Geiger counter*. For normal applications, a G-M tube operating in this region is considered satisfactory if it has a plateau in the graph of count rate against applied voltage with a slope not greater than 2–2.5%/100 V [19]. Above V_5, a continuous discharge occurs, because new ions are continually being formed by the movements of the very highly accelerated charged particles within the chamber.

a. Ionization Chamber. The ionization chamber finds its major use nowadays as a health physics dosemeter for measuring exposure to γ rays, X rays, or neutrons [19]. It can also be used for counting α particles [19], but for counting β particles, proportional and G-M counters are generally used.

b. Proportional Counter. The anode of a proportional counter usually consists of a loop of fine wire enclosed in a chamber that often contains a mixture of 90% argon and 10% methane [22,23]. The gas can be sealed into the chamber at high or low pressures or can be allowed to flow slowly through the chamber under a slight positive pressure. This latter type is shown diagrammatically in Fig. 6a. The gas-flow system has the advantage of avoiding problems associated with the deterioration of the gas, and because the pressure differential used is very low, the end window can be very thin, thereby facilitating the penetration of low-energy β particles, such as those from ^{14}C, into the counter. The gas used must be purified before use to remove corrosive agents, dust particles, etc. A further modification for suitable radionuclides, e.g., ^3H, in gaseous forms is to pass them through the chamber with the flow of counting gas.

Since the pulses produced by these counters are proportional to the energy of the incident radiation, it is possible, by use of suitable electronic equipment, to distinguish between different particle energies and carry out spectroscopic measurements, or count two nuclides separately when both are present, as in double labeling experiments, or simply to reject unwanted pulses [9].

Although proportional counters are most commonly used for counting β particles, they can also be used for counting α emissions if

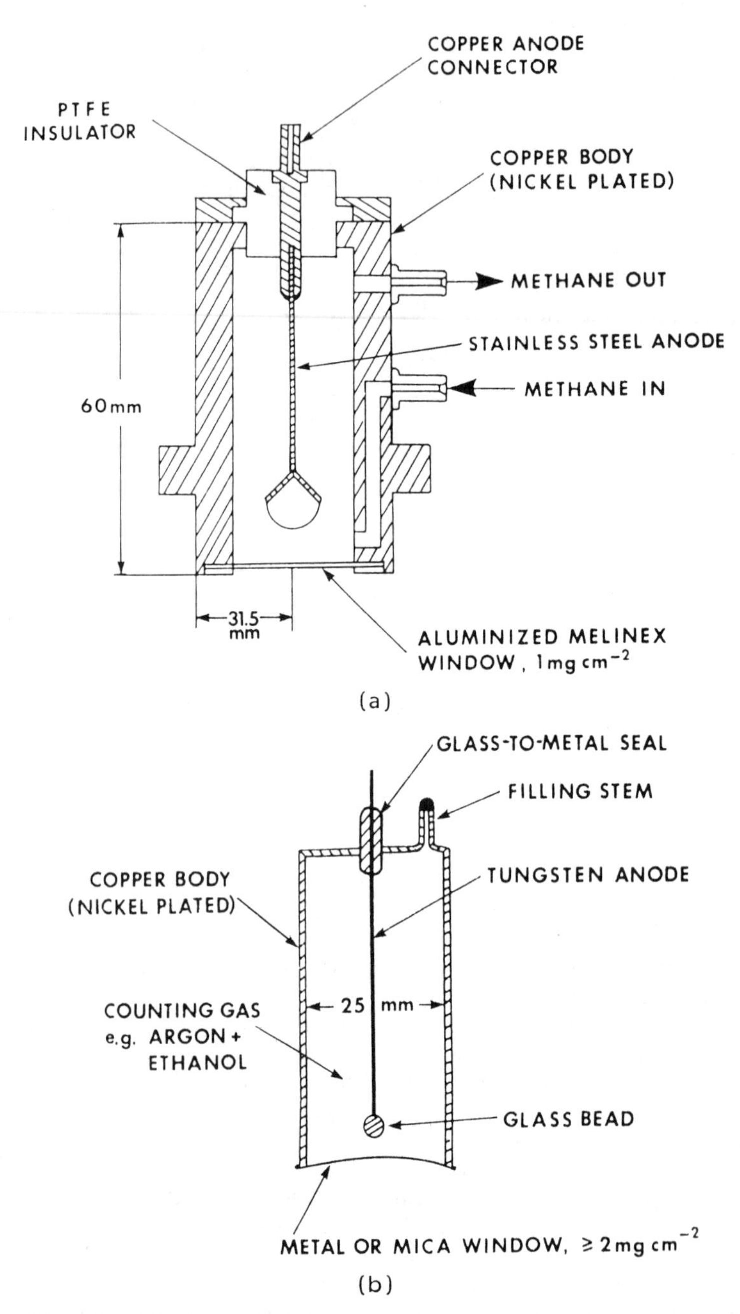

FIG. 6 (a) End-window gas flow proportional counter. (b) Typical G-M counter. (From Ref. 19).

fitted with a very thin window, or if the α source is placed inside the counting chamber [9,19].

Modern laboratory instruments are available incorporating proportional counters with automatic sample changers and printout systems, which are capable of accommodating up to 100 samples.

1. Geiger–Muller Counter

In the Geiger–Muller region, any ionization gives rise to uniform, large pulses. This means that although the G-M detector is unable to distinguish radiations of different energy, it is ideal for use with electronic systems that are simple, cheap, and robust. As a consequence, it has proved very popular as a general-purpose β counter in spite of its other limitations. A typical end window G-M tube is shown in Fig. 6b. The anode is a quite substantial wire sealed into but insulated from the metal body of the tube. The end of the wire is coated with a glass bead to prevent unwanted point discharges. The tube is fitted at one end with a sealed mica window that is as thin as possible. Even so, the higher gas pressure used in G-M tubes means that this window must be thicker than that used in a gas-flow proportional counter, thereby causing a decrease in counting efficiency.

A particularly important part of G-M tube design is the nature of the gas filling. To ensure that each ionization event results in a finite pulse, a quenching agent, usually chlorine or ethanol, is introduced into the counting gas (normally argon) to bring the discharge to an end. During the time taken by the ionization and quenching processes, the tube is unable to detect or respond to any other ionizing radiation. To ensure that this dead time is a constant value, it is usual to superimpose a known, electronically generated dead time, typically 100 μsec [2], triggered by the receipt of a pulse. When the count rate is high, it becomes necessary to perform a correction to make up for lost counts. Although this is not difficult, it is inconvenient, and it is better practice to avoid when possible the need for correction by careful adjustment of sample size and activity and its distance from the tube.

If a dead time correction *is* necessary, then it can be calculated from

$$C_t = \frac{C_0}{1 - C_0 t} \tag{8}$$

where C_t is the true count rate (per second), C_0 the observed count rate (per second), and t the dead time (seconds). This correction is an approximation but is adequate for most routine purposes.

2. *Absorption and Scattering of β Particles*

Solid sources for β counting are commonly obtained by evaporating to dryness aliquots of dilute salt solutions or of slurried precipitates on metal planchets, or by filtering precipitates through suitable papers, membranes, or sinters (see Sec. IV). The resulting layer of solid material on its support is then counted. In these cases the solid material of the source itself gives rise to self-absorption, which reduces the count rate, and scattering, which can either reduce or increase the count rate.

a. Self-Absorption. This problem can be overcome largely by ensuring that the amount of solid material on the planchet, filter, etc. is sufficiently small so as to cause no significant absorption. If this is not possible, it becomes necessary to calibrate the system using standards of known activity and known amounts of solid. Alternatively, if the radionuclide emits γ rays as well as β particles, it may be preferable to count the γ radiation, since this is much less affected by self-absorption.

If increasing weights of a material of given specific activity are made into sources of uniform density and cross-sectional area, and counted, the count rate increases with increasing mass (and thickness), but since self-absorption also increases, the relationship is not linear (Fig. 7). A source thickness is reached that is equal to the range of β particles in this medium. All particles emitted from below this depth are absorbed, and the count rate remains constant; this is known as an "infinitely thick" source. Such sources of equal mass can be counted and compared on the basis of specific activity rather than total activity. This technique can be particularly useful when working with heavy precipitates containing low-energy β emitters, e.g., barium carbonate containing ^{14}C or calcium oxalate containing ^{45}Ca.

b. Back-Scattering. Particles emitted from a source in a direction away from the detector may be scattered into the detector, due to collisions with nuclei in the source itself, the source mount, or the counter shield. Back-scattering increases with the thickness of the source mount and with the increasing atomic number of source and mounting materials. The counting rate of a high-energy β source may thus be raised by 50–60% above that expected purely on the basis of counting geometry [20,25]. Thus, it is important, in order to prevent errors, to prepare and count sample sources and standards under the same conditions.

c. Use of Absorbers. The introduction of an aluminum or polyethylene absorber between a β source and a detector will prevent all β particles below a certain energy from entering the detector. This simple principle has been used for many years to allow the

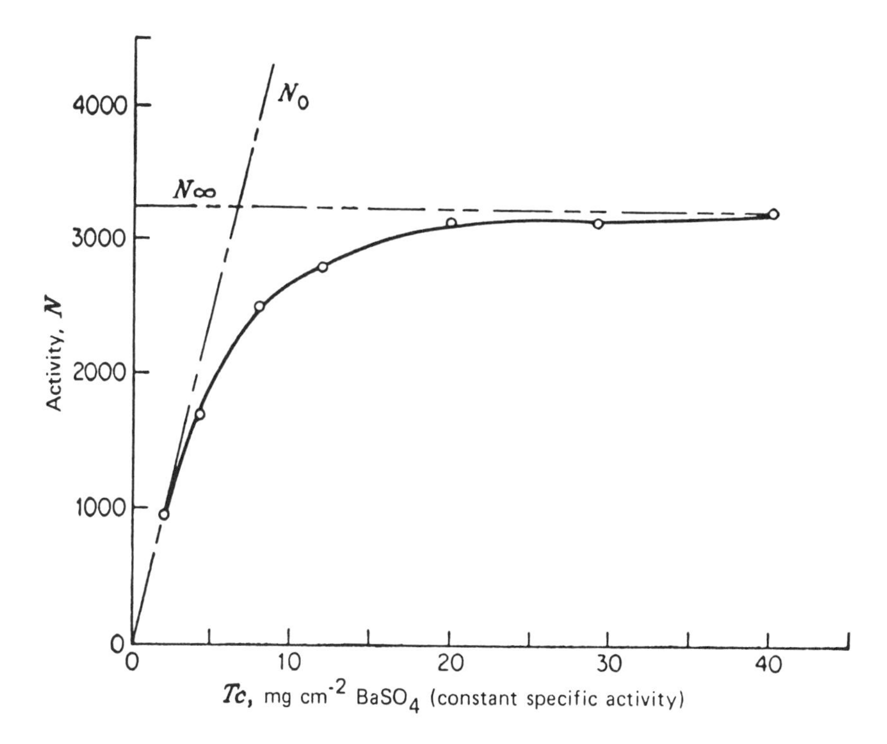

FIG. 7 Self-absorption of β particles. (From Ref. 24.)

determination of two different nuclides, with different E_{max}, present in the same source. Good examples are the determination of ^{89}Sr in the presence of ^{90}Sr, using a 100 mg cm^{-2} absorber [25], and ^{32}P and ^{33}P used in double-labeling experiments [26]. In suitably equipped laboratories, this technique has been superseded largely by liquid scintillation counting.

C. Principles of Scintillation Counting

When an α or β particle, or a γ ray, passes through certain transparent materials, all or part of the energy is converted into a small light pulse, or *scintillation*. These materials may be crystals of inorganic salts, e.g., sodium or cesium iodide (activated with thallium), or bismuth germanate, solutions of fluorescent aromatic compounds (fluors), or *plastic scintillators* (solid solutions of organic fluors in translucent plastic). By recording the number of light pulses emitted, the activity of the source can be determined. Furthermore, as the intensity of each light pulse is proportional to the energy absorbed, if pulses of different intensities can be distinguished, then the energy spectrum of the radiation may be determined. The scintillator is placed in optical contact with the cathode of a photomultiplier (PM) tube and enclosed in a light-proof container. Light photons from the scintillator eject electrons from the photocathode, and the size of the pulse emitted by the photomultiplier is proportional to the light energy received. Detailed accounts of some or all of these processes are given in Refs. 4–9.

Because of the distribution of energy within the β particle spectrum, plus the nature of energy transfer in scintillation systems and poor resolution of pulse size, it is found that the relationship between pulse size and energy is less well defined for β than for γ radiation. This restricts the use of scintillation counting as a spectroscopic technique for β particles, but the resolution of energies is sufficiently good to allow double-labeling experiments to be carried out using two β emitters with significantly different energies (e.g., ^{3}H and ^{14}C).

Plastic scintillators can be cast or machined to shapes, sizes, and thicknesses appropriate for the particular radiation detection task involved and have replaced gas ionization detectors in some commercially available β counters. Such scintillation counters have a very much shorter dead time (~1 μsec) than either G-M counters or proportional counters. Liquid scintillators are discussed in Sec. III.D.

A key component of the scintillation counting system is the photomultiplier tube. When light photons hit the photocathode, a number of electrons are emitted, and this photoelectric current is amplified by a factor of $10^{6}-10^{8}$ by the PM tube. The thermionic emission of electrons from the photocathode or dynodes will give rise to unwanted pulses, and this can lead to an unacceptably high level of background noise or counts. Much of this consists of low-energy pulses that can be rejected by the use of a discriminator or bias in the associated electronics. In older instruments the noise was reduced by refrigerating the tube, to minimize thermal emissions. Now in modern liquid scintillation counters, it is achieved by the use of two PM tubes linked through a coincidence circuit. This is shown schematically in Fig. 8. The coincidence circuit is designed so that an output pulse is produced only when pulses arrive simultaneously (or in practice within a very short time interval, ~1 μsec) from both PM tubes. Pulses due to random thermionic noise are unlikely to be produced simultaneously by both PM tubes and thus nearly all are rejected. On the other hand, photon emissions caused by scintillations within the sample will arrive simultaneously at both PM tubes, giving rise to pulses in coincidence that are recorded.

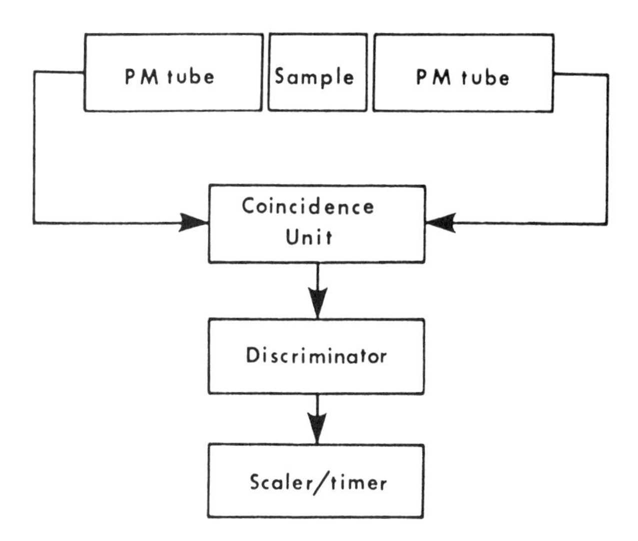

FIG. 8 Coincidence counting system used in scintillation counters.

D. Liquid Scintillation Counting

There are two major reasons for the popularity of liquid scintillation counting in biological studies. First, the technique is ideal for counting the low-energy β radiation emitted by nuclides such as ^{3}H, ^{14}C, and ^{35}S that are used extensively in such studies. Indeed, work with ^{3}H would be almost impossible without this technique. Second, liquid scintillation counting lends itself to automation, and highly sophisticated counters are available that minimize the tedium of manual operation and the need for technical understanding by the operator. More detailed information on the technique than can be included here is available in, e.g., Refs. 4 and 9, and in specialist monographs (e.g., [27]).

In general terms the liquid scintillation source is a sample consisting of radioactive material, one or two organic fluors, and a solvent, contained in a vial. The light flashes produced by the sample are detected by a PM tube and counted in the usual way. The intimacy of the sample and the detector (i.e., the organic fluor) minimizes absorption problems and maximizes counting efficiency. Although the system is simple in essence, errors and problems can arise. The solvents used must be able to dissolve the organic fluor and preferably the radioactive material also (although this is not absolutely essential). In addition, the solvents play an important role in the transfer of energy within the sample. They absorb the energy of the β radiation and then transfer it to the organic phosphor [28]. Solvents that have been found to fulfill these criteria are aromatic compounds such as xylene, toluene, and, to a lesser extent, benzene.

The most commonly used primary fluor is PPO (2,5-diphenyl-oxazole)

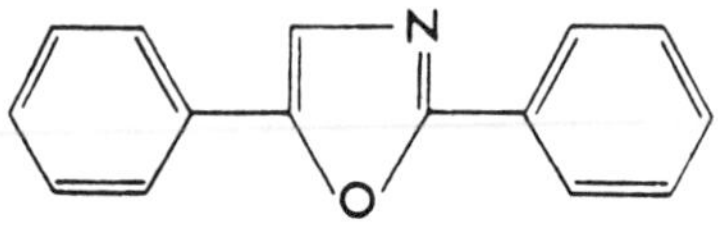

The photons emitted from this fluor peak at 380 nm, which is close to the optimum wavelength of 385 nm for the production of photoelectrons by modern bialkali photocathodes [9,29].

The primary solute PBD 2-(4-biphenylyl)-5-phenyl-1,3,4-oxadiazole

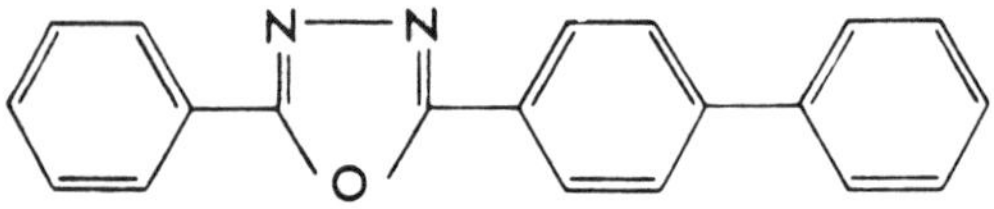

is reported to give greater pulse heights than PPO for the same radiation, but it has the disadvantages of low solubility and high cost [9,28].

In some cases it is advantageous to include a secondary fluor that absorbs the light emitted from the primary fluor and re-emits it at a longer wavelength. This was originally done to produce a better response in the photomultiplier tube, but the greater sensitivity of modern photocathodes at shorter wavelengths now often makes this unnecessary. A secondary fluor is useful, however, in other circumstances, e.g., when the scintillator vial is only partially transparent to the primary solute fluorescence [30]. A compound often used for this purpose is 1,4-bis-(5-phenyloxazol-2-yl) benzene (POPOP)

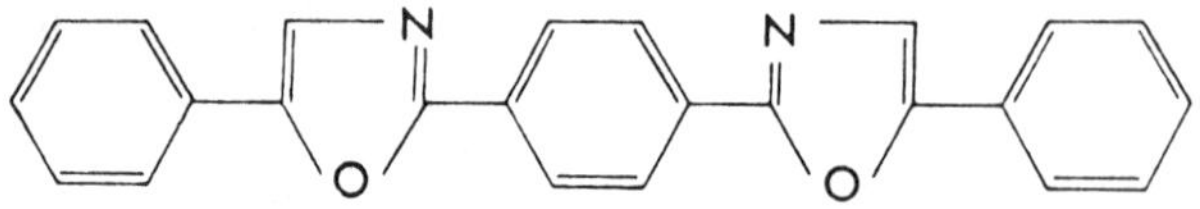

Many of the samples obtained from soil studies are inevitably in the form of aqueous solutions and contain material that is insoluble in nonpolar solvents. The most satisfactory way to count these solutions is to mix them with commercially available water-miscible "cocktails" containing fluors, solvents, and a detergent such as Triton X-100 (octylphenoxypolyethoxy ethanol) [9,31]. The water content of these mixtures can be as high as 30–40% [9].

Apart from electronic considerations (discussed in Sec. III.C) the greatest problem in present-day liquid scintillation counting arises from a phenomenon known as quenching (this should be distinguished from the "quenching" of a G-M tube; see Sec. III.B). Quenching substances lower the efficiency of the energy transfer-light output scintillation process and thus reduce the scintillation pulse heights and the counting efficiency [9]. The main sources of quenching are "chemical quenchers," which are a variety of organic compounds that interfere with energy transfer from solvent to fluor molecules, and colored substances that absorb light in the visible region of the spectrum and thus reduce the number of photons reaching the photocathodes of the PM tubes [9].

A number of methods can be used to overcome or correct for quenching. For instance, the colored material can be removed by treatment with charcoal or some other adsorbent, but this is laborious and not always successful. Another procedure is the *internal standard method*, in which the sample is counted, then a known amount of the same radionuclide is added and the sample recounted. Comparison of the count rate due to the standard added to the sample with the known unquenched count rate will give the loss of efficiency due to quenching. Neither of these methods readily lends itself to the high degree of automation found in modern liquid scintillation counters and other, more convenient techniques have been developed for this purpose.

In the *external standard method* the sample is counted, then an external source is placed alongside the sample, which is then recounted. The external standard is usually a γ-ray source (e.g., ^{137}Cs) that generates Compton recoil electrons within the counting vial (see Sec. III.E) and these interact with the liquid scintillant in the normal way. Once again, comparison of the standard count rate against a calibration curve will enable the degree of quenching in the sample to be determined.

The method of choice for many workers is the *channels ratio* technique. If a plot is made of counts against discriminator setting for β radiation, a curve of the type shown in Fig. 9 is obtained. It can be seen that although quenching reduces the overall count rate, the effect is greater on high-energy pulses. Thus, if counting is carried out in two channels (e.g., channel 1, L_1 to L_2, and channel 2, L_2 to L_3), then the ratio of counts in channel 1 to channel 2 will change as quenching occurs. By comparison with ratio values obtained for known levels of quenching, the count rate can be corrected. The channels ratio count can be performed at the same time as the sample count and the equipment programmed

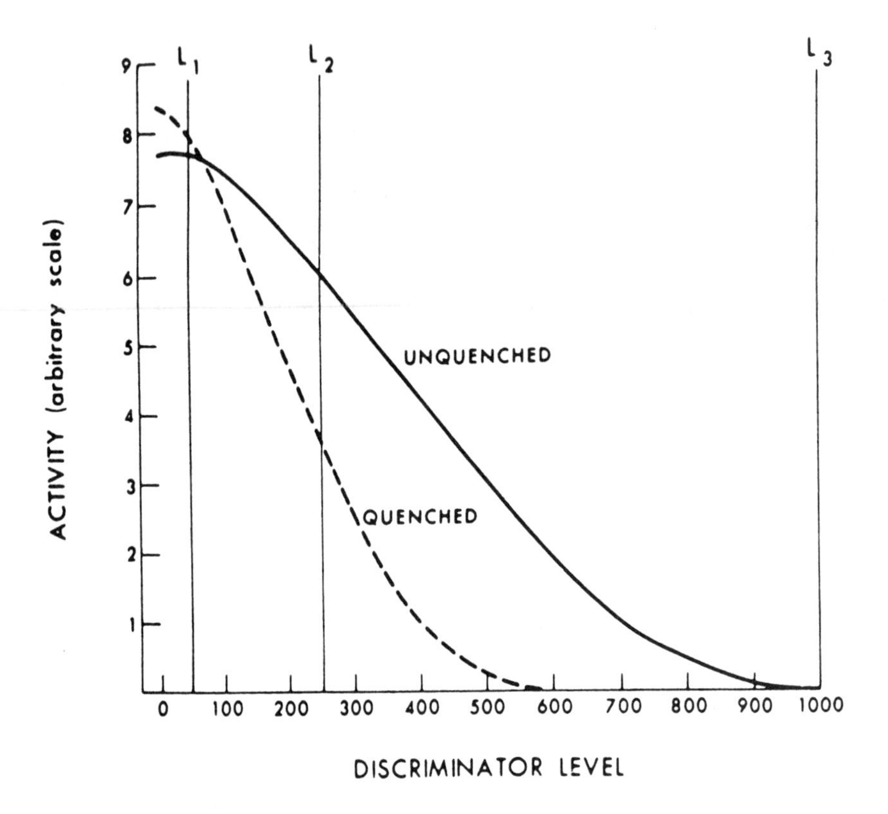

FIG. 9 Unquenched and quenched pulse-height spectra for ^{14}C. (From Ref. 32.)

to correct the observed sample count. The system is therefore ideal for use with automated equipment.

Erroneously high count rates can be caused by chemiluminescence of the sample. Cooling, addition of various compounds, and storage prior to counting are all employed to overcome this problem [4].

The current generation of microprocessor-controlled counters also includes features such as library programs for different counting requirements, automatic calibration, and on-line computing to produce information in the final form required.

E. Other Uses for Liquid Scintillation Counters

1. *Cerenkov Counting*

High-energy β emitters dissolved in water have been found to emit light pulses even with no scintillant in the system. The light is known as Cerenkov radiation and arises when β particles travel at a velocity greater than that of light in the liquid medium [33]. The electron is found to lose some of its energy in the form of photons emitted at an angle to the direction of travel of the particle [34]. The mechanism is analogous to the production of a shock wave when an object travels at greater than the speed of sound in air. Cerenkov radiation is a threshold phenomenon and β-particle energies have to be greater than a certain value (related to the refractive index of the liquid medium) before light emissions occur. For water, E_{max} must be > 0.26 MeV but, in practice, worthwhile efficiencies are not achieved until $E_{max} \geq 1$ MeV. Even so, several nuclides that are very useful in soils and plant work, e.g., ^{32}P, ^{42}K, and ^{86}Rb, can be counted by this method [35–39]. Although the number of photons emitted in this process is relatively small, the light pulses are detected easily by standard liquid scintillation counters.

The Cerenkov counting method has several advantages. Little or no sample preparation is required; the whole sample volume can be occupied by radioactive solutions, thereby decreasing the specific activity required; the sample is unadulterated by the addition of solvent and scintillant and can be used for other purposes after counting. Also, the cost of conventional liquid scintillant is avoided, although small amounts of fluorescent compounds can be added to improve the counting efficiency. Chemical quenching in Cerenkov counting is nonexistent, but color quenching can be greater than that which occurs in conventional liquid scintillation counting. The methods of dealing with color quenching are similar to those outlined in Sec. III.D above and are also discussed fully by L'Annunziata [9].

2. *Counting α and γ Radiation*

Liquid scintillation counting systems are ideal for counting α particles and, because of intimate mixing in the sample, very good efficiencies can be obtained. For γ radiation, however, the efficiency of the system is very low since most γ rays pass through the sample without interacting. Improved counting efficiencies can be obtained for γ rays by incorporating organometallic compounds (e.g., of Sn, Tl, Pb) in the liquid scintillant that improve its stopping power [4,9,40]. Such scintillant liquids are sufficiently expensive to deter the common practice of discarding the liquid after use. This can be overcome by using an Ashcroft vial with a central well for the sample, which is thus kept separate from the scintillant (Fig. 10) [41]. In this way, the scintillant liquid can be used indefinitely. A low-cost alternative to the rather expensive commercially available Ashcroft vial, in which the metal-loaded scintillator cocktail can be changed easily, has been described by Cecchi and Somenzi [42].

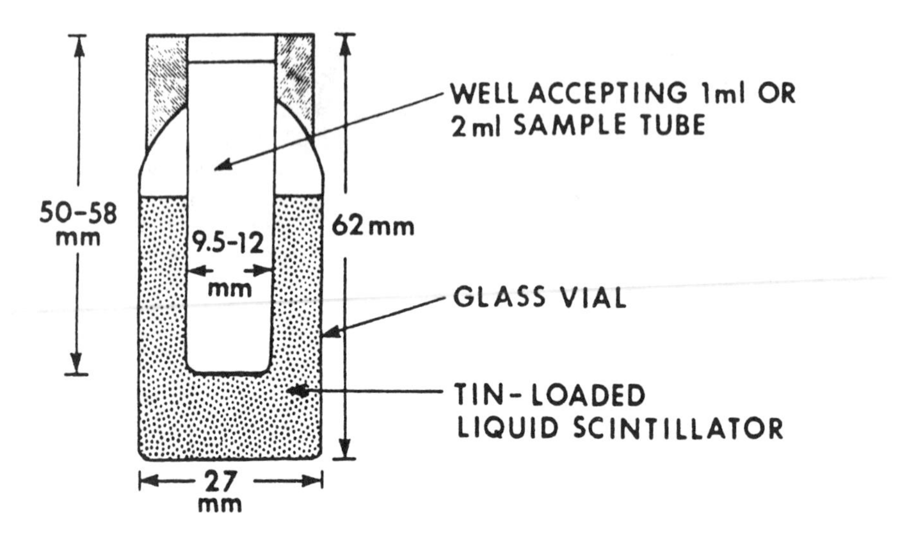

FIG. 10 Vial containing metal-loaded scintillator for gamma counting in liquid scintillation counter. (Courtesy of Koch-Light Laboratories Ltd., Slough, Berkshire, England.)

F. Gamma-Scintillation Counting and Spectrometry

The effectiveness with which γ rays are absorbed increases rapidly with the increasing density and atomic number of the absorbing material. For many years, the most widely used type of γ detector has consisted of a large crystal of sodium iodide (for iodine, Z = 53), activated with a low concentration of thallium, encased in a thin aluminum lightproof can, and mounted in an integral assembly on a PM tube. Crystal sizes commonly range from 25 mm diameter × 25 mm high to 100 × 100 mm or even larger. Some versions of the detector have a reentrant well cut into the crystal into which the source may be lowered to improve counting geometry (Fig. 11). In recent years, some other crystalline inorganic materials that have still greater efficiency for absorption, such as cesium iodide, CsI, and bismuth germanate, $Bi_4Ge_3O_{12}$, have found applications in γ detection.

When γ rays enter the crystal, they lose part or the whole of their energy by transfer to electrons. The size of the output pulse is proportional to the γ energy absorbed. In many studies, the only requirement is to count the single γ-emitting tracer nuclide present. For this purpose, it is sufficient to record, within a single energy channel, all the pulses corresponding to total

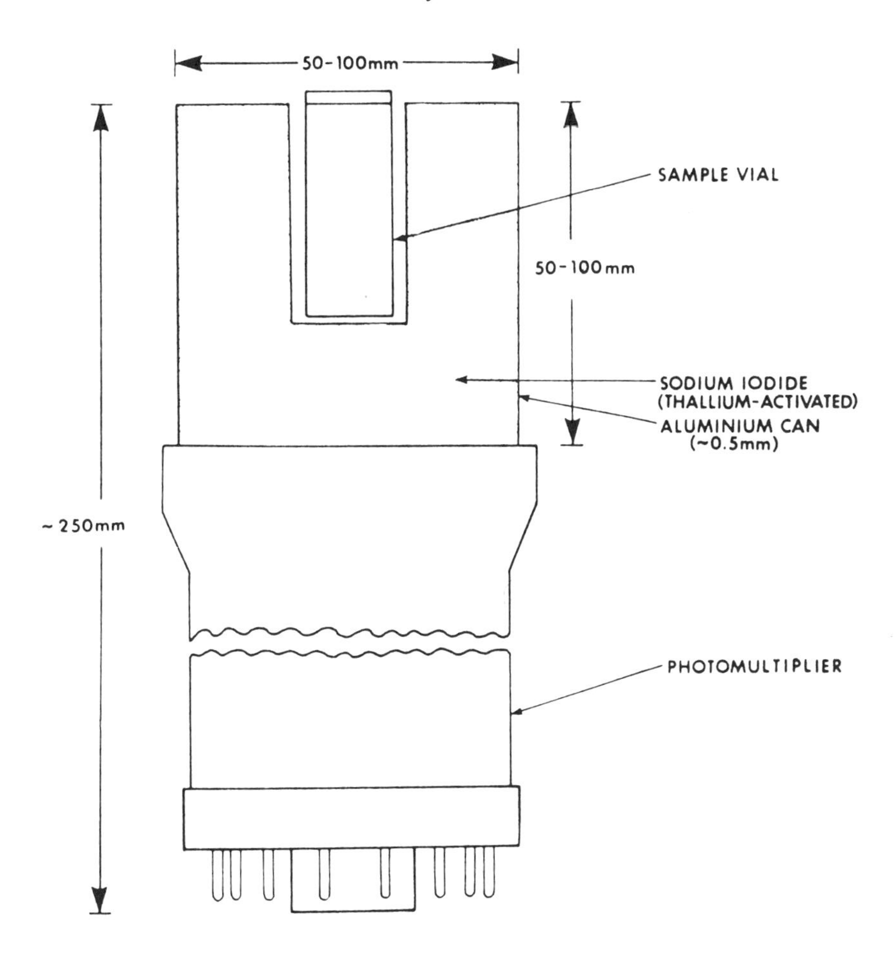

FIG. 11 Sodium iodide (thallium-activated) well crystal-photomultiplier assembly for γ-scintillation counting.

absorption, i.e., those in the photopeak. This is achieved by setting upper and lower discriminator levels to leave a window of the desired width. A conventional electronic scaler is used to record the events.

Large automated γ counters in current use normally include a number of preset channels for counting particular radionuclides: mainly those used widely in medical research, e.g., ^{125}I. However, one or more channels may be adjusted to any desired energy. Crystal sizes are normally 50 × 50 mm or 75 × 75 mm, with wells up to

25 mm in diameter. The sample conveyors can accommodate several hundred samples, and because of the potentially high background activity caused by a large number of samples on the conveyor, heavy lead shielding of the crystal is normal. Typically, the crystal is surrounded by at least 75 mm of lead, with up to 200 mm on the side adjacent to the conveyor. As with liquid scintillation counters, microprocessor control of counting and data output is a standard feature of the latest instruments.

Each γ-emitting radionuclide emits radiation of a characteristic energy. By determining the energy of the radiation, it is usually possible to make an unequivocal determination of the identity of the nuclide; furthermore, by separately recording radiation of different energies, it is possible to determine quantitatively the separate activities present in a mixture. These principles are embodied in the *gamma spectrometer*, in which a scintillation detector or semiconductor detector (see Sec. III.F) is coupled to a pulse-height analyzer. The detector emits a pulse that is proportional to the γ energy absorbed, and this is converted from analog to digital form by a conversion circuit and recorded in the corresponding energy channel of the analyzer. Purpose-built multichannel pulse height analyzers (MCAs) have been replaced to a large extent in the last few years by personal computers. This has resulted in far greater flexibility of data handling at much less cost.

The γ rays that are completely absorbed give rise to the photopeak and those only partially absorbed to the Compton continuum (Fig. 12). In addition, γ rays in excess of 1.02 MeV in energy may result in *pair production*, i.e., an electron-positron pair is created: the reverse of the annihilation process described in Sec. II above. The positrons will inevitably be annihilated in turn by collision with other electrons, resulting in the production of 0.51 MeV annihilation radiation. The probability of this process occurring increases with increasing γ energy, and 0.51 MeV peaks are visible, above the Compton continuum, in the spectrum of the more energetic γ emitters.

The activity of a source is determined normally from the counting rate observed in the photopeak. Two or more γ emitters may be determined in the presence of each other if the photopeaks are resolved adequately. The resolution of detectors is usually expressed as a ratio of the full width of the ^{137}Cs photopeak at half the maximum height (or FWHM, shown as ΔE in Fig. 12) to the energy of the radiation causing the peak (662 keV). Thus, if the FWHM for a scintillation detector = 50 keV, the resolution is (50/662) $\times$ 100 = 7.5%. Spectrometric applications of sodium iodide scintillation detectors are now restricted primarily to measurements involving only one or two radionuclides.

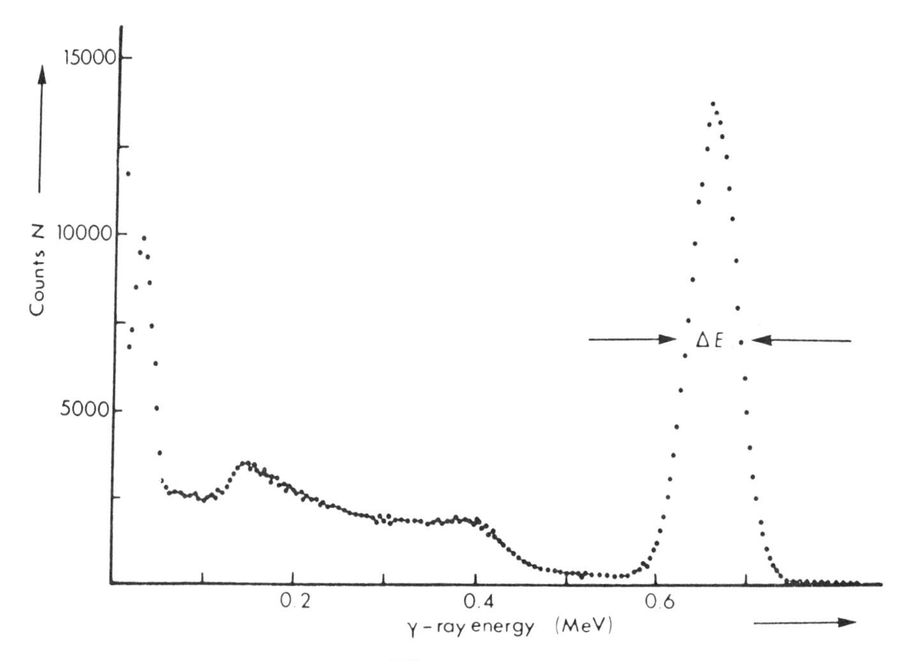

FIG. 12 γ-ray spectrum of ^{137}Cs, obtained using a NaI (Tl) detector, showing photopeak and Compton continuum. (From Ref. 43.)

G. Semiconductor Detectors

Semiconductor radiation detectors can be regarded as the solid-state equivalents of gas ionization detectors. When charged particles or γ rays lose energy by interactions with the detector material, electrons are raised in energy from the valence band to the conduction band. Under the influence of an applied electric field, these electrons move through the material, resulting in a pulse [4,26].

Compared with gas detectors, semiconductor devices have certain advantages. First, the fact that they are solid means that they have a far greater stopping power than a gas-filled detector. Second, excellent energy resolution is obtainable, which may be as much as 20 times better than that of a gas ionization detector or 40 times that of a sodium iodide scintillation detector. Third, the ability to make semiconductor detectors in extremely small sizes greatly increases their possible applications as probes used in biological studies.

The presence in semiconductor materials such as quadrivalent germanium or silicon of pentavalent or trivalent atoms imparts semiconducting properties, and the materials are known as *n*- or *p*-type semiconductors, respectively. One type of detector consists of layers of *p*- and *n*-type materials with a voltage applied so that they are connected to negative and positive poles, respectively. Under this *reverse bias* potential, the holes migrate toward the negative pole and the electrons to the positive, thus creating a depletion zone in the center. Radiation entering this zone induces the formation of electron-hole pairs that migrate to the poles, resulting in a current pulse whose magnitude is proportional to the energy of the radiation. Such a device is known as a noncompensated surface barrier detector and has a depletion zone up to a few millimeters in width, thus making it suitable for counting charged particles or low-energy X rays.

Detectors with much greater depletion zone widths (particularly those made of germanium) have much more "stopping power" and thus are suitable as γ detectors. For some years the best germanium contained $\sim 10^{14}$ *p*-type atoms/cm^3, and this needed careful counter-doping or "drifting" with lithium (an *n*-type atom) to compensate for these residual charge carriers, thus giving rise to the *lithium-drifted germanium*, or Ge(Li), detector [44]. From the mid-1970s, higher-purity Ge with only $\sim 10^{10}$ *p*-type atoms/cm^3 became available, and this material is used now in detectors variously known as "intrinsic," "hyperpure," and "high purity." Lithium compensation is not required, and the impurities may be of either the *p*- or *n*-type [44].

Both types of detector have to be used at liquid nitrogen temperature (77 K) to reduce the thermally induced electronic noise to an acceptable level. A major advantage of the intrinsic germanium detectors is that they are not damaged by warming up to room temperature, whereas the Ge(Li) detectors must be always kept at low temperature. The former are much more costly, however.

In terms of resolution, either type of Ge detector represents an enormous improvement over the sodium iodide detector, as can be seen from Fig. 13, and this allows the measurement of complex mixtures of emitters without prior separation (see Chap. 9).

1. Alpha Spectrometry

The unequivocal identification of α-emitting isotopes can be made normally by determining the energy of the particles. The introduction of solid-state detectors has resulted in greatly improved resolution of α spectra compared with that obtainable with ionization chambers. However, the good resolving power of the detector is useless if the energy spectrum is badly degraded by contamination of the source, and this is frequently the limiting factor in studies concerned

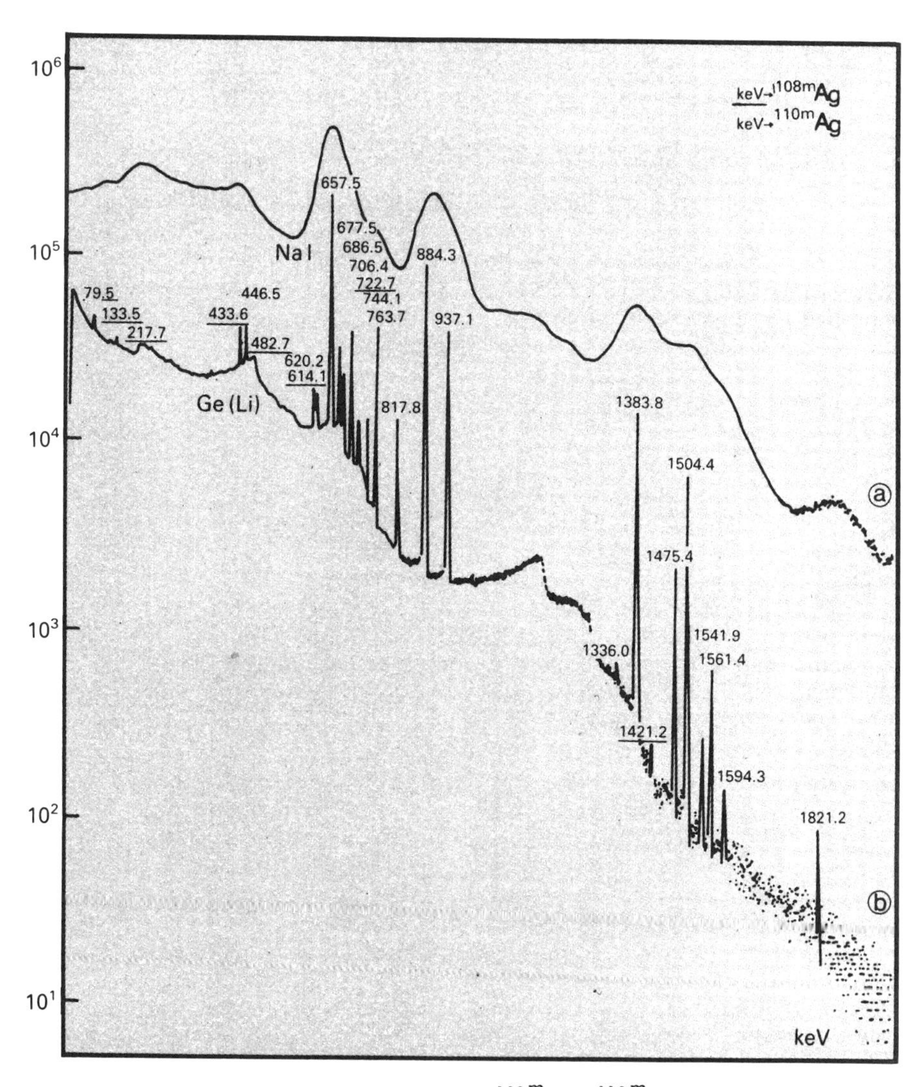

FIG. 13 Pulse-height spectrum of a $^{108m}Ag-^{110m}Ag$ source. (a) NaI(Tl) detector and (b) Ge detector. (From Ref. 44.)

with the analysis of radionuclides present in matrices such as soil and plant ash. The presence of even a few micrograms of material on the source results in particles emerging with a range of energies from the maximum possible virtually to zero. Sources for α spectrometry, therefore, must be formed from carrier-free material after careful radiochemical purification to remove all extraneous matter. Procedures for this include electrochemical deposition, electroplating, and evaporation of solutions onto metal disks [25,45].

H. Autoradiography

Ionizing radiation will interact with and expose the silver halide in a photographic emulsion so that on development the areas subjected to the radiation show up as blackened. Thus, a sheet of photographic paper can be used to give a facsimile, or an *autoradiograph*, of the distribution of radioactivity in a particular object.

Autoradiography is most suitable for use with low-energy β emitters, such as ^{3}H, ^{14}C, ^{33}P, ^{35}S, ^{45}Ca, etc., which produce well-defined images because of the short path length of the particles. High-energy β particles, with longer path lengths, and γ rays give much more diffuse images that are of limited use. Full details may be found in the book and review by Rogers [46,47].

Autoradiography has been used for many years in the study of nutrient uptake from soil by plant roots and to study the distribution of nutrients in plants after uptake [e.g., 48–50]. Another important use is with paper and thin-layer chromatograms. Here it is used to monitor the distribution of activity (usually of ^{14}C) in particular compounds, following synthesis or degradation or some other biochemical process.

Although autoradiography is essentially a qualitative technique, quantitative measurements may also be made using a microdensitometer to scan the film, e.g., with ^{33}P [50]. When quantitative results are required, a more common method is to recover the radionuclide from the sample and count it by one of the methods referred to in previous sections. The autoradiograph can be invaluable in pinpointing the areas of sample to be assayed.

An interesting combination of two techniques is shown in *fluorography*, in which the specimen for autoradiography is impregnated with a scintillation fluor, and the photographic film is affected by the light produced by the fluor as well as by the direct action of the radiation. Fluorography is especially effective with tritium, because the low energy of the β particles from this nuclide results in very poor penetration of the photographic emulsion. However, there is little to be gained by applying the technique to radionuclides that emit more energetic radiation [9,46].

IV. RADIOCHEMICAL METHODS

A. Inorganic Separation Procedures

The manipulations that have to be performed to prepare a radionuclide in a form suitable for counting vary enormously in complexity.

At one end of the scale, a direct nondestructive determination may be carried out on the original material. This is possible not only for γ emitters but also in some circumstances for high-energy β emitters, e.g., in the determination of ^{32}P in plant material. Again, in studies of isotopic exchange between soil and solution (see Sec. V.B), the activity remaining in solution may be determined directly, after separation from the soil by centrifugation.

The development of instrumental neutron activation analysis (INAA; see Chap 9) using high-resolution germanium detectors has made it possible to analyze a complex mixture of γ emitters, but it has been recognized that purely instrumental techniques cannot meet all analytical requirements [13], and separation procedures (including groups of elements as well as complete separations) are still valuable. Radiochemical procedures for use with NAA of environmental and biological samples have been reviewed by Pietra et al. [51]; the extensive literature describing separation procedures for preparing radionuclides in a sufficiently pure state for nonspectrometric counting has been reviewed by a number of authors [e.g., 3,10–13].

1. Precipitation Methods

When chemical separation procedures are required, it is necessary to consider whether separations involving the use of isotopic carriers can be employed or whether carrier-free methods are necessary. If the former situation applies, precipitation of a radionuclide along with a macro-amount of the same (inactive) element is one of the commonest ways of separating the radionuclide from solution. Examples are the precipitation of ^{35}S as $BaSO_4$ and ^{45}Ca as calcium oxalate [24]. In situations where insufficient element is initially present, a quantity usually in the range 10–100 mg is added to the sample. To achieve satisfactory carrying of the radionuclide, it must first be uniformly mixed with the carrier in solution. Elements of variable valence, e.g., iodine and manganese, should also be taken through an oxidation–reduction cycle to ensure that both active and inactive forms are in the same valence state.

When considering the effectiveness of a particular separation, it is necessary to distinguish between *chemical* and *radiochemical* purity. A precipitate of, say, 50 mg of barium sulfate containing 100 Bq (2.7 nCi) of ^{35}S would be radiochemically very impure if it contained half this activity of ^{32}P as a contaminant (particularly as the latter would be counted with much greater efficiency); yet the weight of phosphorus this activity represents is only 5×10^{-12} mg.

It is often difficult to achieve simultaneously both adequate chemical and radiochemical purity *and* a quantitative recovery. It is more desirable to achieve radiochemical purity with a reasonable yield, say in excess of 50%, and subsequently to determine that yield by a suitable analytical procedure, than to achieve 100% recovery and also carry down unwanted activities in the precipitate. Contamination of a precipitate can be greatly reduced often by the addition to the solution of milligram quantities of the inactive element isotopic with the unwanted activity, as a "holdback" carrier. The diluting effect of this addition greatly reduces the amount of unwanted activity that is coprecipitated.

Nonisotopic carrying, i.e., coprecipitation with ions of other elements, may be employed with carrier-free material. Several different processes may be employed: isomorphous replacement, anomalous mixed crystal formation, surface adsorption on the precipitate, or internal adsorption, i.e., adsorption during the growth of crystals that traps the carried ions within the crystals [20]. These carrying reactions may also be employed to remove carrier-free impurities from a solution containing the wanted activity in the presence of its isotopic carrier. This process has become known as "scavenging," and a common example is the precipitation of iron (III) hydroxide to remove the ^{90}Y and ^{140}La daughter products of ^{90}Sr and ^{140}Ba, respectively [52].

2. *Ion Exchange*

This technique has the advantage of being applicable to both carrier-free and carrier-borne radioisotopes. It has been very widely used in radiochemical analysis, particularly for separations of complex mixtures of radionuclides.

The two procedures used most frequently are ion-exchange elution chromatography and selective adsorption [53]. In the former, a mixture of different ionic species is adsorbed on a column, and the individual species are then sequentially eluted with one or more reagents. In the latter, conditions are created so that one species is much more strongly adsorbed than the others. For detailed discussion of the theory of ion-exchange processes and their applications in analytical separations, the reader is referred to one of the specialist books on the subject, e.g., Rieman and Walton [54].

The great majority of ion-exchange separations employ the synthetic cation- and anion-exchange resins based on polymers of styrene and divinylbenzene. Strong-acid cation exchangers (e.g., Dowex 50) contain sulphonic acid functional groups, and strong-base anion exchangers (e.g., Dowex-1) contain quaternary ammonium groups. In general, for solutions under 0.1 *M* the affinity for cation exchanger increases with charge on the cation and decreases with increasing size of the hydrated cation. This results in the following order of affinities [20]:

For cations,

$$Ba^{2+} > Sr^{2+} > Ca^{2+} > Mg^{2+}$$

$Mg^{2+} > Mn^{2+} > Cd^{2+}$

$Cs^+ > Rb^+ > NH_4^+ > K^+ > Na^+ > H^+ > Li^+$

For polyvalent anions, on a quaternary ammonium resin,

citrate > sulfate > oxalate

and for univalent anions,

$I^- > HSO_4^- > NO_3^- > Br^- > Cl^- > HCO_3^- > H_2PO_4^- > OH^-$

Some ions that are only separated from each other with difficulty by elution from cation-exchange columns may be much more readily separated by anion-exchange in a strong acid solution. For example, Fe^{3+} and Co^{2+} form anionic chloro-complexes in strong HCl, and Ni^{2+} does not and is not adsorbed.

The use of complexing agents often greatly improves the separation that can be achieved between two very similar ions, because the ion with the greater affinity for the resin also has the lesser tendency to form complex ions. Thus, Ba^{2+} may be separated from Ra^{2+} by eluting the former ion from a cation-exchange column with 0.01 *M* EDTA at pH 9 [45].

Some ions may be adsorbed selectively from solution by the use of chelating resins, which have a particular affinity for ions of a certain size or charge. An example of the application of one such resin, Chelex 100, to soil analysis is the determination of copper [55]. Certain insoluble inorganic substances can also have a high selectivity for the absorption of particular ions, e.g., ammonium molybdophosphate [25] and the much more recently developed titanium ferrocyanide [56], both for ^{137}Cs, and powdered Sb_2O_5 implanted into a resin matrix, for ^{24}Na [57].

3. *Solvent Extraction*

Solvent extraction, or liquid–liquid distribution, shares with ion exchange the property of being applicable to carrier-free or macroamounts of material. In its most common form, the process involves the extraction of a metal chelate from aqueous solution into an immiscible nonpolar solvent, thus achieving a satisfactory separation from interfering ions. The widespread analytical applications have resulted in a large output of papers and a number of books, e.g., Refs. 58–60.

According to Coomber [53], for satisfactory separation of a substance from others by solvent extraction the following points should be satisfied:

1. The solubility of the extracted compound should be high.
2. The separation factor should be as large as possible.
3. The solvent should be immiscible, or nearly immiscible, with the aqueous phase.
4. The specific gravity of the solvent should differ as much as possible from the aqueous phase, to encourage rapid separation and minimize the formation of emulsions.

If adequate recovery is not achieved with a single extraction, the process may be repeated as many times as is necessary, and the extracts combined. Repeated extraction with small aliquots of solvent is preferable to one extraction with a single large volume.

Analytical separations may be readily carried out in separating funnels or stoppered or capped test tubes and centrifuge tubes. When several successive extractions are needed, or when many samples are to be extracted, the use of commercially available continuous-extraction apparatus is desirable.

Metal chelates are not the only species that can be extracted into organic solvents. Other species include ion-association compounds formed by large organic ions such as tetraphenylarsonium; nitrate and chloride complexes (solvates) formed with neutral solvents such as tri-*n*-butyl phosphate (TBP); and complexes between anions such as $FeCl_4^-$ and long-chain amines [59].

In recent years some very great improvements in traditionally difficult separations have been achieved by utilizing the selectivity of cyclic crown ethers in *emulsion membrane* extraction as well as conventional solvent extraction. For example, Kimura et al. [61] and Mikulaj et al. [62] have used dicyclohexyl-18-crown-6 for the separation of Sr from Ca. The latter workers achieved a separation factor of ~500 with a yield of Sr of 95–98%, even in the presence of a 1000-fold excess of Ca.

B. Preparation of Sources

1. Solid Sources for β Counting

These may be prepared either by (1) transferring a precipitate as a slurry onto a planchet and evaporating the liquid under an infrared lamp, (2) evaporating a solution in the same way, or (3) preparing a flat pad of precipitate by filtration under suction. Efficient preparation techniques were established over 30 years ago [20] but are still perfectly acceptable for present-day applications.

Two alternative ways of applying method (3) are illustrated in Figs. 14a and b. Figure 14a shows a demountable assembly from which the filter paper, together with its pad of precipitate, may be

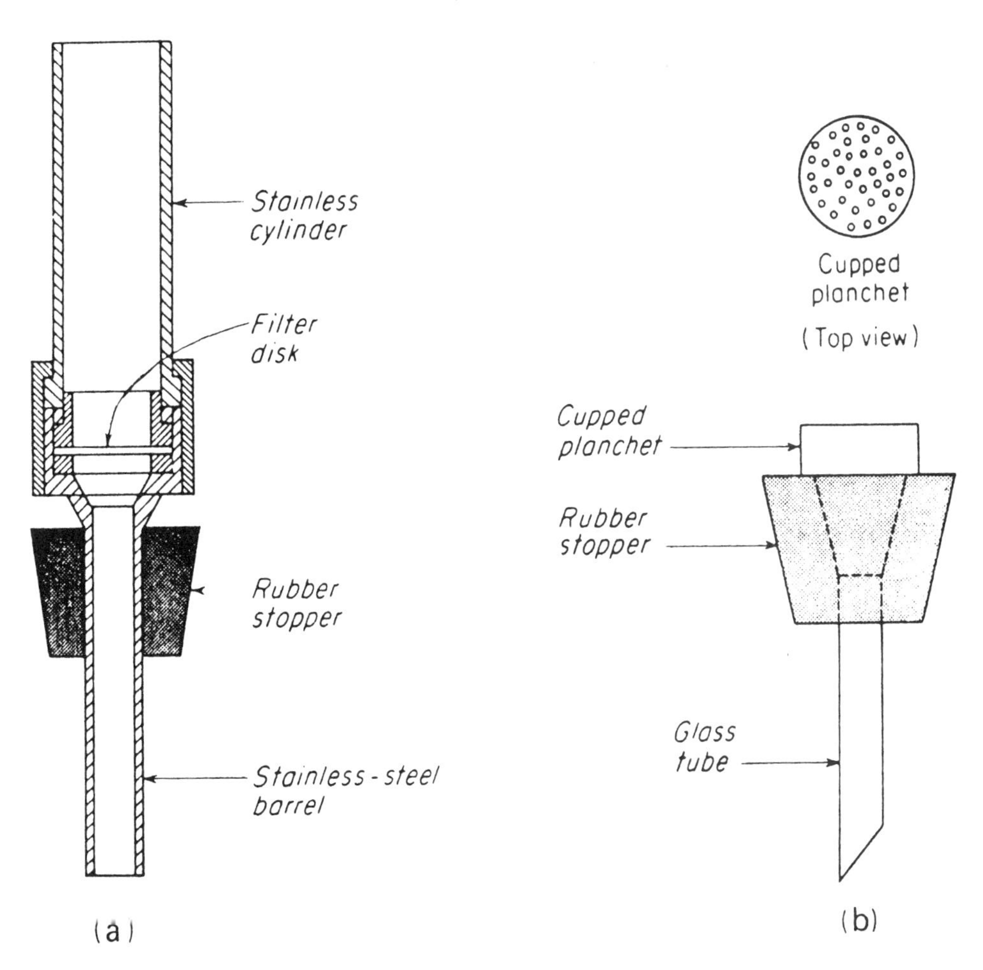

FIG. 14 Filtration devices for preparation of solid sources for β counting. (From Ref. 20.)

transferred to a planchet. Figure 14b shows a perforated planchet that serves as a mounting for the filter paper during filtration. This avoids the risk of damage to the pad during transfer.

Precipitates should be dried slowly, to avoid cracking. Curling of the filter paper and crumbling or flaking (which results in lost activity or contamination of the counter) can be avoided by cementing with a dilute solution of polyvinyl acetate in methanol [52].

2. Other Sources

The lack of self-absorption problems in γ counting, and the consequent ability to count large samples, greatly simplifies source preparation. Either the original sample may be counted directly without processing, or, if concentration of the radionuclide is required to improve efficiency, a simplified procedure may be adopted such as adsorption of the radionuclide on ion-exchange material which is then transferred to a bottle and counted in a well-type sodium iodide scintillation detector [25]. Where a purification procedure is necessary, resulting in a solution of the radionuclide, this solution can be counted directly, thus eliminating the source preparation stages necessary for most β-counting techniques.

Reference has already been made in Sec. III to the preparation of sources for α spectrometry and liquid scintillation counting and will not be discussed further here.

C. Isotope Dilution Analysis

In isotope dilution analysis (IDA), the quantity of an element present in a sample is determined by adding a known amount of a radionuclide that is isotopic with the element. A procedure is followed that ensures the uniform mixing of active and inactive material. Then after a suitable separation, which need not be quantitative but which gives the element in a pure form, the change in specific activity is determined. Thus, if the specific activities before addition to the samples and after recovery are S_a and S_r, respectively,

$$S_r = S_a \frac{W_a}{W + W_a} \tag{9}$$

where W and W_a are the quantities of the element initially present and added to the sample, respectively. Rearranging Eq. (9) gives

$$W = W_a\left(\frac{S_a}{S_r} - 1\right) \tag{10}$$

When the radionuclide added is carrier-free or if the quantity of inactive substance added is very much less than the quantity in the sample ($W_a \ll W$), then

$$W = \frac{A_a}{S_r} \tag{11}$$

where A_a is the activity added to the sample and S_r, as before, is the specific activity of the substance recovered from the sample.

The method may be applied to the determination of naturally occurring radioelements, as well as to inactive ones, provided that it is possible to distinguish between the radiation emitted from the radionuclide and that from the radioelement already present in the sample. For example, an artificial isotope of radium or polonium can be used to measure the recovery of another isotope of the same element because the radiations can be distinguished by α spectrometry [16,45].

With the exception of those heavy elements which have no stable isotopes, it is possible for the added isotopic yield tracer to be carrier-free or to contain added inactive element, as desired. For those separation procedures which work as well with carrier-free materials as with macroscopic amounts, such as solvent extraction and ion exchange, either approach may be employed. If precipitations are involved, milligram quantities of inactive carrier are required.

Reverse isotope dilution analysis involves the addition of an inactive element or compound to a sample containing the element in labeled form. As the name implies, this is the reverse of the procedure described in the previous section.

A common application is in neutron activation analysis (NAA), in which the activity in the sample is induced by irradiation by neutrons from a nuclear reactor or neutron source. (For a full discussion of the neutron activation technique, see Chap. 9 and the references therein, and specialist books such as Ref. 6.) A quantity W_a of carrier is added that greatly exceeds the quantity W in the sample. Typically, 10–50 mg may be added to a sample containing microgram or submicrogram amounts of the element. If the quantities of activity originally present in the sample and recovered at the end of the chemical separation are A_s and A_r, respectively, and the weight of the carrier recovered = W_r,

$$A_s = A_r \frac{W_a}{W_r} \tag{12}$$

The specific activity S_s of the element in the sample is determined by comparison with a standard irradiated under identical conditions. Thus, W may be calculated as

$$W = \frac{A_s}{S_s} \tag{13}$$

Chemical recovery may be determined either by using a radionuclide as yield tracer or "spike" (with or without inactive carrier) or by using an inactive carrier alone. The choice depends very much on the availability of necessary instrumentation and the variability in concentration of the element between samples. The yield of inactive carrier is measured normally by flame emission or atomic absorption spectrometry or a gravimetric method. The use of radioactive yield tracers may be preferred if a multichannel pulse height analyzer is available. Computer processing of the counting data may be employed to determine the individual components of the composite spectra obtained from the mixture of nuclides. However, when element concentrations vary widely, there is a problem of matching the quantity of added tracer to the induced activity, to prevent one activity from being "swamped" by the other. In such situations, superior results may be achievable by determining the recovery of inactive carrier. The latter method is also more appropriate in laboratories that are well equipped for the analysis of inactive elements, but which have only simple γ counters available.

There is a relatively new approach to IDA that is applicable to the determination of elements that have at least two stable isotopes, which may be converted by activation analysis methods to γ-emitting radionuclides. A known amount of the element, enriched in one of the isotopes, is added as a spike, and the relative γ activities of both the spiked sample and the spike material are determined, following activation. If we know the isotopic composition of the natural element and the spike, the amount of the former in the sample can be calculated [63].

1. *Substoichiometric Isotope Dilution Analysis*

The determinations of inactive elements by the procedures described in the two previous sections necessitate chemical analysis as well as measurement of radioactivity. This can cause difficulties if concentrations are very low. A technique known as substoichiometric isotope dilution analysis was pioneered by Růžička and Starý [64], as a means of eliminating the need for a chemical determination.

If the specific activity of the radionuclide added to the sample is

$$S_a = \frac{A_a}{W_a} \tag{14}$$

the specific activity after isotopic exchange with the element in the sample is

$$S = \frac{A_a}{W + W_a} \tag{15}$$

where W is the weight of element in the sample. By adding the same *substoichiometric* quantity of reagent required to bring about a precipitation of the element or to form a complex for solvent extraction, the same weight of element W_r ($< W_a$) can be separated from the radionuclide solution and the sample that has undergone isotope dilution. If the activities of these two fractions are A_r and A_x, respectively,

$$A_r = W_r S_a = W_r \frac{A_a}{W_a} \tag{16}$$

and

$$A_x = W_r S = W_r \frac{A_a}{W + W_a} \tag{17}$$

Combining these two equations, we obtain

$$A_x(W + W_a) = A_r W_a \tag{18}$$

Thus,

$$W = W_a\left(\frac{A_r}{A_x} - 1\right) \tag{19}$$

i.e., the weight of element in the sample can be determined from the weight added and a ratio of two activity measurements.

D. Radiocarbon Dating

Radiocarbon dating is a well-established tool for estimating the age of organic material in archaeological and geological studies and is equally applicable to the dating of organic material found in soils.

The technique depends on the change in the specific activity of ^{14}C in the organic matter with time, due to the decay of the ^{14}C ($t_{1/2}$ = 5700 yr). The nuclide is produced in the upper atmosphere by the reaction $^{14}N(n,p)^{14}C$, the neutrons being generated by the interaction of cosmic rays with other atoms. The ^{14}C is oxidized to CO_2, moves into living tissue by photosynthesis, and becomes incorporated into soil organic matter. When the organic material is no longer living, exchange with the pool of atmospheric CO_2 ceases and, in principle, a comparison of the specific activity of carbon from a soil sample to be dated with a contemporary sample of organic carbon will allow the age of the former to be calculated. In practice, complications arise, e.g., because of changes in the specific activity of atmospheric carbon during the past few decades, due to the combustion of fossil fuels and extra production of ^{14}C as a result of the atmospheric testing of nuclear weapons.

Because of the low specific activity, the achievement of satisfactory measurements requires the counting of samples that contain as much carbon as possible and the use of low-background counters. The organic carbon is combusted to CO_2 that may be counted directly, or after conversion to a hydrocarbon, in an anticoincidence proportional counter. Such gas counters may be operated at high pressure, to incorporate more sample, and the conversion of CO_2 to acetylene, C_2H_2, is an effective way of increasing the amount of carbon in the gas filling the counter. The formation of acetylene is achieved by the reaction of the CO_2 with lithium to form lithium carbide, Li_2C_2, which is then reacted with water. By catalytic conversion of the acetylene to benzene and counting in a liquid scintillation counter, even more carbon can be counted than in a large high-pressure gas counter [65].

E. Preparation and Analysis of Labeled Organic Substances

Study of the synthesis and degradation of simple and complex organic molecules in the soil is facilitated greatly by tracer techniques using radioactively labeled compounds. A wide range of such compounds may be obtained from commercial suppliers, who will also prepare to order compounds that are not catalog items. Uniformly ^{14}C-labeled plant material can be produced by growing plants in a growth chamber in an atmosphere containing $^{14}CO_2$ (the Commissiariat à L'Énergie Atomique, St-Paul-les-Durance, France, is a source of such material). An alternative method is to grow plants in soil or solution culture to which $NaH^{14}CO_2$ has been added [66]. Knight [67] has described solution culture methods for labeling with ^{59}Fe, ^{22}Na, ^{36}Cl, ^{65}Zn, and ^{45}Ca. Humic substances, soil polysaccharides, etc., labeled with ^{14}C or ^{35}S, are not commercially available and also have to be prepared by researchers themselves.

Labeled plant material can be used directly in decomposition studies [e.g., 68] or allowed to decompose in soil to produce labeled organic matter. The specific activity of organic matter fractions extracted after this process tends to be rather low, but can be increased by adding labeled compounds directly to the soil for incorporation by microorganisms [69]. However, it should be noted that with either procedure the added radionuclide is present only in newly formed material and therefore the soil organic matter fractions are not uniformly labeled.

1. *Analysis*

Suitable techniques for the separation and quantitative assay of labeled organic compounds have been described in detail by L'Annunziata [14] and will only be summarized here. Chromatographic methods, including paper and thin-layer, gel filtration, ion exchange and gas chromatography, and electrophoresis, have been used in this way. The separation procedures are the conventional ones used in nonradioactive work with the addition of an appropriate method of measuring the activity present.

The positions of radioactive compounds on a thin-layer chromatographic plate or paper chromatogram may be determined either by autoradiography or by the use of a scanner, which consists of a detector (under which the plate or paper moves) connected to a rate meter and a chart recorder. Alternatively, sections of a paper chromatogram, or the coating of a thin-layer plate, may be counted in a scintillation counter directly, or after digestion.

In liquid chromatography, including gel filtration and ion-exchange methods, the effluent solution may either be collected in a fraction collector, followed by the determination of the activity in each fraction, or passed through a flow cell where the activity is monitored continuously with a suitable detector.

Gaseous or volatile substances may be assayed by gas radiochromatography. In a typical system, a gas chromatograph (see Chaps. 12 and 13) is modified to allow part of the gas stream leaving the column to pass through a combustion/reduction train in which the organic compounds are converted to CO_2 and H_2, and then to a gas-flow proportional counter. The remaining part of the gas stream is monitored with a GC flame ionization detector. The outputs of the two detectors are recorded normally with a twin-pen chart recorder.

V. APPLICATIONS

Applications of nuclear and radiochemical methods to the elemental analysis of soils and biological materials and estimates of nutrient labile pools began as early as the 1940s. The literature covering these aspects and also other applications up until 1979 was reviewed in Chap. 6 of the first edition of this book [70]. This section is restricted to selected illustrations of appplications made over the last two decades, to give an indication of when nuclear and radiochemical methods may have advantages over alternative methods of analysis.

A. Elemental Analysis

1. Neutron Activation Methods

Neutron activation methods, because of their sensitivity, are particularly useful for the determination of certain trace elements. For example, the limit of detection for mercury was 10^{-10} g in 100 mg soil samples, using a procedure involving the solvent extraction and sulfide precipitation of the ^{197}Hg activation product [71].

Copper in biological materials has been determined by NAA combined with extraction chromatography, using LIX-70 (2-hydroxy-3-chloro-5-nonobenzylphenone oxime) supported on Bio-Beads SM-1. This technique is reported to be very selective for Cu, with recoveries > 95% and decontamination factors of 10^4-10^5 [72].

Trace elements in biological materials have been determined by separation into groups after neutron activation and counting with a Ge(Li) detector. The groups were Se, Ag, Au, Sb, Pt; Hg, Co, Fe, Zn, Ni; and Mo, Sn, Cr, Cd, Cu, and As (after separation from Na, K, Rb, Cs, and partially from Br) [73]. Low levels of rare earths in rocks have also been determined by NAA, using rapid ion-exchange and hydroxide precipitation methods, followed by γ counting with Ge(Li) and intrinsic Ge detectors [74].

For elements of higher concentration, the neutron flux available in nuclear reactors often induces excessive levels of activity. Thus, manganese and sodium in geological samples have been determined more satisfactorily by activation with a ^{252}Cf spontaneous fission neutron source, giving 6×10^5 n cm^{-2} sec^{-1}, followed by counting on a NaI(Tl) detector [75].

Iodine in water samples from an area in Pakistan where goitre (iodine deficiency of the thyroid) is endemic has been determined by NAA following preconcentration by solvent extraction, and γ counting of the ^{128}I activation product with a 30 cm^3 Ge(Li) detector and multichannel analyzer [76].

Neutron activation has been applied to the determination of chemical yields when assaying the fission products ^{89}Sr and ^{90}Sr in grass samples. 40 mg of stable Sr were added and the ^{87m}Sr activity determined after a 1 hr irradiation at a neutron flux of 2.5×10^{12} n cm^{-2} sec^{-1} [77].

2. *Isotope Dilution Methods*

Several elements have been determined in soils and other environmental materials by radiochemical isotope dilution procedures. A typical study was that by Brown et al. [78] who used ^{76}As as the yield tracer in the determination of arsenic in soils and rocks. Their procedure involved the separation of arsenic as arsine gas, which was then absorbed in iodine-potassium iodide-sodium bicarbonate solution and determined colorimetrically. This inevitably resulted in variable losses due to leakage, and chemical yields (determined by γ counting with a single-channel analyzer) varied from 40–90%. Branquinho and Robinson [79] similarly used ^{212}Pb in the determination of lead in water and dust samples, in the course of an investigation of lead pollution in Rio de Janeiro, Brazil. This short-lived tracer ($t_{1/2}$ = 10.6 hr) is a decay product of the gaseous nuclide ^{220}Rn, which itself results from the successive decays of ^{228}Th and ^{224}Ra. The ^{212}Pb was conveniently "milked" from a ^{228}Th emanation source by collection on a gold cathode, from which it was removed with concentrated HCl.

The recovery of ^{227}Ac from sediments has been measured by using ^{225}Ac as the yield tracer. The activities of the two nuclides were determined by α spectrometry [80].

3. *Substoichiometric Analysis*

This technique has been applied to, e.g., the measurement of the specific activity of ^{32}P in plants fertilized with labeled phosphate [81] and of mercury in biological materials, using ^{203}Hg as the tracer [58]. Other applications include the determination of iron in plant material [82] and fluorine in plant material and water [83, 84].

Uranium in phosphate rock (50–200 mg kg^{-1}) has been determined by adding ^{237}U tracer, followed by solvent extraction with a substoichiometric amount of TOPO and γ counting in a NaI(Tl) well counter [85].

According to a recent review of the applications of this technique in the Soviet Union [86], the overwhelming majority of studies have been devoted to the analysis of "pure substances, rocks, reference materials, and biological and environmental samples," and the rate of appearance of papers is doubling every 2.5 yr.

4. *Direct Counting Methods*

The natural radionuclides ^{210}Pb and ^{226}Ra have been determined in sediments by direct γ counting, measuring the 46.5 keV peak from ^{210}Pb and the 351.9 keV peak from the ^{214}Pb decay product of ^{226}Ra [87]. Low-background γ spectrometry has been used to determine uranium, radium, thorium, and potassium in soils [88] and sediments [89,90].

B. Measurement of Labile Ions in Soil

The measurement of isotopically exchangeable soil phosphorus and thus the labile pool of P in the soil was first carried out by McAuliffe et al. in the 1940s [91], using a ^{32}P tracer. Counting methods have, of course, become more sophisticated in recent times, and Cerenkov counting of the phosphate in solution is a very satisfactory method [36]. In their studies of available soil phosphorus, Fardeau and Jappe [92] used double labeling with ^{32}P and ^{33}P, counting the former by the Cerenkov method and the latter by liquid scintillation. Available sulfur has been determined, using ^{35}S, in Australian soils [93].

The short half-life of ^{42}K (12.4 hr) necessarily limits the duration of exchange studies. Rubidium, in the form of ^{86}Rb, has been investigated as an alternative tracer, but significant differences occur between the two elements [94]. A more satisfactory method for long-term exchange studies with potassium has been described by Mercer and Gibbs [95], who used the stable isotope ^{39}K free from the other stable isotope ^{41}K, which occurs in natural potassium to the extent of 6.9%. Exchange between a solution containing ^{39}K and a soil results in both ^{39}K and ^{41}K entering the solution, and the increase in ^{41}K concentration is determined by neutron activation analysis (by activation to ^{42}K).

The availability of a number of biologically essential trace elements has been studied by tracer methods, including cobalt (using ^{58}Co) [96], manganese (using ^{54}Mn) [97], and selenium (using ^{75}Se) [98].

C. Transformations of Labeled Organic Matter

The numerous studies of transformations of ^{14}C-labeled organic compounds in soil, and estimations of the age of organic matter, have been reviewed by Stout et al. [99] and Scharpenseel et al. [100]. Recent applications include the pulse labeling of carbon compounds in plants by a brief exposure to $^{14}CO_2$ and subsequent monitoring of their translocation into roots and release as CO_2 in the rhizosphere [101], and at the higher end of the time scale, an investigation of the fate of ^{14}C-labeled barley straw after 20 years of incubation in soil under field conditions [102]. The mineralization of ^{14}C-labeled humic acids and humic acid-bound compounds such as chlorinated and nonchlorinated phenols and anilines has been investigated by Haider and Martin [103].

The dynamics of ^{35}S-labeled organic sulfur compounds have been studied in forest soils [104–106], as has the turnover of carbon, nitrogen, and phosphorus through the microbial biomass by incubating soils with ^{14}C-, ^{15}N-, and ^{32}P-labeled bacterial cells [107]. The absorption by soil and subsequent transformation of atmospheric dimethylselenide have been investigated using ^{75}Se tracer [108].

Carbon-dating measurements have demonstrated different rates of organic matter decomposition in mollisols from three continents [109], and studies of soil organic matter enrichment with ^{14}C from atmospheric nuclear tests in the 1950s and 1960s have allowed estimates to be made of the carbon input rate, decomposition times, and downward diffusivity in the soil profile [110].

REFERENCES

1. H. A. C. McKay, *Principles of Radiochemistry*, Butterworths, London, 1971.

2. R. A. Faires and B. H. Parkes, *Radioisotope Laboratory Techniques*, 3rd ed., Butterworths, London, 1973.
3. D. I. Coomber (ed.), *Radiochemical Methods in Analysis*, Plenum Press, New York, 1975.
4. M. F. L'Annunziata, *Radiotracers in Agricultural Chemistry*, Academic Press, London, 1979.
5. D. Brune, B. Forkman, and B. Persson, *Nuclear Analytical Chemistry*, Verlag Chemie, Weinheim, West Germany, 1984.
6. J. M. Kolthoff and P. J. Elving (eds.), *Treatise on Analytical Chemistry*, Vol. 14, Section K: *Nuclear Activation and Radioisotope Methods of Analysis*, Wiley, New York 1986.
7. W. Geary, *Radiochemical Methods*, Wiley, Chichester, England, 1986.
8. H. J. Arnikar, *Essentials of Nuclear Chemistry*, Wiley, New York, 1987.
9. M. F. L'Annunziata, *Radionuclide Tracers*, Academic Press, London, 1987.
10. G. W. Leddicote, *Anal. Chem.*, *34*:143R (1962); *36*:419R (1964).
11. W. S. Lyon, E. Ricci, and H. H. Ross, *Anal. Chem.*, *38*: 251R (1966); *40*:168R (1968); *42*:123R (1970); *44*:439R (1972); *46*:431R (1974).
12. W. S. Lyon and H. H. Ross, *Anal. Chem.*, *48*:96R (1976); *50*: 80R (1978); *52*:69R (1980); *54*:227R (1982); *56*:83R (1984).
13. W. D. Ehmann and S. W. Yates, *Anal. Chem.*, *58*:49R (1986); *60*:42R (1988).
14. M. S. Smith, M. K. Firestone, and J. M. Tiedje, *Soil Sci. Soc. Am. J.*, *42*:611 (1978).
15. D. A. Barber, M. Ebert, and N. T. S. Evans, *J. Exp. Bot.*, *13*:397 (1962).
16. K. A. Smith, Ph.D. thesis, Reading Univ., Berkshire, England, 1968.
17. M. Vasilaki, L. Salmon, and J. A. B. Gibson, *Geochim. Cosmochim. Acta*, *30*:601 (1966).
18. E. R. Mercer, in *Radioactivity and Human Diet* (R. S. Russell, ed.), Pergamon, Oxford, 1966, p. 489.
19. J. C. Cunninghame, in *Radiochemical Methods in Analysis* (D. I. Coomber, ed.), Plenum Press, New York, 1975, p. 1.
20. R. T. Overman and H. M. Clark, *Radioisotope Techniques*, McGraw-Hill, New York, 1960.
21. G. Friedlander, J. W. Kennedy, and J. M. Miller, *Nuclear and Radiochemistry*, 2nd ed., Wiley, New York, 1964.
22. J. N. Mundy and S. J. Rothman, *Nucl. Instr. Meth.*, *200*:355 (1983).
23. J. N. Mundy and S. J. Rothman, *Meth. Exp. Phys.*, *21*:50 (1983).

24. C. L. Comar, *Radioisotopes in Biology and Agriculture*, McGraw-Hill, New York, 1955.
25. Anonymous, *Methods of Radiochemical Analysis*, WHO, Geneva, 1966.
26. J. R. Robinson, *Int. J. Appl. Radiation Isotopes*, *20*:531 (1969).
27. B. W. Fox, *Techniques of Sample Preparation for Liquid Scintillation Counting*, North Holland, Amsterdam, 1976.
28. G. Laustriat, R. Voltz, and J. Klein, in *The Current Status of Liquid Scintillation Counting* (E. D. Bransome, Jr., ed.), Grune and Stratton, New York, 1970, p. 13.
29. C.-Y. Lin and T.-Y. Chang Mei, *Int. J. Appl. Radiation Isotopes*, *35*:25 (1984).
30. J. B. Birks, in *The Current Status of Liquid Scintillation Counting* (E. D. Bransome, Jr., ed.), Grune and Stratton, New York, 1970, p. 3.
31. K. E. Collins, M. G. Farris, O. A. S. Estrazulas, and C. H. Collins, *Int. J. Appl. Radiation Isotopes*, *28*:733 (1977).
32. C. T. Peng, in *Radiochemical Methods in Analysis* (D. I. Coomber, ed.), Plenum Press, New York, 1975, p. 79.
33. P. A. Cerenkov, *Dokl. Akad. Nauk S.S.S.R.*, *2*:451 (1934).
34. E. H. Belcher, *Proc. Roy. Soc. Lond., Ser. A*, *216*:90 (1953).
35. A. T. B. Moir, *Int. J. Appl. Radiation Isotopes*, *22*:213 (1971).
36. R. E. White and A. W. Taylor, *J. Soil Sci.*, *28*:48 (1977).
37. V. F. Nascimiento Filho, *Int. J. Appl. Radiation Isotopes*, *29*:789 (1977).
38. J.-C. Fardeau, *Fert. Agric.*, *86*:23 (1984).
39. D. D. Lefebvre and A. D. M. Glass, *Int. J. Appl. Radiation Isotopes*, *32*:116 (1981).
40. H. Lundqvist, K. J. Johanson, and G. Jonsson, *Int. J. Appl. Radiation Isotopes*, *27*:233 (1976).
41. J. Ashcroft, *Anal. Biochem.*, *37*:268 (1970).
42. L. Cecchi and E. Somenzi, *Lab. Practice*, *35*:79 (1986).
43. P. Marmier and E. Sheldon, *Physics of Nuclei and Particles*, Vol. 1, Academic Press, New York, 1969.
44. J. D. Hemingway, *Lab. Practice*, *35*:15 (1986).
45. K. A. Smith and E. R. Mercer, *J. Radioanal. Chem.*, 5:303 (1970).
46. A. W. Rogers, *Techniques of Autoradiography*, 3rd ed., Elsevier, Amsterdam, 1979.
47. A. W. Rogers, *Practical Autoradiography*, Review 20, Radiochemical Centre, Amersham, Buckinghamshire, England, 1979.
48. J. M. Walker and S. A. Barber, *Science*, *133*:881 (1961).
49. J. P. Baldwin and P. B. Tinker, *Plant Soil*, *37*:209 (1972).
50. K. K. S. Bhat and P. H. Nye, *Plant Soil*, *38*:161 (1973).

51. R. Pietra, E. Sabbioni, M. Gallorini, and E. Orvini, *J. Radioanal. Nucl. Chem.*, *102*:69 (1986).
52. E. R. Mercer, K. B. Gunn, P. M. Lay, W. Harris, W. Downs, and M. G. Johnson, *UK Agricultural Res. Council Letcombe Lab. Ann. Rept. 1968*, Supplement, HMSO, London, 1969, p. 1.
53. D. I. Coomber, in *Radiochemical Methods in Analysis* (D. I. Coomber, ed.), Plenum Press, New York, 1975, p. 175.
54. W. Rieman and H. F. Walton, *Ion Exchange in Analytical Chemistry*, Pergamon, Oxford, 1970.
55. R. G. McLaren and D. V. Crawford, *J. Soil Sci.*, *24*:172 (1973).
56. J. Narbutt, J. Siwinski, B. Bartos, and A. Bilewicz, *J. Radioanal. Nucl. Chem.*, *101*:41 (1986).
57. A. Bilewicz, B. Bartos, J. Narbutt, and H. Polkowski-Motrenko, *Anal. Chem.*, *59*:1737 (1987).
58. Y. Marcus and A. S. Kertes, *Ion Exchange and Solvent Extraction of Metal Complexes*, Wiley, New York, 1969.
59. J. Starý. in *Analytical Chemistry*, Part 1 (T. S. West, ed.), Physical Chemistry Series I, Vol. 12, Butterworths, London, 1973, p. 279.
60. M. Pimpl and H. Schuettelkopf, Proc. 5th Int. Conf. Nucl. Meth. Environ. Energy Res., CONF-840408 (J. R. Vogt, ed.), NTIS, Springfield, Va., 1984, p. 216.
61. T. Kimura, I. Iwashima, T. Ishimori, and T. Hamada, *Anal. Chem.*, *51*:1131 (1979).
62. V. Mikulaj, H. Hlatky, and L. Vasekova, *J. Radioanal. Nucl. Chem.*, *101*:51 (1986).
63. K. Masumoto and M. Yagi, *J. Radioanal. Chem.*, *79*:57 (1983).
64. J. Růžička and J. Starý, *Talanta*, *8*:228 (1961).
65. G. W. Pearson, J. R. Pilcher, M. G. L. Baillie, and J. Hillam, *Nature (Lond.)*, *270*:25 (1977).
66. H. W. Scharpenseel, in *Soil Biochemistry*, Vol. 2 (A. D. McLaren and J. Skujins, eds.), Marcel Dekker, New York, 1971, p. 96.
67. A. H. Knight, *New Phytol.*, *79*:573 (1977).
68. D. S. Jenkinson and A. Ayenaba, *Soil Sci. Soc. Am. J.*, *41*:912 (1977).
69. K. Haider, J. P. Martin, and Z. Filip, in *Soil Biochemistry*, Vol. 4 (E. A. Paul and A. D. McLaren, eds.), Marcel Dekker, New York, 1975, p. 195.
70. K. A. Smith and R. S. Swift, in *Soil Analysis – Instrumental Techniques and Related Methods* (K. A. Smith, ed.), Marcel Dekker, New York, 1983, p. 229.
71. S. Nakamo, J. Hanboh, and N. Urabe, *Bull. Chem. Soc. Japan*, *49*:2437 (1976).

72. R. Dybczynski, H. Maleszewska, and M. Wasek, *J. Radioanal. Nucl. Chem. Lett.*, *96*:69 (1985).
73. M. Czanderna, *J. Radioanal. Nucl. Chem.*, *89*:13 (1985).
74. G. A. Wandless and J. W. Morgan, *J. Radioanal. Nucl. Chem.*, *92*:273 (1985).
75. G. I. Khalil, *J. Radioanal. Nucl. Chem. Lett.*, *96*:539 (1985).
76. S. Ahmed, A. Mannan, I. H. Qureshi, S. M. Khan, and I. Ahmed, *J. Radioanal. Nucl. Chem.*, *120*:89 (1988).
77. U. Niese and S. Niese, *J. Radioanal. Nucl. Chem.*, *122*:347 (1988).
78. F. W. Brown, F. O. Simon, and L. P. Greenland, *J. Res. U.S. Geol. Surv.*, *3*:187 (1975).
79. C. L. Branquinho and V. J. Robinson, *J. Radioanal. Chem.*, *24*:321 (1975).
80. R. Bojanowski, E. Holm, and N. E. Whitehead, *J. Radioanal. Nucl. Chem.*, *115*:23 (1987).
81. J. Starý and J. Růžička, *Talanta*, *18*:1 (1971).
82. E. Gundersen and E. Steinnes, *Talanta*, *18*:1167 (1971).
83. G. D. Wals and H. A. Das, *Radiochem. Radioanal. Lett.*, *26*:353 (1976).
84. H. A. Das, W. H. Kohnemann, G. D. Wals, and J. Zonderhuis, *J. Radioanal. Chem.*, *25*:261 (1975).
85. N. Suzuki, K. Hanzawa, and H. Imura, *J. Radioanal. Nucl. Chem.*, *97*:81 (1986).
86. G. N. Bilimovich, *J. Radioanal. Nucl. Chem.*, *88*:171 (1985).
87. S. R. Joshi, *J. Radioanal. Nucl. Chem.*, *116*:169 (1987).
88. Z. Huang, X. Li, Z. He, Y. Li, and H. Teng, *Fushe Fanghu (China)* *6*:304 (1986); *Chem. Abs.*, *106*:17470h.
89. Y. Dong and Q. Zhou, *Hejishu (China)* *10*:33 (1987); *Chem. Abs.*, *107*:189661f.
90. G. Just, *Isotopenpraxis*, *23*:187 (1987).
91. C. D. McAuliffe, N. S. Hall, L. A. Dean, and S. B. Henricks, *Soil Sci. Soc. Am. Proc.*, *12*:119 (1948).
92. J.-C. Fardeau and J. Jappe, in *Isotopes and Radiation in Soil-Plant Relationships Including Forestry*, IAEA, Vienna, 1972, p. 499.
93. M. E. Probert, *Plant Soil*, *45*:461 (1976).
94. A. A. R. Hafez and P. R. Stout, *Soil Sci. Soc. Am. Proc.*, *37*:572 (1973).
95. E. R. Mercer and A. R. Gibbs, in *Soil Chemistry and Fertility* (G. V. Jacks, ed.), Trans. Int. Soc. Soil Sci. Comms. II, IV, Aberdeen, Scotland, 1967, p. 233.
96. R. G. McLaren, D. M. Lawson, and R. S. Swift, *J. Sci. Fd. Agric.*, *39*:101 (1987).
97. S. P. Goldberg and K. A. Smith, *J. Sci. Fd. Agric.*, *36*:81 (1985).

98. G. Gissel-Nielsen and A. A. Handy, *Z. Pflanzen. Bodenk.*, *141*:67 (1978).
99. J. D. Stout, K. M. Goh, and T. A. Rafter, in *Soil Biochemistry*, Vol. 5 (E. A. Paul and J. N. Ladd, eds.), Marcel Dekker, New York, 1981, p. 1.
100. H. W. Scharpenseel and H. V. Neue, in *Organic Matter and Rice*, IRRI, Los Baños, Philippines, 1984, p. 273.
101. J. K. Martin and J. R. Kemp, *Soil Biol. Biochem.*, *18*:103 (1986).
102. L. H. Sorensen, *Soil Biol. Biochem.*, *19*:39 (1987).
103. K. M. Haider and J. P. Martin, *Soil Biol. Biochem.*, *20*:425 (1988).
104. T. C. Strickland and J. W. Fitzgerald, *Soil Biol. Biochem.*, *18*:463 (1986).
105. S. C. Schindler and M. J. Mitchell, *Soil Biol. Biochem.*, *19*: 531 (1987).
106. J. W. Fitzgerald and M. E. Watwood, *Soil Biol. Biochem.*, *20*: 833 (1988).
107. J. A. Van Veen, J. N. Ladd, J. K. Martin, and M. Amato, *Soil Biol. Biochem.*, *19*:559 (1985).
108. R. Zieve and P. J. Peterson, *Soil Biol. Biochem.*, *17*:105 (1985).
109. H. W. Scharpenseel, K. Tsutsuki, P. Becker-Heidmann, and J. Freytag, *Z. Pflanzen. Bodenk.*, *149*:582 (1986).
110. B. J. O'Brien, *Soil Biol. Biochem.*, *16*:115 (1984).

9

Instrumental Neutron Activation Analysis

LEONARD SALMON and PETER A. CAWSE *Harwell Laboratory, Harwell, Oxfordshire, England*

I. INTRODUCTION

An analytical method is selected by an investigator for a variety of reasons, and a method that offers fairly accurate results on a wide variety of elements at submicrogram levels, without any need for chemical treatment, is not to be discarded lightly. Indeed instrumental neutral activation analysis (INAA) offers all these features to the soil chemist, yet the literature relating to this application is sparse compared with that, e.g., for related studies of the atmospheric environment and in geochemistry.

Like all good physical methods of elemental analysis, activation analysis depends on the measurement of a specific property that is unique to the element and whose magnitude is related reliably to the quantity of the element present in the sample. The principle is simply to induce radioactivity in the material of the sample and then to measure radiation specific to each element being determined. It happens that, in soil, the major elements of low atomic number are barely susceptible to activation by neutrons, whereas many minor and trace elements do activate readily. In instrumental activation analysis, a radiation spectrometric technique is used to discriminate between different elements by the measurement of photons specific in energy to individual isotopes.

Neutrons are chosen as the irradiation source since the probability of activation is greater than with charged particles, which need to overcome a coulomb barrier, whereas available nuclear reactors provide particularly intense and isotropic sources of neutrons of low enough energy to avoid the complication of interfering secondary reactions. Similarly, γ rays rather than the primary β radiation are chosen for the identification and assessment of the induced activity, since they exhibit discrete energies are are not subject to significant absorption within the sample.

There are many general reviews of activation analysis in the literature including those by Gibbons and Lambie [1], De Soete et al. [2], Kruger [3], and compilations by staff of the University of Ghent [4,5]. Many papers are published in the proceedings of regular international conferences such as *Modern Trends in Activation Analysis* [6] and *Nuclear Activation Techniques in the Life Sciences* [7].

There have been numerous technical developments beyond the use of neutrons. One such development, the subject of much current interest, is the use of high-energy photons as an activation source. Photon activation offers analysis for certain elements not determined easily by neutron irradiation (e.g., Pb, Cd, and Ti) and a similar instrumental approach employing γ-ray spectrometry can be used. However, the application of photons differs sufficiently in practical details from neutron activation to make a single review of the two methods unsuitable. The reader is referred to a general review of the photon activation technique by Hislop [8] and its specific application to soil by Chattopadhyay and Jervis [9].

The present chapter confines its attention to the use of neutrons in soil analysis, the employment of instrumental methods alone, and the consideration of published applications of the technique.

II. PRINCIPLES OF NEUTRON ACTIVATION ANALYSIS

A. Theory

When material is bombarded with neutrons, charged particles, or photons, some may be converted by a nuclear reaction to a radioactive isotope of the same or another element. In the case of neutron irradiation, the conversion is usually to a heavier isotope of the same element. Excess energy is released in the form of prompt γ rays

$$^{w}A_{z} + {}^{1}n_{0} = {}^{w+1}A_{z} + \gamma \tag{1}$$

Usually, although not always, the new isotope is radioactive and the measurement of its activity yields an estimate of the quantity of element *A* present in the irradiated material.

Each isotope of an element (active or nonactive) has a specific probability of capturing a neutron of particular energy. Nuclear reactors are used generally as the source of neutrons; the neutron energy involved in most reactors is low (thermal), and the capture probability remains constant. This capture probability has the dimension of area and is called the *capture cross section*, having units of *barns*, viz. 10^{-24} cm^{-2}. The amount of induced activity can be calculated [10] from the following equation:

$$D_0 = \frac{m\theta\sigma N\phi}{w}\left\{1 - \exp\left[-\ln(2)\frac{t}{T_{1/2}}\right]\right\} \tag{2}$$

where

D_0 = disintegration rate in becquerels (dis sec^{-1})
m = mass of element undergoing irradiation
θ = fractional natural abundance of the relevant isotope
σ = absorption cross section (cm^2)
w = atomic weight of isotope
N = Avogadro's number
ϕ = irradiation flux (n cm^{-2} sec^{-1})
t = irradiation period
$T_{1/2}$ = half-life of active isotope, in same units as t

After a decay period T,

$$D_T = D_0 \exp\left[-\ln(2)\frac{T}{T_{1/2}}\right] \tag{3}$$

A consequence of the relationship shown in Eq. (2) is that for low values of t, the induced radioactivity rises almost linearly with t, but is limited to a saturation activity at an infinite value of t when decay and production rates of activity are balanced. Note that there will be half the saturation activity produced after an irradiation period of one half-life.

The relationship between mass and induced activity is, of course, linear. Some indication of the potential sensitivity of activation analysis can be gained from the consideration of a hypothetical isotope with some typical nuclear parameters. The activity induced in one nanogram of a monoisotopic element of atomic number Z = 100 and a cross section of 0.1 barn, when irradiated for one half-life in a neutron flux of 10^{14} n cm^{-2} sec^{-1}, is 300 becquerels (~8 nCi). Classical chemical methods would permit us to separate this element with a high degree of radiochemical purity, and this level of activity would then be measured with considerable precision and sensitivity.

Indeed, neutron activation analysis has employed radiochemistry as its major discrimination technique almost since its inception (see Chap. 8).

The principle of instrumental neutron activation analysis is to avoid tedious and expensive chemical separation and to use a nuclear property, the energy of associated γ rays, as the discriminating factor. As will be shown in the next section, the resolving power of the detection systems used for this, although considerable, can hardly compete with the separation factors of 10^8 or more that are obtained by chemistry. Instrumental activation therefore relies on limiting its application to matrices in which most of the activation takes place among the minor and trace elements present. Examination of the activation cross sections of the elements in the periodic table shows very little relation between them and chemical properties (unlike parameters normally encountered in chemical discrimination methods), but activation susceptibility does tend to be low for elements of low Z, either because the cross section is small or the resultant isotope has a very short half-life. So far as soil is concerned, C, N, O, H, and Si produce very little activity, whereas Al only produces an isotope with a half-life of 2.3 min.

Although the linear relation between mass and induced activity is shown in Eq. (2) above, the establishment of the proportionality constants requires knowledge of a number of experimental and nuclear parameters. Such data are available, together with calculated values of induced activity, in tabular form in a compilation by Erdtmann [11].

In practice, a comparative method normally is used, in which a known quantity of the element to be determined is irradiated under similar conditions to the sample. Then all irradiation and nuclear parameters can be regarded as constants between the sample and standard and the activity ratios can be considered equal to the mass ratios.

Normalization refers only to correction for differences in decay times between the sample and standard

$$\frac{\text{Mass of element in soil sample}}{\text{Mass of element in standard}} = \frac{\text{normalized activity in soil}}{\text{normalized activity of standard}} \qquad (4)$$

Recently, much attention has been paid to the so-called k_0 method of standardization [12]. Reliance is placed on absolute physical parameters rather than comparative techniques that require the remeasurement of all standards whenever minor changes of irradiation and spectrum measurement are made to the system. The physical parameters concerned are determined once only for each element and consist effectively of the product of neutron absorption cross section, isotopic abundance, atomic mass, and absolute γ ray intensity.

This is the parameter k_0. One or two comparators are irradiated with the samples to monitor the neutron flux and the detector counting efficiency. Computer procedures are described to calculate changes of detector efficiency with changes in sample geometry [13]. The effect of modifications to the neutron energy distribution has also been further explored [14]. A great deal of care is needed to initiate the system and there is much reliance on the adequacy of the monitors used. It does, however, result in a powerful and efficient multielement analysis system.

A judicious combination of irradiation and decay times can optimize the sensitivity with which any particular element can be determined. To do this rigorously in a complex matrix requires a suitable computer program simulating the pulse-height distributions obtained for a variety of irradiation and measurement parameters such as those developed by Zikovsky and Schweikert [15], Guinn [16], Hsia and Guinn [17], and Law [18]. However, a common assumption is that a decay period of approximately one half-life will optimize the signal from a specific isotope. Because of the growth factor in Eq. (2), the irradiation period should be kept low in relation to the half-life, provided sufficient activity is induced to be measured with adequate precision.

At any one time only a few isotopes are visible in a pulse-height spectrum, but by continuing measurements over a considerable period as many as 40 elements may be determined in suitable matrices.

B. Gamma-Ray Spectrometry

The general principles of γ-ray spectrometry have been discussed in Chap. 8 and are therefore only briefly stated here. The technique simply involves energy discrimination by the use of a detector whose response to an impinging γ ray is an output related to the γ-ray energy. Thus, the detector behaves as a "proportional counter." Analog signals from the detector may be translated to digital form and the resulting number used to select a memory channel of a pulse-height analyzer, where the event is recorded. With present-day techniques, the memory of a suitable computer may be used instead. A large number of events received over a period of time will result in the buildup of a frequency histogram or pulse-height spectrum that relates quite closely to the initial distribution of photon energies from the radioactive source.

Currently, the most popular form of detector is the solid-state type constructed of high-purity germanium. This exhibits considerable powers of energy discrimination, although somewhat inconveniently needs to be maintained at the temperature of liquid nitrogen when in use. Germanium detectors suitable for γ-ray spectrometry are available currently as *p*- or *n*-type detectors, depending on their semiconductor properties. The latter type, although slightly more expensive, is capable of measuring very low-energy γ rays and X rays. The lithium drifted germanium, or Ge(Li), detector, which requires permanent cryostatic storage and at one time was in general use, is no longer manufactured.

The principles of operation and other features of the spectrometric technique are described in standard textbooks [19,20].

Solid-state germanium detectors may be conveniently thought of as ion chambers where the number of ion pairs produced is proportional to the absorbed energy. Consideration of the energy required to produce an ion pair and the statistical nature of the ion production process shows that the resolving power, i.e., the width of a total absorption peak, can be theoretically as low as 1 keV at 1000 keV. In fact, detectors giving peaks less than 2 keV in width are routinely available commercially. It is not usual for these detectors to be available with useful volumes greater than 100–200 cm^3, and in some special applications requiring large samples, a scintillation (sodium iodide) detector is required simply to provide sufficient detection efficiency. Such detectors are available with a volume of a liter or more. However, the inefficiency of energy collection and its transfer in the scintillation process is such that the resolving power is some 30–40 times worse than that of a Ge solid-state detector.

The majority of γ rays entering the detector will be scattered due to the Compton effect, when the energy of the secondary electron is given by

$$E_e = \frac{E}{1 + a(1 - \beta)} \tag{5}$$

where

E_e = energy of secondary electron (keV)
E = energy of incident photon (keV)
a = E/511
β = angle of photon scatter

Figure 1 shows the relative probability of compton interaction at different energies and compared with other processes. The dependence on energy is not great. The highest energy imparted to a scattered Compton electron is less than the primary energy of the incident photon and corresponds to a scattering angle of 180°.

$$E_e(\max) = \frac{E}{1 + 0.5/a} \tag{6}$$

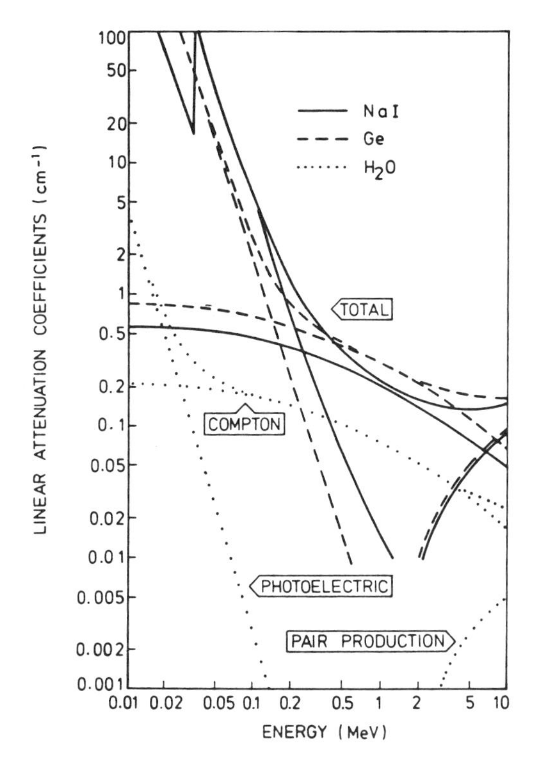

FIG. 1 Absorption coefficients for γ radiation in three media. (Courtesy of the Harshaw Chemical Co., Northants, England.)

Figure 2 illustrates a pulse-height spectrum from ^{24}Na. This isotope has two equally intense γ rays at energies 2754 and 1369 keV. The corresponding Compton continua are shown with maximum energies in accord with Eq. (6) above. The somewhat concave shape of the continua is due to the nonisotropic nature of the scattering process. The differential probability of scattering at specific angles is given by Klein and Nishina [21].

The important photon interaction process, from the analyst's viewpoint, is the photoelectric interaction in which all the energy is converted (albeit indirectly) into the output signal. The photoelectric absorption probability varies as Z^5, hence, germanium ($Z = 32$) is preferred to silicon ($Z = 14$) as material for semiconductor detectors of γ radiation. It appears from Fig. 1 that since the probability for photoelectric absorption falls rapidly with increasing energy, the

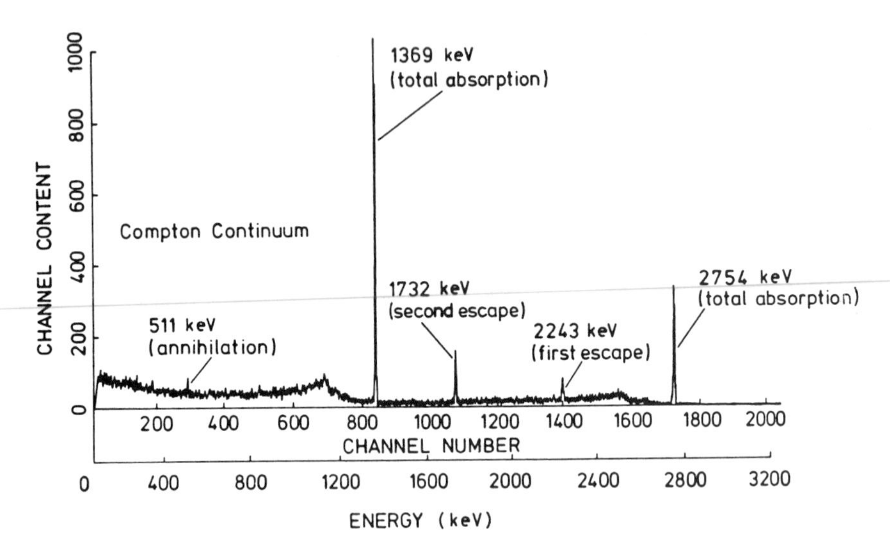

FIG. 2 γ-Ray spectrum from ^{24}Na.

size of the most energetic peaks will be very small. However, the scattering process reduces the initial energy of the photon, hence, if the detector is sufficiently large to permit several interactions within its volume without the escape of the scattered photon, then its eventual photoelectric capture becomes distinctly probable as its energy drops. It is thus more accurate to refer to the observed peak as a "total absorption" peak. The appropriate two peaks are seen clearly in Fig. 2. Total absorption efficiencies for detectors of different sizes and materials are commonly tabulated in the literature either from experimental data such as those compiled by Heath [22] or from calculations using mathematical modeling techniques [23].

Other peaks are observed from high-energy photons due to the "pair production" phenomenon (see Chap. 8) when "escape" peaks occur at $E - 511$ and $E - 1022$ keV (see Fig. 2), where E is the energy of the total absorption peak.

Ideally, the peaks in a pulse-height spectrum should be of a symmetric Gaussian type. However, if the source itself is large or an absorbing material is interposed between the source and detector, then some degradation may take place due to low angle scattering and the low-energy side of the peak may be relatively enhanced. Quite severe degradation may also take place if the amplification circuitry is unable to restore the pulse base line rapidly following the record of an event. Then if event rates are high, pulse heights are in error, resulting in peak broadening or shape distortion. The interpretation of such data is discussed in a later section.

The important relationships in a spectrometer are those between photon energy and peak position and between peak size and source intensity. The first relation is practically linear, although deviations from this are also discussed later. The property of linear addition in a spectrometry system is almost always obeyed at low event rates. At high rates, the total number of events recorded in a peak is subject to linear addition, but this may not be true of individual channels within the peak.

The resolving power of a detector is defined usually in terms of the full width at half the maximum height of a peak (FWHM) at a specific energy, e.g., 1.9 keV at 1330 keV (^{60}Co). The resolution varies as

$$\text{FWHM} = k_1 + k_2 E \tag{7}$$

where k_1 and k_2 are specific constants for the detection system.

In activation analysis, the detector used should be as efficient and of as high resolution as possible. The second factor is the more important, should economic considerations prevail. A useful Ge detector should have a resolving power of less than 2.0 keV at 1330 keV and have a total absorption efficiency at 667 keV of 10% of that of a standard 3 in. (75 mm) sodium iodide crystal. (The parameters are those generally in use commercially.)

Many laboratories find it useful to support an additional low-efficiency detection system to measure sources of high activity, rather than perform additional irradiations to suit the single sensitive detector.

It has been demonstrated by Atkins et al. [24] that, for some limited applications, sodium iodide detectors are quite adequate devices for short lived activation analysis. Also, a nonactivation application, the determination of natural radioactivity (radium, thorium, and potassium) in soil, is also best performed by a large sodium iodide detector [25].

There is probably some advantage in employing detectors particularly sensitive to low-energy photons or X rays. This has been demonstrated particularly in the estimation of a range of rare earths and platinum metals in rocks and lunar soils by Rosenberg and Wiik [26] and Gijbels and Govaerts [27]. Such detectors are virtually windowless and are made from intrinsic *n*-type germanium.

III. PRACTICAL DETAILS OF ANALYSIS

A. Collection and Preparation of Samples

The application of instrumental neutron activation analysis to the determination of trace elements in soils requires special techniques of sample collection and preparation. Although most soil samples are taken by auger or spade, the use of a coring device of known cross-sectional area is preferred for studies on the input of elements from the atmosphere to ground, since the integrated deposition per unit area needs to be known. Thus, Ragaini et al. [28] took soil core samples to a depth of 26 cm by driving a steel cylinder into the ground, employing an interchangeable plastic liner to avoid contamination by metals. A polypropylene corer, 32 mm in diameter, has also been used to avoid metal contamination [29].

Whereas for conventional analyses, samples >10 g are commonly prepared from air-dried soil sieved through a 2 mm mesh, the requirement of the activation method for small samples, usually from 50–200 mg, demands the production of a very homogeneous specimen of soil with a final particle size below 300–500 μm diameter. This problem was also recognized during analysis of the standard Bowen kale powder [30].

For preparation of finely ground or sieved soil, various approaches have been described by different workers. Nevertheless, many reports in the literature do not provide details of the processing of soil samples prior to analysis by INAA. The danger of contamination by metals from conventional disk grinders and mills, and the difficulty of cleaning out such apparatus between samples to avoid cross-contamination, has led authors to the use of an agate ball mill for pulverization [31,32] or an agate pestle and mortar [33].

Tests with a vibrating agate ball pestle mill of 50 cm^3 capacity showed that 15 g of dry soil (<2 mm) were pulverized in 20 min to <200 μm particle diameter. Furthermore, 70% of the sample passed a 150 μm sieve and 18% was <53 μm [34]. Other grinding techniques include the use of diamonite mortars [35,36] and an iron mill [36]. In archaelogical work [37], samples of 20 g of soil were freeze-dried before grinding in a vibrating mill to pass a 1 mm^2 nylon mesh when 200 mg samples were used for analysis by neutral activation. In other work [28], soil samples were air-dried and ground to 75 μm mesh prior to the short-term irradiation of 25 mg and longer irradiation of 200 mg samples.

The dependence of trace element concentrations on soil particle size must be appreciated if specific size fractions are sieved out for analysis without homogenization. Evidence to show the effect of particle size on measured element concentration is mentioned in Sec. V.C.

For the packaging of samples for irradiation, practically all analysts recommend the use of clear silica tubes. Tubing should be cleaned with a powerful oxidant and thoroughly rinsed in water followed by deionized water. After insertion of the soil sample, the tube should be sealed by fusion. For quite short irradiations (a few minutes), the authors have used successfully unsealed silica tubes plugged with aluminum foil. Similarly, polythene vials and sealed bags are useful if selected for their low inorganic contents.

Liquid samples may arise from soil extracts and solutions (Sec. IV.B). These should be carefully evaporated onto suitable matrices (e.g., Whatman 40 filter paper) and the matrix itself packaged and irradiated. It is possible to irradiate, for short periods, liquids sealed in quartz or even polythene tubes. However, the procedure is not recommended nowadays. An air space should be left in the vessel equal to at least half of its volume. Also the solution should not contain substances (e.g., nitric acid) subject to radiolysis.

Samples should always be removed from silica irradiation containers before measurement. However, for the measurement of isotopes having short half-lives, it is quite practicable to measure irradiated soil or filter papers within an irradiated polythene container.

B. Irradiation of Samples

The majority of neutron activation analyses performed on soil have used a nuclear reactor as the source of neutrons. Not only is a high neutron flux available (up to 10^{14} n cm^{-2} sec^{-1}) for maximum sensitivity, but a reasonable degree of irradiation uniformity can be expected due to the isotropic distribution of neutrons within a reactor.

Nuclear research establishments frequently permit access to the larger reactors (e.g., DIDO heavy water reactor at the Harwell Laboratory, U.K.) where irradiation facilities exist that permit uniform, high-intensity thermal neutron fluxes. However, many universities throughout the world provide "swimming-pool" reactors that use a light water moderator with enriched uranium fuel. These may even be designed deliberately for activation analysis, providing special irradiation facilities and operating at low cost. Such reactors have been marketed as TRIGA reactors by General Atomic Inc. [38] and SLOWPOKE by Atomic Energy of Canada Ltd. [39]. Nuclear power reactors rarely have useful irradiation facilities and operate at inconveniently high temperatures.

It is possible to use nonreactor sources of neutrons for the analysis of rocks and soils, but they are so lacking in intensity that they are really only suitable for the measurement of major elements. Plutonium-beryllium and radium-beryllium sources have been used in the past although effective thermal fluxes have not usually exceeded 10^5 n cm^{-2} sec^{-1} [40]. In recent years, however, ^{252}Cf sources have become available, yielding fluxes of up to 10^7 thermal n cm^{-2} sec^{-1}, depending on the quantity and phsyical arrangement of the isotope [41].

An alternative source is the neutron generator that, by the deuteron excitation of a tritium target, produces 14 MeV neutrons from the ${}^{3}_{1}H(d,n){}^{4}_{2}He$ reaction. Fluxes of 10^{10} n cm^{-2} sec^{-1} are possible [42], although again the irradiating beam is not isotropic and the high energy of the neutrons makes the device more suitable for special reactions on elements of low atomic number, rather than for the more usual application of multitrace element analysis. A detailed review of neutron beams in activation analysis is given by De Soete et al. [2].

The relative intensities of thermal and epi-thermal neutrons vary considerably, depending on their source. Generally, it is advantageous to employ as low energy a source of neutrons as possible. The cross section for (n,γ) reactions is highest at low energies, thus yielding maximum sensitivity. Also, a wide variety of alternative reactions, e.g., (n,α), (n,p), $(n,2n)$, occur with generally increasing probability at higher energies. These may well interfere with the determination of one element due to the enhancement of the appropriate activity by an alternative neutron reaction with another element (see Sec. II.A).

It may happen that a particular element can only be conveniently measured by the use of a high-energy neutron reaction. A well-known example is the measurement of nickel by ${}^{58}Ni(n,p){}^{58}Co$, a more useful and sensitive reaction than ${}^{64}Ni(n,\gamma){}^{65}Ni$, whereas oxygen can be determined by ${}^{16}O(n,p){}^{16}N$. In general, however, it is in the interest of the analyst to use as thermalized a flux as possible. For this reason, large heavy water (i.e., well moderated) reactors available at some national laboratories are practically ideal. Elsewhere, a homogeneous thermal flux can usually only be obtained at the cost of source intensity, and a compromise has to be accepted between sensitivity and potential interference.

Another area in which some compromise is frequently necessary is the length of time of irradiation and the subsequent period of decay before measurement. To obtain the maximum level of induced activity would require an irradiation of several half-lives. However, the rate of induction of activity ceases to be linear before an irradiation period of one half-life has elapsed [Eq. (2)]. Thus, inducement of sufficient total activity should be controlled by the selection of the sample weight and neutron flux, whereas the irradiation period should be kept short in relation to the relevant half-life. Also, the decay period should be selected to optimize the relative level of activity it is desired to measure compared with all others present. If all other activities were distributed uniformly in quantity and half-life, then the optimum time to make the measurement would be about one half-life of the activity of interest after irradiation. In most practical cases, however, one interfering activity is often dominant for a particular period of time. Such examples in the case of soil are the activities arising from aluminum, sodium, and scandium. Thus, decay periods between measurements are chosen to avoid the major sources of interference. It is possible to forecast the optimum conditions of irradiation and decay by means of suitable computer programs, which model the γ-ray distribution induced in soil of known approximate composition and hence derive the relative signal of particular interest [9,10, 18].

The instrumental activation technique is normally employed for multielement analysis, when 20–30 elements are estimated. Conditions cannot be optimized for all these, and some general compromise needs to be adopted. In the authors' laboratory, two separate soil samples each of about 50–100 mg are prepared and irradiated in the DIDO reactor, one for a few seconds and the other for 4–8 hr. These periods could be scaled up or down depending on the flux used and weight of the sample taken. The first sample is measured within 5 min of irradiation and thereafter at intervals of 30 min, 3 hr, and (possibly) 14 hr. The second is measured after 2 days and 20 days.

Figure 3 displays the differing form of the spectra as the source decays and indicates the appearance of a variety of peaks from many isotopes.

Ideally, all samples should be irradiated uniformly in an isotropic neutron flux unattenuated by the sample. In the case of small soil samples, the average capture cross section is sufficiently low [$\ll$1 barn (10^{-24} cm^2)] for attenuation to be quite negligible. It is, however, quite important to ensure that neither the irradiation container nor material in the reactor near the sample are of sufficiently high cross section to produce significant local flux changes in the sample. In reactors designed for activation experiments, there are devices to rotate a number of samples in the reactor to ensure uniformity of irradiation [38]. This uniformity is particularly difficult to achieve in nonisotropic irradiation conditions as found in neutron generators and ^{252}Cf sources.

In order to monitor the neutron flux received by the sample, standards of all elements to be determined should be irradiated with the sample and be in a similar physical state to the sample. It is not practicable, however, to irradiate a large set of fresh standards for every irradiation made, and it is assumed that one set of standards can be irradiated and measured to cover a reasonable period of time (perhaps some weeks), during which the energy spectrum of the neutron source will not change. Small differences in irradiation time and absolute flux level can then be monitored by including one or two suitable reference standards with each set of samples irradiated. Such standards should be of high purity and of such concentration and nuclear characteristics that an accurate measurement of the neutron dose received can be made without perturbing that flux. The use of sodium salts (e.g., sodium carbonate) and pure

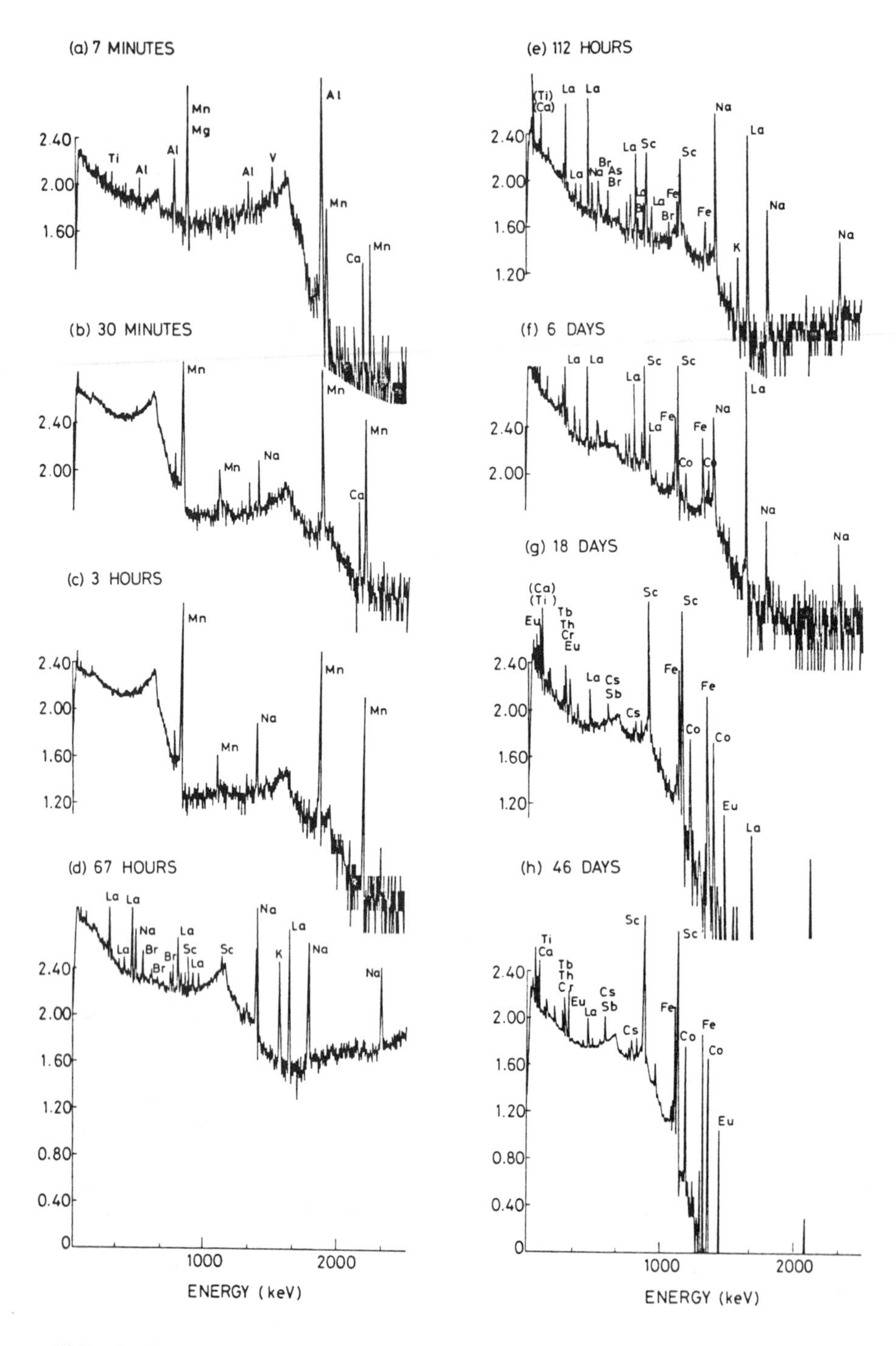

FIG. 3 Sequential γ-ray spectra from irradiated soil.

iron wire is recommended for short and long periods of irradiation, respectively.

Master standards of elements to be determined will, of course, be required. Most workers employ freshly prepared dilute solutions of appropriate compounds. The compound chosen should be a soluble salt of the element concerned containing no other "activatable" elements and available in a chemically pure form. It should be diluted to a concentration comparable with that to be determined and a measured quantity sealed in a container of material such as silica. After irradiation under known conditions, the solution should be removed from the container and a source prepared physically resembling a soil sample.

C. Measurement of Radioactivity

The actual measurement of γ radiation is performed in the conventional manner of radiation detection. The detector is shielded so far as is possible from external radiation sources and the sample itself arranged in a well-defined and reproducible position relative to the detector.

The physical arrangement may be somewhat difficult with solid-state Ge detectors, since the associated cryostat is inevitably bulky (e.g., 0.8 m in height and 0.4 m in diameter), although in activation work it is usually unnecessary to provide a lead shield to cover the entire assembly. Most laboratories performing activation analysis routinely provide some form of sample changing mechanism. For such a system to be useful, it is essential that each sample be moved to a measurement position identical with its predecessors and that the storage system for samples awaiting measurement be sufficiently distant and well shielded for their activity to be insignificant to the detector. Figures 4a and b show such a system with automatically controlled doors to the shield. An accurate sample positioning mechanism driven by stepping motors is controlled by a small local computer and provides options in the measurement sequence.

The minimization of radiation background is not perhaps so important in activation analysis as in some other applications of γ-ray spectrometry. Nevertheless, some pains should be taken to shield the detector effectively with 100 mm of lead (selected for its low activity) and to plug gaps between the shield and cryostat with bags of lead shot. Regular measurements of background should be made to discover occasional contamination and sources of varying external radiation present in some nuclear environments. The lead should be lined with successive layers of cadmium, copper, and aluminum sheet, to absorb interfering characteristic X rays from the lead and subsequent metals due to irradiation by the source being measured. The device supporting the sample should be of lucite or similar material.

(a)

FIG. 4 (a) General view of detector and sample-changing mechanism. (b) Local view of detector and sample-changing mechanism.

If the irradiation and measurement conditions are chosen correctly, then there will be a sufficient event rate registered by the detector to provide an adequate spectrum in a reasonably short time. At too great a rate, there will be spectrum distortion due to overloading of the pulse amplification equipment. If the source is of such activity that it needs to be measured at a considerable distance from the detector, then unwanted scattering effects are observed to distort the continuum of the spectrum. It is essential that the sample exhibits the same geometry as the standards used for comparison. Efficiency is very dependent on the solid angle subtended by a source onto the surface of the detector. It should be noted that the point of greatest sensitivity to small changes for a disk-shaped source appears to be a distance of one disk radius from the detector.

It is equally important that characteristics of the detection system remain unchanged between the measurement of the standard and sample. The position and shape of absorption peaks must remain constant. They should also exhibit the property of linear addition, i.e., twice the activity should produce twice the size of the corresponding

(b)

FIG. 4 (Continued)

peak. The amplification characteristics (zero offset and channel width) should remain fixed with carefully chosen equipment. The detection equipment must be kept at a constant temperature. This will obviously be so for a solid-state detector cooled by liquid nitrogen. A scintillation detector, if used, should not only be thermostatically housed but should have some gain stabilization system fitted, e.g., that used by Keane [43], employing an internal reference standard. The peak shape is maintained best by measuring similar event rates with both sample and standard, but the amplification system should use a baseline restoration system before presenting an analog signal to a digital converter, and thus avoid some of the distortion caused by the overlap of signals that occur during high event rate measurements.

The analog-to-digital converter (ADC) should either be of the Wilkinson type using a 100 MHz oscillator to cope with the high detection rate encountered in activation analysis or employ the successive approximation technique in a suitably accurate fashion. The ADC should provide a digital signal of at least 12 bits.

The pulse-height spectrum is accumulated in a conventional computer memory that may be an attachment to the ADC, an integral part of a computer, or part of a *pulse-height analyzer*, a hardware device devoted to the pulse-height analysis function and controlled by specifically designed circuitry. The spectrum is formed by incrementing the memory address corresponding to each digital signal from the ADC. Each memory item corresponds to an energy interval. Thus, if the detector is capable of discriminating between energies differing by 2 keV and the energy range under consideration covers some 3000 keV, a memory of some 4000 or 8000 words is appropriate, although 2000 is adequate for some applications.

The pulse-height analyzer, although commonly still in use with many laboratories, has now been superseded mostly by computer analyzer systems. When spectrum accumulation, manipulation, display, and storage are carried out using physical controls of limited capability, these operations are performed nowadays by interactive computer procedures with considerable flexibility and power.

The initial incorporation of small computers for the accumulation of spectra was accompanied by the use of large powerful "number crunching" computers to analyze spectra automatically on a "batchwise" basis. A number of computer programs were developed: HEVESEY [44], GASP [45,46], GASPAN [47], HYPERMET [48], GAMANAL [49], and SAMPO [50] are but a few. However, the increasing power of the smaller mini-computers, of which the PDP-11 and later the VAX-11 are prime examples, could accommodate small versions of these programs, although some were developed specifically to provide interaction combined with computational power [51]. Manufacturers have developed and continue to develop systems based on mini-computers that will support simultaneously a number of active detectors and analysis tasks. Some have developed single-user systems based on cheap "personal" computers, notably the IBM-PC and its numerous imitators. This style of development, requiring the addition of but a single circuit card to an already cheap computer, obviously will continue with a consequent sharp reduction in cost per unit analysis.

The operating features of the computer-analyzer system should include good manipulative and spectrum-viewing characteristics. The capability of selecting specific regions or peaks within the spectrum, rapidly identifying such features, and determining their numerical characteristics ought to exist. Any displayed information should, of course, take advantage of known calibration characteristics. A most important aspect, however, is the system's ability to store and archive experimental spectra. On small machines this is accomplished usually by the use of the ubiquitous "floppy disk," although larger systems may employ magnetic tape or on-line Winchester disks providing instant access to hundreds or even thousands of stored spectra.

It goes without saying that stored spectra should be accompanied by necessary identifying information and all calibration parameters.

The analysis of the spectra is usually performed by proprietary software supplied by the manufacturer of the system, although some laboratories maintain their own numerical procedures [51]. The analytical programs available on the smaller computers are usually modified versions of those mentioned earlier, operating in an interactive fashion rather than in the batch mode.

D. Interpretation of Spectra

The activation analyst will find inevitably the need for data evaluation and analysis by computing methods even though subjective judgment must still play a large part in the initial interpretation of activation data. The quantity of digital data produced is so large, even from one measurement, that some aid is necessary. Admittedly, much useful analysis has long been performed by manual examination of data plotted in analog form and with purely graphical methods used to assess the peak energies and their magnitudes. Such methods, however, apart from being laborious. inevitably introduce subjective error and usually fail to utilize all available information. They certainly fail to provide a statistical assessment of the data.

The operations performed in the interpretation of spectra begin with a recognition of the major features, i.e., the total absorption peaks. The energies of the peaks need to be measured with some accuracy using suitable calibration methods. The qualitative analysis of the spectrum constituents is then carried out by recognition of specific photon energies and their relative intensities. In the routine analysis of similar batches of soil samples, this assessment does not need to be made manually in most laboratories.

The quantitative assessment may be made a general one in which all components are included. In this case some routine procedure such as regression analysis, with a large library of standard spectra, must be invoked. Alternatively, if specific elements are sought, then direct comparison between total absorption peaks in the sample and an appropriate standard must be used. If a direct reference standard is not available, then an efficiency calibration may be established with secondary standards of different energies.

The first requirement, that of finding significant peaks, has been subject to considerable investigation by Routti and Prussin [50], Robertson and Kennet [52], Mariscotti [53], Yule [54], Blok et al. [55], among others. The topic has been reviewed by Op de Beeck and Hoste [56] and by Phillips and Marlow [57]. A numerical procedure based simply on the rise and fall of successive channel contents is too crude even when put onto a formal statistical basis. It is found necessary to perform a "convolution" operation (e.g., calculate a smoothed second derivative) on successive groups of channels. Then, although the procedure is one in which statistical fluctuations are smoothed, it also filters out the contributions to the spectrum from continua and other slowly varying changes such as those caused by scatter peaks.

By comparing the convulate with the standard deviation at any point, to give a ratio R, we will locate peaks where R exceeds some arbitrary value. In practice, five standard deviations will detect a peak with considerable certainty. Above this value the test becomes insensitive, and below too many "spurious" peaks will be found.

Although location techniques may indicate the channel corresponding to the center of a peak, a more accurate figure is required to assess the corresponding γ-ray energy. For this, a peak "centroid" measurement is needed. The continuum beneath the peak is first subtracted, making use of located boundaries to the peak or by making assumptions about the peak width and shape. A trapezoidal shape for the continuum is assumed normally. The residual peak may then be treated as a symmetric Gaussian, and a simple least-squares fit is performed on the quadratic shape obtained from the logarithmic transform of the peak. Alternatively, a simple way to calculate the centroid is by moments [56]. In cases where the symmetry assumption is unreasonable, then a more elaborate nonlinear regression analysis needs to be performed using a suitable function describing peak shape, e.g., those by Routti and Prussin [51] and Booker [58].

The conversion of peak position to energy is performed by the measurement of the position of peaks of known energy, either from calibration standards measured independently but under identical conditions, or from known peaks used as internal standards within the spectrum itself. The relation between peak position and energy is very nearly linear

$$E = (a + p)b \tag{8}$$

where E is the energy represented by the peak at position p channels, and a and b are the calibration constants "zero offset" and "channel width," respectively. However, some small deviation from linearity is normal, and quadratic or higher polynomials may be invoked for the calibration line, depending on the number of accurately measured points that are available [56,59].

A calibration line may be drawn alternatively from experimental points using a cubic spline interpolation [60].

The identification of isotopes from energies and relative intensities is performed mostly by manual search. Major compilations of γ-ray energies include those by Lederer et al. [61] and Erdtmann and Soyka [62]. The latter provides a very complete list of all γ rays in ascending energy order and is available in computer-compatible form for incorporation into any suitable program suite. Tables limited to neutron activation products are available (e.g., Adams and Dams [63] and Carder et al. [64]).

Computer methods of isotope identification vary in complexity. The simplest method is just to match a measured energy against that which is arithmetically nearest in a computer-stored table of isotopes. Usually, an arbitrary level of energy uncertainty is chosen such that all isotopes within a probable energy level are offered for consideration. Clearly, the more comprehensive the table stored, the more complex the solutions offered. A degree of sophistication can be incorporated that checks on the half-life of the isotope in relation to the proffered γ-ray energy and the presence of confirmatory γ rays can also be noted [64]. A more stringent procedure by Gunninck and Niday [65], elaborated on by Christensen et al. [66], generates a confidence index of identification by the measurement of all associated γ rays.

In the limited case of activation analysis associated with relatively few isotopes in a relevant half-life range, it might be more practicable to employ a method of inverted search such as that employed in the analysis of mass spectra. Here, rather than identify seprately each peak in the spectrum, a library of candidate isotopes is compared with the spectrum to find whether or not relevant γ-ray energies are present and whether in the right proportions.

The quantitative assessment of identified isotopes may either be performed by individual peak area measurement or by linear regression analysis applied to the whole of the spectrum. The former technique is commonly employed, particularly in small computer systems.

The area of a peak may be calculated simply as the area above a baseline that can either be assumed straight or subjected to some polynomial extrapolation of the continuum on either side of the peak, as in Quittner [67].

It is often difficult to assess accurately the peak boundaries due to local statistical fluctuations (although convolution methods may be used to smooth these effects), and a comparison of a peak with one from an identical standard may be made in relative terms, based on partial peak area measurement. The classical method is due to Covell [68], in which a center channel is chosen and a symmetric range of k channels on either side of the center is selected that does not reach either peak boundary. As modified by Op de Beeck and Hoste [56], this is

$$A_p = \sum_{i=-(k-1)}^{k-1} a_i - (k - 1/2)(a_k + a_{-k}) \qquad (9)$$

The variance on this assessment is

$$V_p = \sum_{i=-(k-1)}^{k-1} a_i + (k - 1/2)^2(a_k + a_{-k}) \tag{10}$$

Maximum estimates for components too low to be measured can therefore be made. Other useful methods of partial peak area measurement are reviewed by Baedecker [69] and by Op de Beeck and Hoste [56].

Where two or more peaks overlap, their area may be assessed simply by linear regression analysis if the position and width of the components are known accurately. This method has been further elaborated on by Christensen et al. [66], who provide an iterative correction to the assessment of peak position.

Nonlinear fitting procedures for peak fitting have been in use for many years; the literature is considerable and has been reviewed by Phillips and Marlow [57], McNelles and Campbell [70], and Booker [58]. Peak shapes suggested vary from simple Gaussian to modifications with exponential tails on one or both sides of the peak and a variety of asymmetric distributions based on the supposed behavior of pulses emerging from the analog amplifier and shaping systems. None of these models is very difficult to describe mathematically, but their fitting to experimental data using the method of least squares depends on the controlled behavior of an iterative minimization procedure. This can be a delicate operation when dealing with experimental data containing information limited by the statistical nature of the nuclear disintegration process. Nevertheless, such numerical procedures are used commonly for resolving multiple peaks of asymmetric shape and are employed by the programs listed earlier in this section.

For the analysis of large numbers of similar spectra, a nonsubjective means of analysis, that of multilinear regression, is suitable. Although developed for sodium iodide detectors in which the majority of component spectra overlap [71,72], it is equally applicable to solid-state detection systems, provided a high degree of system stability is maintained and the computational power to handle fairly large matrices is available.

Other numerical procedures may involve corrections for minor differences in irradiation and measurement parameters between sample and standard. Gibbons et al. [73] have considered the corrections to be made from extended flux level variations during irradiation.

In cases where an isotope has been identified as present in a spectrum but no standard is available, then an estimate can be made of the original elemental mass by applying the equations of

Sec. II.A and using the appropriate nuclear constants and estimates of absolute flux level. Absolute disintegration rates are needed and thus the peak area needs to be estimated from one or more multi-energy standards such as ^{226}Ra or ^{152}Eu. Efficiencies calculated for a series of known energies are fitted to an experimental curve or joined by a series of cubic splines. A suitable general function is a pair of power law curves [59].

Finally, an important area of data processing is the collated storage of analytical results such that they can be validated [74] and presented in a useful form either as analytical or derived data [75]. It is also profitable in such circumstances to have automatic mechanisms for associating analytical data with the sample from which it arose.

E. Sensitivity and Applicability

A simple statement of sensitivity for the determination of individual elements in soil by INAA cannot be made. The relative contributions of constituent elements determine the signal-to-background ratios of absorption peaks in γ-ray spectra, hence, sensitivity varies with each element in every soil type. Figure 5 shows four γ-ray spectra of differing soil types (see Table 3, Sec. IV.A) that illustrate the differing relative magnitude of constituent peaks and hence their degrees of interference in the determination of minor peaks.

Detection (or minimum determination) levels can be expressed in a variety of ways, as discussed in Currie's classical review [76]. However, it is convenient to use the calculated standard deviation of the measured levels in the determination of a particular element as some indication of the minimum levels measurable. To give the reader some approximate minimum determination levels of particular elements in soil, Table 1 provides a list of *average* minimum levels for a group of soils found in the United Kingdom.

The quoted levels are also some measure of the precision or reproducibility of results to be expected in soil analysis by INAA. The data are based on the statistical uncertainties associated with the radiation measurement, i.e., of "counting" statistics and detector resolving power. No account is taken of uncertainties in sampling or in preparation of the standards. Errors related to nonuniformity of irradiation flux may occur, but these should be very small in well-designed facilities. Likewise, errors due to geometrical dissimilarities in the measurement equipment should be small if reasonable care is taken in its design and operation. A more likely source of significant error is that introduced by distortion of spectrum shape due either to large differences in levels of measured radioactivity, or to instabilities in the measuring equipment itself.

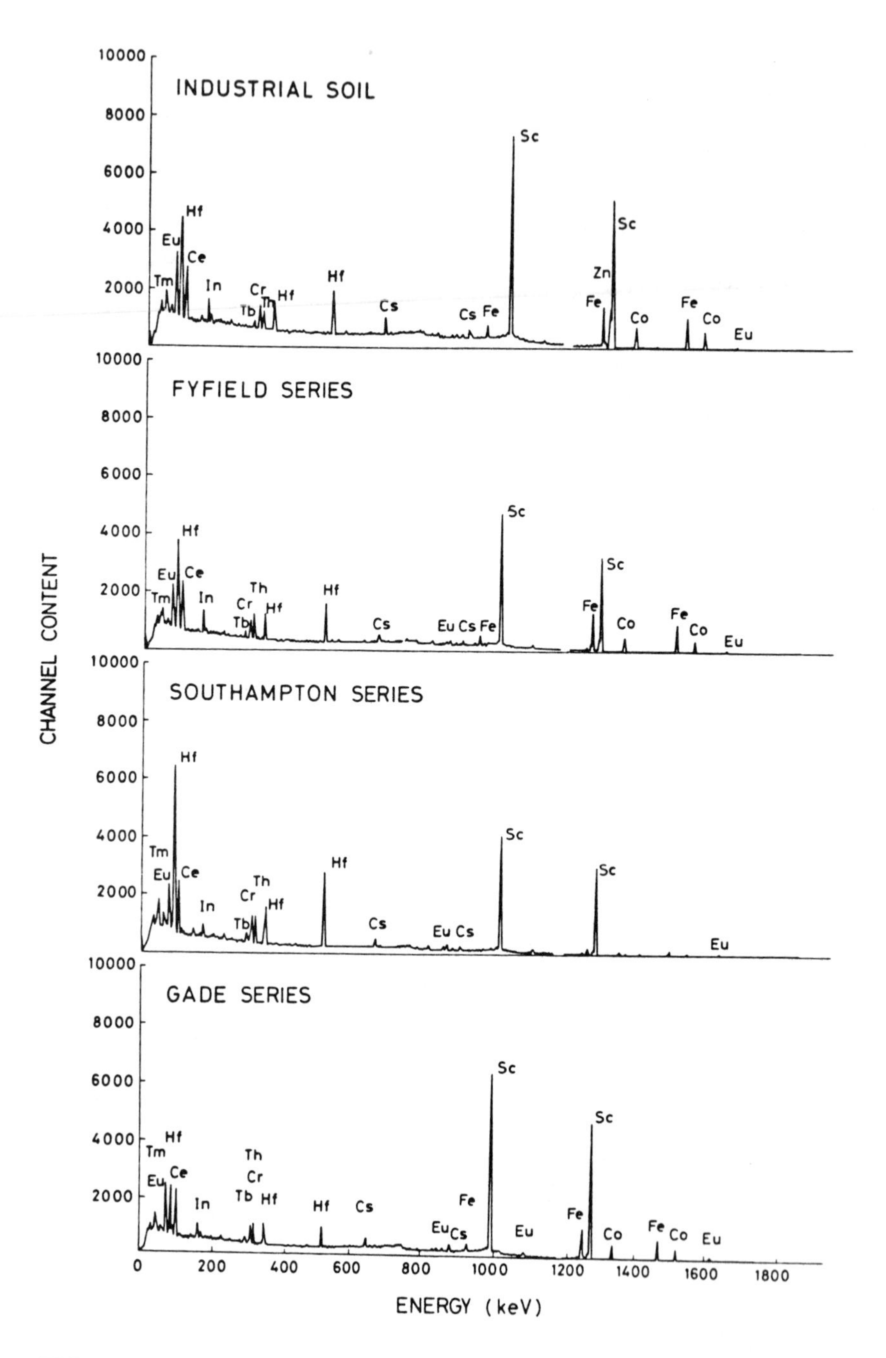

FIG. 5 γ-Ray spectra from irradiated samples of differing soil type.

TABLE 1 Average Minimum Concentrations of Elements Detectable in a Group of U.K. Soils

Element	Concentration (mg/kg^{-1})[a]	Element	Concentration (mg kg^{-1})[a]
Al	0.15%	K	0.15%
Ag	0.15	La	1.5
As	0.5	Mn	20
Au	0.005	Mo	5
Br	2.0	Na	300
Ca	0.4%	Ni	30
Cd	4	Rb	20
Ce	2	Sb	0.5
Co	0.35	Sc	0.25
Cr	2.5	Se	0.4
Cs	0.3	Sm	0.2
Cu	4	Tb	0.1
Dy	1	Th	0.3
Eu	0.02	Ti	1000
Fe	0.10%	V	10
Hf	0.15	W	0.3
Hg	0.6	Yb	0.3
I	10	Zn	4.0
In	0.05		

[a]Except where indicated.

Overall accuracy is not determined easily, but when analysis of similar material was examined using INAA [77], then 10–20% accuracy at least could be expected for most elements where the precision was good.

A review of this form of analysis of soil should include a statement of disadvantages. These are at least threefold:

1. The physics and economics of irradiation procedures demand that samples should not exceed a few grams and preferably be a good deal less. Thus, a problem immediately exists with such a heterogeneous material as soil in the production of a suitably representative sample for analysis.
2. Access to a suitable irradiation source is a very limiting factor, especially if isotopes with a half-life of a few hours or even minutes are to be measured. Furthermore, it is unrealistic to expect a wide range of elemental analyses unless the irradiation source is a nuclear reactor.

3. A wealth of numerical data is provided with every sample measured. The laboratory is faced with either an excessive amount of manual computation on the part of its analytical staff or a fairly heavy investment in computing resources (both hardware and software) beyond the already expensive electronic equipment necessary to measure the samples and store the consequent digital data. For the investment to show a good economic return, the throughput of samples needs to be quite high.

These problems are essentially organizational and economic. They do not detract from the basic scientific worth of the activation technique with its high specificity and sensitivity for a range of environmentally significant elements. It seems a reasonable conclusion, therefore, that in cases where the particular scientific advantages of the method are overwhelming, it is worth the trouble and labor of manually applying the method to a few important samples. At the other end of the scale, if the irradiation, measurement, and data handling resources are available, then this powerful method should be used routinely.

Depending on the element and irradiation and measurement conditions, levels of activity corresponding to submicrogram quantities can be determined.

A feature, not necessarily an advantage, of activation analysis is that the total element, irrespective of its chemical or physical composition, is measured.

De Corte and Hoste [78] presented a survey of the European contributions to activation analysis, including NAA, CPAA, and PAA of environmental samples and standard reference materials. It was concluded that some 90% of activation analysis applications are made with reactor neutron facilities. The authors discuss advances in computerized data reduction techniques and the development of very high γ count-rate systems.

Finally, some advantages of INAA may be summarized thus:

1. It is an economic multielement technique in the appropriate circumstances and well suited to automation and data processing operations.
2. The sensitivity is particularly good for a number of elements of interest both in plant and animal nutrition studies and in toxicity investigations related to pollution.
3. Trace element contamination by reagents and chemical processing is completely avoided.
4. Dissolution and other chemical operations are avoided in total element studies.

IV. BASIC STUDIES

A. Total Element Analysis of Soils

Basic investigations on the induced γ activities in a wide range of neutron irradiated soils have been reported over the last 25 years and have concentrated on details of irradiation, instrumentation, and analysis of γ spectra rather than applications of the technique to various areas of research, which are discussed in detail in Sec. V.

In 1959, McNeilly and Eichler [79] examined the activities induced by the irradiation of five U.S. soils of varied composition with both thermal and 14 MeV neutrons. They presented spectra obtained by a NaI(Tl) detector and identified peaks from ^{28}Al, ^{56}Mn, and ^{24}Na following thermal neutron irradiation, and an additional peak from ^{27}Mg from fast neutron irradiation.

Instrumental NAA of soils and rocks in Norway was achieved by Thoresen [80] in 1964, for Al, Br, Cu, Fe, K, La, Mn, Na, and Sc. In this case, separate irradiations were made at low and high neutron fluxes up to 10^{12} n cm^{-2} sec^{-1}. The analysis of γ-ray spectra was made by application of a spectrum stripping method to reveal the photopeaks of lower energy. Thoresen [80] commented that the amounts of Na in his samples limited detection of shorter-lived isotopes and quoted approximate relative detection limits of 10^{-2} for Cu:Na and 1 for K:Na. For detection of Br, the amount of La was critical; thus, the limit of detection was Br:La ~ 10^{-1}. Also in 1964, Kline [81] was able to obtain the quantitative measurement of INAA of ^{60}Co, ^{51}Cr, ^{59}Fe, ^{140}La, and ^{46}Sc. Interference from ^{24}Na prevented analysis of isotopes with shorter half-lives than 15 hr.

A method for the rapid determination of total Al, Mn, and Na in Norwegian soils by INAA, with a precision of 1–2%, was published by Khera and Steinnes [82] in 1969. The authors gave attention to interfering nuclear reactions by (n,p) or (n,α) processes and found that correction factors were significant only for ^{28}Al.

With the advent of Ge(Li) detectors, Salmon and Creevy [83] developed an on-line computer system for INAA of air, water, and soil samples whereby γ-ray spectra were obtained from a NaI(Tl) detector for short-lived isotopes and a Ge(Li) detector for longer-lived activities. The authors pointed out the inflexibility of pulse-height analyzers for accumulation of data from several detectors in a laboratory handling large numbers of routine determinations of numerous elements in many different media and suggested replacement by a computer analyzer system.

In 1977, Naidenov and Travesi [84] presented INAA results for 43 soils from Bulgaria that were representative of a wide range of soil classes and profile depths. Altogether, 19 elements were determined following irradiation of samples with thermal neutrons and application of an Ortec 459 Ge(Li) semiconductor detector for low-energy photon spectrometry (LEPS) in addition to a 40 cm^3 Ge(Li) detector for energies >200 keV. Use of the LEPS system on irradiated soils permitted analysis of low-energy peaks from ^{141}Ce, ^{177}Lu, ^{153}Sm, and ^{169}Yb, and reductions in interference from high-energy peaks allowed the measurement of isotopes such as ^{147}Nd (91.4 keV peak) and ^{160}Tb (87.0 keV peak). For some elements, a check of the analysis was possible at both low and high energies, using the different detectors. The authors also showed that there was negligible interference from U in soil, in the analysis of elements whose activation products were also fission products.

Trace elements were measured in several Alaskan soils and clay minerals following irradiation with resonance (epithermal) neutrons to extend the analysis to elements with longer half-lives. Samples were sealed in quartz ampoules for irradiation over 40 hr at a total flux of 10^{12} n cm^{-2} sec^{-1} and were shielded by Cd foil to exclude thermal neutrons [85]. Quantitative data were obtained for Co, Cr, Fe, Ni, Ti, and Zn using a Ge(Li) detector. Ten additional elements were identified, including Ba, Cs, Ta, and U. This technique was claimed to represent an improvement on solely thermal neutron irradiation, in which induced activities of elements with a high cross section or concentration in soil may obscure the γ-ray spectra of other elements that are present in trace amounts and have low activation cross sections. Clay minerals from Turkey were analyzed for 13 elements by thermal NAA [86] at a neutron flux of 1.5×10^{13} n cm^{-2} sec^{-1} for 2 hr; comparison with measurements on lignite showed enrichment of the trace elements Ce, Co, Hf, La, Sc, and Sm in the clays.

In 1973, total elemental analysis of soils from eastern Uruguay was reported by Buenafama [87]. Combined techniques of INAA and NAA with radiochemical separation were used, and both thermal and epithermal neutrons were applied for activation. Concentrations of 25 elements were thus determined by INAA, γ-ray spectra being measured by a Ge(Li) detector system; peak areas were evaluated by the Wasson method [63]. After radiochemical separation, Cu, Ga, Mo, Sr, W, and Zn were determined. Some results obtained by Buenafama [87] for some elements of agricultural interest, including potentially toxic pollutants, are listed in Table 2 and represent the means of at least triplicate determinations.

Modern techniques of INAA have been applied by the authors [34] to the analysis of a range of contrasting soil types from southern England. These soils were sampled to 15 cm depth from arable,

TABLE 2 Trace Element Content of a Basaltic Soil Determined by INAA

Element	Total concentration ($mg\ kg^{-1}$ dry soil)	S.D. (± %)
Br	22	23
Ca	8,900	12
Co	33	1.2
Cr	220	4.2
Fe	69,100	0.7
K	5,900	3.4
Mn	1,930	1.2
Na	4,570	1.3
Sb	2.8	10
V	660	5.0

Source: Ref. 87.

grassland, and woodland locations. The concentrations and uncertainties found for 37 elements are shown in Table 3. Compared with other soils, the podzol (Southampton Series) showed considerably lower concentrations of As, Co, Ni, V, and Zn than others, probably the result of leaching under acid conditions. The excellent permeability to water of the brown earth (Fyfield Series loamy sand) was reflected in relatively low concentrations of Br, Cs, Na, and rare earths.

Korr [37] established that the activation of soil samples with thermal neutrons combined with Ge(Li) spectrometry gave quantitative data for Al, Ba, Br, Ca, Cl, K, Mg, Mn, Na, Ti, and V; use of a low-energy photon detector provided additional results for Co, Dy, Eu, Sm, Th, and U. Cadmium shielding for core irradiations was only of value for Si, Sr, and Sm. Following cyclic neutron-activation (30 cycles, 5 sec irradiation, and 5 sec count), Sc, Se, and Hf were also detected, but this technique was the most affected by the production of ^{28}Al by neutron capture, which interferes with the measurement of spectra from other isotopes. INAA of soils and standard rock samples from China was made by Sun and Jervis [88] using the Canadian "Slowpoke" reactor at a neutron flux of 2.5×10^{11} n cm^{-2} sec^{-1} to provide data for 35 elements.

Detection limits by activation analysis were compared with other methods of microanalysis for Cl, Co, Cu, Fe, Mn, Mo, Si, and Zn, and although it was recognized that B, an important trace element,

TABLE 3 Determination of Total Elemental Concentrations in Topsoils from Southern England ($mg\ kg^{-1}$ dry soil or % where stated)[a]

	Fyfield, Brown earth, Arable	Denchworth, Surface-water gley, Woodland	South-hampton Humus-iron podzol, Woodland	Gade, calcareous ground-water gley, Grassland	Icknield, Rendzina, Grassland	Sulham, Peaty gley, Grassland	Westhay, Organic soil, Grassland	S.D. (%)
Al	1.4%	5.3%	1.1%	3.6%	4.5%	3.4%	3.7%	< 5
Ag	0.96	2.4	1.5	1.3	1.8	2.4	1.6	< 10
As	7.3	16.7	4.1	6.6	10.2	13.7	16.5	< 5
Au	0.0036	0.024	0.065	0.0037	0.029	0.018	0.057	< 20
Br	3.1	14.4	6.2	16.2	28	10.0	89	< 10
Ca	0.96%	2.4%	2.1%	5.3%	4.2%	2.1%	3.2%	< 15
Cd	3.9	6.6	<6	<3	<7	5.9	9.9	> 25
Ce	32	59	37	41	59	44	42	< 5
Co	5.3	9.1	0.85	8.0	10.8	7.2	8.3	< 5
Cr	42	57	57	50	38	37	47	< 20
Cs	1.4	6.1	1.3	2.9	2.9	1.7	4.4	< 10
Cu	20	5.9	10.9	3.1	22	29	18.4	> 25
Dy	2.4	6.9	4.4	6.1	8.2	7.1	4.9	< 15
Eu	0.38	0.73	0.40	0.65	1.2	0.83	0.66	< 5
Fe	1.6%	2.8%	2.6%	1.94%	2.4%	2.2%	2.2%	< 5

Hf	6.9	6.9	12.9	5.6	4.7	4.6	3.0	< 5
Hg	<0.4	<0.8	<0.4	<0.6	<0.7	<1.7	<0.6	> 25
I	31	56	12.6	<10	93	<12	24	> 25
In	0.15	0.38	0.20	0.23	0.35	0.49	0.16	< 20
K	0.69%	2.1%	1.1%	0.54%	1.92%	1.06%	2.1%	< 10
La	18.2	35	22	30	59	36	26	< 5
Mn	220	250	510	340	1430	350	490	< 5
Mo	<7	<5	<2	<9	<4	<6	<4	> 25
Na	410	3000	810	1800	1910	1340	1670	< 20
Ni	45	81	20	69	150	107	40	> 25
Rb	< 13	82	24	48	< 24	< 19	47	> 25
Sb	0.65	<0.9	0.98	<0.6	<0.7	1.6	<0.7	< 15
Sc	2.4	7.6	2.3	5.3	6.8	5.3	5.9	< 5
Se	0.92	<0.9	1.1	1.3	<0.8	1.8	1.0	< 20
Sm	2.0	3.6	2.7	3.7	6.1	4.7	3.3	< 5
Th	2.6	6.9	4.3	4.2	5.8	3.9	4.7	< 5
Ti	1520	3900	5900	1300	1770	2200	< 700	> 25
V	51	102	30	58	74	81	71	< 10
W	<0.3	1.8	1.8	0.94	2.0	1.0	1.4	< 20
Yb	0.71	1.4	1.9	1.1	2.2	1.2	0.79	< 20
Zn	59	78	11.6	80	104	52	96	< 5

[a]Names given are recognized soil series and follow the Soil Survey of England and Wales classification [95].

cannot be determined by activation, sensitivity for the other elements was considered adequate [89].

The analysis of Ag in soil and rock standard reference materials (including IAEA soils 5 and 7) was carried out by Geisler and Schelhorn [90]; 200 mg samples were activated in an epithermal neutron flux of 5×10^{11} n cm^{-2} sec^{-1}, Cd shielded, for 95 hr and ^{110m}Ag was measured at 658, 885, and 937 keV. The results from INAA were in close agreement with those obtained by NAA on radiochemically separated samples.

B. Analysis of Soil Fractions

The measurements of trace elements in soil solutions and extracts by modern techniques of INAA are relatively few compared with work on the determination of total element content, although to most agriculturalists, the exchangeable and solution "pools" of individual elements and the interelement ratios in these available fractions are vital in studies of plant nutrition and plant uptake.

Determination of Cl, Fe, K, and Na in neutron-irradiated soil solutions (simulated) was achieved by direct γ-ray spectrometry and counting with a NaI(Tl) detector, but with reliance on chemical separation by ion-exchange columns for the analysis of Ca, Mg, P, and S [91]. The removal of soil solutions by a centrifugation method [92], followed by evaporation of 50–100 ml of sample onto filter paper (Whatman 41 grade) disks for irradiation has been demonstrated as a practical technique [34]; the composition of soil solutions extracted from a calcareous silty loam and analyzed by this method (Icknield Series) is shown in Table 4.

Soil water from lysimeters was analyzed for Al after removal of Cl by an anion-exchange resin, to avoid interference by ^{38}Cl following irradiation of liquid samples by thermal neutrons [93]. The samples contained from 2–173 μg ml^{-1} chloride. It was noted that the high Mg content of the soil water precluded the use of the 1040 keV γ-ray peak of ^{27}Mg as employed in metallurgical analysis. Production of ^{28}Al by an (n,p) reaction on ^{28}Si necessitated correction after the separate colorimetric determination of Si concentrations. A NaI(Tl) detector was used for counting, and it was possible to reduce interference from ^{56}Mn by spectrum stripping.

For measurement of "plant-available" trace elements in soil, 0.5 *M* acetic acid is frequently used as an extractant. This reagent may be freed easily from metal contamination by distillation in quartz apparatus and dilution with quartz-distilled water. An extractant/dry soil ratio of 40 is recommended, with a 2 hr shaking period [94]. As with soil solutions, analysis by INAA can be effected by the evaporation of filtered extracts onto filter paper.

TABLE 4 Elemental Composition of Different Soil Fractions Determined by INAA[a]

Element	Exchangeable fraction[b]		Soil solution [c]		% Uncertainty of determination
Al	69	(0.2%)	0.074	(0.0002%)	<5
As	0.066	(0.7%)	<0.0007	(<0.007%)	<5
Br	10.7	(38%)	0.0048	(0.02%)	<10
Ca	17,200	(41%)	98	(0.2%)	<10
Cd	0.47	(<7%)	<0.0002	n.d.	>25
Cl	<40	n.d.	1.2	n.d.	>25
Co	0.077	(0.7%)	0.00047	(0.004%)	<5
Cr	0.17	(0.5%)	0.0049	(0.01%)	<10
Cu	6.3	(29%)	0.22	(1%)	<10
Fe	9.2	(0.04%)	<0.4	(<0.002%)	>25
Hg	<0.01	n.d.	0.0048	(>0.7%)	>25
K	240	(1%)	6.2	(0.03%)	>25
Mg	520	(>13%)	5.1	(>0.1%)	<20
Mn	49	(3%)	0.0093	(0.0007%)	<5
Mo	0.02	n.d.	<0.0004	n.d.	>25
Na	155	(8%)	10.2	(0.5%)	<5
Ni	1.4	(0.9%)	<0.06	(0.04%)	>25
Sb	0.029	(>4%)	<0.0006%	n.d.	<20
Se	0.041	(>5%)	0.17	(>21%)	<15
V	<0.1	(<0.1%)	0.0019	(0.003%)	<20
Zn	8.1	(8%)	<0.02	(<0.02%)	<5

n.d.: Not determined; results for both soil fraction and total were below detection limit.
[a]Soil was a rendzina from Icknield Series, Chilton Oxon, U.K., sampled from permanent grassland, 0–15 cm depth.
[b]Extracted with 0.5 *M* acetic acid; 10 g of air-dried soil shaken for 2 hr with 400 ml extractant.
[c]Soil solution removed from soil by centrifugation at 41% moisture on a wet weight basis. Figures in parentheses denote percentage of total elemental concentration in soil.

Other reagents frequently used as soil extractants in agricultural advisory work are KCl, NH_4OAc, NH_4Cl, 0.1 *M* HCl, 0.1 *M* HNO_3, or Morgan's reagent (NaOAc and acetic acid). Of these, NH_4OAc is best suited to the analysis of extracts by INAA, unless only long-lived isotopes are sought and radioactive decay is allowed to eliminate interference from ^{24}Na and ^{38}Cl. Moreover, NH_4OAc can be prepared in a very pure state by bubbling ammonia gas through quartz-distilled acetic acid.

Khera et al. [96] determined ^{28}Al in irradiated extracts from 29 Norwegian soils; a well-type NaI(Tl) detector was used to measure the 1780 keV γ ray. The method had a sensitivity of 5×10^{-5} μg Al, and the overall results compared closely with those obtained by atomic absorption and spectrophotometric methods. In this study the extracts were irradiated in liquid form in a thermal neutron flux of about 1.5×10^{13} n cm^{-2} sec^{-1}. Subsequently, Khera and Steinnes [97] determined ^{56}Mn and ^{24}Na (as well as ^{28}Al) after neutron irradiation of similar soil extracts.

The extraction of soil with 0.1 *M* HNO_3, followed by the irradiation of the centrifuged extracts, was used by other authors [31]; for short-lived isotopes, 0.1–0.5 ml of the supernatant was irradiated in a polyethylene vial, whereas for long-lived ones 0.5 ml of the extract was sealed in a quartz vial prior to irradiation. Results were obtained for 20 elements, using a neutron flux of 2×10^{13} n cm^{-2} sec^{-1} for 50 hr, and a 45 cm^3 Ge(Li) detector coupled to an 8192 channel pulse-height analyzer.

The extraction of ^{46}Sc and ^{140}La from thermal neutron-irradiated soil by agitation with 0.1 *M* HCl has been used in pedological studies [98]; the extractable forms of these elements were compared with their total concentrations to examine relative differences between soil-mineral associations.

The availability of Cu for plant uptake was examined in surface and subsurface soils by the irradiation of samples with thermal neutrons, followed by the extraction of a diffusible fraction, with dilute $CuSO_4$ solution [81]. Soils were extracted further with an acidified (0.1 *M* HCl) $CuSO_4$ solution, and finally the residual Cu fraction was also determined from the induced ^{64}Cu activity.

Clay mineral fractions were prepared for INAA by the sedimentation of soil samples in distilled water in a lucite cylinder [85]. Glassy fractions from volcanic ash soils were obtained by centrifugation in high-purity bromoform-bromobenzene and through ultrasonic scrubbing, centrifuging, and freezing procedures [99].

Vanadium in clays has been determined by paper chromatography of acid extracts, followed by neutron-irradiation of the relevant portion of the chromatogram [100]. This enabled the measurement of ^{52}V with little interference from the induced activities of Al, Mn, and Na, and the method achieved better sensitivity than atomic absorption.

V. APPLICATIONS

The overall range of elements analyzed in soils by INAA has been very extensive; Table 5 indicates some of the main areas to which the technique has been applied. The cited papers are discussed in Sec. V.A to V.E.

A. Plant and Animal Nutrition

The determination of minor and trace elements by INAA in life sciences research has been reviewed by Bowen [101] and Leddicotte [102]. It is evident that INAA may be used to determine all trace elements, with the exception of B, that are essential to plants. Rahman [89] has reviewed the importance of this method of micronutrient analysis in research on soil-plant relationships in Bangladesh.

The essentiality of Co in plant and animal nutrition led Johansen and Steinnes [103] to apply INAA to routine determinations. Neutron-irradiated soils and plant ashes were stored for 4 weeks before the measurement of ^{60}Co, using a well-type NaI(Tl) detector. The relative standard deviation was 7.2% for soils and 6% for plant material, based on the duplicate determination of 14 samples in each case. The authors point out that if immediate analytical data are not essential and long postirradiation cooling times can be tolerated, the method is very well suited to handling large numbers of samples.

The identification of minerals and the reserve of major and trace elements present in soils on the Matadi rock formations in Zaire was determined with the assistance of INAA with respect to Al, K, Mn, Na and Si [104]. However, chemical separation was made for Ba, Cu, and Sr prior to γ-ray spectrometry with a Ge(Li) detector.

In Thailand, the routine analysis of rice soils for Al, Mn, and Na has been accomplished by INAA [105]. Other elements were measured following the chemical separation of thermal neutron-activated soils and included As, Cu, Mn, and Zn.

Activation of plant and fertilizer samples with 14 MeV fast neutrons to determine the macronutrient elements Ca, Cl, K, Mg, N, P, and Si has been carried out on a large scale in the Soviet Union by Srapenyants and Saveliev [106]. A throughput of 250–500 plant and fertilizer samples per 8 hr shift has been achieved, and the system has accomplished the analysis of more than 50,000 such specimens. Although these authors did not describe any application of their technique to the analysis of soils or soil extracts for diagnostic purposes, other workers in the Soviet Union [107] have applied INAA to the analysis of cotton plant tissues and soils

TABLE 5 Application of INAA to Analysis of Soil; References to Research Areas

Element positively measured	Plant and/or soil	General environmental studies	Environmental pollution	Forensic science	Archaeology
Ag	[90]	[132]	[28]		
Al	[88,105,107,108]	[127,137]	[32,115,117,122]	[145,146]	
As	[88]	[127,132,137]	[28,32,115,117,122]		
Au	[108]		[28,32,115]		
Ba	[86,88,108]	[132]	[32,115,117]		[152,153]
Br	[88,108]	[127,130,132,137]	[32,116,122]		
Ca	[88,107,108]		[115,122]		
Cd			[28]		
Ce	[86,88,108,150]	[127,132,137]	[32,116,117,122]	[146]	[152,153]
Cl	[88,107,108,150]		[32]		
Co	[86,88,108]	[127,132,137]	[32,115,117,122]	[146,146]	[152,153]
Cr		[127,132,137]	[32,122]		[152,153]
Cs		[127,132,137]	[32,115,116,122]		[152,153]
Cu		[127]	[28,115]		
Dy	[88,108]		[117]		
Eu	[88,108]	[132]	[32,115,116]	[145]	[152,153]
Fe	[86,88,108]	[127,132,137]	[32,115,117,122]	[145,146]	[152,153]
Ga		[132]	[116,117]		
Hf	[86,88,108]	[132]	[32,115,117]		[152,153]
Hg			[28]		
I	[88]		[130]		
In			[28,117]		
K	[88,108]	[132]	[117]	[146]	[153]
La	[86,88,108,150]	[127,132]	[32,115—117,122]	[145,146]	[153,153]

Lu	[88,108]	[132]			
Mg	[88,108]	[127]	[32,115]		
Mn	[88,105,107,108]	[127,137]	[32,115,117,122]	[146]	[153]
Mo	[107]	[132]			
Na	[86,88,105,107,108]	[127,132,137]	[32,115,117,122]	[146]	[153]
Nd	[88,150]				
Ni	[108]		[28]		
Rb	[88]	[132]	[32,115,117,122]		[152,153]
Sb	[86,88,108]	[132,137]	[28,32,115—117,122]		
Sc	[86,88,108]	[132,137]	[28,32,115,122]	[145,146]	[152,153]
Se		[137]	[28,32,115]		
Sm	[86,88,108]	[132]	[32,116,117]	[145]	
Sr	[88]		[117]		
Ta	[86,88,108]	[132]	[32,115,117]		[152,153]
Tb	[88]	[132]	[32,116]		
Th	[88,108]	[127,132,137]	[115—117,122]	[146]	[152,153]
Ti	[88,108]	[127]	[32,115,117]		
U	[88,108]		[115—117]		[153]
V	[88,108]	[127,137]	[32,115,117,122]		
W	[88]		[122]		
Yb	[88]		[32]		
Zn	[107,108]	[132,137]	[28,32,122]		
Zr			[32]	[146]	

for seven elements. Thus, by irradiation with a flux of epicadmium (epithermal: Cd shielded) neutrons the following isotopes were measured: ^{28}Al, ^{49}Ca, ^{38}Cl, ^{42}K, ^{56}Mn, ^{101}Mo, and ^{24}Na.

Diagnostic measurements on 30 trace elements in Egyptian soils and sugar cane grown on them were reported by Awadallah et al. [108], including the examination of changes in trace element concentrations during processing to extract sugar; a neutron flux of 7.5×10^{13} n cm^{-2} sec^{-1} for 24–48 hr was used for long-lived nuclides. A similar study was made on 13 crop plants grown in Egypt [109].

B. Environmental Pollution

Research to assess the extent and impact of soil pollution by operations such as power generation, smelting and refining of metals, and disposal of industrial waste has made extensive use of INAA. Frequently, such work has been carried out on an interdisciplinary basis to include atmospheric, terrestrial, and aquatic measurements.

The application of sewage sludge to agricultural land for disposal and/or as fertilizer can cause problems because of the content of associated trace metals. The accumulation of metals in the soil may be toxic to plants or render vegetables and fruits unacceptable for human consumption. Contamination of ground water supplies by leaching of metals through soil may also require investigation. These problems demand a multielement analysis technique. The determination of metals in sludges supplied from six cities in North Carolina has been achieved by INAA [110]. Results were obtained for Co, Cr, Fe, Hg, Sb, and Se, and it was suggested that Ag, Sn, and Zn could also be measured after a long decay time. The authors used both X-ray and γ-ray detectors.

The analysis of soil from grassland used for sewage sludge disposal was carried out by INAA, with the results shown in Fig. 6 [34]. Comparison with uncontaminated soil showed increases of two orders of magnitude in total concentrations of Ag, Cd, and Hg. For Cu, Sb, and Zn, an order-of-magnitude increase was observed.

In Japan, the pollution by Cu of rivers used as sources of irrigation water has led to harmful effects on the growth of crops: INAA of Cu in plant tissue, soil, and soil solution was made using a neutron flux of 5×10^{11} n cm^{-2} sec^{-1} and a 50 hr cooling period prior to γ-ray spectrometry [111].

The use of methyl bromide for soil fumigation to eradicate pests increases the uptake of Br by plants, with the result that marketed crops for human consumption may contain more than the acceptable level of Br. Although INAA has been used recently for the determination of ^{80}Br and ^{82}Br in vegetables by Suess and Staerk [112], it has so far found little application in the analysis of Br in soil or

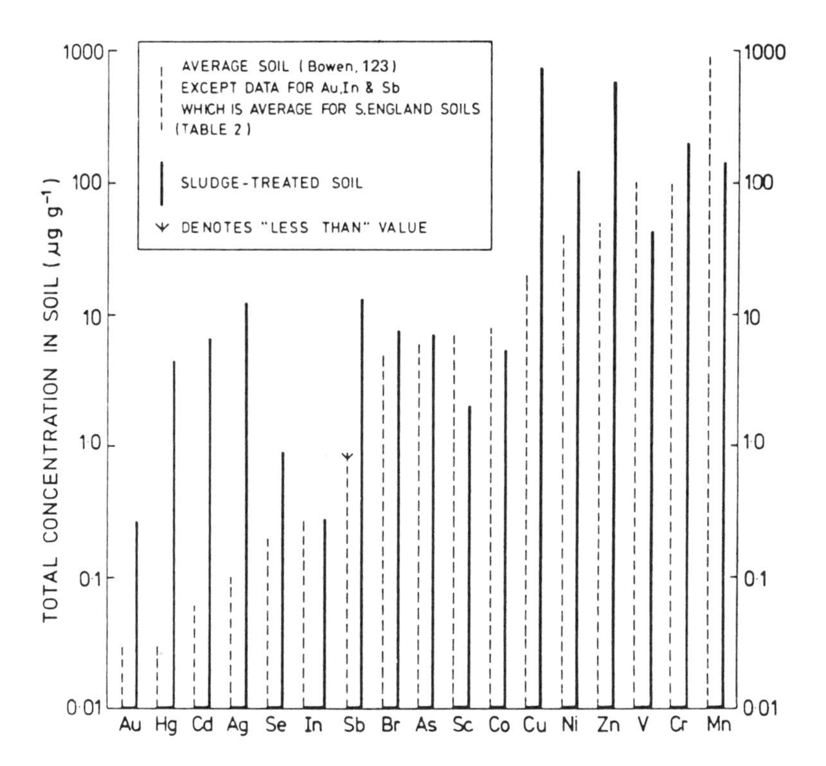

FIG. 6 INAA of trace elements in sandy loam subject to long-term disposal of sewage sludge.

soil components. However, INAA would appear to be a promising technique to assess plant uptake factors for Br under varied conditions of soil management.

The use of arsenical pesticides in agriculture has led to a requirement for the determination of trace elements of As in plants and soils. Among the various techniques discussed by Talmi and Feldman [113], INAA is included. The induced γ activity of ^{76}As is measured with a Ge(Li) detector, and good sensitivity can be achieved in the absence of excessive ^{24}Na in the samples. The contamination of soils with chlorinated hydrocarbon pesticides has been examined by the neutron irradiation of organic extracts of the soils to produce ^{38}Cl for measurement by the 2160 keV peak [114].

Emissions from coal-fired power stations into the terrestrial environment have been investigated by several workers in the United States. Lyon [115] determined 23 trace and major elements by INAA in soil profiles around a power plant in Memphis, Tennessee. Potential marker elements for fly-ash were identified and soil concentrations of Cu and Zn were 5 to 10 times higher than levels in average "world" soil. In Aiken, South Carolina, the analysis of 29 elements was made in samples of fly-ash, soil, and vegetation collected at varying distances from a coal-fired power station [116]. Neutron activation was used to supplement atomic absorption analysis, plasma source emission spectrometry, and conventional spectrometry; 10 g samples were irradiated for 2 weeks with a ^{252}Cf source (100 mg), giving a flux of 10^7 n cm^{-2} sec^{-1}. Elements determined by γ-ray spectrometry were Br, Ce, Cs, Eu, Ga, La, Sb, Sc, Sm, Tb, Th, U, and Yb. Kucera and Soukal [117] used INAA to determine 27 elements in ashes from Czechoslovakian power stations, parallel with the analysis of standard reference soils (IAEA numbers 5 and 7) and standard coal fly ash (NBS 1633a).

The use of Sr as an indicator for pollution around coal-fired power plants in the western United States prompted a comparison of techniques for the analysis of soil and vegetation by Gladney et al. [118]. INAA, epithermal neutron activation (IENAA), and atomic absorption (AAS) methods were applied. In the case of IENAA, higher-energy (0.5 MeV) nonthermal neutrons were used to produce ^{87m}Sr, and although the detection limit was superior to INAA, the lowest detection limit was given by AAS; however, effort is required for sample dissolution in the latter method.

Crecelius et al. [119] measured As, Sb, and Pb contamination of soils at varying distances from a copper smelter: It was found that As and Sb were the best tracers of smelter stack dust. Dried soil was irradiated for 1–2 hr in a thermal neutron flux of 10^{12} n cm^{-2} sec^{-1}, and the ^{76}As and ^{122}Sb γ-ray peaks were counted after a cooling time of 2–4 days.

Near to gold mining and refining operations at Yellowknife, Canada, the analysis by INAA of soil showed relatively high concentrations of As, Au, Co, Cr, and Sb [36]. The more organic soils showed higher levels of Co and Cr than mineral soils from the same general area. The ratios of As to Sb in soil were close to those reported for the mined ore.

More recently, the same gold refinery complex at Yellowknife was the subject of an epidemiological study by Jervis and Tiefenbach [120], but specifically on As pollution. INAA was used on samples of water, hair from children and workers, and on soil and plant specimens, using neutron irradiation at 10^{12} n cm^{-2} sec^{-1}; ^{76}As was measured after a 24 hr decay using a Ge(Li) detector. The authors commented that soils were the most dramatic indicator of local arsenic contamination and within 1–2 km of the smelters they measured two to three orders of magnitude more arsenic than normal terrestrial concentrations.

Pollution caused by urban metal refineries in Toronto, Canada was examined by Jervis et al [121]. Soil, vegetation, airborne particulates, dustfall, and human hair were all analyzed. These authors used INAA to obtain information on pollution by As, Cd, Fe, and Sb, as these were associated with Pb emissions; in addition, data were obtained for Br (from vehicle exhaust and combustion sources) and Al, Ca, V, and Ti from soil and nonsmelter sources.

Pollution from metal mining, smelting, and refining industries in Kellogg, Idaho was assessed by the multielement analysis of air particulates, soil and vegetation, employing INAA together with X-ray fluorescence analysis, since information on Pb levels was required [28]. High enrichments in soil and grass were reported for Ag, As, Au, Cd, Cu, Hg, In, Ni, Pb, Sb, Se, and Zn; Sc was chosen as the normalizing element because it was considered that the only likely source of this element was the soil. An irradiation time of 2 min at 2×10^{13} n cm^{-2} sec^{-1} was used for short-lived isotopes, and 12 hr at 4×10^{12} n cm^{-2} sec^{-1} for long-lived ones. In this work, the depth distribution (profile) of metals in soil was measured to confirm that the parent material at the soil sampling points did not contribute to high levels of metals.

An interdisciplinary study of atmospheric and terrestrial pollution in the Swansea/Neath/Port Talbot industrial area of South Wales was initiated in 1971–1972 and the analysis of soil and smelter waste by INAA was included [122]. The local soils had significantly higher concentrations of As, Cu, Ni, Se, and Zn than uncontaminated soil [123], and the importance of resuspension and transport of waste deposits was examined by a comparison of element ratios with those in atmospheric samples.

The contamination of soil at the site of an old tin smelter in Lancashire, England was assessed by multielement analysis using INAA, supplemented by gamma-photon analysis for Pb and Sn [124]. Compared with the average analysis of surface soils [123], Ag, Sn, and W were enriched by three orders of magnitude, whereas an increase of 350- to 700-fold was detected for As, Cu, and Pb. Radish (roots) from the contaminated soil showed at least an order-of-magnitude increase in As, Br, Cd, Sn, Sb, and W concentrations above crops from normal soil, and As exceeded the statutory limit for foodstuffs.

The practical value of INAA for the determination of 32 elements in soil beneath a landfill (refuse) site has been described by Van der Klugt et al. [32], and the heavy metals Ag, As, Co, Cr, and Zn were detected easily. The effect of leaching on downward migration of metals, and the potential for pollution of ground water, could be explored by an extension of this particular application.

In Canada, Kuja and Hutchinson [25] have used INAA in combination with atomic absorption to assess the toxicity to vegetation of mine wastes, with a view to the eventual selection of tolerant species of grasses and sedges. In this work, elements analyzed in soil and plant specimens included Al, As, Cd, Cu, Ni, Pb, Se, and Zn.

The pollution of urban soils in the Glasgow area of Scotland by Br from automobile exhaust was used by Farmer and Cross [29] as a qualitative indicator of Pb pollution from the same source, using INAA to determine ^{82}Br. The soil samples were irradiated for 24 hr at a flux of 6×10^{12} n cm^{-2} sec^{-1} in the Harwell reactor DIDO. Natural concentrations of Br in soil were enhanced up to fivefold by pollution in urban areas. In a similar study [126], analysis was made of surface dust and vegetation close to highways, and also vehicle exhaust; altogether, 14 elements were measured including Br, which again proved a useful indicator of contamination levels.

C. General Studies on the Terrestrial Environment

In recent years, greater attention has been given to the cycling of both major and trace elements in the natural environment. Inputs of major elements from the atmosphere to plants and to the ground are of interest to agriculturists and hydrologists, whereas the influence of resuspension of soil dust on the qualitative and quantitative composition of the atmosphere is the concern of atmospheric chemists. The provision of data on the multielement composition of soil, plant, water, and atmospheric samples by the application of INAA has proved particularly suitable in this research area, as shown by the following examples. In all cases the total element content of soil has been measured.

The multielement patterns of wet and dry (total) deposition and of grass (*Festuca*) and surface soil from southern England were compared [127]. In this case, simple ratios of 23 elements to Sc rather than Al or Fe were used for comparison, because plants show negligible uptake and translocation of Sc; the results are shown in Fig. 7. It follows that any increase in element:Sc ratios relative to either soil or total deposition will demonstrate uptake by the roots or leaves. In fact, the results showed the significant uptake of Ce, Cl, Mn, and Rb by grass, whereas for other elements analyzed, their ratios to Sc were similar in both total deposition and grass (Fig. 7). It was concluded that atmospheric deposition retained by the grass is a major factor influencing the composition of leaf tissues. Furthermore, it was shown that the ratios to Sc of As, Br, Cl, Co, Cr, Mg, Na, Ni, Sb, V, and Zn in the total deposition were greatly in excess of those in local soil, indicating industrial and maritime influences [127].

In Japan, Tanizaki and Nagatsuka [128] studied the distribution of 21 elements in lateritic soil, rainwater, and streamwater from the Ogasawara Islands by thermal NAA. The resuspension or scattering

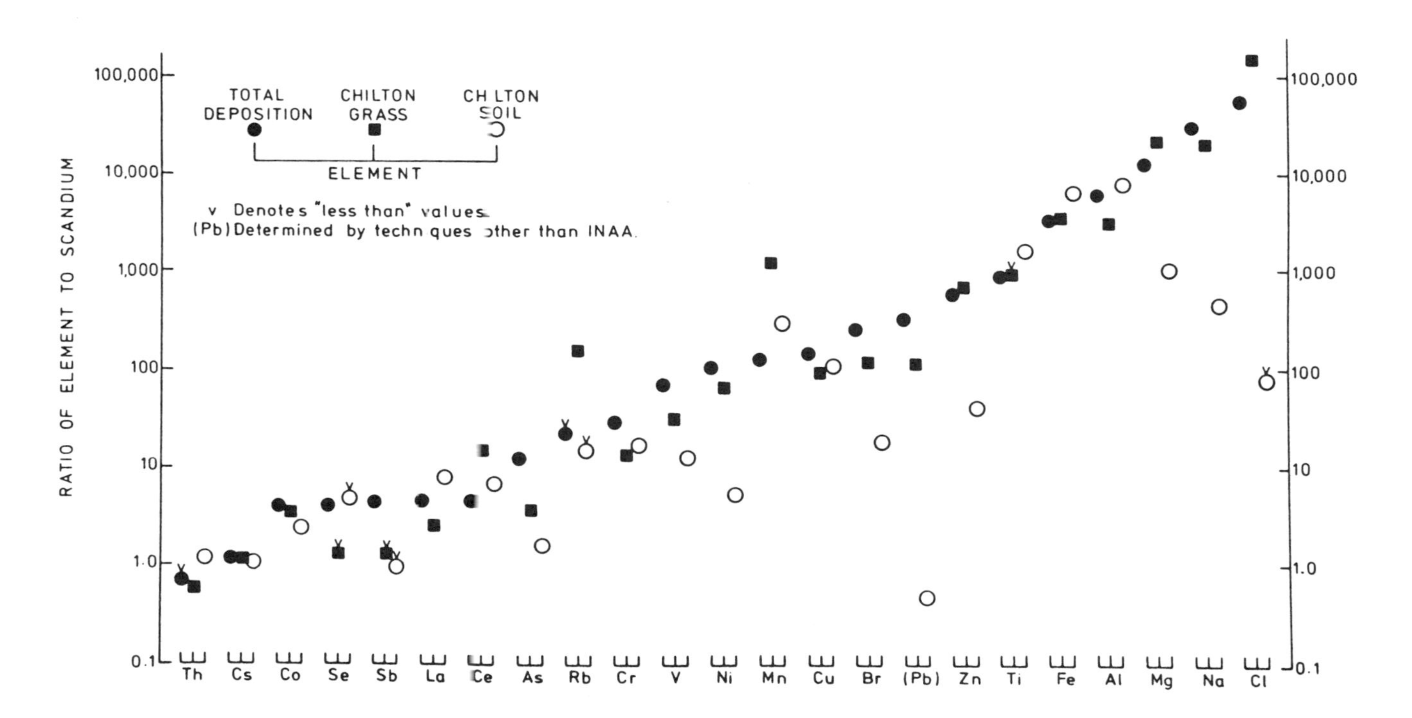

FIG. 7 Ratios of elements to scandium in total deposition, grass and soil at Chilton, Oxon, England. (From Ref. 127.)

of soil was believed responsible for the occurrence of Al, Co, Cr, Cs, Fe, Rb, Sc, and V in rainwater and streamwater; however, the concentrations of As, Sb, W, and Zn found appeared to originate not from lateritic soil but rather from an industrial source.

The distribution of Cl, Br, and I in Norwegian forest soils has been studied by INAA [129] as part of a wider research program on trace elements of geochemical and geomedical interest. However, the presence of large quantities of ^{56}Mn after the irradiation of soils with <80% organic matter content precluded the determination of I unless radiochemical separation was used. Yuita [130] applied INAA to the analysis of Br and Cl in soil solutions, plants, and rainwater; iodine could be included if chemical separation was carried out after radioactivation.

Studies on the inorganic composition of the atmosphere have often been extended to include analysis of local surface soil, a potential contributor to airborne particulate matter. The enrichment of air particulates relative to soil may be examined by double normalization to an element that is not used extensively in industry and is universally distributed in soil. Thus, an enrichment factor is derived, usually based on one of the normalizing elements Al, Ce, Fe, Sc, or Ti. For example, Gordon [131] reported the analysis of 28 surface, subsoils, and deep soils from Maryland as part of an extensive research project on aerosol composition, enrichment, and sources. Some 20 elements were measured by INAA.

The concentrations of rare earth elements (REE) in Sahara soils and aerosols have been determined by INAA to examine their relative abundances, normalized to chondrites [132]. Light rare earth elements (Ce, Eu, La, Nd, Pr, and Sm) are enriched in urban aerosols relative to the heavier REE (Dy, Er, Gd, Ho, Lu, Tb, Tm, and Yb), and such enrichment is found mainly on the smaller particles in the atmosphere [132]. Other studies include the investigation of the origin of arctic haze bands that occur in the troposphere over Alaska [133], and measurement of the chemical composition of aerosols in the Sudan Gezira [134], northern Canada [135], and northern Nigeria [136].

In a survey of trace and major elements in air particulate, rainwater, and dry deposition at seven nonurban sites in the United Kingdom [136], the analysis of the major soil type within a 5 km radius of each sampling station was made. Surface soils (to 80 mm depth) were analyzed by INAA for total concentrations of 16 elements, supplemented by other methods for Ni and Pb. It was found that at certain locations, the total deposition from the atmosphere of As, Br, Co, Cr, Na, Pb, and Zn was significant in relation to the total soil concentrations.

Particle size fractions in a surface profile of a buried loess soil from southeast England [138] were analyzed by INAA, using the

Harwell reactor DIDO for activation at a flux up to 10^{14} n cm^{-2} sec^{-1} [34]. Preferential concentration was found on particles <2 μm in diameter for the elements As, Ce, Cr, Co, Cs, Fe, La, Mn, Rb, Sc, Se, Th, and Zn. The reverse was true for Al, Na, and Ti, which were present at higher concentrations in the particle size range 2–50 μm diameter.

D. Forensic Science

The potential values of INAA in forensic investigations in which the multielement composition of materials must be known for comparison and identification purposes, and a nondestructive method of analysis is preferred, was realized some 30 years ago. In addition, few alternative techniques could analyze extremely small specimens obtained from the scene of a crime and accomplish this without dissolution of the sample, which may introduce trace elements as contamination from reagents. In a historic paper by Pro et al. [139], evidence was provided for the first time by activation analysis for use in the U.S. Federal Court. Their investigation was concerned with illicit distilleries and involved the analysis of whisky, soil, and paint. These materials were irradiated at 10^{12} n cm^{-2} sec^{-1}, followed by measurement of short- and long-lived γ-ray spectra with a NaI(Tl) detector. In eight soil samples, 13 isotopes were identified, including ^{198}Au, ^{131}Ba, ^{51}Cr, ^{239}Np, and ^{185}Os. Applications of INAA to forensic science have been reviewed by Bate et al. [140], Bryan and Guinn [141], Smythe [142], and Jervis [143]. Smythe [142] commented that the occurrence of trace elements in various drugs (plant-based) can often indicate the soils where they were grown, or the country of origin. Jervis [143] emphasized the improved resolution obtained by the Ge(Li) detector and gave examples from the analysis of hair, with comments on the extension of the technique to other media including soil.

In 1964, Bryan and Guinn [141] described developments in the United States of high flux neutron activation analysis, and differences in the elemental constituents of various media of forensic interest, including soil, were characterized. Furthermore, instrumental NAA was preferred to methods using chemical separation prior to γ-ray spectrometry on grounds of speed, preservation of specimens that might be required in court, and the ease of the graphical presentation of results in legal proceedings.

In a comparison of forensic soil specimens by INAA [33,144], samples from Georgia, Texas, and Washington were irradiated in a thermal neutron flux of about 7×10^{12} n cm^{-2} sec^{-1}, for measurement of Ba, Ce, Cl, Cr, Cs, Hf, La, Na, Sc, and Sm using a NaI(Tl) detector. Additional elements were measured by atomic absorption, and with combined results for some 20 elements there was no difficulty in establishing the origin of the soils.

On the other hand, Kline and Brar [145] doubted the capability of INAA with NaI(Tl) detectors and 400-channel analyzers to produce spectra that would characterize soils without other supporting evidence. Their conclusions were reached after the analysis of 12 elements in soils from a worldwide collection. Samples were irradiated at a thermal flux of 2×10^{13} n cm^{-2} sec^{-1}. Short-lived isotopes ($T_{1/2} < 15$ hr) could not be measured owing to interference from ^{24}Na. The authors recognized that the use of low flux or Cd-shielded irradiations and higher-resolution detectors for γ-ray spectrometry would improve the capability of INAA. Bate [146] discussed problems of comparison of soil analytical data for forensic work and reported large variations in the content of some major and trace elements according to soil depth, with the exception of Co, La, Se, and Th.

In conclusion, it seems essential that a wide multielement spectrum is required before individual soil specimens can be characterized. It also appears desirable that results for more heavy metals are obtained for forensic work, since these elements are expected to be more variable in soils than the rare earths, Fe and Sc, and may be able to characterize particularly well urban locations.

E. Miscellaneous Studies

Further applications of INAA to soil research have been made in several areas including pedology, archaeology, and the measurement of tracer isotopes added to soil.

1. *Pedological Studies*

Garrec et al. [147] described the determination of Al, Fe, and Si in some French soils by INAA using 14 MeV neutrons. For the determination of Al, the reaction $^{27}Al(n,\alpha)^{24}Na$ was used, taking account of interference by $^{24}Mg(n,p)^{24}Na$. Silicon was determined by the production of ^{28}Al, and Fe by the induced activity of ^{56}Mn, both from (n,p) reactions. The results compared well with data obtained by activation of soils with thermal neutrons and separation of isotopes by ion-exchange resins.

The total concentrations of La and Sc in three soils from western Wisconsin were measured by INAA, for research on the effect of pedogenic processes on their translocation in depth profiles and on the relationships of the two elements with clay, organic C and Fe content [98]. The authors concluded that since the redistribution of Sc in soil development processes is less complex than that of La, it may be a useful marker for research on both soil morphogenesis and soil correlation.

In many parts of the world, soils have been derived from deposits of volcanic ash, and their identification and physico-chemical properties are of interest to pedologists, agriculturists, and archaeologists. Thus, the determination of trace elements in volcanic ash soils, pumice, and glassy fractions of ash has received attention. Borchardt et al. [148] applied INAA to measure 19 elements in samples from the Cascade Mountains area of the United States, using a flux of 7×10^{11} n cm^{-2} sec^{-1} and a 30 cm^3 Ge(Li) detector. The elements Ce, Fe, Hf, Na, Sc, Sm, and Th were determined with <5% relative standard deviation. In a subsequent paper, Borchardt and Harward [35] established trace element correlations of volcanic ash soils from data obtained by INAA: The composition of volcanic glass from C horizons agreed with known volcanic ejecta from Mt. Mazama (Crater Lake, Oregon), and 15 stable elemental ratios unique to Mazama ash, e.g., Sc/Fe, La/Hf, Ce/Sc, Cs/Yb, were used to assist in the correlation of weathered soil horizons.

Iridium contents of soils were measured in the range 0.27–0.64 $\mu g\ g^{-1}$ from the reaction $^{191}Ir(n,\gamma)^{192}Ir$ [149] for geochemical studies. A relatively short irradiation time of 30 hr at a flux of 5×10^{12} n cm^{-2} sec^{-1} was chosen to avoid the excessive activation of major elements. Gamma-ray spectrometry was performed with a NaI(Tl) detector after 2–3 weeks cooling time. Interest in the exploitation of Brazilian lateritic deposits for metals prompted a comparison of INAA with XRFS for the analysis of Ba, Ce, La, and Nd [150]. Although INAA provided additional data on other rare earths, the long cooling period of 1–5 weeks after 7 hr irradiation at 1.1×10^{12} n cm^{-2} sec^{-1} was a disadvantage for routine analysis and the sensitivity for Ba was inadequate.

For research on the geochemical behavior of Ti, the analysis of soil humic ash samples was achieved by INAA using 14 MeV neutrons [151]. An irradiation time of 20 hr was followed by 42 hr cooling before γ-ray spectrometry. The reaction $^{47}Ti(n,p)^{47}Sc$ was used for quantitative analysis, and spectra were obtained with a Ge(Li) detector. The results showed close agreement with a colorimetric method, concentrations of 7 mg Ti g^{-1} ash being present.

2. *Archeological Research*

INAA of clays and pottery from locations throughout the world has shown that the measurement of Co, Cs, Hf, Sc, Ta, and Th could provide satisfactory clues on the origins of articles [152]. Altogether, 15 isotopes were measured with a 30 cm^3 Ge(Li) detector coupled to a 4096-channel pulse-height analyzer, and the grouping of the analytical data was studied statistically by cluster analysis.

Clay and shale from 20 sites in Egypt, and ancient pottery, were analyzed for 16 major and trace elements by instrumental thermal neutron activation [153]. Fluxes of $1–4 \times 10^{14}$ cm^{-2} sec^{-1} were used for the irradiation of these samples, and after appropriate cooling times the spectra of short- and long-lived activities were measured with a 35 cm^3 Ge(Li) detector. Differences were particularly marked in the concentrations of Ce, Co, Fe, Hf, Mn, Na, Rb, Ta, and Th.

Kerr [37] applied INAA to research at archaeological sites on the detection of buried soils, particularly the location of buried surfaces, by measuring variations in the concentration of elements through the depth profile. Soil samples were obtained from Hadrian's Wall, near Carlisle, England, and were characterized by their relative levels of ^{28}Al, ^{29}Al, ^{139}Ba, ^{165}Dy, ^{42}K, ^{27}Mg, ^{56}Mn, ^{24}Na, ^{51}Ti, and ^{52}V following neutron activation.

3. *Tracer Studies*

The use of stable isotopes as tracers of transfer mechanisms in the terrestrial environment may be preferred to radioactive species, to avoid problems of safety in certain situations. In this connection, it has been demonstrated that the behavior of I in soils can be studied by means of a stable ^{127}I tracer, with subsequent analysis by neutron activation at a flux of 2×10^{11} n cm^{-2} sec^{-1} to yield ^{128}I ($T_{1/2}$ = 25 min) [154]. Benes et al. [154] also analyzed five of the most important soil types in Czechoslovakia by INAA and established that increasing Mn and Al contents reduced the sensitivity of ^{128}I measurements.

The kinetics of Cs transport in a lake ecosystem was determined by the introduction of $^{133}CsCl$ as a tracer, and the concentrations in sediment and other media were measured by neutron activation to ^{134}Cs; absolute quantities of 10^{-8} g of ^{133}Cs were detectable [155]. This technique is promising for research on the behavior of ^{137}Cs fallout from atmospheric tests of nuclear weapons.

VI. CONCLUDING REMARKS

The value of INAA has been recognized and applied in a wide range of scientific disciplines that require soil analysis, including agriculture, research on natural and polluted environments, forensic science, and archaeology. Recent progress in the techniques of INAA resulting from improvements in general instrumentation and computer technology have permitted an extension of the range of elements that can be determined on a quantitative basis and have allowed increases in sensitivity.

According to the published work throughout the world, it appears that research on environmental pollution has made greater use of modern techniques in INAA than other fields such as pedological investigations and research into problems of soil/plant and soil/plant/

animal interrelationships. No doubt the remoteness of many agricultual research laboratories from reactor facilities and the complexity and cost of equipment generally required for INAA have been largely responsible. However, with an increase in the nuclear power program and demands for research on the fate of artificial isotopes in the terrestrial and marine environments, the availability of equipment for γ-ray spectrometry and data processing is also likely to increase. Such equipment can have a dual role, i.e., the analysis of stable elements by INAA, given access to reactor facilities.

The application of INAA to problems of plant and animal toxicology has been relatively slow, and there are few integrated studies concerned with the analysis of soils, plants and animal tissues from specific areas unless there has been massive industrial pollution. Examples of the latter type of study are discussed in Sec. V.B.

The interrelationships of trace elements that are essential in plant and animal nutrition are such that an excess of one element may induce a deficiency of another (e.g., Mn/Fe, Cu/Mo). In this context, the comparison of multielement spectra obtained by INAA of soil and tissues would assist diagnosis. The deliberate addition of trace element fertilizers to land may also create problems of imbalance unless care is taken to avoid excessive use. Furthermore, modern methods of INAA could be applied usefully to the analysis of soils and vegetation from regions where problems with livestock have been encountered, to detect the presence of As, Cu, Mo, Se, and Zn in particular.

ACKNOWLEDGMENTS

The authors wish to express their gratitude to E. M. R. Fisher for providing much experimental detail and most of the analytical data published in this chapter, and to M. Paton for considerable help in literature searches.

REFERENCES

1. D. Gibbons and D. A. Lambie, in *Comprehensive Analytical Chemistry*, Vol. IIc (C. L. Wilson and D. W. Wilson, eds.), Elsevier, Amsterdam.
2. D. De Soete, R. Gijbels, and J. Hoste, *Neutron Activation Analysis*, Wiley-Interscience, New York, 1972.
3. P. Kruger, *Principles of Activation Analysis*, Wiley-Interscience, New York, 1971.
4. R. Dams, F. De Corte, J. Hertogen, J. Hoste, W. Maenhaut, and F. Adams, in *Analytical Chemistry*, Part I (T. S. West,

ed.), Physical Chemistry Series II, Vol. 12, Butterworth, London, 1976, p. 1.
5. R. Cornelis, J. Hoste, A. Speecke,C. Vandecasteele, J. Versieck, and R. Gijbels, in *Analytical Chemistry*, Part I (T. S. West, ed.), Physical Chemistry Series II, Vol. 12, Butterworth, London, 1976, p. 71.
6. Proceedings of the 7th International Conference on Modern Trends in Activation Analysis, Cophenhagen, June 23–27, 1986 (also earlier meetings at College Station, Texas, 1962 and 1965, at Gaithersburg, Md., 1968, Paris, 1973, Munich, 1976, and Toronto, 1981).
7. *Nuclear Activation in the Life Sciences*, Proceedings of the International Conference Vienna, 1978, IAEA, Vienna, 1979 (also earlier meetings at Amsterdam, 1965, and Bled, 1972).
8. J. Hislop, *Proc. Anal. Div. Chem. Soc.*, *15*:193 (1978).
9. A. Chattopadhyay and R. E. Jervis, *Anal. Chem.*, *46*:1630 (1974).
10. H. J. M. Bowen and D. Gibbons, *Radioactivation Analysis*, Oxford Univ. Press, Oxford, 1963, p. 8.
11. G. Erdtmann, in *Kernchemie in Einzeldorstellungen* (K. H. Lieser, ed.), Weinhem, New York, 1976.
12. L. Moens, F. De Corte, A. De Wispelaere, J. Hoste, A. Simonitis, A. Elek, and J. Szabo, *J. Radioanal. Nucl. Chem.*, *82*:385 (1985).
13. L. Moens, J. De Donder, L. Xi-lei, F. De Corte, A. De Wispelaere, A. Simonitis, and J. Hoste, *Nucl. Instr. Meth.*, *187*:451 (1981).
14. F. De Corte, A. Simonitis, A. De Wispeleare, and J. Hoste, in *Proceedings of the 7th International Conference on Modern Trends in Activation Analysis*, Copenhagen, 1986, p. 581.
15. L. Zikovsky and E. A. Schweikert, *Nucl. Instr. Meth.*, *155*:279 (1978).
16. V. Guinn, unpublished work privately communicated, 1978.
17. H. S. Hsia and V. P. Guinn, in *Proceedings of the 7th International Conference on Modern Trends in Activation Analysis*, Copenhagen, 1986, p. 181.
18. D. V. Law, unpublished work, AERE Harwell, 1979.
19. Anonymous, NCRP Rept. 58, *Handbook of Radioactivity Measurement Procedures*, 2nd ed., NCRP, Bethesda, Md., 1985.
20. G. F. Knoll, *Radiation Detection and Measurements*, Wiley, New York, 1979.
21. O. Klein and Y. Nishina, *Z. Physik.*, *52*:823 (1929).
22. R. L. Heath, USAEC Rept. IDO 16880, 1964.
23. B. J. Snyder, *Nucl. Instr. Meth.*, *46*:173 (1967).
24. D. H. F. Atkins, E. M. R. Fisher, and L. Salmon, UKAEA Rept. AERE-R6734, HMSO, London, 1972.

25. M. Vassilaki, L. Salmon, and J. A. B. Gibson, *Geochim. Cosmochim. Acta, 30*:601 (1966).
26. J. Rosenberg and H. B. Wiik, *Radiochem. Radioanal. Lett.*, *6*:45 (1971).
27. R. Gijbels and A. Govaerts, *J. Radioanal. Chem.*, *16*:7 (1972).
28. R. C. Ragaini, H. R. Ralston, and N. Roberts, *Environ. Sci. Tech.*, *11*:773 (1977).
29. J. G. Farmer and J. D. Cross, *Water Air Soil Pollut.*, *9*:193 (1978).
30. H. J. M. Bowen and P. A. Cawse, in *Activation Analysis* (J. M. A. Lenihan and S. J. Thomson, eds.), Academic Press, New York, 1965.
31. D. McKown, M. Kay, D. Gray, A. Abu-Samra, E. Eichor, and J. Vogt, in *Nuclear Methods in Environmental Research* (J. R. Vogt, T. F. Parkinson, and R. L. Carter, eds.), Univ. Missouri, Columbia, 1971, p. 150.
32. N. van der Klugt, P. Poelstra, and E. Zwemmer, *J. Radionanal. Chem.*, *35*:109 (1977).
33. R. L. Brunelle, C. F. Hoffman, K. B. Snow, and M. J. Pro, *J. Assoc. Off. Agric. Chem.*, *52*:911 (1969).
34. L. Salmon and P. A. Cawse, unpublished work, 1978.
35. G. A. Borchardt and M. E. Harward, *Soil Sci. Soc. Am. Proc.*, *35*:626 (1971).
36. J. J. O'Toole, R. G. Clark, K. L. Malaby, and D. L. Tranger, in *Nuclear Methods in Environmental Research* (J. R. Vogt, T. F. Parkinson, and R. L. Carter, eds.), Univ. Missouri, Columbia, 1971, p. 172.
37. S. A. Kerr, Ph.D. thesis, Part II, Univ. Surrey, England, 1978.
38. H. Yule and V. Guinn, in *Radiochemical Methods of Analysis*, Vol. 2, Proceedings of the Symposium, Salzburg, 1964, IAEA, Vienna, 1965, p. 111
39. R. E. Jervis, R. G. V. Hancock, D. E. Hill, and K. Isles, *J. Radioanal. Chem.*, *37*:463 (1977).
40. A. K. De and W. W. Meinke, *Anal. Chem.*, *30*:1474 (1958).
41. Anonymous, *Radiation Sources 1977/78*, Radiochemical Centre, Amersham, Buckinghamshire, England, 1977, p. 60.
42. D. W. Downton and J. D. L. H. Wood, in *Proceedings of the 1968 Conference on Modern Trends in Activation Analysis* (J. R. DeVoe and P. D. LaFleur, eds.), NBS Special Pub. 312, National Bureau Standards Washington, D.C., 1968, p. 1059.
43. J. R. Keane, Direct Information 1/65, Karlsruhe, 1965.
44. H. P. Yule, in *Proceedings of the 1968 Conference on Modern Trends in Activation Analysis* (J. R. DeVoe and P. D. LaFleur, eds.), NBS Special Pub. 312, National Bureau of Standards, Washington, D.C., 1968, p. 881.

45. L. Salmon, in *Radiochemical Methods of Analysis*, Vol. 2, Proceedings of the Symposium, Salzburg, 1964, IAEA, Vienna, 1965, p. 125.
46. L. Salmon and D. V. Booker, UKAEA Rept., AERE-R8870, HMSO, London, 1979.
47. V. Barnes, UKAEA Rept. PG-834(W), HMSO, London, 1968.
48. G. W. Phillips and K. W. Marlow, NRL Memo. Rept. 3198, Washington, D.C., 1976.
49. R. Gunnick and J. B. Niday, *Computerized Quantitative Analysis by Gamma-Ray Spectrometry*, Vols. II and III, Lawrence Livermore Lab., Univ. Calif. Rept. UCRL 51061, 1971.
50. J. T. Routti and P. G. Prussin, *Nucl. Instr. Meth.*, *72*:125 (1969).
51. L. Salmon and M. M. Davies, in *Proceedings of the 5th International Conference on Nuclear Methods in Environmental and Energy Research* (J. R. Vogt, ed.), Puerto Rico, 1984, p. 72.
52. A. Robertson and T. J. Kennet, *Nucl. Instr. Meth.*, *100*:317 (1972).
53. M. A. Mariscotti, *Nucl. Instr. Meth.*, *50*:309 (1967).
54. H. P. Yule, *Anal. Chem.*, *40*:1480 (1968).
55. H. P. Blok, J. L. DeLange, and J. W. Schottmann, *Nucl. Instr. Meth.*, *50*:309 (1967).
56. J. P. Op de Beeck and J. Hoste, in *Analytical Chemistry*, Part 1 (T. S. West, ed.), Physical Chemistry Series II, Vol. 12, Butterworth, London, 1976, p. 151.
57. G. W. Phillips and K. W. Marlow, *Nucl. Instr. Meth.*, *137*: 525 (1976).
58. D. V. Booker, *J. Radioanal. Chem.*, *48*:83 (1979).
59. Anonymous, *SCORPIO/SPECTRAN Version 2 User Manual*, CI-SE-905, Canberra Industries Inc., Meriden, Conn., 1977.
60. Anonymous, *System ND6600 Operation Instruction Summary*, Nuclear Data Inc., Schaumberg, Ill., 1978.
61. C. M. Lederer, J. M. Hollander, and I. Perlman, *Table of Isotopes*, 6th ed., Wiley, New York, 1967.
62. G. Erdtmann and W. Soyka, *Nucl. Instr. Meth.*, *121*:197 (1974).
63. F. Adams and R. Dams, *J. Radioanal. Chem.*, *3*:100 (1969).
64. W. Carder, J. D. Macmahon, and A. Egan, *Talanta, 25*:21 (1978).
65. R. Gunninck and J. B. Niday, *Computerized Quantitative Analysis by Gamma-Ray Spectrometry*, Vol. I, Lawrence Radiation Lab., Univ. Calif. Rept. UCRL-51061, 1972.
66. G. C. Christensen, M. J. Koskelo, and J. T. Routti, CERN, Geneva Rept. HS-RP/015, 1977.
67. P. Quittner, *Nucl. Instr. Meth.*, *76*:115 (1969).
68. D. F. Covell, *Anal. Chem.*, *31*:1785 (1959).
69. P. A. Baedecker, *Anal. Chem.*, *43*:405 (1971).

70. L. A. McNelles and J. L. Campbell, *Nucl. Instr. Meth.*, *127*: 73 (1975).
71. L. Salmon, *Nucl. Instr. Meth.*, *14*:193 (1961).
72. L. Salmon, in *Application of Computers to Nuclear and Radiochemistry* (G. D. O'Kelley, ed.), Proceedings of the Symposium, Gatlinburg, Tenn., 1962, USAEC Rept. NAS-NS-3107, 1963, p. 165.
73. D. Gibbons, L. E. Fite, and R. E. Wainerdi, in *Proceedings of the 5th Conference on Nuclear Reactor Technology*, Gatlinburg, Tenn., USAEC Rept. TID-7629, 1961.
74. C. D. Carter and D. V. Booker, UKAEA Rept. AERE-PR/EMS1, HMSO, London, 1974.
75. C. D. Carter, AERE, Harwell, unpublished document, 1974.
76. L. A. Currie, *Anal. Chem.*, *40*:586 (1968).
77. P. A. Cawse, AERE Harwell Rept. R8191, HMSO, London, 1976.
78. F. De Corte and J. Hoste, in *Proceedings of the 5th International Conference on Nuclear Methods in Environmental and Energy Research* (J. R. Vogt, ed.), Puerto Rico, 1984, p. 2.
79. J. H. McNeilly and E. Eichler, Tech. Rept. CWLR 2299, U.S. Army Chemical Center, Md., 1959.
80. P. Thoresen, *Acta Chem. Scand.*, *18*:1054 (1964).
81. J. R. Kline, Ph.D. thesis, Univ. Minnesota, Minneapolis, 1964.
82. A. K. Khera and E. Steinnes, *Agrichim.*, *13*:524 (1969).
83. L. Salmon and M. G. Creevy, in *Nuclear Techniques in Environmental Pollution*, IAEA, Vienna, 1971.
84. M. Naidenov and A. Travesi, *Soil Sci.*, *124*:152 (1977).
85. R. P. Murmann, R. W. Winters, and T. G. Martin, *Soil Sci. Soc. Am. Proc.*, *35*:647 (1971).
86. S. Aksoyoglu, H. Göktürk, and H. N. Erten, *J. Radioanal. Nucl. Chem.*, *104*:97 (1986).
87. H. D. Buenafama, *J. Radioanal. Chem.*, *18*:111 (1973).
88. J. Sun and R. E. Jervis, in *Proceedings of the 5th International Conference on Nuclear Methods in Environmental and Energy Research* (J. R. Vogt, ed.), Puerto Rico, 1984, p. 2.
89. S. M. Rahman, *Nucl. Sci. Applications*, *6*:42 (1972).
90. M. Geisler and H. Schelhorn, *J. Radioanal. Nucl. Chem. Lett.*, *96*:567 (1985).
91. W. Zmijewska and J. Minczewski, in *Plant Nutrient Supply and Movement*, Tech. Rept. Series No. 48, IAEA, Vienna, 1965.
92. H. J. M. Bowen and P. A. Cawse, *Soil Sci.*, *98*:358 (1964).
93. U. K. Misra, E. R. Graham, W. J. Upchurch, and D. M. McKown, *Soil Sci. Soc. Am. Proc.*, *37*:193 (1973).
94. R. L. Mitchell, in *Trace Elements in Soils and Crops*, MAFF Tech. Bull. 21, HMSO, London, 1971.
95. B. W. Avery, *J. Soil Sci.*, *24*:324 (1973).

96. A. K. Khera, E. Steinnes, and A. Øien, *Acta Agric. Scand.*, *20*:33 (1970).
97. A. K. Khera and E. Steinnes, *Geoderma*, *5*:251 (1971).
98. J. R. Kline, J. E. Foss, and S. S. Brar, *Soil Sci. Soc. Am. Proc.*, *33*:287 (1969).
99. G. A. Borchardt, Ph.D. thesis, Oregon State Univ., Corvallis, Oregon, 1970.
100. H. Jaffrezic, A. Decarreau, J. P. Carbonnel, and H. Deschamps, *J. Radioanal. Chem.*, *18*:49 (1973).
101. H. J. M. Bowen, in *Nuclear Activation Techniques in the Life Sciences*, IAEA, Vienna, 1967.
102. G. W. Leddicotte, *Isotope Radiation Tech.*, *5*:200 (1968).
103. O. Johansen and E. Steinnes, *Acta Agric. Scand.*, *22*:104 (1972).
104. G. Van Compernolle, M. Krivanek, C. Masozera, and F. R. Lumu, in *Peaceful Uses of Atomic Energy in Africa*, IAEA, Vienna, 1970.
105. M. L. A. Nilubol, P. Sington, and D. Chamnirokasarnt, *Soil Analysis by Neutron Activation Method*, Thailand Atomic Energy Commission Rept. AEC-36, Bangkok, 1970.
106. B. A. Srapenyants and I. B. Saveliev, *J. Radioanal. Chem.*, *38*:247 (1977).
107. R. Rustamov, Sh. Khatamov, I. I. Orestova, and A. A. Kist, *Atomnaya Energiya*, *34*:476 (1973).
108. K. M. Awadallah, M. K. Sherif, A. E. Mohamed, and F. Grass, *J. Radioanal. Nucl. Chem.*, *98*:49 (1986).
109. K. M. Awadallah, M. K. Sherif, A. H. Amrallah, and F. Grass, *J. Radioanal. Nucl. Chem.*, *98*:235 (1986).
110. J. N. Weaver, A. Hanson, J. McGaughey, and F. J. Steinkruger, *Water Air Soil Pollut.*, *3*:327 (1974).
111. M. Shibuya and Y. Kamemoto, *Mizushori Gijutsu*, *4*:17 (1963).
112. A. Suess and H. Staerk, *J. Radioanal. Chem.*, *37*:905 (1977).
113. Y. Talmi and C. Feldman, in *Arsenical Pesticides* (E. A. Woolson, ed.), Am. Chem. Soc. Symp. Series 7, ACS, Washington, D.C., 1975.
114. H. Gounchev and T. Dimchev, *Pochvozn. Agrokhim.*, *8*:77 (1973).
115. W. S. Lyon, *Trace Element Measurements at the Coal Fired Steam Plant*, CRC Press, Cleveland, Ohio, 1977.
116. J. H. Horton, R. S. Dorsett, and R. E. Cooper, Savannah River Lab. Rept. DP-1475, Aiken, South Carolina, 1977.
117. J. Kucera and L. Soukal, *7th Czechoslovak Spectroscopic Conference Abstracts*, Vol. 3, Sec. 5, 1984, p. 170.
118. E. S. Gladney, L. E. Wangen, and R. D. Aguilar, *Anal. Lett.*, *10*:1083 (1977).

119. E. A. Crecelius, C. J. Johnson, and G. C. Hofer, *Water Air Soil Pollution, 3*:337 (1974).
120. R. E. Jervis and B. Tiefenbach, in *Nuclear Activation Techniques in the Life Sciences*, IAEA, Vienna, 1978.
121. R. E. Jervis, J. J. Paciga, and A. Chattopadhyay, *Trans. Am. Nucl. Soc., 21*:95 (1975).
122. N. J. Pattenden, AERE Harwell Rept. R7729, HMSO, London, 1974.
123. H. J. M. Bowen, *Trace Elements in Biochemistry*, Academic Press, New York, 1966.
124. P. A. Cawse, I. S. Jones, and G. W. Cox, AERE Harwell Rept. R8103, HMSO, London, 1975.
125. A. J. Kuja and T. C. Hutchinson, in *Annual Report Slowpoke Nuclear Reactor* (W. Paul and R. G. V. Hancock, eds.), Univ. Toronto, Canada, 1978.
126. T. W. Oakes, A. K. Furr, D. J. Adair, and T. F. Parkinson, *J. Radioanal. Chem., 37*:881 (1977).
127. P. A. Cawse, in *Inorganic Pollution and Agriculture*, Proc. ADAS Conf., London, 1977, HMSO, London, 1980.
128. Y. Tanizaki and S. Nagatsuka, *Nippon Kagaka Naishi*, 5:667 (1977).
129. J. Lag and E. Steinnes, in *Isotopes and Radiation in Soil-Plant Relationships Including Forestry*, IAEA, Vienna, 1972.
130. K. Yuita, Nogyo Gigutsu Kenkyusho Hokoku, Part B35, 1983, p. 35.
131. G. E. Gordon (project director), *Atmospheric Impact of Major Sources and Consumers of Energy*, Progress Rept. 1975, Univ. Maryland Dept. Chemistry, College Park, Md.
132. K. A. Rahn, *The Chemical Composition of the Atmospheric Aerosol*, Tech. Rept. Univ. Rhode Island Dept. Oceanography, Kingston, R.I., 1976.
133. K. A. Rahn, R. D. Borys, and G. E. Shaw, *Nature (Lond.)*, *268*:713 (1977).
134. S. A. Penkett, D. H. F. Atkins, and M. H. Unsworth, *Tellus, 31*:295 (1979).
135. W. I. Gizyn and T. C. Hutchinson, in *Annual Report Slowpoke Nuclear Reactor* (W. Paul and R. G. V. Hancock, eds.), Univ. Toronto, Canada, 1978.
136. F. Beavington and P. A. Cawse, *Sci. Tot. Environ., 10*:239 (1978).
137. P. A. Cawse, AERE Harwell Rept. R7669, HMSO, London, 1974.
138. A. H. Weir, J. A. Catt, and P. A. Madgett, *Geoderma*, 5:131 (1971).
139. M. Pro, H. Schlesinger, and M. Cohan, *J. Assoc. Off. Agric. Chem., 48*:459 (1956).

140. L. C. Bate, J. F. Emery, G. W. Ledicotte, W. S. Lyon, and M. J. Pro, *Int. J. Appl. Radiation Isotopes, 14*:549 (1963).
141. D. E. Bryan and V. P. Guinn, *Application of Neutron Activation Analysis in Scientific Crime Detection*, USAEC Rept. GA-5556, San Diego, Calif., 1964.
142. L. E. Smythe, *Atomic Energy Australia, 9*:2 (1966).
143. R. E. Jervis, *Radiation Tech., 6*:57 (1968).
144. C. M. Hoffman, R. L. Brunelle, K. B. Snow, and M. J. Pro, in *Proceedings of the 1968 Conference on Modern Trends in Activation Analysis*, Vol. 1 (J. R. DeVoe and P. D. LaFleur, eds.), National Bureau Standards, Special Pub. 312, Washington, D.C., 1968.
145. J. R. Kline and S. S. Brar, *Soil Sci. Soc. Am. Proc., 33*:234 (1969).
146. L. C. Bate, *J. Radioanal. Chem., 15*:193 (1973).
147. J. P. Garrec, A. Fer, and A. Fourcy, *C. R. Acad. Sci. Paris, 268*:3021 (1969).
148. G. A. Borchardt, M. E. Harward, and R. A. Schmitt, *Quatern. Res., 1*:247 (1971).
149. G. Stefanov and L. Daieva, *Isotopenpraxis, 4*:146 (1972).
150. J. J. LaBrecque, J. M. Beusen, and R. E. Van Greiken, *X-Ray Spectr., 15*:13 (1986).
151. P. Bornemisza-Pauspertl and M. Szilagyi, *Radiochem. Radioanal. Lett., 12*:271 (1972).
152. O. Birgul, M. Diksic, and L. Yaffe, *J. Radioanal. Chem., 39*:45 (1977).
153. S. K. Tobia and E. V. Sayre, in *Recent Advances in Science and Technology of Materials*, Vol. 3 (A. Bishay, ed.), Plenum Press, New York, 1974.
154. J. Benes, J. Frana, and A. Mastalka, *Collect. Czech. Chem. Commun., 39*:2783 (1974).
155. T. E. Hakonson and F. W. Whicker, *Health Phys., 28*:699 (1975).

10

Analysis of Nitrogen, Carbon, and Oxygen Isotope Ratios by Optical Emission Spectrometry

VICTOR MIDDELBOE* and HENRIK SAABY JOHANSEN *Royal Veterinary and Agricultural University, Copenhagen, Denmark*

I. INTRODUCTION

The ideal tracer for nitrogen, carbon, or oxygen would, of course, be an appropriate radioisotope of the element in question. Nitrogen-13 (^{13}N), with a half-life of only 10 min, has been used in isolated instances in experiments related to soil (see, e.g., Gersberg et al. [1]). Apart from ^{14}C, the longest lived radioisotopes of carbon and oxygen have half-lives of the same order of magnitude as that of ^{13}N. Thus, *stable* isotopes of the three elements mentioned and/or carbon-14 are indispensable in N-, C-, or O-tracer experiments lasting more than an hour or two.

All naturally occurring nitrogen consists of two stable isotopes (^{14}N, ^{15}N), whereas natural carbon consists of two stable and one long-lived radioactive isotope (^{12}C, ^{13}C, and ^{14}C), and natural oxygen consists of three stable isotopes (^{16}O, ^{17}O, ^{18}O). The isotopic abundance ratios in nature are very nearly constant. In principle, therefore, nitrogen, carbon, or oxygen having any significantly unnatural mixture of isotopes can be used as a tracer for the element considered. Furthermore, in the case of carbon the supplementary use of ^{14}C renders double-tracer experiments possible.

---

*Professor emeritus.

In tracer experimentation employing stable isotopes, the unit "percentage isotope abundance (mol %)" is the recognized standard. In the case of stable nitrogen, which is comprised of two isotopes only, the relationships between isotope abundance and isotopic ratio are as follows:

$$a_h = \frac{100}{r^{-1} + 1} \text{ mol \%} \tag{1}$$

$$a_l = \frac{100}{r + 1} \text{ mol \%} \tag{2}$$

where

a_h = percentage abundance of heavy isotope (^{15}N in sample nitrogen)
a_l = percentage abundance of light isotope (^{14}N in sample nitrogen)
r = heavy/light isotopic ratio ($^{15}N/^{14}N$ ratio in sample nitrogen)

Similar relationships hold for $^{13}C/^{12}C$ and $^{18}O/^{16}O$ provided due allowance is made for ^{14}C and ^{17}O, respectively.

In the case of nitrogen, the *natural* heavy/light isotopic ratio is 1:272, which is equivalent to about four nitrogen-15 atoms in every 1000 nitrogen atoms. Consequently, a supply of nitrogen containing, say, 10 nitrogen-15 atoms per 1000 nitrogen atoms (i.e., 1.0 mol % ^{15}N) would be considered "labeled." The practical use of nitrogen labeled in this way depends on the accuracy with which small differences in $^{15}N/^{14}N$ ratios can be determined. For $^{13}C/^{12}C$ or $^{18}O/^{16}O$ the situation is quite similar. Up to now, the commonly used methods of analysis have been (1) optical emission spectrometry (OES) and (2) mass spectrometry (MS); see Chap. 11. The subject of the present chapter is the measurement of $^{15}N/^{14}N$, $^{13}C/^{12}C$, and $^{18}O/^{16}O$ ratios in soil, plant, and other environmental samples using OES, including a discussion of the procedure for liberating N_2 or CO_2 from the sample, which is frequently the most difficult part of the complete analysis.

II. BASIC PRINCIPLES AND TECHNIQUES

A. Discovery of Nitrogen-15

The existence of the rare, stable isotope of nitrogen (^{15}N) was finally proven by Naudé in 1929 [2]. A tube containing nitric oxide was prepared and white light was transmitted through the tube parallel to the axis. In the absorption spectrum thereby obtained, isotopically shifted (i.e., wavelength displaced) bandheads of low intensity were observed at certain spectral positions, which coincided accurately with those that could be predicted by assuming the existence of nitrogen-15.

Isotopic shifts take place in molecular spectra, mainly because the vibrational energy of a molecule (in any given quantum state) is a function of the masses of its constituent atoms. In fact, to a first approximation, the vibrational energy is inversely proportional to the square root of the reduced mass. The reduced mass μ for a diatomic molecule is defined as follows:

$$\mu = \frac{m_1 m_2}{m_1 + m_2} \tag{3}$$

where m_1 and m_2 are the masses of the two constituent atoms.

In the case of the ultraviolet (UV) bands studied by Naudé, wavelength (λ) shifts of 0.1–0.2 nm occur when the ^{14}N atom in the NO molecule is replaced by a ^{15}N atom. The observed isotopic shifts correspond to about 10 times the resolving power of the spectrographic equipment used by Naudé. However, he was not able to measure the natural $^{14}N^{18}O/^{15}N^{16}O$ intensity ratio with any degree of accuracy, but this was accomplished two years later by Urey and Murphy [3]. The result of the latter determination, combined with Bleakney and Hipple's value for the natural $^{16}O/^{18}O$ ratio [4], led to a natural $^{14}N/^{15}N$ ratio of 275, which is within 1% of the presently accepted value [5].

B. Analysis of $^{15}N/^{14}N$ Ratios by OES

In 1930, several bands in the emission spectrum of the N_2 molecule were analyzed isotopically by Herzberg [6], whereby the existence of the newly discovered ^{15}N isotope was verified. However, the value obtained for the natural $^{15}N/^{14}N$ ratio proved, with the advent of determinations by MS, to be rather inaccurate. Nevertheless, 20 years later in 1950, the reasonably accurate semiroutine analyses of $^{15}N/^{14}N$ ratios in N_2, by means of OES, were reported by Hoch and Weisser [7]. Two hours were needed per optical analysis, due to the use of photographic emulsion for recording the spectra. The quantity of nitrogen required for one analysis was about 50 μg. A series of ^{15}N-abundances, ranging from natural (0.37 mol %) to 99.8 mol %, were analyzed in samples supplied in part by Clusius [8]. This was the beginning of the routine analysis of $^{15}N/^{14}N$ ratios in N_2 by OES. The next major steps forward were (1) the introduction of photoelectric light detection by Broida and Chapman [9] and (2) the introduction of the monochromator by Meier and Müller [10]. As a result of these improvements, the optical analysis time could be reduced to about 10 min, and the quantity of N_2 required for an analysis was only 5–10 μg (in a 2 cm^3 discharge tube).

The main components in an instrument designed for the analysis of $^{15}N/^{14}N$ ratios by OES are shown in Fig. 1. Power from a microwave generator (also called a high-frequency transmitter) excites N_2 molecules in the discharge container, resulting in the emission of light. The light-dispersive unit is usually a prism or line grating. The photomultiplier tube transforms light quanta into electronic signals, which subsequently are amplified and registered.

The analysis of $^{15}N/^{14}N$ ratios by OES is based on an orbital electron transition ($C^3\Pi$-$B^3\Pi$), which gives rise to a group of molecular bands belonging to the so-called 2nd positive N_2 system situated in the UV spectral region and, as can be seen in Fig. 2, most of the bands appear in triplicate. The three replicates, with

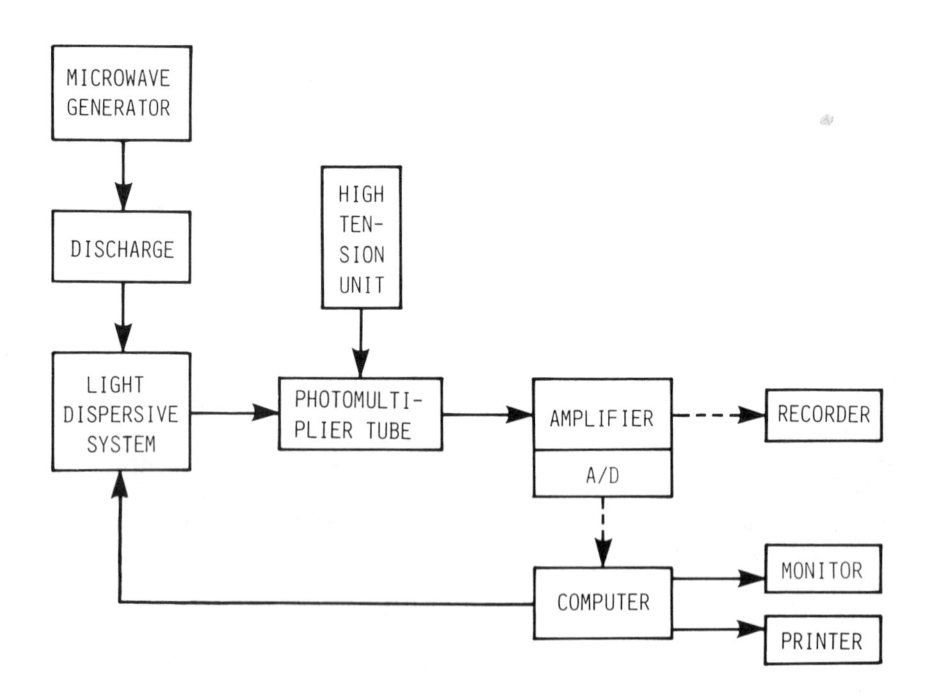

FIG. 1 Block diagram of an optical emission spectrometer and ancillary equipment used for the analysis of stable-isotope ratios. The broken arrows leading from "amplifier A/D" indicate alternatives.

diminishing intensities toward the 0 - 0 transition, correspond to emission by the three isotopic N_2 molecules $^{14}N^{14}N$, $^{14}N^{15}N$, and $^{15}N^{15}N$. These three isotopic molecules will be referred to as N_2-28, N_2-29, and N_2-30 (or when the molecular composition is obvious, simply 28, 29, and 30), according to the particular mass numbers involved. In Fig. 2, each pair of numerals, separated by a dash, indicates the unique set of vibrational quantum numbers (upper

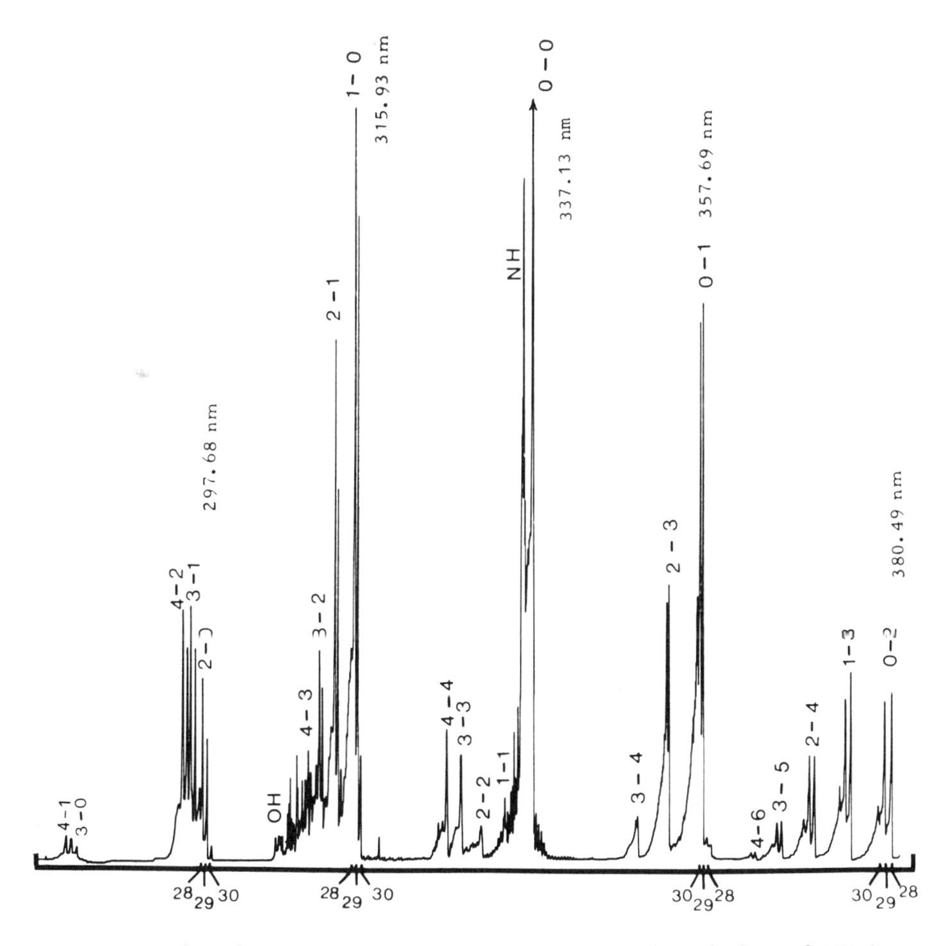

FIG. 2 $C^3\Pi$-$B^3\Pi$ spectrum of N_2 (obtained by dissociation of NH_3) containing 27 mol % ^{15}N. Light intensity versus wavelength is shown in the region 280–380 nm. The discharge tube was cooled strongly with liquid nitrogen.

and lower) pertaining to a given band and, consequently, can be used as a "label" for that band.

Until recently, except when dealing with very high ^{15}N abundances, the band at λ = 298 nm (2977 Å), arising from the 2 - 0 transition, was found to give the best results [10–12]. The main reason for this is that spectral overlap by the 3064 Å system of OH bands (OH being present because of the usual traces of water vapor) is of relatively low intensity around the 2 - 0 band, as can be seen by inspection of Fig. 3. However, the introduction of an

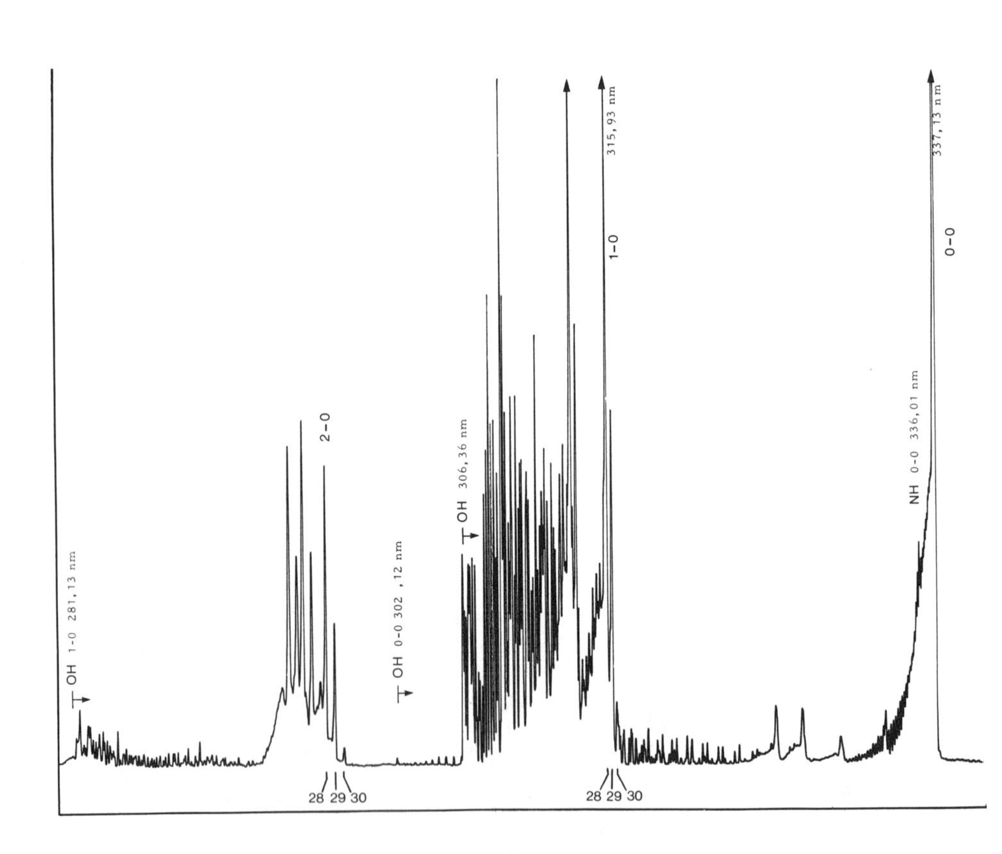

FIG. 3 Spectrum of N_2 containing 20 mol % ^{15}N. Analytical procedure as before (compare left-hand half of Fig. 2) except that (1) customary routine used in preparing sample and (2) discharge tube cooled only slightly by a current of air at ambient temperature.

improved sample-preparation technique (see Sec. III.D) and strong cooling of the discharge plasma make the considerably more intense 1 - 0 band preferable (compare the OH backgrounds in Figs. 2 and 3). Meanwhile, owing to the reduction in isotopic shift, as a result of this choice, a spectrometer with better resolving power is needed (see end of this section).

During discharge the ^{14}N and ^{15}N atoms rapidly become randomly distributed among the three types of N_2 molecule, since these molecules continuously undergo dissociation and recombination. Thus, according to the statistics of the binomial distribution, the following relations will be valid shortly after ignition of the discharge:

$$r^{-1} = 2R \quad (4)$$

$$r = 2R' \quad (5)$$

where r = $^{15}N/^{14}N$ ratio, R = N_2-28/N_2-29 ratio, and R' = N_2-30/N_2-29 ratio.

It is seen by substituting Eq. (4) into Eq. (1) that the percentage ^{15}N abundance, a_{15}, in a nitrogen sample may be calculated as follows:

$$a_{15} = \frac{100}{2R + 1} \text{ mol \%} \quad (6)$$

where R is the population density of the $^{14}N^{14}N$ molecules relative to that of the $^{14}N^{15}N$ molecules.

A detailed spectrum of the 2 - 0 transition in the 2nd positive N_2 system is shown in Fig. 4. In this transition the isotopic shift from the N_2-28 band to the N_2-29 band is 0.61 nm, and the shift from the N_2-29 band to the N_2-30 band is 0.57 nm. The N_2-28/N_2-29 peak-height ratio, assumed to be equal to the corresponding population density ratio R, is directly observable. For the particular sample from which the spectral section shown in Fig. 4 was obtained, it will be seen that R = 95/26 = 3.65, which according to Eq. (6) corresponds to 12.0 mol % ^{15}N.

In the case of the 1 - 0 transition, the isotopic shift is approximately one-half that mentioned above for the 2 - 0 transition. Having taken this and other factors into consideration, we recommend the use of a spectrometer that can operate at a resolving power R of 6000–8000 (i.e., a reciprocal resolving power, $R^{-1} = \Delta\lambda/\lambda$, corresponding to 0.05–0.04 nm in the vicinity of λ = 300 nm). A suitable instrument would be a plane-grating, symmetrical Czerny–Turner-type spectrometer with a focal length of 0.6–0.1 m. An added advantage may be gained by fitting the instrument with properly curved entrance and exit slits.

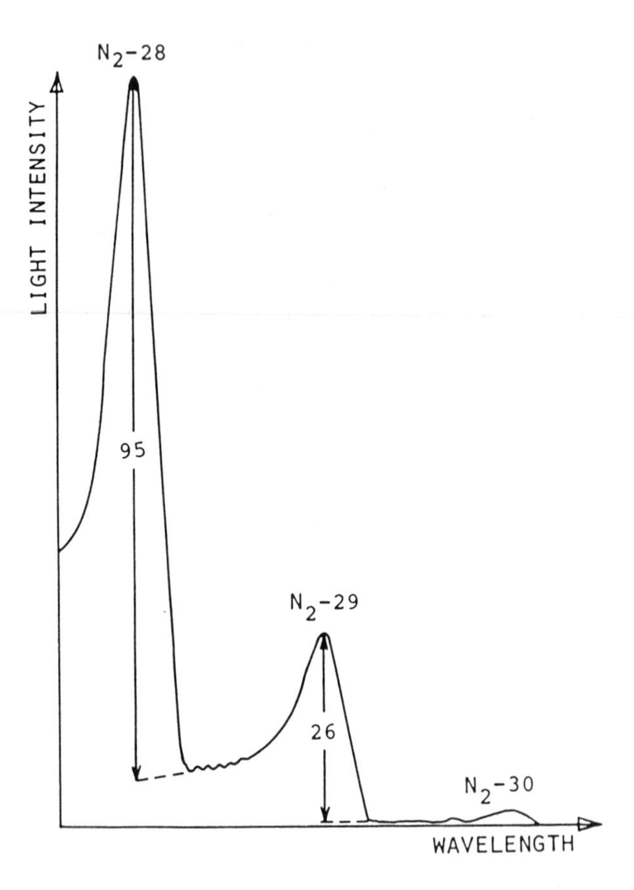

FIG. 4 Detailed spectrum of N_2 ($C^3\Pi - B^3\Pi$, 2 - 0 transition) in the wavelength region 297.5–299.0 nm.

C. Analysis of $^{13}C/^{12}C$ and $^{18}O/^{16}O$ Ratios by OES

In natural carbon and oxygen the abundances of the heavy isotopes ^{13}C and ^{18}O are 1.108 mol % and 0.205 mol %, respectively. Thus, by means of enrichment in one or both of these isotopes, tracers may be obtained for carbon and/or oxygen.

For the purpose of measuring isotopic ratios in singly or multiply labeled tracers, the optical emission spectrum of ionized carbon monoxide (CO^+) has been analyzed for the stable isotopes ^{13}C and ^{18}O. These, together with the most abundant isotopes ^{12}C and ^{16}O, produce quadruplicate bandheads due to the following isotopic molecules: $^{12}C^{16}O$, $^{13}C^{16}O$, $^{12}C^{18}O$, and $^{13}C^{18}O$, the mass numbers of which are 28, 29, 30, and 31, respectively.

Naturally, the radioactive isotope ^{14}C, with a half-life of about 5700 yr, is used frequently as a convenient tracer in environmental research. By the combination of ^{13}C and ^{14}C, double-tracer experiments can be performed, when required.

In principle, ^{13}C and ^{18}O can be analyzed by MS or OES in much the same way as described for ^{15}N. An early prototype ^{13}C-analog of the optical emission ^{15}N-analyzer was introduced by Middelboe and Saaby Johansen in 1978 [13].

D. Comparison with Mass Spectrometry

In a review [14] comparing the use of MS and OES in ^{15}N-aided research, the following two main conclusions were drawn. MS is indispensable for the analysis of the extremely small variations in ^{15}N abundance that occur in nature, and OES is uniquely applicable to the isotopic analysis of nanomole quantities of nitrogen. These circumstances are due largely to the following two facts: (1) The accuracy of a mass spectrometer is superior to that of an optical spectrometer in the analysis of $^{15}N/^{14}N$ ratios. (2) The sample nitrogen is consumed during analysis by MS, but conserved (in the discharge tube) during analysis by OES.

Actually, the above-mentioned conclusions refer to two rather limiting cases, namely, minute differences in isotopic composition or very limited amounts of available nitrogen. However, the situation is often intermediate and the choice of method not so clear-cut. We will consider below a few typical situations that occur in practice (Sec. III).

The same general procedures for the isolation of N_2 from the N-containing sample are applicable, irrespective of the particular method of $^{15}N/^{14}N$ measurement used. Consequently, the application of a certain sample-preparation procedure has relatively little influence on the respective advantages and limitations of OES and MS.

The optical emission method has a distinct advantage in certain multitracer applications, i.e., when either two isotopic molecules with the same molecular mass number or two different but isobaric molecules (e.g., either $^{14}C^{16}O$ and $^{12}C^{18}O$, or $^{13}C^{16}O$ and $^{14}N^{15}N$) are present in significant amounts in the gas to be analyzed. Such molecular pairs are readily distinguishable in OES but inseparable in MS designed for the routine analysis of isotopic ratios.

III. PRACTICAL CONSIDERATIONS

A. Introduction

In this section ^{15}N is taken as an example. However, most of the considerations are equally applicable to other stable tracers such as ^{13}C and ^{18}O.

Invariably, the two most important considerations are the cost of the labeled compound and the cost of the isotopic analyses (instrumentation and operation). The expense per gram of labeled test element (e.g., N) depends progressively on the extent to which the abundance of the heavy isotope (^{15}N) has been altered relative to its natural abundance. The degree of alteration is usually expressed in terms of heavy-isotope enrichment (or depletion), i.e., the difference between the abundance of the heavy isotope in a sample of labeled test element and the abundance of the heavy isotope in a sample of natural test element. The heavy-isotope enrichment (depletion) is given in units of mol % excess (deficiency).

B. Field versus Laboratory or Greenhouse Experiments

Relatively large quantities of labeled nitrogenous compounds are needed in carrying out field experiments, and until about 30 years ago the level of ^{15}N enrichment considered to be desirable was rather high, in fact, on the order of 10 mol % excess. Consequently, the cost of the labeled material often proved to be a serious deterrent. Meanwhile, the Joint FAO/IAEA Division (Vienna), who pioneered the widespread field application of ^{15}N labeling in agronomic research, demonstrated in many instances that the level of ^{15}N enrichment mentioned above could be reduced by at least one order of magnitude without any appreciable loss of useful information [15]. However, the exploitation of *low-level* ^{15}N *enrichment* makes it necessary, in most cases, for the isotopic analyses to be performed by MS, which is considerably more expensive than OES. Nevertheless, experience has shown that the overall savings on the labeled material, throughout a period of years, are substantial enough to justify the use of MS in field experiments on a broad scale.

The amount of nitrogen required for most laboratory or greenhouse experiments is small compared to that needed in field experiments. Therefore, the cost of the labeled material is usually relatively modest, even if the desired level of ^{15}N excess is high. Consequently, OES is likely to be the analytical method of choice for pure laboratory or greenhouse experimentation, since the necessary optical equipment is considerably less costly to acquire, operate, and maintain than a mass spectrometer.

C. Applications of Nitrogen Depleted in ^{15}N

At present, kilogram quantities of almost pure ^{14}N compounds are being produced commercially and these products are offered for sale at a somewhat lower price than tracer material enriched in ^{15}N to an equivalent degree, i.e., N compounds containing 0.36 mol % excess ^{15}N [16]. However, if the place where the tracer material is to be used is situated far from the isotope-separation plant and shipping expenses are taken into account, the lesser cost of the ^{15}N-depleted compound can be more or less matched by transporting rather highly enriched material, which is less bulky, and then diluting it locally with the corresponding natural N compound. Whether or not the use of OES will suffice, when *^{15}N-depleted material* is applied, will depend largely on the degree to which the tracer nitrogen becomes diluted with natural nitrogen during the course of the experiment; if the isotopic dilution factor involved is expected to be at least 10, and the relative error is to be kept below about 25%, no alternative to a mass spectrometer exists.

D. Preparation of Samples

The preparatory treatment of a soil, plant, or other N-containing sample is aimed at getting its chemically bound N liberated as N_2 gas. In principle, the sample N should be liberated completely and the N_2 produced should be uncontaminated by alien N or foreign gases. The original chemical procedure designed for use in the analysis of ^{15}N-labeled samples by OES was developed by Faust [17, 18] in the 1960s. More recently, sample ammonia has been admitted directly into the discharge tube and N_2 liberated, together with H_2, by microwave interaction.

Classically, there are two main approaches to N_2 liberation. In one procedure, called the Rittenberg method, the sample N is all converted to the ammonium form by conventional Kjeldahl digestion of the sample [19], and N_2 is liberated from the solution containing the ammonium ions by mixing it, in a so-called Rittenberg flask, with a sufficiently freshly made solution of sodium hypobromite. In the alternative procedure, called the Dumas method, N_2 is liberated directly from the sample by heating it in the presence of copper oxide.

The Rittenberg method was originally designed for use in MS, and the procedure is described more closely in Chap. 11. The chief modification encountered, in applying the method to OES, is a reduction of the sample size by about an order of magnitude.

The apparatus originally developed for "Probenchemie" by Faust [18], which was designed primarily for the preparation of discharge tubes containing N_2 liberated by the Dumas method, is shown schematically in Fig. 5. The Rasotherm glass discharge tube containing copper oxide and calcium oxide is attached by the aid of black wax (picein) to one end of a T piece, which can be rotated about a horizontal axis (the greased joint in Fig. 5). With the T piece in the horizontal position, the sample is placed just inside the free end of the T piece and this end is then closed by a greased cap with a ground glass joint. By opening the top valve (see Fig. 5) the T

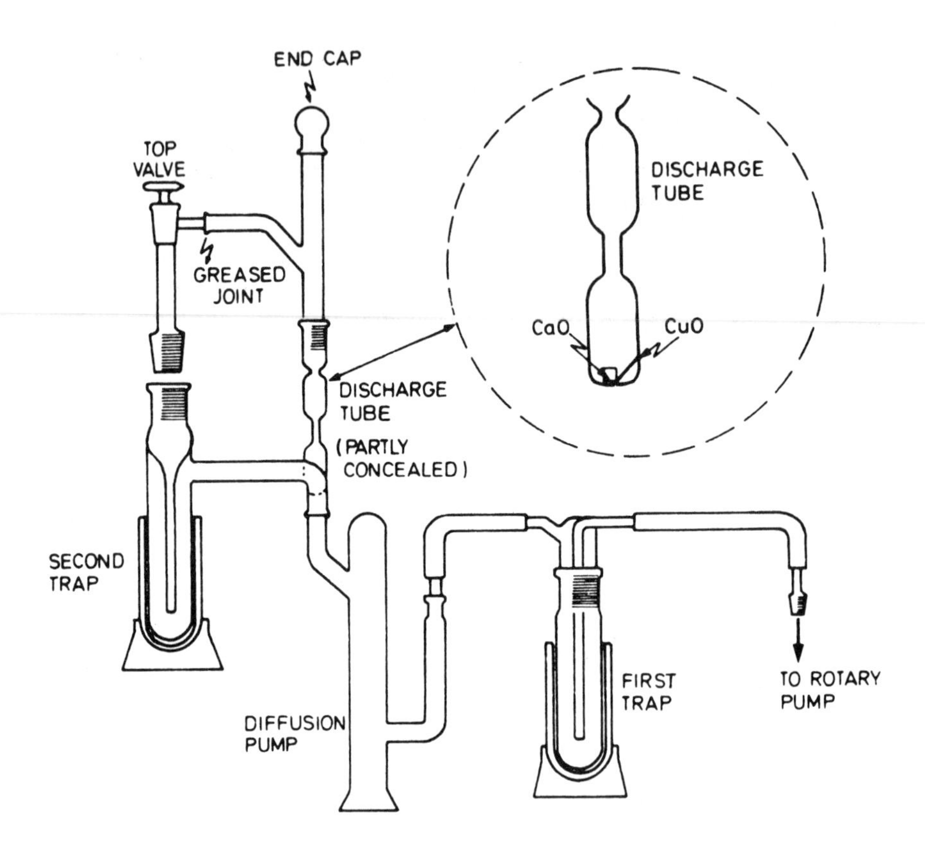

FIG. 5 Faust's apparatus for the preparation of discharge tubes. (From Ref. 20.)

piece and discharge tube are evacuated, via a cold-trap, by means of a diffusion pump, backed by another cold-trap, and a rotary pump. Natural N_2 adsorbed on the surface of the chemicals and the inner wall of the discharge tube is eliminated by heating under a vacuum. Upon cooling, the sample is dropped into the discharge tube by turning the T piece to the vertical position. Finally, the discharge tube is sealed by flame and pulled off the vacuum line.

Incidentally, it should be mentioned that a few years ago an all-metal vacuum line, designed for the preparation of discharge tubes containing N_2, was constructed at IAEA's Seibersdorf Laboratory (near Vienna). This vacuum line has now stood the test of time and it has been introduced in developing countries. The main advantage of the all-metal vacuum line over the glass vacuum line is that the latter is much more vulnerable to accidental breakage, and the replacement/repair of any broken glass part often takes an unduly long time in countries with a less-developed infrastructure.

The closed discharge tube containing the sample and chemicals is placed in a furnace at 550°C for 3 hr. The N_2 liberated from the sample should exert a pressure in the discharge tube that falls in the range of 0.2–0.5 kPa (2–4 torr) in order to facilitate the production of a bright discharge. Most of the alien gases, such as water vapor and carbon dioxide, liberated together with N_2 are absorbed by the calcium oxide in the discharge tube during a predischarge rest period of about 12 hr. However, one gas that remains unabsorbed is carbon monoxide, the spectrum of which overlaps and interferes somewhat with the section of the N_2 spectrum normally used in the determination of $^{15}N/^{14}N$ ratios by OES.

Many applicable variations and further details in sample preparation are described in a comprehensive review by Fiedler and Proksch [21]. For instance, the copper oxide in the discharge tube is replaced by metallic copper, when the sample N is present in the form of nitrates. Furthermore, in the case of very restricted amounts of sample N, on the order of 1 μg or less, isotopic analysis can be effectuated by the use of one or more noble gases to sustain the discharge, as demonstrated by Goleb and Middelboe [22].

Lazeeva et al. [23] introduced a novel sample preparation procedure. The sample is pyrolyzed under a vacuum and the combustion products are passed across nickel oxide at a temperature of 1000°C. Carbon dioxide, water vapor, and other condensible by-products are frozen out in a liquid-nitrogen trap, and the N_2 (together with any remaining contaminants) flows through a purifying unit containing copper at 600°C and copper oxide at 300°C. Finally, the N_2 is admitted to the discharge tube via a second cold-trap.

Faust and Reinhardt [24] have investigated the use of molecular sieves, in place of calcium oxide, for the absorption of unwanted gases produced during the preparation of N_2 by the Dumas method. With 10 μg of N_2 in a 2 cm³ discharge tube, the most promising results were obtained using 10–15 mg of molecular sieve grains (ca. 2 mm diameter) having a pore diameter of 0.3–0.5 nm. The apparently improved quality of the N_2 discharge, brought about by the use of this technique, is probably partly due to the absorption of carbon monoxide (CO), as well as carbon dioxide (CO_2), by the molecular sieve and partly due to a reduction of the residual water vapor in comparison with that left unabsorbed by calcium oxide.

A modification of the Dumas method that permits the use of frozen aqueous samples, as opposed to the normal dehydrated samples, was developed in 1977 by Middelboe [25]. The frozen sample is placed in the usual position inside the T piece (see Figs. 5 and 6) and is

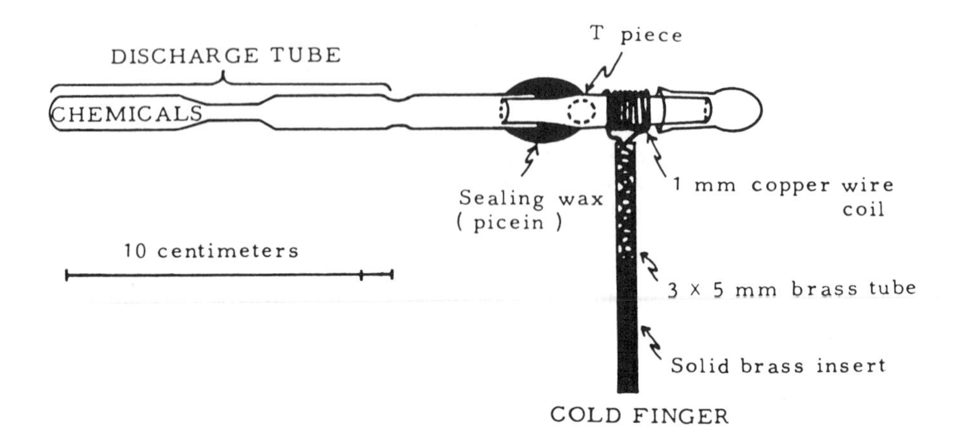

FIG. 6 Part of sample-preparation apparatus (see Fig. 5) modified for use of the "frozen sample" technique. (From Ref. 25.)

kept at about -100°C during the evacuation and degassing of the discharge tube. Finally, having tipped in the frozen sample, the discharge tube is sealed before the sample begins to thaw. Naturally, the quantity of calcium oxide in the discharge tube must be sufficient to absorb all the water (H_2O) present, e.g., 30 mg of calcium oxide are found to be adequate in the case of a 1 μl aqueous sample.

Finally, Burridge and Hewitt [26] have shown that the discharge tube may be filled directly with ammonia, instead of using the Rittenberg method to release N_2 from the ammonium solution. In their original procedure, a frozen aqueous solution containing ammonium ions, to which caustic soda has been added, is attached to a line, evacuated, and allowed to thaw. The ammonia gas (NH_3) liberated in this way is expanded into a permanently installed discharge tube, while H_2O vapor is retained by a carefully balanced cooling system. Upon high-frequency excitation of the NH_3 at 2450 MHz, the emission spectrum of N_2 is obtained with some interference due to CO and other gases. Contemporaneously, the present authors developed a simplified and improved ammoniacal technique [27,28]. A discharge tube is filled with sample NH_3 to a pressure of about 2.5 kPa (ca. 20 torr) and then cooled with liquid nitrogen (ca. -190°C), whereby the NH_3 is solidified together with unavoidable traces of H_2O and CO_2. Intense N_2 light is emitted soon after subjection of the residual NH_3 vapor to high-frequency transmission; see Fig. 2. Traces of OH and NH are to be seen in the spectrum, but no sign of CO is discernible.

The precision of all the commercially available optical ^{15}N-analyzers is very much the same. The standard deviation (SD) on duplicate analysis is, at best, approximately 0.01 mol % in the case of samples in the range from 0–1 mol % ^{15}N. In the case of higher ^{15}N abundances the relative SD is about 1%. However, if the analytical uncertainty in the complete sample-preparation procedure is included, the relative overall SD may be 2–5% [21].

E. Use of a Calibration Curve

It is important to note that the $^{15}N/^{14}N$ ratios obtained by the use of MS are accepted directly as "true" values, whereas those observed in OES have thus far been considered apparent values, i.e., the latter are corrected, usually by the aid of a calibration curve, in order to obtain true values. Experience shows that care must be taken in establishing the calibration curve. Spectroscopically pure N_2 samples (with known ^{15}N abundances) are not suitable for this purpose, presumably due to the absence of the small amounts of gaseous impurities normally present after the preparation of samples of soil, plant, or other environmental material [29]. More consistent results are obtained by isotopic analysis of a set of ammonium chloride standards (with known ^{15}N abundances), samples of which are put through the same preparatory procedure as the experimental samples.

A few years ago, Saaby Johansen [28] found that exploitation of the transition 1 - 0 (rather than the usual 2 - 0), in particular with strong cooling of the discharge plasma (initially N_2 or NH_3), makes the use of a calibration curve superfluous in the determination of $^{15}N/^{14}N$ ratios by OES, provided a spectrometer of sufficient resolving power is employed.

IV. ^{13}C—^{18}O-ANALOG OF OPTICAL ^{15}N-ANALYZER

A. Methodology

Analysis for ^{13}C by OES was first achieved in the 1950s, using either the Swan bands of C_2 (Ferguson and Broida [30]) or the Ångström system of CO bands (Zaidel and Ostrovskaya [31]). However, owing to certain practical problems, neither of these methods attained widespread use. In the 1970s, preliminary investigations in our laboratory [13,32] demonstrated that the development of a practicable optical emission method for the determination of $^{13}C/^{12}C$ ratios was feasible. The method, which now also incorporates the determination of $^{18}O/^{16}O$ ratios, is analogous to that described above for the determination of $^{15}N/^{14}N$ ratios.

Carbon dioxide would seem to be the most suitable C-containing gas for isotopic analysis, since (1) this gas is frequently the direct object of interest, and (2) organic compounds are converted readily to CO_2. However, a high-frequency activated discharge in pure CO_2 produces a spectrum that is dominated by emission bands due to CO, and the diffuseness of these bands renders them ill-suited for isotopic analysis. In contrast, CO^+ produces slender bandheads, almost as sharp as atomic spectral lines, and bandtails fortuitously low in intensity.

The spectrum of CO^+ is obtained by using helium (He) as carrier gas, at about 1000 Pa (ca. 10 torr), mixed with CO_2 at a partial pressure, after strong cooling, which is two or three orders of magnitude lower than the He pressure. One of the most useful parts of the spectrum is that called the 1st negative system of CO^+ bands situated in the UV region. In particular, the 0 - 1 transition produces a band headed at 230 nm. This band undergoes an isotopic shift of 0.3 nm (to lower λ) when the light is emitted by $^{13}C^{16}O^+$ instead of $^{12}C^{16}O^+$. The isotopic shift is of the same magnitude (in reverse) as the N_2-28 to N_2-29 λ shift of the 1 - 0 band in the 2nd positive N_2 system, so in emission spectrometric analysis the optical dispersion and resolution required for ^{13}C are essentially the same as those for ^{15}N.

Singly labeled ^{13}C samples can be analyzed isotopically on the basis of either the 0 - 1 or 1 - 4 transition in the He-sustained CO^+ spectrum (1st negative system). Dual-labeled (^{13}C and ^{18}O) CO^+ can be analyzed by utilizing the same transitions, given some information on the $^{13}C/^{18}O$ ratio. Depending on the strength of the labeling, two different methods of measurement may be applied, as follows:

1. When the enrichment is *low*, the 1 - 4 transition is analyzed optically, since, in the spectrum of this transition, it is possible to resolve the bandheads from CO^+-28, CO^+-29, and CO^+-30 (forthwith 28, 29, and 30; see Fig. 7) in addition to which the background, especially that due to the 4th positive CO system, has been found to be the lowest among those transitions in which the bandheads from 29 and 30 are well resolved.

2. When the enrichment is *higher*, the more intense 0 - 1 transition is preferred, because the bands from 28 and 31, in conjunction with the amalgamated bands AB from 29 plus 30, contain all the necessary information (see Fig. 8). Nevertheless, the best method is to analyze the 1 - 4 transition as well as the 0 - 1 transition and, moreover, to evaluate the background effects by inspection of the CO spectrum.

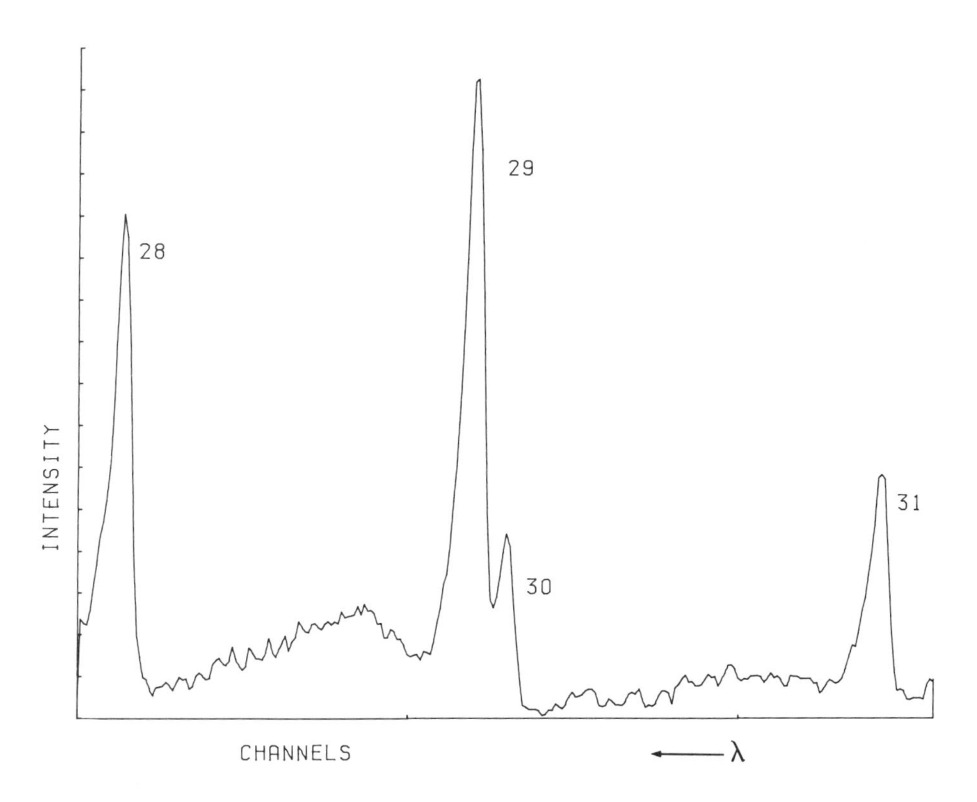

FIG. 7 Spectrum of 1st negative CO^+ system, 1 - 4 transition, in the region λ = 258–255 nm, showing all four bandheads. The bandheads from 29 and 30 (at 256.75 and 256.66 nm) are seen resolved. The ^{13}C abundance was 54 mol % and the ^{18}O abundance 26 mol %.

B. Determination of ^{13}C and/or ^{18}O Abundances

1. *Singly Labeled Samples*

The procedure introduced for deriving ^{13}C- *or* ^{18}O-abundance values from the CO^+ spectrum (0 - 1 transition) is the following: (1) The interval 230.2–229.6 nm is registered in order to monitor both the peak of the band from 28 (headed at 229.96 nm) and the peak of the AB from 29 plus 30 (headed at 229.68 and 229.65 nm, respectively). (2) The data, in the case of ^{13}C-labeling, are corrected for the contribution from the natural content of ^{18}O and vice versa. (3) The observed isotope abundance *a* is calculated

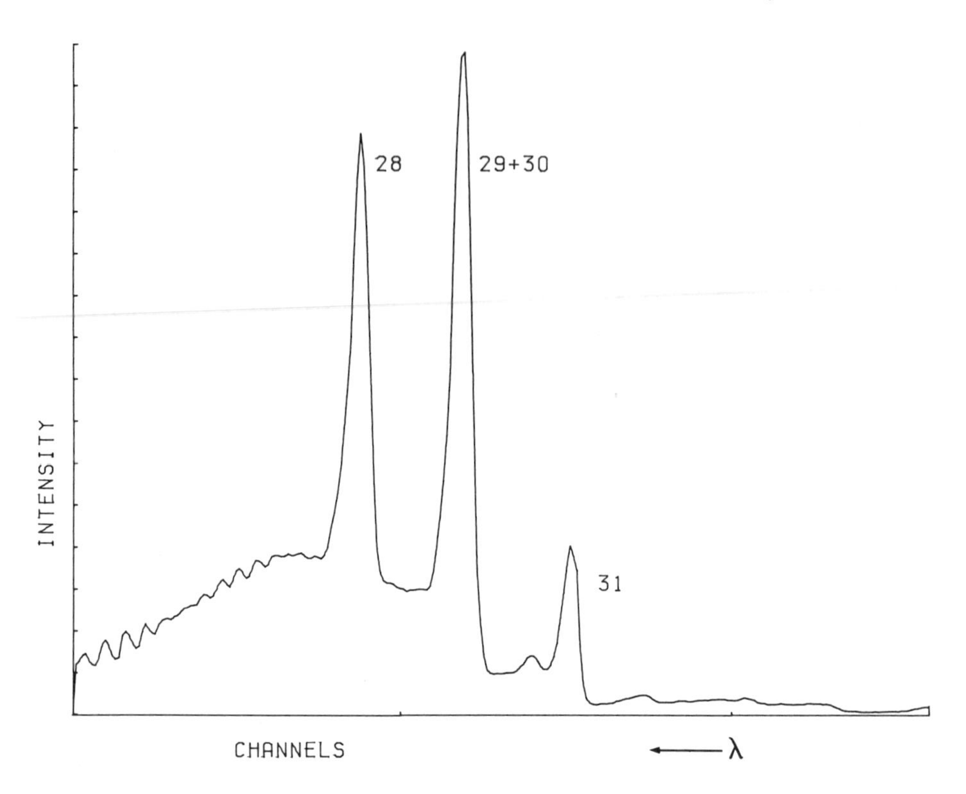

FIG. 8 Spectrum of 1st negative CO^+ system, 0 - 1 transition, in the region λ = 230.8–228.6 nm. The amalgamated bands (AB) from 29 and 30 (headed at 229.68 and 229.65 nm) are seen in the center. The sample is the same as that used for obtaining the spectrum shown in Fig. 7; however, the actual intensity of the 0 - 1 bands is nearly ten times that of the 1 - 4 bands.

from a ratio r between bandhead intensities (peak heights) by applying the appropriate one of Eq. (7).

$$a_{13} = \frac{100}{r_C^{-1} + 1} \text{ mol \%} \quad \text{or} \quad a_{18} = \frac{100}{r_O^{-1} + 1} \text{ mol \%} \tag{7}$$

in which

$$r_C = \frac{(13)}{(12)} = \frac{(13)\cdot(16)}{(12)\cdot(16)} = \frac{(29)}{(28)}$$

$$r_O = \frac{(18)}{(16)} = \frac{(12)\cdot(18)}{(12)\cdot(16)} = \frac{(30)}{(28)}$$

where (12) and (13) represent C-isotope abundances, (16) and (18) represent O-isotope abundances, (28), (29), and (30) represent CO^+ isotopic abundances. The ratio r (i.e., r_C or r_O) is obtained from measured peak heights H as follows:

$$r = \frac{(H_{AB})_{corr}}{H_{28}}$$

where "corr" refers to step (2) above.

2. *Dual-Labeled Samples*

The procedure worked out for the simultaneous determination of both abundances in samples from dual-tracer research is somewhat more complicated. In certain sections of the CO^+ spectrum four bandheads, viz. those from 28, 29, 30, and 31, have to be optically analyzed individually.

In the spectrum of the 1 - 4 transition (see Fig. 7), all four bandheads are resolved when the reciprocal resolution (R^{-1}) of the spectrometer is 0.04 nm at λ = ca. 250 nm, but CO spectral overlap at the 29 bandhead can prevent the determination of a reliable value for the ^{13}C abundance. This problem notwithstanding, measurements on spectral recordings (257.8–256.6 nm) taken from very stable discharges give reasonable results. We found the first three bandhead λ's to be 257.77 nm (from 28), 256.75 nm (from 29), and 256.66 nm (from 30). The observed ^{13}C and ^{18}O abundances are determined by the measurable (29)/(28) and (30)/(28) peak-height ratios; see Eqs. (7).

In the spectrum of the 0 - 1 transition (see Fig. 8) with a band headed at 229.96 nm (from 28), emission by the heavier isotopic CO^+ molecules produces bands headed at 229.68 nm (from 29) and 229.65 nm (from 30), which are barely resolved (i.e., AB). Meanwhile, the simultaneous determination of ^{13}C and ^{18}O abundance in a sample is possible, when the enrichment in one or both of these heavy isotopes of C and O, respectively, is high enough for the 31 bandhead not to be overwhelmed by CO emission. In this case, the interval 230.2–229.2 nm is recorded, and the mutual interference in the AB (between 29 and 30) is subjected to spectral analysis. Finally, the ^{13}C and ^{18}O abundances are each calculated on the basis of measured peak-height ratios.

The peak-height ratios $r_{AB} = H_{AB}/H_{28}$ and $r_{31} = H_{31}/H_{28}$, and the isotope-abundance ratios r_C = (13)/(12) and r_O = (18)/(16), where the notation is as defined above, bear the following relationships to one another:

$$r_{AB} = \frac{x \cdot (13) \cdot (16) + y \cdot (12) \cdot (18)}{(12) \cdot (16)} = x \cdot r_C + y \cdot r_O \qquad (8)$$

where

x = fractional intensity of AB due to $^{13}C^{16}O$ (i.e., 29)
y = fractional intensity of AB due to $^{12}C^{18}O$ (i.e., 30)

and

$$r_{31} = \frac{(13) \cdot (18)}{(12) \cdot (16)} = r_C \cdot r_O \qquad (9)$$

The applicable values of x and y for any given sample are found by spectral analysis. The solution of Eqs. (8) and (9) yields a double pair of abundances [33], the accuracy of each being dependent on the $^{13}C/^{18}O$ ratio. Selection of the best set of results can be done, either by the aid of a test measurement on the spectrum of the 1 - 4 transition, or by an exact determination of the distance in nm units between the peak of the 28 bandhead and that of the AB. This distance is a function of the $^{13}C/^{18}O$ ratio, due to the small difference in λ (0.03 nm) between the 29 and 30 bandhead positions.

C. Automation and Data Processing

Our basic instrument designed for automation and data processing is a microcomputer (personal computer). An instrument of this type constitutes a powerful tool in the analysis of stable isotopes by OES, since it can be used both for control of the measuring system and treatment of the primary observational values. Furthermore, its use increases the rate of data accumulation and enhances the precision of measurement.

Figure 1 (with the alternative arrow vertical) shows a schematic diagram of our experimental set-up. An A/D converter, i.e., a digital multimeter, enables different types of spectrometer to be coupled to the computer system. The multimeter and spectrometer can be programmed, via an interface bus, by means of the computer. The program serves the following twofold purpose: (1) control of the measuring system and (2) analysis of data obtained by the observation of specified spectral regions.

Figure 9 shows a flow diagram of the program that has been developed for the determination of stable-isotope abundances. Corrections for certain technical effects (such as detector hysteresis and spectral irregularities) are incorporated in the program. The working time needed for measurement and calculation, based on a double

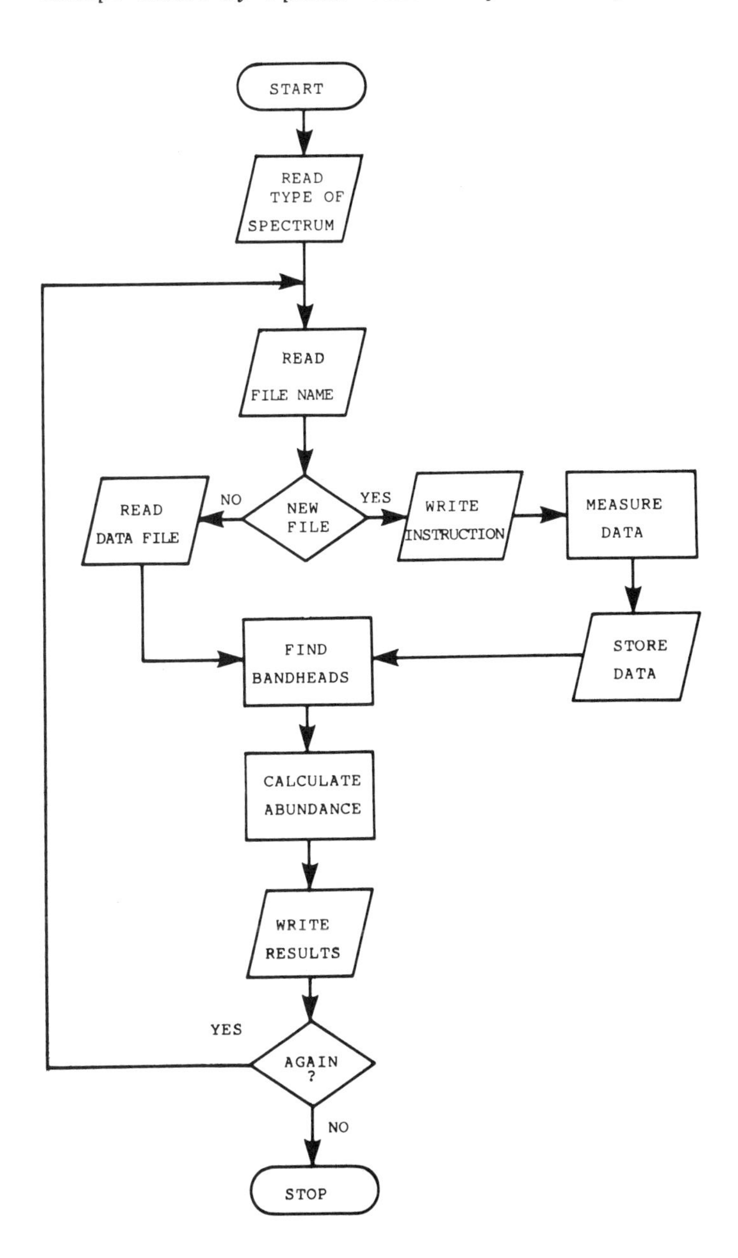

FIG. 9 Flow diagram of the program used for control of the optical measurements and calculation of the results.

scanning of the bandheads of interest, is between 0.5 and 2 min, dependent on the transition(s) selected. The amount of working time required is governed by the time constant of the particular photomultiplier in use as well as the time constant of the indispensable electronic amplifier.

D. Example of Results Obtained

Some samples designed for ^{13}C and ^{18}O analysis by OES were produced using standard procedures [27], and the optical measurements were carried out by the use of a spectrometer with a reciprocal resolution (R^{-1}) of 1.6×10^{-4} and a reciprocal dispersion of 0.8 nm per mm (i.e., a slit width of 0.05 mm). The ^{13}C and/or ^{18}O abundances were determined by peak-height measurments in spectra of the 0 - 1 and 1 - 4 transitions.

In the case of low ^{13}C abundances, the results obtained from the 1 - 4 transition were not very accurate due to overlay by a CO bandhead at 256.78 nm (4th positive, 9 - 20 transition). This interference gave rise to a correction of about 1.1 mol % and an increased SD. A few results obtained by the analysis of mono- and dual-labeled samples are given in Table 1.

E. Conclusion

The method described above (Sec. II.B) has long since proved its worth in analysis for ^{15}N and with time, undoubtedly, the analogous method for determination of ^{13}C and/or ^{18}O abundances will find its

TABLE 1 Abundance Values Found by OES Analysis of Labeled CO_2

Label isotope	Vibrational transitions	^{13}C (mol %)	SD (mol %)	^{18}O (mol %)	SD (mol %)
^{13}C[a]	0 - 1	11.78	0.09	0.2	0.2
	1 - 4	12.1	0.6	0.3	0.3
^{18}O[b]	0 - 1	1.04	0.10	12.06	0.16
	1 - 4	0.2	0.7	12.33	0.10
^{13}C and ^{18}O	0 - 1	11.7	0.7	4.1	0.3
	1 - 4	11.1	0.9	4.6	0.3

[a]Natural abundance of ^{18}O.
[b]Natural abundance of ^{13}C.

place in tracer research related to soils, crops, and the environment in general. The precision of the analytical results need only be comparable to, or preferably somewhat better than, the SD carried by samples from field experimentation before they are subjected to isotopic analysis. Consequently, OES often represents a fairly simple and inexpensive alternative to other techniques used for stable-isotope analysis, such as MS, laser absorption, nuclear magnetic resonance, and infrared spectrophotometry.

We believe that the optical emission method could be improved by the introduction of a UV-sensitive diode array, this being a type of detector well suited for monitoring a narrow spectral region. Furthermore, the working time required per analysis would be reduced. Thus, after further development, OES is likely to become not only a cheap and simple, but also a reasonably accurate, method for analysis of ^{15}N, ^{13}C, ^{18}O, and other stable isotopes applied as tracers in various types of environmental investigation.

VI. EXAMPLES OF ^{15}N, ^{13}C, AND ^{18}O APPLICATION

A. Introduction

In this section attention is focused on problems related to the practical application of stable-isotope analysis by OES. The emphasis here is placed mainly on the limitations and advantages of the technique.

Up to now many thousand papers have been published on agronomical experiments in which ^{15}N has been used as a tracer, and in recent years an increasing fraction of the isotopic analyses involved has been performed by use of the optical emission method [34]. A few examples will serve to illustrate some of the more important tracer problems that are encountered frequently. The examples have been selected, not primarily for the intrinsic value of the results, but as a basis for discussing certain points of tracer methodology associated with the application of isotopically labeled nitrogen, carbon, or oxygen.

The following examples include the determination of A-values in a greenhouse experiment, analysis for total N by isotopic dilution in submilliliter samples of soil water, evaluation of asymbiotic dinitrogen fixation in soil or sediment, photosynthesis of carbon by soybean, and plant uptake of oxygen-labeled phosphate.

B. Evaluation of Soil-N Fertility

The well-known A-value was introduced by Fried and Dean [35] as a concept by which soil fertility with regard to a given nutrient element could be assessed quantitatively. According to this concept, the amount of available soil-N is mathematically defined as follows:

$$A_N = B_N \frac{1 - y}{y} \text{ kg ha}^{-1} \tag{10}$$

where

A_N = A-value for nitrogen
B_N = rate of fertilizer N applied
$y = \dfrac{\text{mol \% excess N-15 in plant N}}{\text{mol \% excess N-15 in fertilizer N}}$
= fraction of plant N derived from fertilizer

Note that mol % excess of a stable isotope is analogous to the specific activity of a radioisotope. Evidently, the fertility of the soil is expressed in terms of the equivalent rate of the applied fertilizer.

A greenhouse experiment was carried out in pots containing nine different soils, and the respective A-values were determined at three rates of ammonium sulfate application (50, 100, 200 kg N ha^{-1}). There were three replicates for each combination of soil and treatment, and the fertilizer N was labeled with 1.0 mol % excess ^{15}N. The result was, essentially, that for each soil the A-value appeared to be independent of the fertilizer rate [36]. The total quantity of labeled fertilizer used in the experiment contained approximately 0.1 g excess ^{15}N, which cost about $10(U.S.). The samples of plant N were prepared by the Rittenberg method and the $^{15}N/^{14}N$ ratios were measured by MS. Thus, ^{15}N-depleted fertilizer could have been used, but the consequent cost reduction (on $10) would have been extremely small.

The determining factor in the calculation of the A-value is the fraction y in Eq. (10). In this experiment the fraction of plant N derived from the fertilizer was never less than one-tenth. Therefore, the mol % excess ^{15}N in the plant was not less than one-tenth of that in the fertilizer. Consequently, if the initial ^{15}N enrichment of the fertilizer had been 5 mol % excess (instead of 1 mol % excess), none of the samples to be analyzed would have contained less than 0.5 mol % excess ^{15}N, and their enrichments could presumably have been determined with a relative SD of 2–3%, if OES had been used (instead of MS). The cost of the labeled fertilizer would still have been very modest.

C. Analysis of Subsoil Water for Total N

Hevesy's isotope-dilution technique may be applied to the determination of an unknown quantity of natural nitrogen [25], as follows:

A known amount of labeled nitrogen, called the spike, is mixed with the unknown nitrogen, which, being natural, contains 99.6 mol % ^{14}N. Next, the isotopic ratio in the uniform mixture is analyzed and, finally, the result is calculated according to the equation

$$n_x = n_s \frac{a_m - a_s}{99.6 - a_m} \text{ mol} \quad (11)$$

where

n_x = moles of N in unknown
n_s = moles of N in spike
a_s = ^{14}N abundance in spike (determined by MS)
a_m = ^{14}N abundance in mixture of spike and unknown

A high sensitivity in total N determination is achieved when (1) the spike contains practically pure ^{15}N, and (2) the isotopic ratio in the nitrogen mixture is analyzed by the use of OES. However, high sensitivity requires the determination of low ^{14}N abundances. Consequently, in the 2 - 0 transition (see Fig. 4), the N_2-30 band will be the most intense and its tail will overwhelm the less intense N_2-29 band, whereby accurate measurement of the N_2-29 peak height becomes impracticable. This problem is overcome by basing the analysis on the 0 - 2 transition, in which the isotopic shift is reversed with respect to λ (see Fig. 2). According to the previous Eqs. (2) and (5), the ^{14}N abundance in the mixture a_m may be calculated as follows:

$$a_m = \frac{100}{2R' + 1} \text{ mol \%} \quad (12)$$

where R' is the N_2-30/N_2-29 peak-height ratio observed. Finally, the value obtained from Eq. (12) is entered into Eq. (11).

The isotope-dilution technique has been applied to trace analysis for total N in a number of subsoil-water samples [25]. The samples contained various concentrations of NH_4-N (0–77 μg ml^{-1}), NO_3-N (0–16 μg ml^{-1}), and organic N (0–3 μg ml^{-1}).

The pH of the sample was adjusted to approximately 4.5, and an aliquot of 100 μl was used per analysis for total N. The spike was added and the aliquot was concentrated by evaporation, under a vacuum at 50°C, to a volume of 1 μl. Evaporation to complete dryness had to be avoided in order to prevent loss of nitrogen, either in the form of nitric acid fumes or as ammonia gas, toward the end of the drying process.

Discharge tubes containing N_2 were prepared according to the "frozen sample" technique described in Sec. III.C (see Fig. 6).

Contamination by alien nitrogen, e.g., natural N_2 dissolved in the water or adsorbed on the surface of the ice, becomes significant when dealing with nanomole (nmol) quantities of sample N. A number of blanks were analyzed and the contamination was found to be 0.11 ± 0.03 μg N per discharge tube.

Due to the statistical uncertainty of the blank value, the lower limit of detection* is about 2 nmol N (i.e., 0.3 μg total N per ml in this example). For total N contents above 10 $\mu g\ ml^{-1}$, the relative accuracy of the analysis is 2–3%. The amount of water needed for a duplicate analysis is only 0.25 ml.

D. Assessment of N_2 Fixation in Soil or Sediment

The rate of asymbiotic dinitrogen (N_2) fixation, or nitrogenase activity, in soil or sediment can be assessed by the acetylene-reduction method. However, the appropriate conversion factor, according to Hardy et al. [37], has been observed to vary between 2 and 25. The "true" conversion factor for any given system can be evaluated by the simultaneous application of acetylene and ^{15}N-labeled N_2.

An investigation, carried out by Potts et al. in the 1970s [38], provides an illustrative example. N_2 fixation rates were determined (by acetylene reduction as well as ^{15}N labeling) at an intertidal mud- and sandflat on the coast of an island off Germany. Sediment cores, 5 mm deep, were transferred to 7 ml serum bottles and sea water was added to make a final volume of about 2 ml. Gases were introduced or removed by the use of a syringe inserted through the rubber seal of the serum bottle. Labeled N_2, with an abundance of about 80 mol % ^{15}N, was generated in the field by application of the Rittenberg method.

Ethylene (C_2H_4) and acetylene (C_2H_2) concentrations were measured by gas chromatography (GC). Total N contents and ^{15}N abundances were determined by the use of a CHN-analyzer in conjunction with a combination of GC and MS. The conversion factor, i.e., the ratio of the C_2H_2-reduction rate [nmol C_2H_4 (mg N)$^{-1}$ hr^{-1}] to the N_2 fixation rate [nmol N_2 (mg N)$^{-1}$ hr^{-1}] was found to be 5.4 ± 0.8 This observed value is significantly greater than the theoretical ratio of 3. The uncertainty of the observed value corresponds to an overall relative error of 15%.

Whether or not essentially the same result could have been obtained using OES (instead of MS) remains an open question. In

*The "lower limit of detection" is defined here as an amount that carries a (single measurement) SD equal to that amount.

the experiment performed, the order of magnitude of the critical ^{15}N measurement was 0.1 mol % excess. Hence, if the analytical error in optical ^{15}N analysis were assumed to be 0.01 mol %, the relative analytical error of the observed conversion factor would be on the order of 10%, which compares favorably with the overall relative error mentioned above.

E. Photosynthetic C Assimilation by Soybean

The radioactive isotope ^{14}C has been used extensively for carrying out short-term experiments on the distribution of photo-assimilated C in plants. Nevertheless, for various reasons the information gained has generally proved to be of a qualitative nature. Kouchi and Yoneyama [39], however, used the stable isotope ^{13}C as the tracer in work on the "carbon economy" of nodulated soybeans, and they obtained quantitative results in relatively long-term experiments by keeping the CO_2 concentration and ^{13}C abundance reasonably constant.

The plants were cultivated in a closed assimilation chamber, in which a constant CO_2 partial pressure as well as a constant $^{13}C/^{12}C$ ratio could be maintained rather well by means of special mass-flow controllers. For a period of 8 hr the plants were exposed to CO_2 labeled with ^{13}C at an abundance around *30 mol %*. Subsequently, tissue samples were taken from various organs and analyzed for (1) ^{13}C enrichment and (2) total C. The tissue material was combusted at 900°C in O_2, and the $^{13}C/^{12}C$ ratio of the resultant CO_2 was determined by infrared absorption spectrometry (IRA) at approximately 4 μm. The method they used is described in detail by Kumazawa and Yanagisawa [40]. The optical part of their isotopic analysis was, in principle, developed originally by McDowell [41], who exploited the spectrum of CO rather than that of CO_2 subsequently used.

The results were expressed primarily as the fraction x of tissue C derived from the CO_2 in the atmosphere of the growth chamber. The value of x is given by the relationship

$$x = \frac{\text{mol \% excess } ^{13}\text{C in tissue C}}{\text{mol \% excess } ^{13}\text{C in CO}_2 \text{ supply}} \tag{13}$$

The amount of tissue-assimilated C was calculated simply as the product of x and total C in the tissue.

Typical values for x were on the order of 0.07, which corresponds to a ^{13}C abundance in plant tissue of about *3 mol %*, as can be deduced from Eq. (13); according to Refs. [40] or [41], this entails a relative error of 3% or 6%, respectively.

The ^{13}C analysis could have been done equally well by OES (instead of IRA), since ^{13}C abundances around 3 mol % can be determined by the use of OES with a relative error of about 3% (see Table 1).

F. Plant Uptake of ^{18}O-Labeled Phosphate

Radioactive ^{32}P has, of course, been very widely applied in a variety of investigations related to plant uptake of fertilizer phosphate. However, in practice, the 15-day half-life makes it impossible to carry out ecological experiments of extended duration. Hence, Axente et al. [42] conducted a basic study of the application of ^{18}O, as an alternative to ^{32}P (or ^{33}P, $t_{1/2}$ = 25 d), for tracing the uptake of phosphate by *Lolium multiflorum*. The necessary fundamental assumption that there is no significant isotopic exchange at ambient temperatures between oxygen in phosphate ions and that in water containing dissolved CO_2 was shown to be valid by Teys et al. in 1958 [43], and Larsen et al. in 1989 [44].

The plants, grown from seed for about 14 days, were cultivated in equal proportions of sand and glauconite in order to avoid the usual loss of phosphate supply by soil fixation. The authors [42] concluded that only one plant (out of four) contained sufficient ^{18}O for a satisfactory determination of isotope abundance by MS. That plant was fertilized with KH_2PO_4 containing 1.2 mol % excess ^{18}O. At the end of the growth period, CO_2, liberated by pyrolysis (at 230–250°C) of the aerial plant parts, was analyzed by MS and found to contain 0.075 ± 0.010 mol % excess ^{18}O (a dilution factor of 16).

In the analysis of fairly low ^{18}O abundances by OES, the SD is about 0.2 mol % (see Table 1). Consequently, the same precision could have been obtained using OES (instead of MS), if the initial ^{18}O enrichment of the fertilizer phosphate had been 25 instead of 1.2 mol % excess (if we assume the same dilution factor).

In conclusion, the remark at the end of Sec. II.D is worth bearing in mind, when comparing OES and MS as methods of analysis for stable isotopes of importance in environmental research.

VI. SUMMARY

The use of optical emission spectrometry for the analysis of stable-isotope ratios has become an established technique in tracer experiments with nitrogen, carbon, and oxygen in soils and crops. Commercial instrumentation is available for sample preparation as well as the optical analysis, and many studies of pertinent technical details have been reported in the literature. The list of references given below is not exhaustive, since reference is made mainly to pioneer work on the one hand and selected recent publications on the other. However, the latter will enable the interested reader to find many further references.

The advantages and limitations of optical emission spectrometry, as opposed to mass spectrometry, may be summarized as follows. The optical technique is cheaper and simpler, and it permits the routine isotopic analysis of considerably smaller amounts of the element under investigation. Conversely, the precision is limited, which excludes, e.g., the use of depleted ^{15}N as a tracer in most cases. Furthermore, natural variations in stable-isotope abundances cannot be detected by optical emission spectrometry. Nevertheless, much worthwhile tracer research, related to the environment in the broadest sense, can be accommodated adequately using the optical emission method for isotopic analysis of nitrogen, carbon, and/or oxygen.

Naturally, the automation and computerization mentioned above in connection with the description of optical emission analysis for ^{13}C and ^{18}O may be applied equally well to the analysis for ^{15}N and, it is hoped, other stable isotopes in the future.

ACKNOWLEDGMENT

We are grateful to Dr. Günther Ewald, Statron (Fürstenwalde/Spree, GDR); Dr. Sung-Kil Yang, Niels Bohr Institute (Copenhagen); and Dr. Ivan Katic, The Danish Veterinary and Agricultural Library (Copenhagen), for kindly providing the translation of Refs. 23, 40, and 11 plus 43, respectively. Gratitude is also due to the Royal Veterinary and Agricultural University for harboring one of us, now retired. Finally, we wish to thank Annelise S. Knudsen for care taken in processing the manuscript and Hanne Sprogø for painstakingly rendered artwork.

REFERENCES

1. R. Gersberg, K. Krohn, N. Peek, and C. R. Goldman, *Sci.*, *192*:1229 (1976).
2. S. M. Naudé, *Phys. Rev.*, *34*:1498 (1929).
3. H. C. Urey and G. M. Murphy, *Phys. Rev.*, *38*:575 (1931).
4. W. Bleakney and J. A. Hipple, *Phys. Rev.*, *47*:800 (1935).
5. G. Junk and H. J. Svec, *Geochim. Cosmochim. Acta*, *14*:234 (1958).
6. G. Herzberg, *Z. Phys. Chem.*, *B* 9:43 (1930).
7. M. Hoch and H.-R. Weisser, *Helv. Chim. Acta*, *33*:2128 (1950).
8. K. Clusius, *Helv. Chim. Acta*, *33*:2134 (1950).

9. H. P. Broida and M. W. Chapman, *Anal. Chem.*, *30*:2049 (1958).
10. G. Meier and G. Müller, *Isotopenpraxis*, *1*:53 (1965).
11. A. I. Gorbunov and P. A. Sagorec, *Ž. Fiz. Chim.*, *29*:1442 (1955).
12. J.-P. Leicknam, V. Middelboe, and G. Proksch, *Anal. Chim. Acta*, *40*:487 (1968).
13. V. Middelboe and H. Saaby Johansen, *Appl. Spectrosc.*, *32*: 511 (1978).
14. V. Middelboe, in *Soil Organic Matter Studies*, Vol. 2, IAEA, Vienna, 1977, p. 205.
15. D. A. Rennie and M. Fried, in *Proceedings of the International Symposium on Soil Fertility Evaluation*, Vol. 1, Indian Agric. Res. Institute, New Delhi, 1971, p. 639.
16. F. E. Broadbent and A. B. Carlton, in *Nitrogen in the Environment*, Vol. 1 (D. R. Nielsen and J. G. MacDonald, eds.), Academic Press, New York, 1978, p. 1.
17. H. Faust, *Isotopenpraxis*, *1*:62 (1965).
18. H. Faust, *Isotopenpraxis*, *3*:100 (1967).
19. J. M. Bremner and A. P. Edwards, *Soil Sci. Soc. Am. Proc.*, *29*:504 (1965).
20. V. Middelboe, *Tracer Manual in Crops and Soils*, IAEA, Vienna, 1976, p. 108.
21. R. Fiedler and G. Proksch, *Anal. Chim. Acta*, *78*:1 (1975).
22. J. A. Goleb and V. Middelboe, *Anal. Chim. Acta*, *43*:229 (1968).
23. G. S. Lazeeva, A. A. Petrov, and E. P. Stolbova, in *Primenenie stabil'nogo izotopa* ^{15}N *v issledovanijach po zemledeliju*, Kolos, Moscow, 1973, p. 16.
24. H. Faust and R. Reinhardt, *Isotopenpraxis*, *11*:321 (1975).
25. V. Middelboe, *Int. J. Appl. Radiation Isotopes*, *29*:753 (1978).
26. J. C. Burridge and I. J. Hewitt, *Anal. Chim. Acta*, *118*:11 (1980).
27. H. Saaby Johansen and V. Middelboe, *Appl. Spectrosc.*, *36*: 221 (1982).
28. H. Saaby Johansen, *Int. J. Appl. Radiation Isotopes*, *35*:1039 (1984).
29. L. Karlsson and V. Middelboe, in *Isotopes and Radiation in Soil Plant Relationships Including Forestry*, IAEA, Vienna, 1972, p. 211.
30. R. E. Ferguson and H. P. Broida, *Anal. Chem.*, *28*:1436 (1956).
31. A. N. Zaidel and G. V. Ostrovskaya, *Opt. Spectrosc.*, *9*:78 (1959/60).
32. H. Saaby Johansen and V. Middelboe, *Appl. Spectrosc.*, *34*: 555 (1980).

33. H. Saaby Johansen, *Appl. Radiat. Isot.*, *39*:1059 (1988).
34. R. D. Hauck and J. M. Bremner, *Advan. Agron.*, *28*:219 (1976).
35. M. Fried and L. A. Dean, *Soil Sci.*, *73*:263 (1952).
36. Ž. Aleksic, H. Broeshart, and V. Middelboe, *Plant Soil*, *29*: 474 (1968).
37. R. W. F. Hardy, R. C. Burns, and R. D. Holsten, *Soil Biol. Biochem.*, *5*:47 (1973).
38. M. Potts, W. E. Krumbein, and J. Metzger, *Environ. Biogeochem. Geomicrobiol.*, *3*:753 (1978).
39. H. Kouchi and T. Yoneyama, *Ann. Bot.*, *53*:875 (1984).
40. K. Kumazawa and K. Yanagisawa, *Nippon dozyô-hiryô-gakw zassi*, *52*:74 (1981).
41. R. S. McDowell, *Anal. Chem.*, *42*:1192 (1970).
42. D. Axente, M. Abrudean, A. Baldea, L. Calancea, and D. Stroescu, *Isotopenpraxis*, *15*:312 (1979).
43. R. V. Teys, T. S. Gromova and S. N. Kochetkova, *Doklad. Akad. Nauk SSSR*, *122**:1057 (1958).
44. S. Larsen, V. Middelboe, and H. Saaby Johansen, *Plant Soil*, *117*:143 (1989).

---

*This volume number was misprinted (as *126*) in the literature list at the end of Ref. 42.

11

Analysis of Nitrogen Isotope Ratios by Mass Spectrometry

DAVID ROBINSON *Scottish Crop Research Institute, Dundee, Scotland*

KEITH A. SMITH *The Edinburgh School of Agriculture, Edinburgh, Scotland*

I. INTRODUCTION

The function of the mass spectrometer in nitrogen isotope analysis is in most cases identical with that of the optical emission spectrometer (Chap. 10), and although each instrument has unique advantages, in many instances either would perform the task equally well. Both instruments exploit a physical property of the nitrogen molecule, N_2, to determine the relative amounts of each of the three possible species, $^{14}N_2$, $^{14}N^{15}N$, and $^{15}N_2$. In optical emission spectrometry, N_2 molecules are separated on the basis of their vibrational properties, whereas in mass spectrometry, charged ions are separated according to their mass to charge (m/e) ratio. The terms *mass spectrometer*, *mass spectroscope*, and *mass spectrograph* all refer to instruments performing this function, the difference being limited to the way in which the ions are detected and measured.

A further similarity between nitrogen isotope analysis using a mass spectrometer and analysis using an optical emission spectrometer lies in the procedures used to convert soil or biological material to elemental nitrogen. Since N_2 is usually the material analyzed in both instruments, techniques developed for one can frequently be adapted for the other.

Revision of Chapter 9 by Anthony Haystead in first edition (1983).

Historical aspects of the discovery of the isotopic nature of nitrogen and the properties of the various isotopes are described in some detail in Chap. 10 and are not discussed further here. This chapter describes the principles on which the most commonly encountered mass spectrometers operate and the advantages that isotope mass spectrometry has over optical emission spectrometry, and discusses briefly those applications to which mass spectrometry is best suited. Detailed reviews of the practical aspects of nitrogen isotope analysis have been published [1–5] and Hauck and Bremner [6,7] have discussed the applications of tracer nitrogen work in soil science and plant analysis.

II. MASS SPECTROMETERS

The development of the mass spectrometer followed the discovery in 1897 of "positive rays" (in fact, positively charged ions) that were seen to emerge from the perforated cathode of a partially evacuated discharge tube (Fig. 1). Instruments built before 1920 contained all the fundamental elements present in modern instruments. Figure 2 is a schematic diagram of a mass spectrometer. In the *source*, positively charged ions are generated in a gas or vapor and restricted into a narrow beam. The beam of ions is further restricted and accelerated electrically into the *analyzer* where the beam is separated into its components in electric and/or magnetic fields. The separated components of the beam are allowed to impinge on the *collector* assembly that, in turn, is connected to some sort of voltage-measuring device.

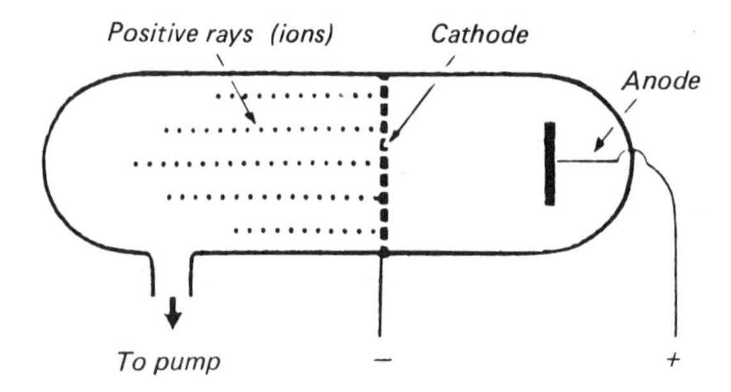

FIG. 1 Positive rays in a partially evacuated discharge tube.

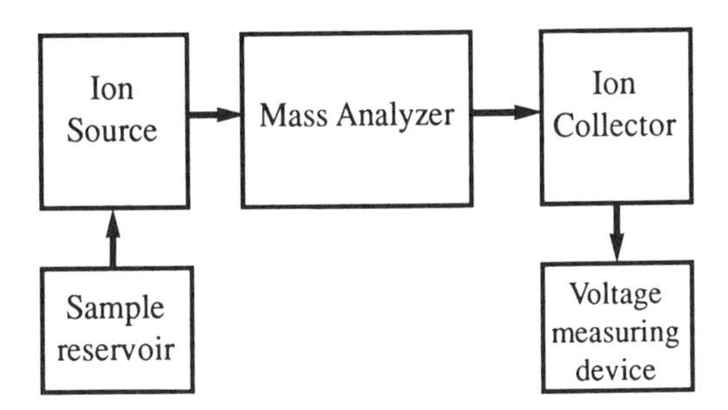

FIG. 2 Schematic diagram of a mass spectrometer.

It is not within the scope of this chapter to discuss in detail the historical development of the isotope mass spectrometer; interested readers are directed to review material by Howe et al. [8] and Chapman [9].

A. Operating Principles

Mass spectrometer development, particularly over the last 25 years, has been extremely rapid and now a number of types of mass spectrometer are available. Not all types are suitable for routine isotope analysis, however, In the main, single-, double-, and triple-collector magnetic sector instruments and, to a lesser extent, quadrupole mass spectrometers have been used in isotopic tracer studies. A detailed discussion of mass spectrometer types can be found in Chapman [9]. All isotope mass spectrometers have certain features in common; they all operate under high vacuum conditions and have the basic ion source-analyzer-collector arrangement shown in Fig. 2. Major differences between magnetic sector, double focusing, quadrupole, cycloidal focusing, and time of flight mass spectrometers lie in the analyzer section, in that the ion beam is separated into its components in different ways.

Several types of source are used in commercially available instruments. Chemical ionization sources in particular are used extensively in the high-resolution mass spectrometry of labeled and unlabeled organics and metabolic intermediates. The isotopic ratio analysis of inorganic gases, however, is almost invariably carried out using electron impact ionization.

The resolution and precision of an isotope ratio mass spectrometer to a large extent determine how well it performs. Resolution, the ability of the instrument to separate the ion spectrum, is expressed in terms of the peak width (ΔM in atomic mass units, $^{16}O^+$ =

16 AMU) at 0.1 or 0.5 of the peak height h. Resolving power R is often described as

$$R = \frac{M}{\Delta M_{0.1 \text{ or } 0.5}} \tag{1}$$

where M is the value of the peak in atomic mass units and ΔM the width of the peak. Unfortunately, there are several other definitions of resolution and resolving power, so a prospective mass spectrometer purchaser is well advised to check exactly how R has been calculated for a particular mass analyzer system. It should also be borne in mind that resolution is affected by the collector aperture or slit width of a magnetic sector instrument and by field parameters in a quadruple. In both cases, resolution can be increased or decreased to a certain extent with reciprocal changes in sensitivity.

Precision defines the ability of a mass spectrometer to differentiate reliably between samples of varied isotopic abundance. Quantitatively, precision or internal reproducibility is expressed either as a relative standard deviation for a set of measurements or a minimum detectable difference in isotopic abundance. Both of these numerical estimates of precision are affected by sample size (relative to the optimum for the inlet system) and also the isotopic abundance range of the samples. The figures quoted by most manufacturers refer to an optimum sample size close to natural abundance; only rarely is it possible to determine how the instrument will perform at high enrichment.

B. Single-Collector Mass Spectrometers

1. *Magnetic Sector Mass Spectrometers*

Figure 3 illustrates diagrammatically how one of the simplest types of mass spectrometer, a 180° single-collector instrument, operates. The whole of the source-analyzer-collector assembly is contained within an evacuated stainless steel enclosure that is continuously pumped, so that during sample admission the pressure equilibrates between 10 and 10^{-2} mPa. Low pressure must be maintained in the mass spectrometer so that the mean free path of the ions (the distance an ion can move between collisions) is sufficiently large with respect to the geometry of the analyzer assembly. If this requirement is not met, then scattering due to ion-ion and ion-molecule collisions prevents sharp beam focusing. In the instrument shown in Fig. 3, positive ions are generated in the material to be examined by electrons that are accelerated from a heated filament into a space called the ion chamber. Ions emerge under the influence of voltages on plate E and the ion repeller and are further accelerated

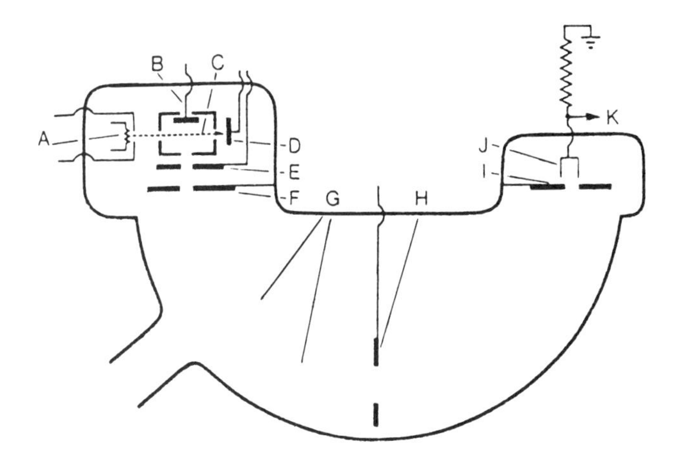

FIG. 3 180° magnetic mass spectrometer. *A*, filament; *B*, ion repeller, *C*, electron beam; *D*, electron trap; *E* and *F*, beam focusing and ion accelerating plates; *G*, ion beams; *H*, alpha slit; *I*, collector slit; *J*, collector plate; *K*, output.

and collimated by a much larger electric field applied between *E* and *F*. A permanent magnet encloses the analyzer such that the homogeneous magnetic field is perpendicular to the diagram and the ion beam is deflected through 180° onto a collector plate *J*. In a constant magnetic field, the radius of the flight path of an ion is a function of its kinetic energy and its mass. The relationship between the radius of flight r, the accelerator voltage V between B and C, the mass m, charge e, and velocity v of an ion is derived as follows:

$$\frac{1}{2} mv^2 = eV$$

Therefore,

$$v^2 = \frac{2eV}{m} \qquad (2)$$

If the flux density of the magnetic field through which the ions pass is B and the source, alpha, and collector slits are arranged so that only a fraction of an ion beam arrives at the collector, the system will select ions with a narrow range of flight radius, the relationship between r and B being

$$r = \frac{mv}{Be} \tag{3}$$

Combining Eqs. (2) and (3) and eliminating v, we obtain

$$\frac{e}{m} = \frac{2V}{B^2r^2} \tag{4}$$

An analysis of a mixture of ions of different m/e values (in fact, m values since ions with a single positive charge predominate) can hence be made by varying V or B and measuring the current passing from source to collector as different parts of the ion spectrum enter the collector aperture, allowing the analyzer to "scan" a range of m/e values. In practice, the voltage produced by the ion current across a very high resistance ($10^{10}-10^{12}$ ohms) is measured using a dc amplifier and voltmeter or potentiometric recorder.

In modern instruments, the ion current is often first amplified by an electron multiplier and subsequently by an operational amplifier using field effect transistors. These first stages of signal amplifications are highly susceptible to outside influences, and to reduce noise the components are usually enclosed in a metallic, often evacuated, enclosure.

2. *Quadrupole Mass Spectrometers*

Figure 4 is a diagrammatic representation of a quadrupole mass spectrometer, from which it is immediately apparent that the ion source and collector sections are identical to those in the magnetic sector mass spectrometer; only the analyzer is different. The quadrupole field into which the ions are accelerated is generated by applying dc and radio frequency (rf) voltages to four metal rods held equidistant and parallel to each other [8,9]. The resolution of the mass analyzer depends to a large extent on the precision of the quadrupole assembly, which in consequence is manufactured to very rigorous standards. Singly charged ions entering the quadrupole field undergo oscillations that are dependent on field parameters and the mass of the ion. Under any given set of conditions, only ions having a particular mass will emerge and impinge on the collector plate. Mass spectra are obtained by varying the dc and rf components of the voltage supply to the quadrupole assembly and so changing the mass filter characteristics of the quadrupole. Conventional quadrupole mass spectrometers have the unique advantage of extremely high scanning speeds, but for isotopic analysis do not have the precision of magnetic sector mass spectrometers [10].

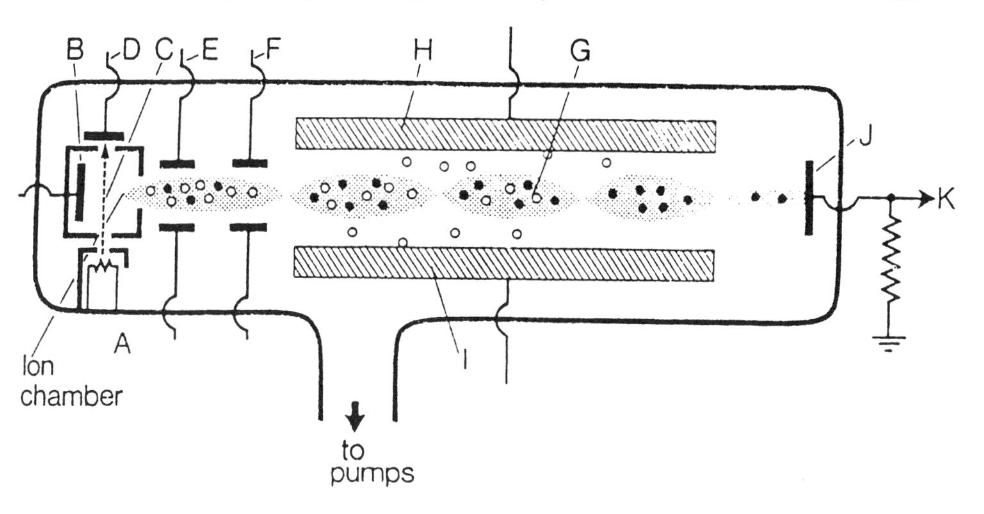

FIG. 4 Quadrupole mass spectrometer. *A*, filament; *B*, ion repeller; *C*, electron beam; *D*, electron trap; *E* and *F*, ion beam focusing plates; *G*, ion beams; *H* and *I*, two of the four quadrupole rods; *J*, collector; *K*, output.

C. Multiple-Collector Systems

In the magnetic sector instruments described in Sec. II.B, ion beams are focused separately onto the collector electrode. If two ion currents, say, *m/e* 28 ($^{14}N_2$) and *m/e* 29 ($^{14}N^{15}N$), are to be compared, they must be measured sequentially. A loss of precision can occur in this procedure because the conditions under which ions are generated, the pressure and composition of the sample, and the characteristics of the measurement system are all susceptible to fluctuations during analysis. In a multiple-collector instrument such effects are minimized because two or three ion beams are collected and measured simultaneously. The two ion beams in a dual collector instrument can be compared electronically and the ratio computed directly. Modern instruments designed for this specific purpose frequently make a series of automatic comparisons of sample with a reference gas and are able to measure differences in ^{15}N abundance down to 0.00001 atom % ^{15}N.

This type of mass spectrometer has been developed considerably during the 1980s in conjunction with automated direct combustion systems of sample preparation and elemental analysis (see Sec. III.B and Chap. 6). The main applications have been in the determination of ^{15}N (or ^{13}C) and total N (C) in biological materials. Manufacturers have recognized that the precision expected of a mass spectrometer by an analytical chemist far exceeds that possible when analyzing any biological or environmental sample. For such applications, therefore, it is possible to sacrifice some instrumental precision without compromising the statistical accuracy of the result. This has allowed modern mass spectrometers to be designed to more robust specifications than had previously been possible.

The high vacuum conditions essential to the operation of single-collector instruments are not as critical in those designed specifically to analyze biological samples. This is because the operating pressure of the mass spectrometer is ca. 500 μPa in the ion source, compared with ca. 10^{-2} to 10^{-1} μPa in single-collector instruments. Consequently, the pumping system can be smaller and no special inlet manifold is required. Smaller mass spectrometers mean that bench top or even portable instruments can now be built.

Dedicated mass spectrometers usually have three collectors, preset to a certain configuration and capable of analyzing only a restricted *m/e* range. All modern instruments are controlled by computer. Although this limits the extent to which the mass spectrometer can be set up and adjusted manually, the use of menu-driven software to control amplifier settings, voltages, sample handling, and data collection greatly speeds up routine analysis. It means also that relatively little knowledge of mass spectrometry per se is required. Other advantages of such systems to soil scientists and biologists are given in Sec. III.B.

Further improvements in precision are pointless for the analysis of samples from most tracer experiments, given the aforementioned sample variability. Any useful future improvement will be in sample handling and processing. (See Sec. III.) For some applications, such as determination of differences in natural ^{15}N abundances, high precision mass spectrometer will still be needed.

D. High-Vacuum Technology

Three main factors dictate the maximum operating pressure of the mass spectrometer. First, as discussed in Sec. II.A, analyzer resolution is dependent on the main free path of the ions being sufficiently large with respect to the geometry of the flight path in the analyzer. Second, at pressures greater than about 10 mPa, space charge effects upset the linearity of the ionization process. Third, the sensitivity of the mass spectrometer at any particular point in the mass spectrum is dependent on the ion current due to residual gases in the source. Ideally, the background ion current at any particular value of *m/e* should be negligible with respect to the sample ion current. It is not, however, sufficient simply to attach to the mass spectrometer a pumping system capable of removing the gas in question extremely quickly, since this would necessitate high rates of sample admission to maintain a pressure in the source sufficient for an analysis to be made. In some mass spectrometers a pumping system is used that pumps efficiently enough to give reasonable background ion currents but slowly enough to maintain instrument sensitivity, so that small samples can be admitted to the source. In other instruments, pumping system efficiency is maximized, and the sample is admitted directly into the ion chamber so that the pressure at the point of ion generation can be one or two orders of magnitude greater than that in the analyzer. With such a system, the objective of negligible backgrounds for the larger residual ion currents (e.g., at *m/e* 28, the $^{14}N_2$ peak) is attained.

The low pressures required in the mass spectrometer cannot be attained using mechanical, oil-sealed, rotary pumps. Combined rough and high-vacuum pumping systems must be used. The vapor (or diffusion) pump backed by a mechanical rotary pump is the type most commonly used for isotope mass spectrometers. Other systems, such as ion pumps backed by sorption pumps, have been used, but not widely. Vapor pumps operate by creating jets of oil or mercury vapor that transfer momentum to residual gas molecules, compressing the gas in the region of the rotary pump outlet. The ultimate vacuum in a continuously pumped system is limited by (1) leaks into the system from outside, (2) backstreaming of vapor from the pump into the mass spectrometer, (3) the rate at which substances are desorbed from the inside of the evacuated system (outgassing), and (4) the efficiency of the pumping system. Leak tightness is obtained by reducing demountable joints to a minimum and by using precision flanges with soft metal gaskets (copper, gold, or aluminum) when joints are essential. Ingress of oil vapor (backstreaming) is reduced by incorporating baffles (which may be refrigerated) or cold traps between the vapor pump and analyzer assembly. Outgassing at room temperature can be a very slow process, depending on the nature of the gases being evolved and the material from which the evacuated system is constructed. The rate of outgassing increases with temperature, and for this reason the whole source-analyzer-collector assembly can usually be heated (baked) to 200–300°C. In this way, the ultimate vacuum of the system is attained as quickly as possible.

A number of isotope ratio mass spectrometers currently available employ turbomolecular high-vacuum pumps rather than vapor pumps. In fact, the operating principle of the two is analogous. In the vapor pump, momentum is transferred to the residual gases by a supersonic vapor jet, whereas in the turbomolecular pump, a turbine rotating at 30,000–60,000 rpm is used. The turbomolecular pump has the advantage that there is no pump fluid backstreaming. It is, however, more expensive to purchase and has a higher maintenance requirement.

E. Sample Inlet Systems

In most isotope mass spectrometers the sample is admitted continuously into the pumped system. Clearly, if the mass spectrometer is to be maintained at a low pressure during analysis, a means must be devised to admit sample gas (1) at a constant rate and (2) in extremely small amounts. This is done usually by introducing a flow restrictor (or leak) into the inlet of the mass spectrometer. The way in which gas flows through such a leak can be either *molecular* or *viscous*, depending on the pressure at the entrance side of the leak and the construction of the leak itself.

Molecular flow occurs when the pressure at the entrance to the leak is low and the mean free path of the sample molecules is large relative to the dimensions of the leak. Under molecular flow conditions, lighter molecules pass into the source more readily and the sample reservoir becomes enriched in the heavier components of the mixture. In the source itself, however, movement of gases is also molecular and lighter gases are pumped away more quickly. Hence, at any point in time, the relative numbers of chemically identical ions generated in the source reflect accurately the partial pressures of the parent molecules in the sample reservoir. Isotope analysis using a typical molecular leak requires the sample reservoir to be large and sample pressure to be low, usually between 10 and 100 Pa. Molecular leaks are manufactured by sealing a plug of porous material into the sample inlet tube close to its connection with the mass spectrometer. Sintered silicon carbide, quartz, and glass are often used as refractive, chemically inert, microporous materials for this purpose.

At higher pressures, gas moves along a tube in much the way that water flows in a pipe, and the flow is said to be viscous. A viscous leak is usually constructed by crimping a capillary inlet tube close to its point of entry to the mass spectrometer. Clearly, as the pressure falls across a constriction of this type, gas flow changes from viscous to molecular, so the overall characteristics of the leak depend on its construction and operating pressure. For example, a relatively long and narrow connection to the sample reservoir will restrict the backflow of heavy molecules accumulated at the constriction, reducing changes in the sample reservoir. Viscous inlets are generally preferred for isotope ratio measurements, because it is more convenient to handle gaseous samples in small volumes at relatively high pressures (5–10 kPa), but many commercially available isotope mass spectrometers use molecular leaks. Two gas-handling systems are shown in Fig. 5, one for a double

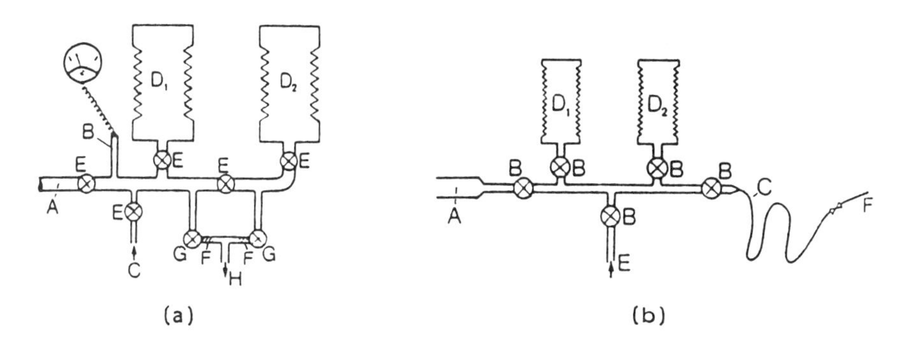

FIG. 5 (a) Inlet and gas-handling system suitable for mass spectrometer with a molecular leak. *A*, outlet to vapor pump; *B*, Pirani gauge; *C*, sample inlet; D_1 and D_2, stainless steel bellows reservoirs (200–1000 ml); *E*, toggle-operated bellows gas-handling valves; *F*, porous leaks; *G*, diaphragm inlet valves (zero dead space); *H*, mass spectrometer inlet. (b) Inlet and gas-handling system suitable for mass spectrometer with a viscous leak. *A*, outlet to rotary pump; *B*, toggle-operated bellows gas-handling valves; *C*, capillary leak with crimp (0.5 mm × 0.1 mm ID); D_1 and D_2, stainless steel bellows adjustable-volume reservoirs (50–150 ml); *E*, sample inlet; *F*, connection to mass spectrometer.

molecular leak suitable for isotope ratio comparisons (Fig. 5a) and one suitable for a viscous inlet into a dual-collector mass spectrometer (Fig. 5b). Gas-handling systems can be constructed from glass, using greaseless high-vacuum stopcocks, but all-metal construction using packless (bellows) or diaphragm valves gives longer trouble-free service and has the advantage that these systems can be baked periodically.

F. Interpretation of Ion Current Measurements

When a sample of combined nitrogen is converted to N_2 (see Sec. III), the proportions of the three molecular species $^{14}N_2$, $^{14}N^{15}N$, and $^{15}N_2$ will be distributed binomially. In a sample containing fractional amounts *a* and *b* of ^{14}N and ^{15}N atoms, respectively, by definition $a + b = 1$, and the relative amounts of the three species will be

$$(^{14}N_2):(^{14}N^{15}N):(^{15}N_2): = a^2:2ab:b^2 \tag{5}$$

Since there is no significant difference in the ionization potential of the three species, the ratio of molecular ions with *m/e* values

28, 29, and 30, and hence the ion currents i_{28}, i_{29}, and i_{30} measured at the collector, will be in the same proportion as the original diatomic molecules. By measuring any two of the three ion currents, the third can be calculated from Eq. (5). The isotopic composition of a sample of nitrogen is usually expressed in terms of its content of the minor isotope, ^{15}N, either in absolute terms as a ^{15}N-abundance in atom percent (%) ^{15}N

$$\text{Atom \% }^{15}\text{N} = \frac{\text{number of }^{15}\text{N atoms}}{\text{total number of N atoms}} \times 100 \tag{6}$$

or, alternatively, with respect to the ^{15}N-abundance of atmospheric N_2 (0.3663 ± 0.0004) as an enrichment or depletion from this norm

$$|\text{atom \% }^{15}\text{N of sample} - 0.3663| = \text{atom \% }^{15}\text{N excess or depletion}$$

Small variations in ^{15}N abundance, e.g., variations in the ^{15}N content of naturally occurring soil components, are often measured in terms of parts per thousand differences ($\delta^{15}N$) from a reference gas, usually atmospheric N_2

$$\delta^{15}\text{N} = \frac{R_{\text{sample}} - R_{\text{reference}}}{R_{\text{reference}}} \times 1000 \tag{7}$$

where $R = {}^{15}N/{}^{14}N$.

With this system, one $\delta^{15}N$ "unit" is equivalent to 3.7×10^{-4} atom % ^{15}N. Some authors [11,12] use a slightly different definition of $\delta^{15}N$ in that they use ^{15}N concentrations in their calculation rather than R. Close to natural abundance, the difference in $\delta^{15}N$ values calculated using the two procedures is very small. Equation (6) can be written in the form

$$^{15}\text{N abundance} = \frac{(^{14}\text{N}^{15}\text{N} + 2\,^{15}\text{N}_2) \times 100}{2\,^{14}\text{N}_2 + 2\,^{14}\text{N}^{15}\text{N} + 2\,^{15}\text{N}_2} \tag{8}$$

or in terms of ion currents

$$^{15}\text{N abundance} = \frac{(i_{29} + 2i_{30}) \times 100}{2(i_{28} + i_{29} + i_{30})} \tag{9}$$

Samples of very low or high ^{15}N content produce low ion currents at i_{30} and i_{28}, respectively, that are more difficult to measure accurately. Depending on the ^{15}N content of the sample, either the i_{28}/i_{29} or i_{29}/i_{30} ratio is measured and the ^{15}N abundance calculated from the equations

$$\text{Atom \% }^{15}\text{N} = \frac{100}{2R_1 + 1} \tag{10}$$

where $R_1 = i_{28}/i_{29}$, and

$$\text{Atom \% }^{15}\text{N} = \frac{200}{R_2 + 2} \tag{11}$$

where $R_2 = i_{29}/i_{30}$.

In practice, the determination of either of these ratios depends on the characteristics of the mass spectrometer that is being used. For example, if residual gases in the source give rise to significant ion currents, then background values must be subtracted from the sample currents. Similarly, if atmospheric contamination is likely to be significant, corrections must be made to the sample ion currents to allow for the components due to atmospheric N_2. In critical studies, sample pressure may affect the measured ratios. If it is not possible to standardize sample pressures, then this phenomenon must be taken into account. Mook and Grootes [13] and Kerven and Saffigna [14] discuss corrections that must be applied to measured ion current ratios in the high-precision determination of isotopic abundances. Statistical procedures for high-precision determination of ^{15}N values are described by Steele et al. [15].

G. Analytical Precision

Table 1 shows the analytical precision possible with various mass spectrometers that are currently available. For samples prepared by direct combustion (Dumas) procedures (see Sec. III), the standard deviation for the analysis of samples containing ^{15}N at natural abundance is $\geq 1 \times 10^{-4}$ atom %, depending on sample type. Analysis of samples prepared using traditional digestion-distillation-oxidation (Kjeldahl–Rittenberg) procedures (see Sec. III) are potentially more precise. However, the figures quoted in Table 1 refer to the ^{15}N determination of standard N_2 gas and so do not include errors associated with sample preparation. Pruden et al. [4] reported a standard deviation of 5×10^{-4} for the analysis of a standard grass sample enriched in ^{15}N (0.4156 atom % excess), prepared using the Kjeldahl–Rittenberg method. In practice, the analytical precision of the two preparation procedures are comparable for many types of plant and soil samples.

III. SAMPLE PREPARATION

A. Introduction

It would be possible to determine the ^{15}N abundance of soil- or plant-derived nitrogen by converting it into any stable, low-

TABLE 1 Analytical Precision of Isotope Mass Spectrometers at Natural Abundances of ^{15}N

Manufacturer	Instrument	Sample preparation system	Sample type	Analytical precision (standard deviation) Atom % ^{15}N	$\delta\, ^{15}N$ (0/00)
Europa	Tracermass	Dumas	Clover	1.3×10^{-4}	0.34
			Grass	5×10^{-4}	1.4
			Urea	1.2×10^{-4}	0.31
			Ammonium sulfate	5×10^{-4}	1.4
VG Isogas	Isomass	Dumas	Grain	2.4×10^{-4}	0.65
			Urea	3.6×10^{-4}	0.97
VG Isogas	Sira	Dumas	Soil	4×10^{-4}	1.1*
			Crop	4×10^{-4}	1.1*
			Ammonium sulfate	2×10^{-4}	0.54*
Finnegan MAT	Delta	Dumas	Acetanilide	2×10^{-4}	0.60
VG Isogas	Sira II	Kjeldahl-Rittenberg	Pure N_2	7×10^{-6}	0.02
VG Isogas	Prism	Kjeldahl-Rittenberg	Pure N_2	2×10^{-6}	0.006

Data obtained from manufacturers, except statistics marked with a * which were obtained from J. M. Bracewell, G. W. Robertson, and M. E. Reid, unpublished results. All Dumas data are for actual sample analyses ("external" precision); those for Kjeldahl–Rittenberg systems are for the analyses of standard gases ("internal" precision).

molecular-weight compound of nitrogen. Ideally, this compound should be gaseous at room temperature and contain only one or two nitrogen atoms. Elemental nitrogen N_2 is such a compound, and consequently it is used almost universally (but see Sec. III.C). N_2 is very stable in conventional ion sources, the contribution from atomic ions at m/e 14 and 15 being negligible. In addition, since only nitrogen atoms are present, the mass spectrum of N_2 is not confused by ions originating from other isotopic elements, whereas N_2O, for example, produces two species at *m/e* 46: $^{15}N^{15}N^{16}O$ and $^{14}N^{14}N^{18}O$, which are not separated in single-focusing mass spectrometers.

The principal requirements of procedures used to prepare N_2 from experimental material are as follows:

1. Quantitative conversion of sample nitrogen to pure N_2.
2. Random pairing of nitrogen atoms during formation of N_2.
3. Removal of contaminant gases and volatiles, particularly water vapor, CO_2, and CO.

B. Sample Preparation Procedures

Two principal sample preparation procedures are used in the determination of ^{15}N in plant and soil material: (1) digestion-distillation-oxidation (Kjeldahl–Rittenberg) and (2) direct combustion (Dumas). To a large extent, the one that is available dictates how particular types of sample should be collected and handled; the options are summarized in Fig. 6.

1. *Digestion-Distillation-Oxidation (Kjeldahl–Rittenberg) Procedures*

Organic nitrogen is converted by Kjeldahl digestion to ammonium-N, which is then recovered by steam distillation or microdiffusion in the presence of a strong base (NaOH), the ammonia liberated being trapped as ammonium in dilute acid. Forms of inorganic nitrogen such as nitrate or nitrite present in extracts of plants or soils can be converted to ammonium without digestion by reduction with Devarda's alloy, and then distilled or diffused in the presence of a weak base (e.g., MgO). The ammonium solutions are usually dried, at temperatures of <80°C to avoid volatilization, and in an ammonia-free atmosphere, to prevent adsorption of ammonia. These methods have been described fully by Hauck [5] and Reeder [16].

Usually, it is essential to determine the total nitrogen content of samples whose isotopic composition is of interest. This is done by titrating an aliquot of the distillate or diffusate against standard acid [5,17]. Pruden et al. [4] showed that when samples of different ^{15}N enrichment are processed sequentially, a significant isotopic "memory" is retained on the distillation apparatus, caused by the adsorption of ammonia. A convenient way to eliminate this is to distill an aliquot of the digest or extract and to use this distillate for the determination of the sample's total nitrogen content. This will wipe out any memory of the previous sample, allowing a second aliquot to be distilled safely, and it is this distillate that is used for the ^{15}N analysis. Alternatively, any adsorbed ammonia

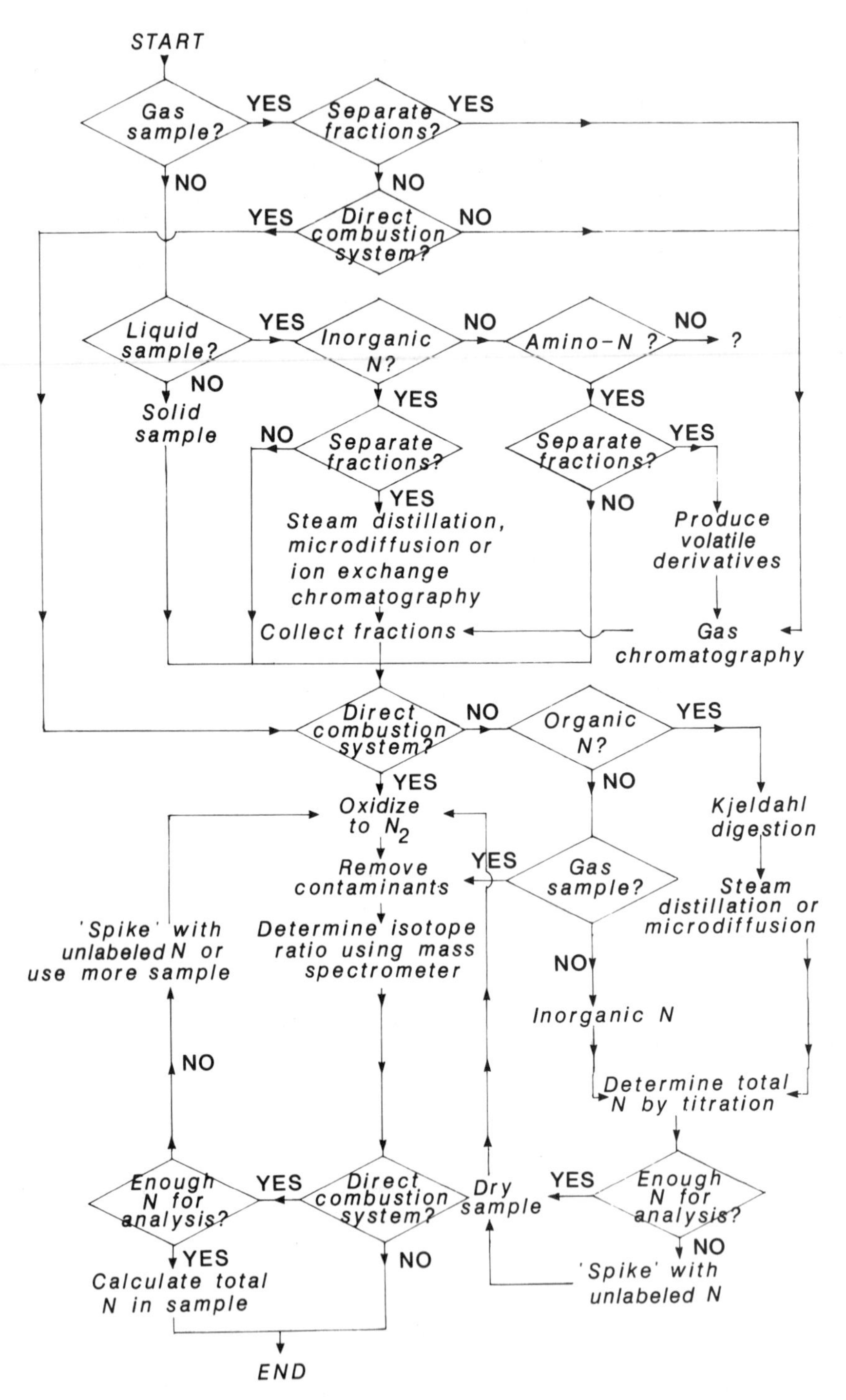

FIG. 6 Flowchart showing the preparative steps required for the ^{15}N determination of different types of sample.

can be removed between samples by distilling ethanol or flushing the apparatus with steam while the coolant supply to the condenser is switched off [4].

The ammonium-N thus prepared is oxidized to N_2 by reaction under a vacuum with a degassed solution of alkaline lithium or sodium hypobromite, the Rittenberg method [18]. This can be done by placing the dried ammonium salt into one arm of a Rittenberg flask and the hypobromite into the other (Fig. 7). The flask is attached to an inlet manifold on the mass spectrometer. As the flask is rotated, an excess of hypobromite is brought into contact with the ammonium, liberating N_2 that is then admitted to the mass spectrometer via a cold trap to remove water vapor and contaminants such as CO_2, Br_2, and N_2O [5].

The original Rittenberg procedure has now been superseded by methods based on that described by Ross and Martin [19], but the chemistry involved is the same. Ammonium solutions are dried in glass vials [16]. Each vial is connected to a stainless steel inlet system (Fig. 8) connected to the mass spectrometer. Hypobromite is delivered to the vial from a reservoir located above the apparatus. Ross and Martin recommended freezing the vial in liquid N_2 to entrap the contaminants before allowing N_2 into the mass spectrometer, but Pruden et al. [4] showed that this is not necessary. Pruden et al. modified Ross and Martin's method principally by separating the hypobromite reservoir from the inlet system. This prevents accidental spillages of hypobromite onto the instrument. The hypobromite is delivered to the sample through a rubber septum using

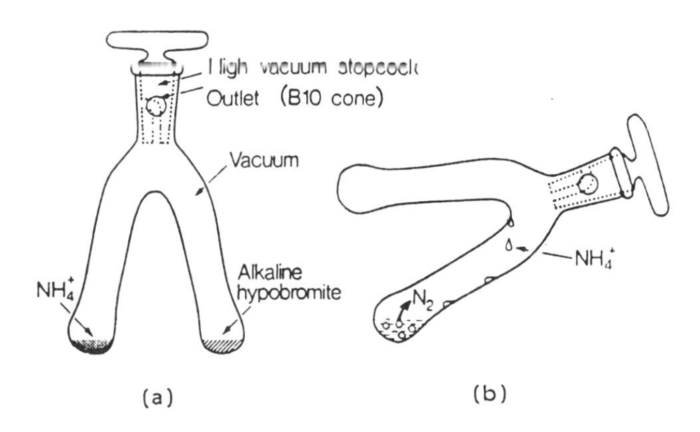

FIG. 7 Rittenberg flask. (a) Upright position for reagent addition and degassing. (b) Tilted to add NH_4^+ to the alkaline hypobromite.

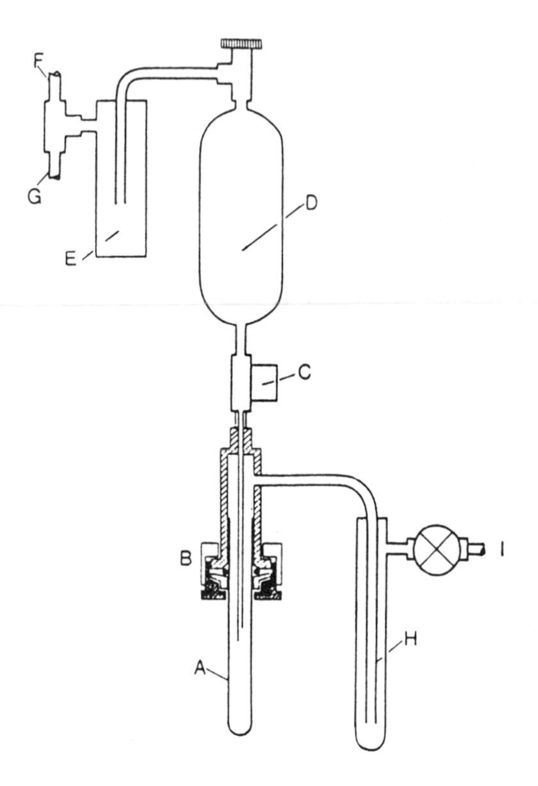

FIG. 8 Stainless steel N_2 generation system based on Ross and Martin's [19] glass apparatus: *A*, 10 mm glass tube; *B*, rapid vacuum seal; *C*, PTFE diaphragm valve (Nupro Corp., Cleveland, Ohio); *D*, hypobromite reservoir; *E*, trap; *F* and *G*, pump and helium outlets; *H*, cold trap; *I*, inlet to mass spectrometer gas-handling system.

a syringe. Two possible sources of error with this procedure are the absorption of atmospheric N_2 into the hypobromite solution via the syringe needle, and leakages of atmospheric N_2 into the inlet system when the needle is inserted or removed from the septum. With care and experience, however, these errors are small.

The most commonly used reference gas in Kjeldahl–Rittenberg systems is atmospheric N_2. Marriotti [20] has demonstrated the remarkable stability of the natural abundance of ^{15}N in the atmosphere. The reference gas is used to calibrate the mass spectrometer on a daily or twice-daily [4] basis. During the analysis of a sample, the ion beam current generated by the reference gas is compared with that of the sample, allowing the ^{15}N enrichment of the sample to be calculated. However, it is essential to include in any batch of samples some internal standards. These may be ammonium salts of known ^{15}N abundance (available from U.S. Department of Energy, Los Alamos, New Mexico) that have been subjected to the same procedure as the samples, or, if available, plant or soil material of uniform, known ^{15}N abundance.

All manual methods that involve the Rittenberg method are laborious and messy. B. B. McInteer and J. G. Montoya (cited by Hauck [5] and Bremner and Hauck [7]) have developed an automated Rittenberg system using an x-y autosampler controlled by a computer. Such techniques must now be regarded as obsolete since automated direct combustion systems became available commercially.

2. Direct Combustion (Dumas) Procedures

The principle underlying Dumas system is self-explanatory: Sample nitrogen is combusted and converted directly to N_2. The combustion of the sample is achieved in a stream of pure O_2 on an oxidant, e.g., NiO, at a temperature of ca. 1000°C. A catalyst, the sealed tin cup that holds the sample, increases the combustion temperature to ca. 1700°C. The combustion products are swept by a carrier gas (He) over a reductant (Cu) at ca. 700°C. Water vapor and CO_2 are removed by an absorbent column. The N_2 is carried via a GC column (which separates the N_2 from trace contaminants) and stainless steel capillary into the mass spectrometer. An example of a Dumas system is shown schematically in Fig. 9. In modern instruments, the mass spectrometer is a dedicated triple-collector system (see Sec. II.B). Details of the development of such systems are given in Refs. 21–24.

In direct combustion systems no reference gas is used. Instead, internal standards are used to calibrate the instrument. These standards are nitrogen compounds of known total nitrogen and ^{15}N contents that are close to those expected in the samples. The standards' atom % ^{15}N and % total nitrogen values are entered into the computer before analysis. When standards are analyzed, the ion beam current and current ratios generated are matched with the standards' ^{15}N and total nitrogen contents. During the automated analysis of a batch of samples (see below), standards are located at designated intervals (ca. between every five or ten samples) within the batch. A common feature of such systems is software that allows results to be corrected at the end of a batch analysis for any instrument drift that might have occurred during the analysis. It is good practice to include some standards and blanks (unlabeled standards) as "samples" in the batch to check that their ^{15}N and total nitrogen contents are being determined accurately.

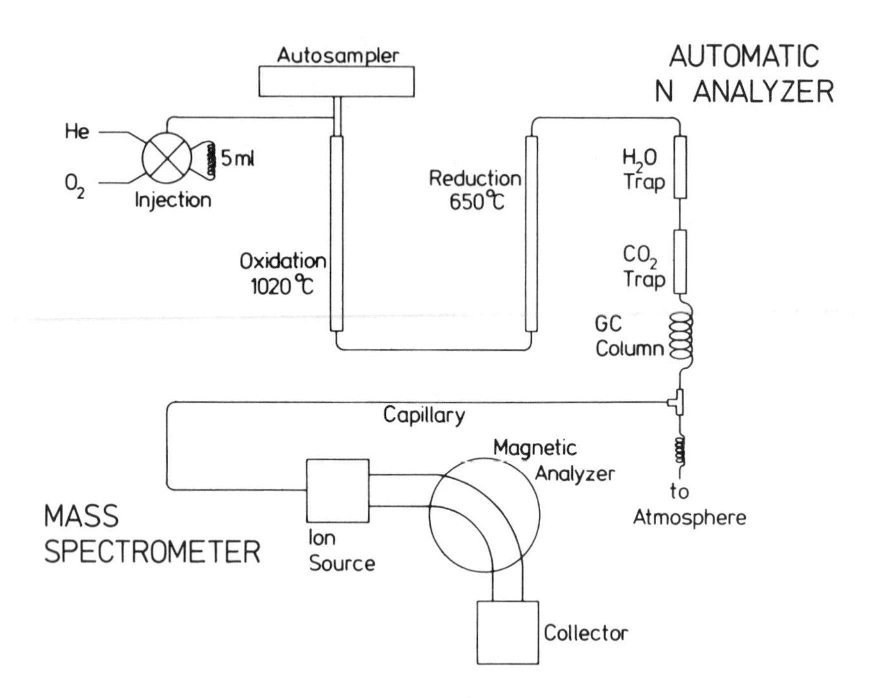

FIG. 9 Schematic diagram of automatic nitrogen analyzer with direct combustion system, interfaced with isotope mass spectrometer.

Dumas combustion methods have many important advantages over Kjeldahl–Rittenberg systems for the determination of ^{15}N in soil and plant materials. The time-consuming digestion procedures are eliminated. Distillation or diffusion is necessary only if certain forms of nitrogen are to be separated and analyzed individually (Fig. 6). Conversion of sample nitrogen to N_2 is performed automatically and completely; the problem of nonquantitative conversion of nitrate-N to ammonium-N does not exist. This feature of direct combustion systems tends to give slighly higher total N results compared with Kjeldahl digestion-distillation, as the latter methods do not include nitrate and nitrite, unless a predigestion reduction step is included [17]. Samples can be analyzed automatically using an autosampler that, typically, carries 50 or 60 samples including blanks and standards. Analysis time is rapid, ca. 5 min per sample, allowing the many samples generated in agronomic or ecological experiments to be handled easily. The sample's total nitrogen content is determined, as well as its ^{15}N enrichment, using the total ion beam current generated in the mass spectrometer. Sometimes a thermal conductivity detector (TCD) is placed between the GC column and atmospheric vent. The TCD output is another means by which the sample's total nitrogen content can be determined; this serves as a useful cross-check on that produced by the mass spectrometer.

Direct combustion systems do, however, have a number of disadvantages compared with the conventional digestion-distillation-oxidation method. Some of these are discussed below as they are important for certain types of sample. The major disadvantage applies to all, that of "blind" analysis. Because total nitrogen content is determined on the same material as the ^{15}N enrichment, there is no way of knowing before the sample is introduced into the instrument if there is enough nitrogen in the sample to achieve an acceptable ion beam current inside the mass spectrometer. This can be a problem if enough sample is available for only one analysis. When large quantities of material are available, the answer is simply to repeat the analysis using more sample. If the amount of material needed is greater than the capacity of the sample cup, the material can be "spiked" with a known quantity of a concentrated solution of an unlabeled nitrogen compound. If this is done, the effect of isotopic dilution on the precision of the ^{15}N determination (which may be serious if the sample's ^{15}N enrichment is close to natural abundance) and the need to allow for the spike in subsequent calculations must be considered. The only other possibility, unfortunately, is to digest and/or distill the sample, as in a Kjeldahl–Rittenberg system, and to concentrate the distillate.

Generally, the advantages of direct combustion systems far outweigh their disadvantages. Dumas systems must now be the standard choice for ^{15}N work in most areas of plant and soil science. Nonetheless, particular applications might be better served by conventional preparative and analytical techniques; some of these will be referred to below.

C. Types of Sample

1. Solids

Many agronomic or ecological experiments require only that total nitrogen and ^{15}N are determined in whole pieces of plant tissue or, less commonly, in whole soil. Direct combustion systems originally were developed for this application. The combustion of solid samples, particularly soils, causes a gradual build-up of residue on top of the oxidation column. Loose residue can be removed, after shutting down the inlet system and isolating the mass spectrometer, by vacuuming it away [25]. Alternatively, a perforated quartz tube can be placed inside a newly packed oxidation column [26]; the residue accumulates in the insert tube that can be removed easily and replaced with an empty one. Experience has shown this to be a cleaner and quicker method, which considerably prolongs the life of the oxidation column.

A more difficult problem with the analysis of solids by direct combustion is the achievement of sample homogeneity, in order that the small quantity that can be introduced into the direct combustion system (no more than 10–20 mg of dry plant tissue, or 30–50 mg of dry soil) is adequately representative of the larger sample. Experiments at the Edinburgh School of Agriculture have shown that a degree of uniformity adequate for most purposes may be achieved with plant samples by grinding in a hammer mill and then grinding a 5 g subsample for an additional 10 min in an agate ball mill to a fine "flour." Both the variability between subsamples and that between replicate 10 mg portions of the same subsample (Table 2) were very much less than that commonly observed between replicate field plots.

2. *Liquids*

Into this category fall extracts of soluble nitrogen (e.g., ammonium, nitrite, nitrate, amino acids) from soils or plants, or samples taken from water courses or drainage outlets.

Separating the various forms of nitrogen from each other is straightforward. Inorganic nitrogen ions are separated using steam distillation, microdiffusion or an ion chromatograph equipped with a fraction collector. Amino acids are separated by HPLC [27] or by GC following their conversion to volatilizable derivatives [28]. If integrated GCMS facilities are available and high precision is not essential, it is worth considering using them for ^{15}N analyses of amino acids. Rhodes et al. [28] give full details of the procedures and examples of their application. Other organic compounds of nitrogen (proteins, nucleic acids, etc.) have rarely been considered in ^{15}N studies of the type with which most soil or plant scientists are familiar. Standard techniques for the separation of such nitrogenous compounds—HPLC, fast protein liquid chromatography (FPLC), electrophoresis, immunoprecipitation—are documented fully in the biochemical literature. They may be adapted readily for application to soil or plants.

The main difficulty with all liquid samples is the low concentration of nitrogen relative to the amount needed for reliable analyses. It is often necessary to concentrate the sample by drying [16], or to spike it with a known amount of unlabeled nitrogen. In digestion-distillation-oxidation systems, drying is, in any case, a routine step in sample preparation (see Sec. III.B).

TABLE 2 Variation in ^{15}N and Total N Content of Dried, Ground Plant Material (Barley) as Determined by a Direct Combustion Method

Subsample for ball-milling (5 g)[a]	Replicate portion of subsample after ball-milling (ca. 10 mg)	^{15}N (atom %)	Total N (% dry wt.)
1	A	0.4822	1.135
	B	0.4833	1.114
	C	0.4831	1.119
2	A	0.4813	1.083
	B	0.4820	1.057
	C	0.4818	1.065
3	A	0.4844	1.156
	B	0.4845	1.189
	C	0.4839	1.137

[a]Taken from 60 g sample retained from total quantity of 580 g of dried material and passed through a hammer mill. (K. A. Smith and I. C. Crichton, unpublished data.)

The analysis of liquid samples by direct combustion is possible by adsorbing the liquid onto an inert carrier (e.g., Chromosorb) contained in the sample cup. The maximum volume of solution that can be accommodated in such a cup without spillage is only 5 μl. This volume must contain at least 20 μg nitrogen for analysis. No naturally occurring liquids contain any form of nitrogen at this concentration (4 μg μl^{-1} or 286 m*M*), so a concentration step is an essential preparative step for the direct combustion analysis also. The dried salt is taken up into a small volume of dilute acid (e.g., 5 m*M* H_2SO_4): 50 μl is enough to dissolve all the salt obtained from the distillation of a KCl extract of 10 g of fresh soil. A 5 μl aliquot of this is then used for direct combustion, which means that ten times as much nitrogen must be extracted and concentrated than that actually required for analysis.

An additional complication with direct combustion analysis is that most systems presently available require that the dry weight of the sample be entered into the microcomputer before analysis. This is used to calculate the total nitrogen content, but not the ^{15}N enrichment. If the total nitrogen content is unimportant, the same arbitrary weight can be entered for all samples. Of course, the resulting total nitrogen determination will be meaningless, but the ^{15}N determination will be correct. If an accurate determination of total nitrogen is needed, one way to do this is to concentrate the sample in a preweighed glass vial. After reweighing, the dry weight of the sample, and of that contained in a 5 μl aliquot, can be calculated. However, it is likely that the increase in weight of the vial caused by the deposition of salt will be virtually undetectable and any value for the sample's dry weight wildly inaccurate. The obvious (if tedious) solution to the problem is to make an independent determination of the sample's total nitrogen content by distilling and titrating an aliquot, as for a Kjeldahl–Rittenberg analysis.

3. *Gases*

Gas samples are obtained in most experiments on denitrification and ammonia volatilization, and in some on N_2 fixation. In most cases, the gas of interest must be separated from others in the sample, for which purpose GC is the only method. Again, if a GC-mass spectrometer system is available, this may be a better means of analyzing the isotopic enrichment of a gas sample than a dedicated stable isotope instrument. It is more usual, however, to separate the gases using a stand-alone GC, collect the fractions, and introduce that containing the gas of interest into the mass spectrometer [29].

Gases can be collected directly from the GC into evacuated containers [30]. These are then taken to the mass spectrometer and injected manually using a gas-sampling syringe via a sampling port. Such ports are not standard features on most stable isotope mass spectrometers. They must be custom-built and plumbed into the existing inlet system. On a direct combustion system, if a gas-sampling port is installed before the oxidation column, all nitrogenous gases will be converted to N_2. The mass spectrometer can therefore remain tuned to the normal *m/e* of 29. If so desired, a mixture of gases, obtained from the headspace of a soil cover, for example, could be injected to give an integrated determination of the mixture's total nitrogen content and ^{15}N enrichment. Alternatively, if the port is placed after the water vapor and CO_2 traps, there is no conversion of the sample gas to N_2. Consequently, the mass spectrometer must be returned to an appropriate *m/e*, 45/46 to analyze $^{15}N_2O$, for example. In modern systems that normally run samples in batches, modifications to software might be required to allow individual samples to be analyzed.

Successful analysis of gas samples using a mass spectrometer that is not interfaced with a direct combustion system depends on the removal of contaminants either before injection (e.g., by GC: see above) or afterwards (e.g., water vapor and CO_2 can be removed by a liquid N_2 trap in the inlet line). If evacuated containers are used, these can be purged before collection with a pure gas that would not interfere with the isotopic analysis of the sample. So if $^{15}N_2O$ is to be determined, it can be collected safely in containers filled with pure N_2 [30].

IV. APPLICATIONS OF ISOTOPE MASS SPECTROMETRY

A. Introduction

Qualitative descriptions of nitrogen cycles in natural and agricultural ecosystems are common in the literature. There is no doubt, however, that the absence of a radioisotope of nitrogen that can be used conveniently in tracer studies has slowed down the quantification of such models. High-precision isotope mass spectrometers that can be operated easily and maintained by biologists and soil scientists are now available. In consequence, the rate of publication of isotopic tracer studies of the N cycle has increased dramatically.

The unique advantage of the mass spectrometer is its ability to measure extremely small differences in isotopic abundance between experimental material and a nitrogen standard, and it is when this advantage is being exploited that the additional expense of mass spectrometry is justified. Examples of such applications are the following:

1. Analysis of low enrichment material derived from experiments in which the tracer has been diluted extensively or was initially of low enrichment (e.g., in fertilizer uptake experiments).
2. Analysis of material from experiments in which ^{15}N-depleted material is used.
3. Measurement of variations in natural abundance.
4. Determination of isotope fractionation in soil-N transfers.
5. Determination of N-isotope abundances in soil gases other than N_2, e.g., N_2O.

The fact that conventional mass spectrometers are able to measure routinely differences in isotope ratio down to 0.05 $\delta\,^{15}N$ units does not, however, mean that differences of this order can be interpreted simply as a bulk movement of N from one pool to another. At this level of precision, differences in isotopic abundance can result from the fact that chemically identical but isotopically different species do not react at identical rates, that is to say, a fundamental assumption underlying all tracer studies is invalid.

A comprehensive review of the application of mass spectrometry to soil analysis is beyond the scope of this chapter. Detailed information of this type can be obtained in the reviews of Hauck [5] and Hauck and Bremner [6,7]. The remainder of this chapter will be devoted to a discussion of a few selected topics of particular interest, in which mass spectrometry is an indispensable analytical tool.

B. Isotopic Fractionation

From a practical point of view, the limit of precision in ^{15}N-tracer studies is set by a variation in ^{15}N abundance occurring between soil-N fractions of different chemical composition, from different depths in the soil profile, and from different locations. These variations in ^{15}N abundance that usually fall in the range ±10 $\delta^{15}N$ units (±0.0037 atom % ^{15}N) originate from the isotopic fractionation that occurs when nitrogenous components of the soil undergo chemical and biological transformations. Fractionation occurring as a result of such processes is expressed numerically as a fractionation factor (Table 3). In the simple case of the nitrogen fixation reaction, $N_2 \rightarrow 2NH_4^+$, in which substrate supply is essentially infinite and not subject to isotope depletion, the fractionation factor β is given by [31,32]

$$\beta = \frac{^{15}N/^{14}N \text{ in air}}{^{15}N/^{14}N \text{ in fixed N}} \tag{12}$$

Reactions occurring in the soil solution, e.g., denitrification, in which the substrate is limited and subject to isotope depletion, require a more complicated analysis, and a kinetic fractionation factor must be derived [31]. Table 3 shows that fractionation during N_2 fixation occurs to a lesser extent than during nitrification and denitrification; this is the reason why soil organic and mineral N fractions are, in general, enriched relative to atmospheric N_2.

The importance of isotopic fractionation depends on the application to which measurements of $^{15}N/^{14}N$ ratios are applied. If they are used to determine precisely the amount of nitrogen derived from a particular source (e.g., fertilizer, atmospheric N_2), the effects of certain fractionation processes could be significant. This is especially so if the ^{15}N abundances of the material(s) are close to background [32]. On the other hand if, for example, balance sheets of soil-derived and fertilizer-derived nitrogen present in whole crops are being drawn up, the results are not likely to be affected seriously

Table 3 Reactions Resulting in Fractionation of Nitrogen Isotopes

Process		Fractionation factor[a]
Ammonia-ammonium equilibrium	$NH_3 + H^+ \rightleftarrows NH_4^+$	1.02
Ammonia volatilization	$NH_4^+ \rightarrow NH_3$	1.005 (equilibrium conditions)
Ammonia volatilization	$NH_4^+ \rightarrow NH_3$	1.032 (nonequilibrium conditions)
Ammonium exchange	$NH_4^+ (aq) \rightleftarrows NH_4^+ (s)$	1.0014
Ammonium fixation	$NH_4^+ (aq) \rightarrow NH_4^+ (s)$	~zero
Nitrate exchange	$NO_3^- (aq) \rightleftarrows NO_3^- (s)$	~zero
Denitrification	$NO_3^- \rightarrow N_2O \rightarrow N_2$	1.03
Nitrification	$NH_4^+ \rightarrow NO_2^- \rightarrow NO_3^-$	1.04
N_2 fixation	$N_2 \rightarrow NH_4^+$	1.004
Ammonium assimilation	$NH_4^+ \rightarrow$ organic-N	~zero

Source: Ref. 32.
[a]The fractionation factors are not constants for the reactions; they are the maximum values that have been measured experimentally.

by whatever isotopic fractionation has occurred during the experiment.

The conclusion has been drawn from measurements of $\delta^{15}N$ (and $\delta^{18}O$) values for atmospheric N_2O that denitrification is an important mechanism for the release of this gas by soils [33,34].

C. Uptake of Labeled Fertilizer

This is one of the most widely used applications of ^{15}N in plant and soil science, a vast literature now being devoted to it. The principle involved is exceedingly simple. Fertilizer enriched or depleted in ^{15}N, relative to natural abundance, is applied to soil in which plants are grown. The recovery of ^{15}N by the plants is an indication of the extent to which fertilizer nitrogen and native soil nitrogen have contributed to nitrogen uptake. In more sophisticated experiments, the ^{15}N enrichment of the various soil fractions (microbial biomass, dead organic matter, soil solution, atmosphere) can indicate the fate of nitrogen derived from fertilizer following its transport and transformation in the soil.

There is no comprehensive review of the practical aspects of such experiments. One of the most thorough accounts is by Powlson et al. [35]. Their technique, applying ^{15}N-enriched fertilizer to small (2×2 m) microplots contained within a larger plot to which unenriched fertilizer had been applied at the same rate, has been used widely. The size of the microplot is dictated by the requirements of the experiment (especially the frequency and intensity of destructive sampling) and by the cost of the ^{15}N-labeled fertilizer. The lower the enrichment of ^{15}N, the cheaper it is and the greater the area of land that can be fertilized at a given rate. However, the lower the enrichment, the greater is the dilution of ^{15}N derived from the fertilizer by native soil nitrogen. An enrichment of ca. 5 atom % is usually adequate to give detectable ^{15}N abundances in soil fractions by the end of a growing season in the United Kingdom. If only plant fractions are to be analyzed, a doubling of the natural ^{15}N abundance (i.e., to ca. 0.7 atom %) is sufficient [36].

Agricultural scientists are concerned usually with annual crops over a timescale of a single growing season. Those with an interest in perennial crops or natural ecosystems may wish to follow the fate of nitrogen over several years. Nonagricultural systems do not normally receive large inputs of nitrogen fertilizer. In tracer experiments, the most satisfactory method is to use a small amount of ^{15}N-labeled fertilizer that contains ^{15}N at a high enrichment, so that the total input of nitrogen will not perturb the system too much. Again, the constraint imposed by cost usually dictates whether this aim is realized or not. Clark [37], for example, applied ^{15}N at an enrichment of ca. 80 atom % to microplots only 20 cm in diameter in order to quantify the cycling of nitrogen within a short-grass prairie ecosystem during four subsequent years.

It is useful to include at this point some basic equations that are used in the manipulation of data from ^{15}N tracer experiments. To calculate the total weight of ^{15}N in a sample,

$$^{15}N_{tot} = \frac{N_{tot} \times 15a^*}{15a^* + 14(100-a^*)} \tag{14}$$

where $^{15}N_{tot}$ is the total weight of ^{15}N (from *all* sources) in the sample, N_{tot} the total weight of nitrogen (again from all sources), and a^* the atom % abundance of ^{15}N in the sample as measured by the mass spectrometer.

If the sample was spiked with unlabeled nitrogen (see Sec. III), the ^{15}N present naturally in it must be deducted from $^{15}N_{tot}$.

$$^{15}N_{spike} = \frac{N_{spike} \times 15a_{spike}}{15a_{spike} + 14(100-a_{spike})} \tag{15}$$

where $^{15}N_{spike}$ and N_{spike} are the weights of ^{15}N and total nitrogen, respectively, added to the sample as a spike. The atom % abundance of ^{15}N in the spike, a_{spike}, may be approximated by the natural abundance of ^{15}N in atmospheric N_2 (0.3663 atom %), but it is possible that a_{spike} could differ significantly from this and should be checked analytically.

It is also necessary to deduct from N_{tot} the weight of ^{15}N present naturally in the sample as background and which is not derived from the fertilizer, $^{15}N_0$.

$$^{15}N_0 = \frac{a_0[^{15}N_{tot}(1400 + a^*) - {}^{15}N_{tot}a^*]}{a_0(1400 + a^*) - [a^*(1400 + a_0)]} \tag{16}$$

where N_{sample} is the weight of nitrogen in the sample, i.e., N_{tot} minus N_{spike}, if appropriate, and a_0 is the natural abundance (atom %) of ^{15}N in the sample. Again, a_0 may be approximated by the natural abundance of ^{15}N in atmospheric N_2, but strictly speaking, a_0 is the natural abundance of ^{15}N in the same material as the sample. Isotopic fractionation (see above) means that the natural abundance of ^{15}N in one soil or plant fraction can differ significantly from that in another fraction and from that in atmospheric N_2. Further, natural abundances of ^{15}N in soils and plants can differ spatially and temporally [32].

To calculate how much fertilizer nitrogen, N_{fert}, is represented in the sample by the amount of ^{15}N derived from that fertilizer, a rearrangement of the above equation is used

$$N_{fert} = \frac{^{15}N_{fert} \times [14(100-a_{fert}) + 15a_{fert}]}{15a_{fert}} \tag{17}$$

where $^{15}N_{fert}$ is $^{15}N_{tot} - (^{15}N_{spike} + {}^{15}N_0)$. The atom % abundance of ^{15}N in the original fertilizer is given by a_{fert}. Ideally, this would be the nominal enrichment chosen when the experiment was planned, but inevitably, errors occur during the formulation of the labeled fertilizer, so a_{fert} should be checked analytically, preferably before application. Equation (17) is essentially the same as Fried and Broeshart's [11] concept of "A values."

Techniques such as these are now being used to investigate in detail the processes involved in nitrogen cycling in soils and plants. These studies [38–43] rely jointly on high-precision tracer data and on mathematical models of varying degrees of sophistication.

The interpretation of the results of experiments using ^{15}N as a tracer is not easy: The guide published by the IAEA [44] and papers by Jenkinson et al. [38] and Hart et al. [39] on the subject are essential reading. The basic message from these is that although ^{15}N tracers are invaluable tools, they are not the answer to everything. The information they provide should be viewed with the perspective provided by other techniques, such as comparing the uptake of nitrogen by plants growing on unfertilized soil with that of plants in soil to which fertilizer has been applied.

D. Organic Matter Turnover

The use of ^{15}N tracers in studies of the transformations of nitrogen in soil was mentioned in Sec. IV.C. Another technique has been used to investigate these processes that uses instead the variations in natural abundance of ^{15}N in decomposing organic matter.

The natural abundance of ^{15}N is generally lower in leaf litter than in root debris, which, in turn, is lower than in mineral soil [45,46]. Typically, the ^{15}N abundance of organic matter is increased by isotopic fractionation during microbial decomposition [47]. The result of this is a gradual increase in ^{15}N abundance with depth in the soil profile [48], corresponding to progressive reductions in the organic matter content of the soil. However, this depends on soil particle size and degree of cultivation [49].

Although this technique may be a way of assessing the long-term transformation of nitrogen in certain soils, it cannot reveal in detail the sequence of events leading to the observed variations in natural ^{15}N abundance.

E. Movement of N Between Soil and Atmosphere

1. Nitrogen Fixation

The incorporation of elemental nitrogen into biological systems is most simply measured by exposing the test material to an atmosphere containing ^{15}N-labeled N_2. Subsequent conversion of the isotopically enriched products of fixation to N_2 for analysis in a mass spectrometer permits a quantitative determination of N_2 fixation to be made.

If the results from this sort of experiment are to be meaningful, however, the enclosed test system must be maintained under conditions identical to those from which the system was taken. N_2-fixing organisms that can be tested in this way, e.g., aqueous suspensions of microorganisms, incrustations of blue-green algae, and lichens tend not to be those of general agricultural significance. The method has, however, been used to determine N_2 fixation by detached nodules of legumes and nonlegumes and by free-living soil bacteria. When this approach is used to measure N_2 fixation by whole plants growing in soil, it is difficult to maintain constant partial pressures of CO_2 and O_2 for sufficiently long periods to measure ^{15}N incorporation (see, e.g., Ross et al. [50]). The use of $^{15}N_2$ in enclosed systems to measure nitrogen fixation in a range of biological systems has been reviewed by Bergersen [51].

In most situations in which N_2 fixation is to be measured on whole plants growing in undisturbed soil, either the acetylene reduction technique (see Chap. 12) or an indirect isotopic method must be used [52–55]. It is possible indirectly to estimate N_2 fixation by adding labeled fertilizer to a nonfixing test crop (often an ineffectively nodulated or uninoculated legume) and a nitrogen-fixing species growing in the same soil. The test crop derives all its nitrogen from the soil, so the nitrogen it takes up will reflect the ^{15}N abundance of the available soil-N pool integrated over the growth period. The nitrogen fixer, on the other hand, will contain less ^{15}N because a proportion of its nitrogen is derived from the atmosphere. The extent of this isotope dilution can be used to determine the amount of nitrogen fixed during an increment of growth as follows: If ΔN_f, ΔN_{nf} are the increases in total plant nitrogen for the fixing and nonfixing plants, respectively,

$$\text{N fixed} = \Delta N_f(1 - X) \tag{18}$$

where

$$X = \frac{^{15}\text{N enrichment of } \Delta N_f}{^{15}\text{N enrichment of } \Delta N_{nf}}$$

If plants are grown from seed, calculating ΔN presents no problem. If this method is used to determine N_2 fixation during regrowth of a defoliated or perennial crop, then calculating ΔN is clearly more difficult and subject to greater errors [32]. Fried and Broeshart [11] transform the proportions of nitrogen derived from fertilizer, soil, and atmosphere calculated from the measured isotopic abundances into fertilizer equivalents ("A values" in their terminology), to simplify the arithmetic in cases where the nitrogen-fixing and control plants have received different amounts of fertilizer nitrogen. The technique has been used to determine N_2 fixation and also in attempts to measure N transfer from legume to grass in pasture [53]. An estimate of fixed-N transfer is obtained by comparing the enrichment of the test crop growing alone with the same crop growing in close association with a legume. Among the possible sources of error are (1) plant roots of different species may discriminate differently between isotopes of nitrogen; (2) the N_2 fixer and control species may exploit different rooting zones, so N available to one may not be available to the other; (3) root and shoot turnover in the two species may differ; (4) if only aboveground organs are sampled, it is possible that the partitioning of labeled metabolites between root and shoot may be different in the two species; (5) uptake of nitrogen by the two species may occur at different times during the growth period if their patterns of growth are markedly different [32,55].

An alternative to using ^{15}N labeled fertilizer is to use the variations in natural abundance of ^{15}N to estimate the fractional contribution made by N_2 fixation to nitrogen acquisition by plants. The theoretical and methodological details of this technique have been reviewed comprehensively by Shearer and Kohl [32]. They concluded that it compared favorably with others, given an appropriate sampling regime to account for spatial and temporal variations in ^{15}N abundances of N_2 fixing plants and appropriate nonfixing reference species. Whether these requirements can be met depends on the conditions occurring at each site, and empirical checks are necessary to determine if the technique is likely to be successful or not.

It should be noted that to use the natural abundance technique reliably, a prerequisite is a mass spectrometer of high analytical precision. Rennie and Rennie [52] recommended a precision of better than 0.1%. This is possible only with instruments that are not interfaced with a direct combustion sample preparation system (Sec. II.C and III.B). Because direct combustion systems, by definition, do not operate at low source pressures, it is difficult to see how their precisions may be improved for such applications in the future.

It may prove that it is essential to have a mass spectrometer dedicated either to natural abundance determinations or to tracer work (this applies, in any case, to other items of equipment used in the preparation of unlabeled samples [32]). The introduction of tracers enriched (or depleted) in ^{15}N into a high-precision mass spectrometer may render it incapable subsequently of reaching its peak analytical performance. Such "memory" effects could be due to isotopic exchange between the sample gas and ^{15}N adsorbed from a previous sample onto the walls of the inlet system. This problem can be overcome by baking out the inlet between analyses of batches of labeled and unlabeled samples, an inconvenient and disruptive procedure at the best of times. Another source of isotopic memory effect in direct combustion systems is the formation of nitrides in the heated copper reduction furnace (Fig. 9). These are retained on the copper and, if enriched in ^{15}N, will contaminate unlabeled samples analyzed subsequently and give incorrect values of ^{15}N abundance. A way of eliminating this problem is to analyze numerous unlabeled "blanks" after a batch of tracer samples to flush the system with unlabeled N_2, or to replace the copper in the reduction furnace [25]. The most obvious solution to the problem, having one high-precision isotope mass spectrometer dedicated to natural abundance measurements and another operating as a direct combustion system for labeled material, is beyond the means of most laboratories.

2. *Ammonia Volatilization*

The volatilization of ammonia may occur following urination by grazing livestock, the spreading of manures or slurries, or when urea- or ammonium-based fertilizers are applied to dry, calcareous soils or flooded soils [56,57].

In most studies, the primary aim has been to measure total fluxes, not to identify how much ammonia has been derived from a particular source of nitrogen such as fertilizer or urine. In only a few experiments [58,59] has ammonia labeled with ^{15}N derived from a particular source been trapped and its isotopic composition determined. For example, Whitehead and Bristow [59] applied ^{15}N-labeled cattle urine to a microplot of grass enclosed by a wind tunnel to draw the air above the plot through an acid trap. Ammonia was collected continuously over several days to provide enough ^{15}N for analysis. In such experiments, some sort of soil cover is necessary to concentrate gases emitted from soils and plants; to use an open micrometeorological technique [60] for the same purpose would to to risk enormous dilution and dispersal of the ^{15}N-labeled ammonia and a low probability of recovering detectable amounts of the tracer.

In many other ^{15}N-based studies, ammonia losses have been inferred rather than measured directly by isotopic analysis, after making allowance for other losses such as leaching and denitrification [61,62]. In view of the considerable amount of work involved in such experiments, and the possibilities for sampling errors, it is surprising that they seem to have been favored over those involving the direct trapping of volatilized ammonia. No critical comparisons between the two approaches have been made to assess which is the more reliable indicator of volatilization losses. However, a ^{15}N-based approach is needed if resorption of volatilized ammonia (or of another gaseous form of nitrogen) by vegetation downwind of the treated microplot is to be measured [63]. This is an aspect of nitrogen cycling between soil, atmosphere, and plants that is often overlooked.

3. Denitrification

In denitrification, nitrate is reduced anaerobically to nitrous oxide, N_2O, and molecular nitrogen, N_2. Quantitative measurement of this process is difficult, particularly in the field, but also in some circumstances in the laboratory. The most widespread method involves the inhibition of the reduction of N_2O to N_2 by acetylene, C_2H_2, so that N_2O is the only gaseous product, and measurement of the N_2O by gas chromatography (GC) (see Chap. 12). As the natural atmospheric concentration of N_2O is only ca. 0.3×10^{-6} ml ml^{-1}, the sensitivity of the method is very high.

The GC-acetylene method has the virtue that it is simpler to operate and uses cheaper apparatus than the mass spectrometer methods (and, of course, requires no ^{15}N). However, in very wet or impermeable soils, there can be no certainty of complete permeation by C_2H_2 and therefore no certainty of complete inhibition of N_2O reduction [64].

The alternative is to use fertilizer highly enriched in ^{15}N and to measure the $^{28}N_2$, $^{29}N_2$, and $^{30}N_2$ released by mass spectrometry, and any N_2O by the same method or GC. The high enrichment is necessary because any labeled N_2 has to be detected against a background of $\geq 78\%$ N_2 in the soil air.

The ^{15}N-mass spectrometric method requires a different calculation from that used for samples in which all the N_2 gas present has been liberated from ammonium-N or organic-N by direct combustion or digestion-distillation methods. Air samples will contain two pools of N_2 between which isotopic equilibrium has not been achieved: (1) atmospheric N_2 in which the ^{15}N is at natural abundance and with the corresponding binomial distribution (see Sec. II.F) of ^{14}N and ^{15}N atoms between molecules of masses 28, 29, and 30, and (2) N_2 from denitrified nitrate, with a binomial distribution corresponding to the ^{15}N content of that nitrate.

Procedures for calculating the amount of denitrification from measurements of N_2 at masses 28, 29, and 30 with such a nonrandom distribution were first published by Hauck et al. [65] and Hauck and Bouldin [66], but there was little application of the technique until triple-collector mass spectrometers became generally available in the 1980s.

Following Hauck and Bouldin [66], the calculation may be summarized as follows. If the fractions of N_2 from denitrification and the atmosphere are d and $(1-d)$, respectively, then

$$^{28}x_f = (1-d)(^{14}x_a)^2 + d(^{14}x_s)^2 \tag{19}$$

$$^{29}x_f = 2(1-d)(^{14}x_a\ ^{15}x_a) + 2d(^{14}x_s\ ^{15}x_s) \tag{20}$$

$$^{30}x_f = (1-d)(^{15}x_a)^2 + s(^{15}x_s)^2 \tag{21}$$

where x represents atom or mol fraction; 14, 15, 28, 29, and 30 represent atomic or molecular masses; and subscripts a and s represent atmospheric and substrate (nitrate) nitrogen, respectively.

Rearranging equations (20) and (21) gives

$$d = [^{28}x_f - 2(^{14}x_a\ ^{15}x_a)]/[2(^{14}x_s\ ^{15}x_s) - 2(^{14}x_a\ ^{15}x_a)] \tag{22}$$

$$d = [^{30}x_f - (^{15}x_a)^2]/[(^{15}x_s)^2 - (^{15}x_a)^2] \tag{23}$$

By substituting $(1-^{15}x)$ for ^{14}x, A for $^{15}x_a$, and S for $^{15}x_s$, and combining the two equations, a quadratic relation can be devised for S

$$\begin{aligned}&[^{29}x_f + 2(^{30}x_f) - 2A - 4A^2]S^2\\ &\quad + [2A^2 - 2(^{30}x_f)]S\\ &\quad + [4A^2 - 2(^{30}x_f) - ^{29}x_f]A^2\\ &\quad + 2A(^{30}x_f) = 0\end{aligned} \tag{24}$$

In order to be able to perform the calculation of $^{15}x_s$ and d, it is necessary to determine $^{29}x_f$ (mol fraction of $^{15}N^{14}N$ in the final mixture), $^{30}x_f$ (mol fraction of $^{15}N^{15}N$ in the final mixture), and $^{15}x_a$ (atom fraction of ^{15}N, in the original atmosphere, assumed to be at isotopic equilibrium)

$$^{29}x_f = {^{29}n_f}/(^{28}n_f + {^{29}n_f} + {^{30}n_f})$$

= ion current at m/e 29/sum of ion currents at m/e 28, 29, and 30

$$^{30}x_f = {^{30}n_f}/(^{28}n_f + {^{29}n_f} + {^{30}n_f})$$

= ion current at m/e 30/sum of ion currents at m/e 28, 29, and 30

$${^{15}x_a}^2 = {^{30}n_a}/(^{28}n_a + {^{29}n_a} + {^{30}n_a})$$

= ion current at m/e 30/sum of ion currents at m/e 28, 29, and 30

A limitation of this method is that it has to assume a single nitrate pool. In many situations, the overwhelming proportion of soil nitrate comes from the applied labeled fertilizer, and so errors are small. However, if a substantial fraction of the nitrate comes from a separate pool that is not in equilibrium with the fertilizer nitrate, e.g., mineralization of soil organic matter or organic manure, then the isotopic measurement will give an incorrect result [67,68].

Dual-inlet ratio mass spectrometers have also been used for denitrification measurements. For example, Siegel et al. [69] used one to measure the ion current ratios i_{29}/i_{28} and $i_{30}/(i_{28} + i_{29})$ and applied calculations similar (but slightly simplified) to those above.

An alternative approach was suggested by Craswell et al. [70] to convert the N_2 into a common pool where isotopic equilibrium could be achieved, by oxidation in an electric arc to NO_x, followed by reduction to ammonium that is then analyzed conventionally by a mass spectrometer.

Lloyd et al. [71] and Boddy and Lloyd [72] have described a method by which denitrification products (and other gases) can be measured directly in sediment cores. They used a quadrupole mass spectrometer coupled to a gas-sampling probe, the open end of which was covered by a gas-permeable membrane. The probe is inserted into the sediment, gases diffuse through the membrane and into the inlet system of the mass spectrometer for analysis.

ACKNOWLEDGMENTS

The help of the following in the preparation of this chapter is acknowledged with thanks: J. R. M. Arah, A. Barrie, L. Boddy, K. S. Killham, C. Quarmby, M. Riding, G. W. Robertson, R. J. Thomas, and D. C. Whitehead.

REFERENCES

1. R. Fiedler and G. Proksch, *Anal. Chim. Acta, 78*:1 (1975).
2. R. Fiedler, in *Isotopes and Radiation in Agricultural Sciences*, Vol. 1 (M. F. L'Annunziata and J. O. Legg, eds.), Academic Press, London, 1984, p. 233.
3. R. J. Buresh, E. R. Austin, and E. T. Craswell, *Fert. Res., 3*:37 (1982).
4. G. Pruden, D. S. Powlson, and D. S. Jenkinson, *Fert. Res., 6*:205 (1985).
5. R. D. Hauck, in *Methods of Soil Analysis*, Part 2, 2nd ed. (A. L. Page, ed.), Am. Soc. Agron., Madison, Wis., 1982, p. 735.
6. R. D. Hauck and J. M. Bremner, *Adv. Agron., 28*:219 (1976).
7. J. M. Bremner and R. D. Hauck, in *Nitrogen in Agricultural Soils* (F. J. Stevenson, ed.), Am. Soc. Agron., Madison, Wis., 1982, p. 467.

8. I. Howe, D. H. Williams, and R. D. Bowen, *Mass Spectrometry, Principles and Applications*, 2nd ed., McGraw-Hill, New York, 1981.
9. J. R. Chapman, *Practical Organic Mass Spectrometry*, Wiley, Chichester, 1985.
10. D. D. Focht, in *Nitrogen in the Environment*, Vol. 2 (D. R. Neilsen and J. G. MacDonald, eds.), Academic Press, New York, 1978, p. 429.
11. M. Fried and H. Broeshart, *Plant Soil, 62*:331 (1981).
12. M. Fried and V. Middelboe, *Plant Soil, 47*:713 (1977).
13. W. G. Mook and P. M. Grootes, *Int. J. Mass Spectrom. Ion Phys., 12*:173 (1973).
14. G. L. Kerven and P. G. Saffigna, *Int. J. Mass Spectrom. Ion Phys., 23*:333 (1977).
15. K. W. Steele, M. J. Grinsted, A. T. Wilson, and C. B. Dyson, *Int. J. Mass Spectrom. Ion Phys., 28*:313 (1978).
16. J. D. Reeder, *Soil Sci. Soc. Am. J., 48*:695 (1984).
17. D. R. Keeney and D. W. Nelson, in *Methods of Soil Analysis*, Part 2, 2nd ed. (A. L. Page, ed.), Am. Soc. Agron., Madison, Wis., 1982, p. 643.
18. D. B. Sprinson and D. Rittenberg, *J. Biol. Chem., 180*:707 (1949).
19. P. J. Ross and A. E. Martin, *Analyst (Lond.), 95*:817 (1970).
20. A. Marriotti, *Nature (Lond.), 303*:685 (1983).
21. T. Preston and N. J. P. Owens, *Analyst (Lond.), 108*:971 (1983).
22. A. Barrie and C. T. Workman, *Spectrosc. Int. J., 3*:439 (1984).
23. R. Marshall and J. Whiteway, *Analyst (Lond.), 110*:867 (1985).
24. A. Barrie and M. Lemley, *Int. Labmate, 14*:27 (1989).
25. G. W. Robertson and M. E. Reid, personal communication, 1987.
26. B. Mary, personal communication, 1987.
27. W. S. Hancock (ed.), *CRC Handbook of HPLC for the Separation of Amino Acids, Peptides and Proteins*, Vol. 1, CRC Press Inc., Boca Raton, Fla., 1984.
28. D. Rhodes, A. C. Myers, and G. Jamieson, *Plant Physiol., 68*:1197 (1981).
29. D. E. Rolston, M. Fried, and D. A. Goldhamer, *Soil Sci. Soc. Am. J., 40*:259 (1976).
30. K. S. Killham, personal communication, 1989.
31. J. Y. Tong and P. E. Yankwich, *J. Phys. Chem., 61*:540 (1957).
32. G. Shearer and D. H. Kohl, *Aust. J. Plant Physiol., 13*:699 (1986).

33. N. Yoshida and S. Matsuo, *Geochem. J.*, *17*:231 (1983).
34. T. Yoshinari and M. Wahlen, *Nature (Lond.)*, *317*:349 (1985).
35. D. S. Powlson, G. Pruden, A. E. Johnston, and D. S. Jenkinson, *J. Agric. Sci. (Camb.)*, *107*:591 (1986).
36. K. A. Smith, A. E. Elmes, R. S. Howard, and M. F. Franklin, *Plant Soil*, *76*:49 (1984).
37. F. E. Clark, *Ecology*, *58*:1322 (1977).
38. D. S. Jenkinson, R. H. Fox, and J. H. Rayner, *J. Soil Sci.*, *36*:425 (1985).
39. P. B. S. Hart, J. H. Rayner, and D. S. Jenkinson, *J. Soil Sci.*, *37*:389 (1986).
40. D. Barraclough, E. L. Geens, G. P. Davies, and J. M. Maggs, *J. Soil Sci.*, *36*:519 (1987).
41. D. Barraclough and M. J. Smith, *J. Soil Sci.*, *38*:519 (1987).
42. D. D. Myrold and J. M. Tiedje, *Soil Biol. Biochem.*, *18*:559 (1986).
43. S. Bjornsen, *J. Soil Sci.*, *39*:393 (1988).
44. International Atomic Energy Agency, *A Guide to the Use of Nitrogen-15 and Radioisotopes in Studies of Plant Nutrition. Calculations and Interpretations of Data*, Tecdoc 288, IAEA, Vienna, 1983.
45. K. J. Nadelhoffer and B. Fry, *Soil Sci. Soc. Am. J.*, *52*:1633 (1988).
46. R. E. Karamanos and D. A. Rennie, *Can. J. Soil Sci.*, *60*:365 (1980).
47. C. C. Delwiche and P. J. Steyn, *Environ. Sci. Technol.*, *4*:299 (1970).
48. S. F. Ledgard, J. R. Freney, and J. R. Simpson, *Aust. J. Soil Res.*, *22*:155 (1984).
49. H. Tiessen, R. E. Karamanos, J. W. B Stewart, and F. E. Selles, *Soil Sci. Soc. Am. J.*, *48*:312 (1984).
50. P. J. Ross, A. E. Martin, and E. F. Henzell, *Nature (Lond.)*, *204*:444 (1964).
51. F. J. Bergersen (ed.), *Methods for Evaluating Biological Nitrogen Fixation*, Wiley, New York, 1980.
52. R. J. Rennie and D. A. Rennie, *Can. J. Microbiol.*, *29*:1022 (1983).
53. P. M. Chalk, *Soil Biol. Biochem.*, *17*:383 (1985).
54. S. K. A. Danso, *Soil Biol. Biochem.*, *18*:243 (1986).
55. J. Witty, *Soil Biol. Biochem.*, *15*:631 (1983).
56. R. J. Haynes, *Mineral Nitrogen in the Plant-Soil System*, Academic Press, London, 1986.
57. D. W. Nelson, in *Nitrogen in Agricultural Soils* (F. J. Stevenson, ed.), Am. Soc. Agron, Madison, Wis., 1982, p. 327.
58. L. Overrein, *Soil Sci.*, *106*:280 (1968).

59. D. C. Whitehead and A. W. Bristow (in preparation).
60. J. C. Ryden and J. E. McNeill, *J. Sci. Food Agric.*, *35*:1297 (1984).
61. D. R. Keeney and A. H. Macgregor, *N.Z. J. Agric. Res.*, *21*: 443 (1978).
62. G. A. Rodgers and G. Pruden, *J. Sci. Food Agric.*, *35*:1290 (1984).
63. R. J. Thomas, personal communication, 1989.
64. K. A. Smith, *Mitteil. Deutsch. Bodenk. Gesell.* (in press).
65. R. D. Hauck, S. W. Melsted, and P. E. Yankwich, *Soil Sci.*, *86*:287 (1958).
66. R. D. Hauck and D. R. Bouldin, *Nature (Lond.)*, *191*:72 (1961).
67. D. D. Focht, *Soil Sci. Soc. Am. J.*, *49*:786 (1985).
68. C. W. Boast, R. L. Mulvaney, and P. Baveye, *Soil Sci. Soc. Am. J.*, *52*:1317 (1988).
69. R. S. Siegel, R. D. Hauck, and L. T. Kurtz, *Soil Sci. Soc. Am. J.*, *46*:68 (1982).
70. E. T. Craswell, B. H. Byrnes, L. S. Holt, E. R. Austin, I. R. P. Fillery, and W. M. Strong, *Soil Sci. Soc. Am. J.*, *49*:664 (1985).
71. D. Lloyd, K. J. P. Davies, and L. Boddy, *FEMS Microbiol. Ecol.*, *38*:11 (1986).
72. L. Boddy and D. Lloyd, in *Nutrient Cycling in Terrestrial Ecosystem* (A. Harrison, P. Ineson, and O. W. Heal, eds.), Elsevier, London, 1990, p. 139.

12

Gas Chromatographic Analysis of the Soil Atmosphere

KEITH A. SMITH and JONATHAN R. M. ARAH *The Edinburgh School of Agriculture, Edinburgh, Scotland*

I. INTRODUCTION

The soil atmosphere may be modified significantly in its composition, with respect to the air above the soil, as a result of the respiratory activity of plant roots and soil microorganisms. The concentration of oxygen decreases, and the concentration of carbon dioxide increases, particularly when gaseous exchange with the air above the soil is impeded. Furthermore, other substances such as oxides of nitrogen (produced by microbial transformations of nitrate and ammonium), gaseous hydrocarbons, and volatile sulfur compounds may be present. The release of some of these gases from the soil into the atmosphere is believed to make a significant contribution to such environmentally important processes as the "greenhouse effect," acid deposition, and changes in both stratospheric and tropospheric ozone concentrations. Conversely, soils are a sink for many gaseous atmospheric pollutants of natural and anthropogenic origin. The development of gas chromatographic methods, and their application to the study of these processes, are considered in this chapter.

Historically, the first analysis of the free air in the soil appears to have been conducted by Boussingault and Lewy in 1853 [1]. They inserted a pipe into the soil to a depth of 0.3–0.4 m, aspirated 2.5–10 liters of soil air through barium hydroxide solution, and weighed the carbonate formed, to determine the carbon dioxide content of the soil air. For the next 100 years or more, the measurement of soil gases continued to rely on absorptiometric/manometric procedures, and in spite of the experimental problems involved, many of the important chemical and biochemical processes in the soil involving gaseous substances had been investigated long before the advent of modern instrumental methods of analysis. However, there were often limitations due to lack of sensitivity that could not be overcome.

The availability, since the end of the 1950s, of gas chromatographic methods for the analysis of permanent gases and volatile compounds has resulted in very considerable progress in research in this area. The change from traditional chemical methods of analysis, which were not specific for all gases, to a system with which most, if not all, of the substances of interest could be separated readily, and the accompanying increases in sensitivity obtainable by gas chromatography, have made possible a wide range of new investigations in soil science, as in so many other disciplines.

Gas chromatographic methods are applicable both to investigations of the soil atmosphere as it occurs in the field, and also to laboratory studies of the evolution of gases from, and their uptake by, soil. Such studies form an integral part of many investigations that have as their ultimate objective the understanding of natural processes in the field.

II. GAS CHROMATOGRAPHY

A. Principles

A brief outline of the principles involved in gas chromatography (GC) is given below, but for a more comprehensive discussion, the reader is referred to books solely devoted to this subject [2–5] and a number of important theoretical papers [6–11].

Chromatography is essentially the separation of the components of a mixture resulting from differences in their partition between a stationary phase with a large active surface area and a moving phase that flows over the stationary phase. Depending on the state of the moving phase, a distinction is made between liquid and gas chromatography. Separations that would be very difficult to achieve by any other means are possible by chromatographic methods because even small differences between the components in their partition between the phases are multiplied many times during their passage through the chromatographic system.

GC may be conveniently classified into two types: gas-liquid and gas-solid chromatography. In the former, the stationary phase is an involatile liquid coated either on an inert support material in a packed column, or on the internal wall of an open tubular (capillary) column. In the latter type, the column is packed with an adsorbent of small particle size, or the inner surface of a capillary column is coated with a thin adsorbent layer. For separations of gaseous mixtures, solid adsorbents (see below) have been used more widely than liquid stationary phases.

At constant temperature, pressure, and carrier gas velocity, the rate at which a component travels along the column is related to the partition coefficient K

$$K = \frac{C_s}{C_g} \tag{1}$$

where C_s and C_g are the concentrations in the stationary and gaseous phases, respectively. The more strongly bound components stay for a longer time in the stationary phase and thus travel along the column more slowly than the more weakly bound components. In an actual column the zone occupied by a component broadens as it moves along, due to the diffusion of molecules in the gaseous phase.

If K is independent of concentration and the adsorption isotherm is as in Fig. 1a, then the elution peak has a symmetrical, Gaussian shape. However, if the adsorption isotherm is of the Langmuir type, which is generally the case for solid adsorbents (Fig. 1b), then the chromatographic peaks have sharp fronts with trailing backs, although the degree of asymmetry may often be quite small.

By analogy with the theory developed to explain the separation of components in a distillation column, the resolving power of a chromatographic column was described formerly in terms of the *number of theoretical plates* to which it was equivalent. By dividing the length of the column by the number of theoretical plates, *the height equivalent to a theoretical plate* (HETP) could be determined. It is now considered to be more appropriate to replace the term "theoretical plate number" by *efficiency* and "plate height" by *relative peak broadening* [3].

The separation of the components in a sample depends on the relative values of the partition coefficients. The closer the values are to each other, the longer is the column required to give an adequate separation. Reducing the temperature has the effect of increasing the separation. In practice, the variables that are most commonly exploited to achieve satisfactory separations are

1. The material used as stationary phase
2. The temperature
3. The length and diameter of the column
4. The flow rate of the moving phase (the carrier gas)

The essential parts of all gas chromatographs consist of

1. A carrier gas system, generally in the form of a cylinder supply at pressures up to 20 MPa (200 bar), with a two-stage regulator to reduce the pressure and a flow-rate regulating valve

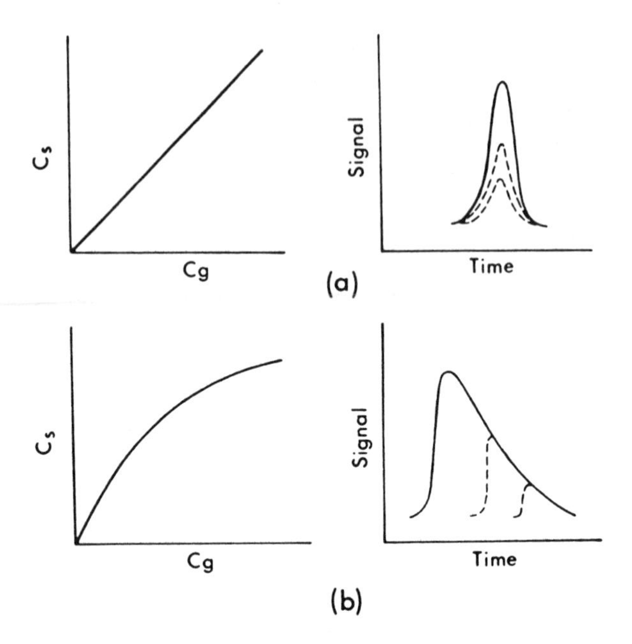

FIG. 1 (a) Linear adsorption isotherm symmetrical elution peak (position of a peak maximum constant). (b) Nonlinear (Langmuir) isotherm, distorted elution peak "tailing" (position of peak maximum depends on quantity of sample). (From Ref. 3.)

2. A device for introduction of the sample into the carrier gas stream
3. A column to separate the components of the sample
4. A detector to indicate the elution of each component and produce a response proportional to the quantity of the component present
5. A recording system to provide a permanent record of the detector response

The column and detector (and sometimes the injection device) are enclosed normally in separate thermostats that can be controlled individually. This is because the optimum column temperature for achieving a desired separation is often different from the optimum detector temperature.

B. Commercial Instruments

A wide range of commercial instruments is available from manufacturers in several countries. For many applications the choice of instrument is not critical, because the substances to be measured are easily within the detection capabilities of any commercial instrument. In this situation, the decision is more likely to be made on the basis of price and the availability of after-sales service than on technical features. However, in those applications where the common detectors have inadequate sensitivity, only one or two manufacturers may offer the particular device that is required. Bayer [12] has reviewed the commercial instruments available and gives details of the various features included in the current generation. For information on the latest developments, manufacturers' literature should be consulted.

It frequently happens that the chromatographer has a perfectly serviceable gas chromatograph, with a single detector, which has been used for some years on one application, but which the chromatographer now wishes to use for a new analysis that requires a different type of detector. In principle, it is perfectly feasible to fit a detector made by one manufacturer to a chromatograph made by another (or a detector produced for a current range of instruments to an earlier model of the same make), to provide a very cost-effective solution to the new analytical requirement. However, in practice, difficulties arise because of manufacturer's reluctance to sell detectors rather than complete instruements, and because of differences in complex microprocessor control systems. Skilled technical assistance with electronics may well be required in such an operation.

C. Columns

Generally, gas chromatographic methods commonly used for the measurement of the constituents of the soil atmosphere have been adopted from methods originally developed for other purposes. There has been little or no basic research into new types of column for this particular application. As was mentioned earlier, solid adsorbents have been used more widely than liquid stationary phases for separations of permanent gases. The most widely used adsorbents are

1. Molecular sieves 5A and 13X, which are synthetic zeolites with channels in the crystal lattice capable of occluding small molecules. They are the only commonly used packing materials that will separate oxygen from nitrogen at ambient or higher temperatures.
2. Porous polymer beads, first developed by Hollis [13] and marketed under the name Porapak, and followed by similar materials such as Polypak, Phasepak, Chromosorb, and Hayesep. These materials are available in different grades that vary in their retention characteristics for particular substances. Porapak Q has been used widely for separating carbon dioxide and nitrous oxide from nitrogen and oxygen in air samples.

3. Alumina. This has been used widely for the separation of hydrocarbon gases, often after partial deactivation by the addition of water or silicone oil [14] or inorganic salts [15], to reduce tailing of the elution peaks. Alumina is a polar adsorbent on which the order of separation depends on the molecular structure. Thus, for example, unsaturated hydrocarbons are retained more strongly than saturated ones with the same number of carbon atoms.

4. Silica gel and activated charcoal. These materials have been used for the separation of gases such as carbon dioxide and nitrous oxide. However, the properties of both these adsorbents have been found to vary considerably with the source of supply and method used in preparation, making it very difficult to reproduce retention times from one column to another [4]. They have now been superseded largely by the porous polymers, which have given much more reproducible results.

5. Graphitized carbon black materials, such as the Carbopack range. Carbopack B was developed to separate hydrogen sulfide, sulfur dioxide, and methanethiol at concentrations in the range 10^{-9}–10^{-6} ml ml^{-1} [16].

6. Nongraphitized carbon molecular sieve. This material, first prepared by Kaiser [17,18] by pyrolysis of polyvinylidene chloride, has proved very useful for the separation of complex mixtures of inorganic gases and hydrocarbons [19,20].

1. Separation of Oxygen from Argon

Although the separation of the two major components of air, oxygen and nitrogen, from each other is accomplished easily on molecular sieves, the separation of oxygen from argon is much more difficult, and at ambient temperature or above the two gases elute as a single peak. Their separation involves the use of molecular sieve 5A or porous polymer columns operated at subambient temperatures, which necessitates either a chromatograph with a suitable subambient capability or (less satisfactorily) the use of cooling baths or Dewar flasks containing dry ice-acetone or similar mixtures. On molecular sieve 5A at -72°C, argon is eluted first, followed by oxygen, and the nitrogen is retained on the column [21]. In contrast, on Porapak Q nitrogen is eluted first, followed by oxygen and then argon [22]. Recently, a modified N-type porous polymer has been synthesized on which this separation can be achieved at only −10°C on an 8 m column [23].

2. Nitrogen Compounds

Trade concentrations of *nitrous oxide*, N_2O, are determined normally with a selective electron capture detector (ECD) (see Sec. II.D). The primary requirement, therefore, is separation of N_2O from other gases to which the ECD is sensitive, e.g., oxygen water vapor, acetylene, and halogenated compounds. This is achieved easily using a material such as Porapak Q, and analysis times can be reduced by using backflushing techniques [24]. In situations in which the nonselective and less sensitive thermal conductivity detector is employed, N_2O must also be separated from carbon dioxide. Satisfactory procedures for this separation on Porapak Q [25,26], Porapak Q-S [27], and Carbosieve B [28,29] have been published. If the carbon dioxide concentration is very high relative to that of nitrous oxide, as is frequently the case, it is important to ensure that these peaks are well separated on the chromatogram. Otherwise, the nitrous oxide peak will be obscured by the tail from the carbon dioxide peak. Passage of the sample through a precolumn of sodalime [26] or ascarite [30] will either remove carbon dioxide completely or so reduce the peak size that this ceases to be a major problem.

a. Nitric Oxide. NO only occurs in soils, or the air above, at very low concentrations, and nowadays it is measured usually with a chemiluminescence analyzer [31] rather than by GC. The determination of NO by GC is complicated by its rapid oxidation to nitrogen dioxide, NO_2, in the presence of oxygen, and a difficult separation from nitrogen and carbon monoxide. Molecular sieves have been used for separating NO, but problems with the tailing of the peak can occur [4]. Quinlan and Kittrel [32] compared several adsorbents and concluded that Chromosorb 102 and Carbosieve B, both at subambient temperatures, were the most satisfactory.

b. Nitrogen Dioxide. NO_2, like NO, is more satisfactorily measured by a chemiluminescence analyzer (e.g., [31]) than by a chromatographic method. One reason is its reactivity, e.g., it reacts with the aromatic rings of the polymer comprising Porapak Q to form nitric acid, water, and aromatic nitration products [33].

c. Ammonia. This is another nitrogen compound that has not been determined very successfully in soil atmosphere samples by GC, but which is readily determined by nonchromatographic methods. Smith and Clark [30] reported that of many packings tried, only Carbowax 600 on firebrick washed with sodium hydroxide had proved successful. The strong retention of ammonia on some gas-solid chromatographic packings is indicated by the work of Burford and Bremner [34,35] who could not obtain a peak for the gas after injecting it onto Porapak Q columns at 25° and 100°C, even though phosphine, hydrogen sulfide, and (at the higher temperature) water were all eluted.

3. *Sulfur Compounds*

Reactive sulfur compounds, sulfur dioxide and hydrogen sulfide in particular, have proved difficult to determine by GC at concentrations of the order of 10^{-6} ml ml^{-1}, due to adsorption on column walls or solid supports or irreversible reaction with the stationary phase. Glass, stainless steel, and even Teflon (polytetrafluoroethylene) tubing have all been found to exhibit undesirable retention of SO_2 below 10^{-5} ml ml^{-1}, but FEP Teflon (fluorinated ethylene-propylene) is free of this effect [36]. Powdered Teflon appears to be the one material sufficiently inert to serve as support, and one stationary phase has been identified that combines the two qualities of inertness and the ability to separate low-molecular-weight sulfur compounds: polyphenyl ether (five-ring polymer) to which orthophosphoric acid had been added [36,37]. Several solid adsorbents have also been used to achieve a number of difficult separations of sulfur gases: Carbopack B, to which reference has already been made; a specially treated silica gel, Chromosil 310; and Supelpak S, which is a modified Porapak Q-S [38]. A typical separation on the latter material is shown in Fig. 2.

D. Detectors

Gas chromatograph detectors are devices that indicate the presence of a sample component eluting from the column in the carrier gas stream and produce an electrical signal proportional to the concentration of the component. A wide variety of physical properties has been employed as the basis of these devices. For satisfactory performance, detectors should have high sensitivity to the components of the sample, a rapid response that increases linearly with concentration, and a high signal-to-background-noise ratio. They should be easy to operate and reliable in constant use. A useful property associated with certain types is that of selectivity, only responding to components with particular properties.

Some of the detectors in current use for the determination of gaseous substances are listed in Table 1 and are described below. However, as the properties of the electron capture and flame photometric detectors are also discussed in Chap. 13 because of their great importance in pesticide analysis, their consideration in this section is restricted mainly to aspects particularly relevant to soil atmospheric analysis.

1. Thermal Conductivity Detector (TCD)

In this detector, the carrier gas passes through channels drilled through a metal block and flows over heated resistance filaments or thermistors suspended in the channels. Pure carrier gas passes over the filaments in one channel of the detector, and the carrier gas containing the sample constituents passes over the filaments in the other channel. Changes in composition of the gas result in

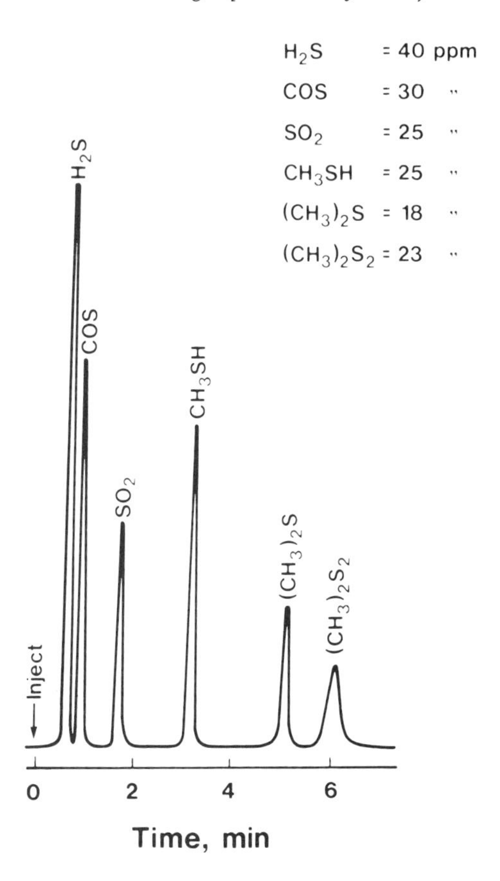

FIG. 2 Typical chromatogram of sulfur gases obtained with Supelpak-S. (From Ref. 38, reproduced from the *Journal of Chromatographic Science*, by permission of Preston Publications, a division of Preston Industries Inc.)

changes in thermal conductivity and thus in the temperature of the filaments and hence their resistance. The filaments are connected in a Wheatstone bridge circuit that is balanced when pure carrier gas flows through both channels. As a sample constituent elutes from the column into the detector, the bridge will be unbalanced and result in an output signal. The sensitivity of the detector depends on the difference between the thermal conductivities of the carrier gas and sample. As hydrogen and helium have much higher thermal conductivities than all other gases, these are the

TABLE 1 Some Detectors Used for Determination of Gases

Detector	Application
Thermal conductivity (Katharometer)	all gases (major constituents of mixtures)
Flame ionization	hydrocarbons (trace concentrations)
Helium ionization	all gases (trace concentrations)
Electron capture	nitrous oxide; oxygen; halogenated compounds
Ultrasonic	all gases (wide concentration range)
Flame photometric	sulfur and phosphorus compounds

most appropriate carrier gases. In view of the possible hazards associated with emission of hydrogen into the laboratory atmosphere, helium is to be preferred in spite of its higher cost.

Modern TCDs have limits of detection an order of magnitude lower than those achievable with earlier devices. This is partly due to improvements in detector geometry and partly due to improved electronics. Instruments are now available with different modes of operation: constant current, constant voltage, and constant temperature. According to Wells and Simon [39], the constant temperature mode is 7–10 times more sensitive than the others, but the signal-to-noise ratio is no better.

The response of a typical four-filament TCD can be expected to become nonlinear when the change in filament resistance caused by a sample approaches the same magnitude as the original resistance in the absence of the sample. To overcome this nonlinearity, Patterson et al. [49] devised the "dynamic current drive" method of powering the TCD, whereby the TCD current is made a weak positive function of the change in filament resistance. Consequently, a small increase in TCD current at high sample concentrations compensates for the loss in detector sensitivity that would otherwise occur. This method results in a linear response for sample concentrations from 10^{-6}–1 ml ml^{-1} [40].

2. *Flame Ionization Detector (FID)*

The flame ionization detector is probably the most widely used of the selective detectors. It consists of a hydrogen-air burner and two electrodes at a potential difference of 100–300 V. Usually, the burner jet is one electrode and the other, the collector electrode, commonly takes the form of a cylindrical tube surrounding the flame. At the high temperature of the flame, some ionization of hydrogen occurs, producing an ion current between the electrodes. This is the standing current of the detector. Gases eluting from the column pass into the jet, and the combustion of any organic compounds present greatly increases the ion current, which, after amplification, is recorded. Baseline drift during temperature-programmed analysis is overcome by the use of dual detectors with opposite polarities, one of which receives carrier gas and column effluent and the other carrier gas only.

For studies of hydrocarbons and other volatile organic species in the soil atmosphere, the FID has great advantages over other detectors. Apart from a small response to high concentrations of oxygen and nitrous oxide, the detector is insensitive to the presence of inorganic gases and water vapor, and in principle any inorganic carrier gas can be used. Nitrogen and argon are the gases most commonly used in practice. Modern detectors have a limit of detection for hydrocarbon gases of 10^{-9}–10^{-8} ml ml^{-1}, i.e., in the parts per billion range, with a linear dynamic range of 10^{6}–10^{7}. Thus, the upper limit of concentration satisfactorily determined with the FID is of the order of 10^{-2} ml ml^{-1}, and samples containing higher concentrations have to be diluted before analysis.

3. *Helium Ionization Detector (HID)*

The helium ionization detector depends on the increase in conductivity resulting from the ionization of gases by collision with metastable helium atoms. These atoms are excited to the metastable state by β-radiation from a high-activity tritium source. The excitation potential of helium, 19.8 eV, is sufficient to ionize all gases except neon, making the HID a universal detector. It also has great sensitivity, giving adequate responses to concentrations in the 10^{-9}–10^{-8} ml ml^{-1} (parts per billion) range, with a linear range of about four orders of magnitude [41]. Although, in theory, the detector should give a positive response to all gases except neon, in practice the response to hydrogen, argon, oxygen, and nitrogen is sometimes negative or bipolar; the reasons are not wholly understood. However, the HID may be operated satisfactorily in either mode for these gases [41]. Brazell and Todd [42] have shown that operation of the HID in a pulsed mode gives increased sensitivity, linearity, and stability.

4. *Electron Capture Detector (ECD)*

In the electron capture detector, β-radiation from a radioactive source (usually nickel-63) ionizes the carrier gas (nitrogen or an argon-methane mixture), and an applied potential causes current to flow. Electrophilic molecules, e.g., halogen compounds, "capture" electrons produced by the ionization and the current is reduced. This provides the basis for a highly sensitive and selective detector.

The response of the ECD is inherently nonlinear, but provided the detector current is not reduced by more than about 20%, an acceptably linear relationship between response and sample concentration can be achieved [43]. In modern equipment the constant current mode of operation is often preferred to the constant pulse frequency mode (see Chap. 13), because of its claimed superior linear response range [43]. The demonstration that at 350°C the ECD also shows a sensitive reponse to nitrous oxide [44] has led to its widespread adoption for the measurement of this gas in studies of nitrification (e.g., [45]) and denitrification (e.g., [24,45]). The ECD will detect readily nitrous oxide even at less than its normal atmospheric concentration (about 0.3×10^{-6} ml ml^{-1}). Another valuable application of this detector is the determination of low concentrations of oxygen, particularly in aqueous solution (see Sec. III.C).

5. Flame Photometric Detector (FPD)

The specificity and high sensitivity of this detector for sulfur compounds have been a major asset in studies of sulfur-containing gases in the soil atmosphere and environment generally. The sensitivity for sulfur compounds is 10^4 times as great as that for other substances, so only the major constituents of the soil atmosphere are likely to produce a response comparable with that from trace levels of sulfur gases. However, this detector also has an inherently nonlinear response, the signal-concentration relationship being given by the expression

$$I = I_0[S]^n$$

where I is the signal, $[S]$ is the sulfur concentration, and I_0 and n are constants under given experimental conditions. In general, the values of n reported have been between about 1 and just greater than 2 [46], but n varies with the flame conditions and is thus affected by flow rates of hydrogen and fuel gases. All the factors affecting the flame conditions must be kept constant to achieve any degree of analytical accuracy. The application of constant or variable n-value electronic linearizers to the FPD output signal may be useful over very limited concentration ranges and operating conditions [46].

6. Ultrasonic Detector

The principle on which the ultrasonic detector operates is the change in the speed of sound through the carrier gas as a sample gas elutes from the column. This detector has the potential advantage that, like the TCD, it responds to all gases, but with greater sensitivity and with a linear range from 10^{-6}–1 ml ml^{-1}. However, when it was introduced, it was very expensive compared with, say, an ECD, and only a few reports of its application to soil atmosphere studies have appeared [47–49]. The ultrasonic detector is much less sensitive than the HID and is therefore not competitive with it for studies of trace gases in the parts per billion concentration range.

E. Calibration

The concentrations of the various constituents in a sample are conveniently determined by comparisons of peak height (or peak area) with that obtained from a known volume of a gas mixture of known composition, having first established the relationship between peak height (or area) and detector response. Cylinders of compressed gases of the desired composition may be obtained from commercial suppliers, together with a certificate of analysis. This is normally the most satisfactory method for gas concentrations of 10^{-2} ml ml^{-1} or more, but an alternative is the use of gas-proportioning pumps, which can produce accurate dilutions of one gas in another, down to concentrations of about 3×10^{-3} ml ml^{-1} [4]. For trace levels, commercial mixtures may also be obtained, but it is possible to achieve reasonably satisfactory results at negligible cost by injecting small volumes of the pure gas into larger vessels of known volume. For example, mixtures of hydrocarbons in nitrogen at concentrations of the order of 10^{-6} ml ml^{-1} can be prepared readily by injecting 1 ml of pure hydrocarbon into a 1 liter flask filled with nitrogen and, after equilibrium has been reached, injecting 1 ml of the diluted gas into a second flask. (The exact volumes of the flasks may be determined simply, prior to this operation, by filling with water and weighing.) However, an advantage of using compressed gases as standards is that the cylinder may be connected directly to a gas-sampling valve fitted with a loop of suitable volume.

Better reproducibility of results is usualy possible with a sampling valve than with a syringe, particularly if the carrier gas pressure is much above atmospheric. Improved results with syringe injection may be achieved by injecting the sample into a chamber forming part of the sample loop of a sampling valve [4].

F. Automation

1. Sample Injection

When large numbers of samples have to be analyzed, a degree of automation can greatly reduce operator time. In principle, the automation of gas sample injection and analysis is not difficult, and such systems have become widely used in industrial process control applications and in the monitoring of the levels of air pollutants. In these situations, the supply of sample gas is unlimited and is frequently at pressures above atmospheric. Thus, a continuous flow of gas can be maintained through a gas-sampling valve, and the contents of the sample loop can be injected automatically at any desired time interval. In contrast, the samples obtained from gas-sampling probes buried in the soil are normally too small to allow the purging of a sampling valve to be carried out in the normal way, and each one (usually in a syringe) must be kept separated from the others. Automated systems for the injection of such samples have been described, in which the loop of a motor-driven gas-sampling valve is evacuated by a pump and then coupled, by the operation of a rotary switching valve, to each of a series of sample syringes in turn. The gas in the syringe expands into the loop, and the contents of the loop are then injected [15,50]. A new, microprocessor-controlled version of this auto-injector, which can accommodate up to 32 samples (with scope for further expansion), has been developed in the authors' laboratory. The apparatus is illustrated schematically in Fig. 3. The original concept has been developed similarly by Brooks and Paul [51] for the automatic sampling of the atmospheres in soil incubation jars.

An alternative design for an auto-injection system has been published by Robertson and Tiedje [52]. This is based on (1) a shuttle base from a fraction collector carrying gas sample vials sealed with rubber septa, (2) a pneumatic piston that drives a hypodermic needle through the septa, and (3) a six-port pneumatically driven valve that connects a sample loop with either the hypodermic needle or carrier gas stream (Fig. 4). This system requires the vials to be pressurized above atmospheric pressure, either with additional sample or by the introduction of some nonsample gas.

2. General Chromatograph Operation

Whereas automated equipment such as the sample injection systems described above has to be purpose-built, automation of the operation of the gas chromatograph itself is now a normal feature of commercial instruments. The incorporation of microprocessors or personal computers allows parameters such as oven temperature and carrier gas flow rate, and the timing of events such as the switching of a valve, to be set to values stored in the memory and called

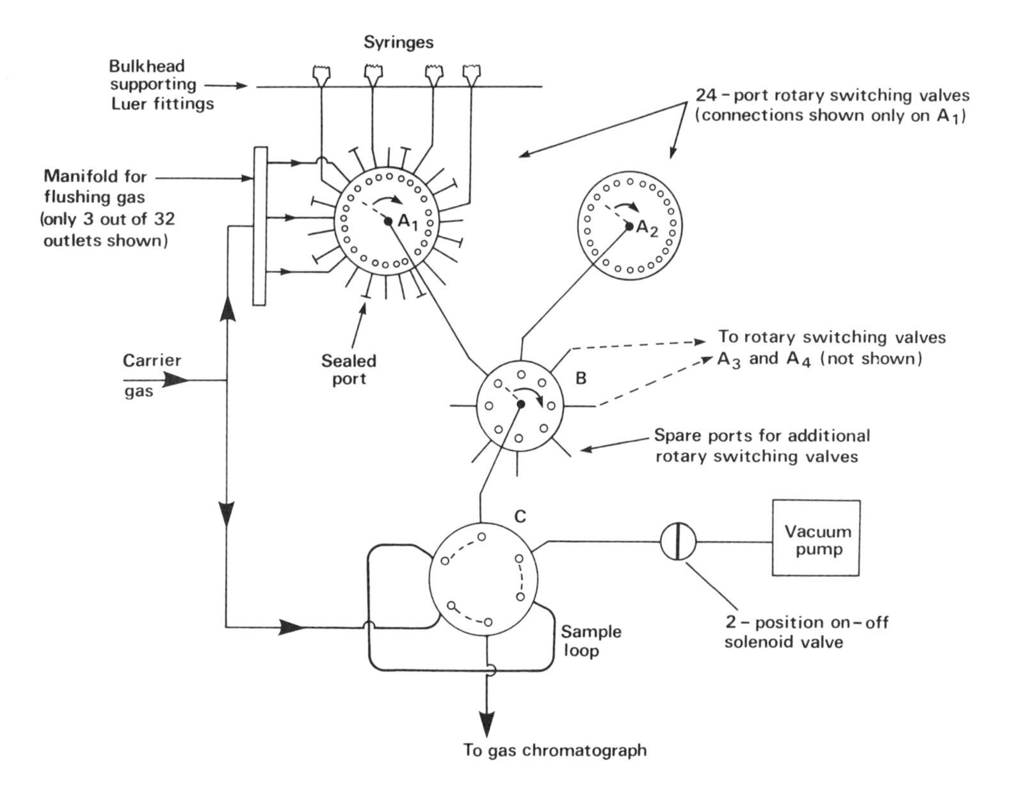

FIG. 3 Schematic diagram of auto-injector system for gas samples contained in syringes. (J. R. M. Arah and K. A. Smith, unpublished.)

up by means of a control progam. Visual display units and keyboards or touch pads allow interactive operation. Questions can be displayed on the screen that prompt the operator to enter the required instructions to program the instrument for a particular analysis. A whole series of complete programs can be stored in the instrument's memory, or on floppy disks. The range of facilities offered is expanding steadily, and readers are urged to consult manufacturers' literature to extablish what features are currently available, when a purchase is contemplated.

3. *Data Acquisition and Processing*

Automatic computation of chromatographic data can be achieved either by using purpose-built digital integrators or by passing the analog

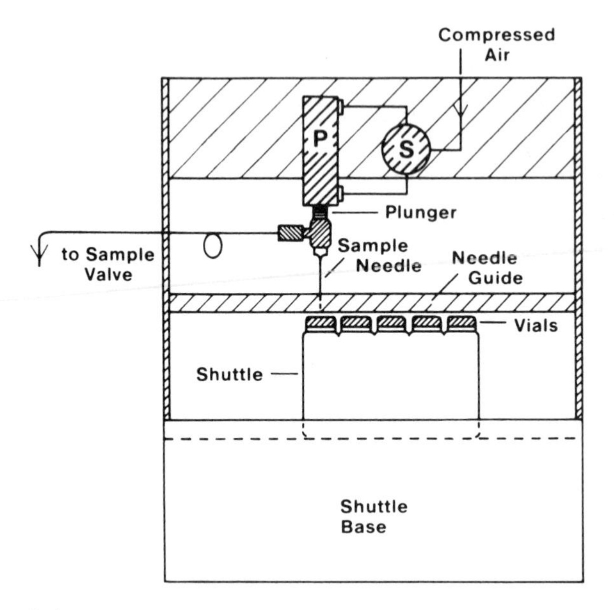

(a)

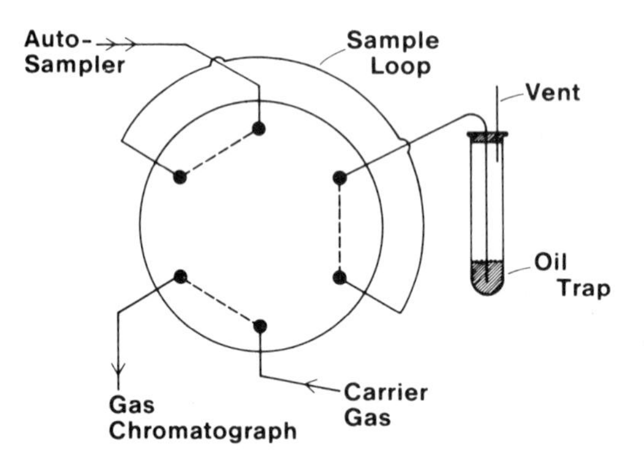

(b)

FIG. 4 (a) Autosampler for gas collection vials, built on a modified fraction collector. (b) Associated sample value in "fill" position. (From Ref. 52, G. P. Robertson and J. M. Tiedje, *Plant and Soil*, *83*:453 (1985), published by Martinus Nijhoff Publishers BV, The Hague, The Netherlands.)

output signal of the chromatographic detector via an analog-to-digital converter to a computer (which may also be used to control the chromatograph itself; see above). Dedicated integrators incorporate several standard programs that will allow, for example, the calculation of either peak heights or areas, and the percentage of the total sample represented by each peak; by incorporating response factors or normalizing to internal or external standards, results can be printed out in suitable concentration units. Low-cost instruments such as the Spectra-Physics SP 4270 and Hewlett-Packard 3396 integrators conveniently provide as output both an analog chromatogram and the quantitative numerical data on the same paper chart. Integrators are able normally to cope with sloping baselines and unresolved or tailing peaks, and spikes on the baseline due to noise can be ignored.

When the chromatograph is interfaced to a computer, it becomes possible, in addition to performing the same operations as can be undertaken by the integrator, to undertake statistical evaluation of replicate analyses and store results for subsequent retrieval without the manual transfer of data at any stage. This has the benefit of eliminating errors, as well as saving operator time.

III. ANALYSIS OF SOIL ATMOSPHERES

A. Sampling

Reference has already been made to the nineteenth century work of Boussingault and Lewy [1]. Their technique of withdrawing a sample via a pipe inserted to the required depth is fundamentally the same as the methods in use today; the only real difference is that now, with the availability of GC, small samples of the order of 1–5 ml are adequate for most analyses. Such small volumes are, in fact, to be preferred for GC, both for ease of collection and quality of analysis (there is less dispersion in the column and, consequently, sharper peaks than when larger samples are used).

Simplicity of sampling procedure and low-cost sampling equipment are important for most field studies of the soil atmosphere, because of the enormous spatial variability commonly encountered. From a statistical point of view, it is acceptable to sacrifice some accuracy in respect to each individual measurement if by doing so, a much greater number of replicate samples can be analyzed.

Two main types of gas-sampling device have been used. For instantaneous sampling, a tube is driven into the soil so that it fits tightly against the soil, and gas samples are withdrawn via an aperture at the lower end up through the tube into a syringe or some other receptacle. The alternative method involves the creation of a cavity in the soil, usually by the insertion of a small container, that is connected to the soil surface by a capillary tube. The gas in the container is allowed to come to equilibrium with the surrounding soil atmosphere by diffusion, and thus a sample withdrawn from the container is representative of the soil atmosphere. Examples of both types of sampling device are shown in Fig. 5.

The former type of probe has been used for an investigation of the levels of natural gas in subsoils above gas-bearing strata. A steel tube was driven 2–3 m into the ground by a sledge hammer, then withdrawn 0.2–0.3 m so as to expose part of the hole, then gas samples were sucked from the subsoil and taken for chromatographic analysis [53]. This method was very similar to that used as long ago as 1915 by Russell and Appleyard [54]. Tackett [55] used a copper tube 1.6 mm in diameter that could be pushed into the soil. To the upper end was attached a tubing union sealed with a septum. The gas entered the tube through a slot in the side. The side opening has the advantage of being less susceptible to plugging by soil when the probe is inserted. A modification of Tackett's probe has been described in which the gas sample is sucked into a glass ampoule, which is then sealed off by melting with a portable gas burner [56]. In this condition, the sample can be stored for several days, if required, before analysis by GC or mass spectrometry.

Several devices based on the diffusion reservoir principle have been developed to provide samples for gas chromatographic analysis. Reservoirs have been made from small open-ended metal cans [57], pipe unions [55], glass tubing [58], sintered bronze filter cups [59], and plastic pipe with the ends covered by fine-mesh plastic screens [60]. In all these designs, the reservoir is connected to the soil surface by a metal, plastic, or glass capillary tube. To install the reservoir, it is necessary to make a hole with an auger to the required depth, insert the apparatus, and then backfill the hole around the sampling tube with soil or some other material, tamping down firmly to provide a seal against the downward diffusion of air.

A straight tube such as that used for instantaneous sampling [55], to which reference is made above, can also be used as a simple diffusion reservoir. An effective method used by the present authors to insert such tubes and prevent them from becoming blocked with soil is as follows. A length of metal rod of slightly smaller diameter than the tube bore is bent at right angles about 10 cm from one end. The longer section of the rod is pushed through the tube until the shorter section rests against the top of the tube; the rod is then cut so as to leave a few mm protruding from the lower end. The tube is inserted into the soil by pressing on the shorter section of rod. After insertion, the rod is withdrawn and

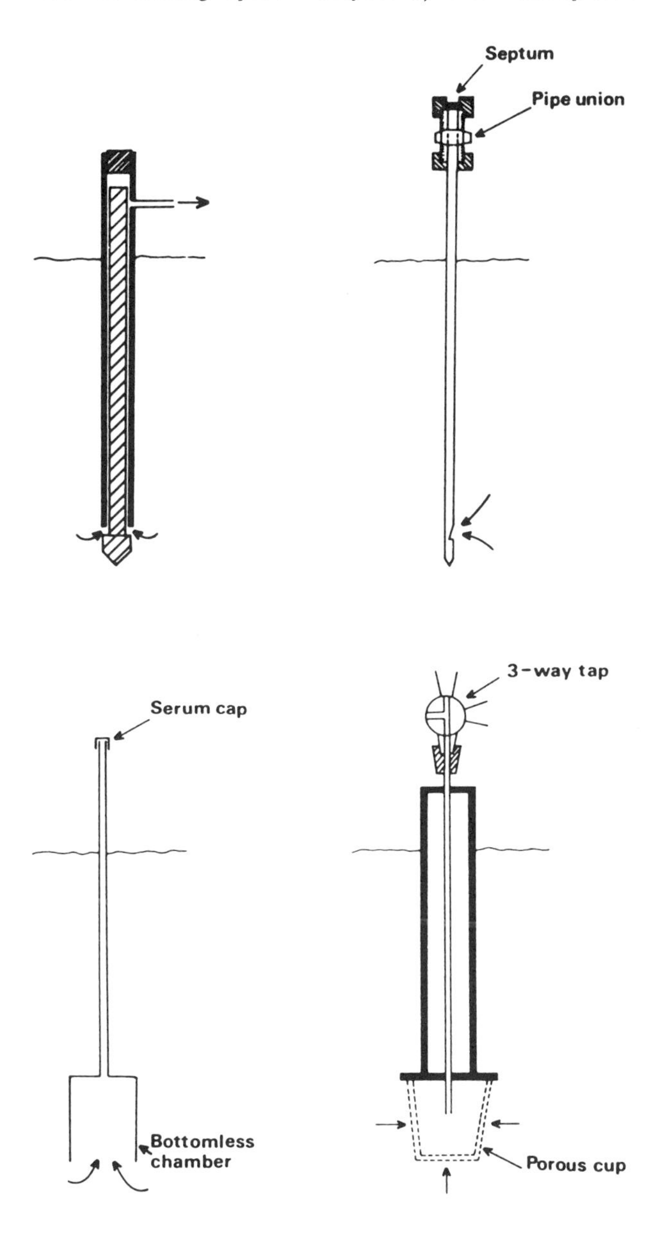

FIG. 5 Devices used for sampling the soil atmosphere. Arrows indicate the direction of gas flow. A variety of different combinations of the above- and below-ground parts shown here have been used by different workers. (From Ref. 61.)

the tube itself (which is then capped with a septum or tap with luer-fitting) forms the sample reservoir.

Gas samples taken in the field may not be analyzed for some hours, or even days, and considerable care is required to ensure that the composition does not change significantly during the intervening period. Sealing the gas in a glass ampoule [56] would seem to be by far the most reliable method, but the risk of contamination by gases from a blow-torch would necessarily limit the range of constituents that could be determined in this way. The high affinity of hydrocarbons for stopcock greases and also their rapid adsorption by the rubber diaphragms of disposable plastic syringes [25] makes necessary the use of all-glass syringes lubricated with glycerol or glycerol-mannitol-starch mixtures [5,25], when these gases are being measured. The alternative is to use gas-tight glass syringes with Teflon plungers manufactured specifically for gas chromatographic work. However, this is likely to be an extremely expensive solution when several dozen syringes are needed to sample a well-replicated field experiment. Some adsorption of inorganic gases by, and diffusion through, plastic syringes was reported by Burford [62], who had to make corrections for these effects. Glass syringes are to be preferred if samples have to be stored for considerable periods before analysis.

An alternative to transporting field samples in syringes is to inject them into vials. Parkin [63] described a method in which 5 ml samples taken by syringe were injected into vials of only 3.5 ml volume, thus creating a pressure above atmospheric, and stored for the later analysis of O_2, CO_2, and N_2O. Parkin's data showed that variability only increased slightly over 28 days. Slight adsorption of N_2O by the rubber stoppers occurred; due to a combination of this effect and sample dilution by residual air in the vials, the recovery of standards was 82%.

Instead of extracting gases from sampling probes inserted into soil in situ, intact cores of soil may be taken, sealed as rapidly as possible, and the entrapped gases extracted later for analysis [64, 65].

In studies of gaseous emissions from the soil surface, it may be desirable to collect an integrated sample over a period of time, or (to increase sensitivity) to concentrate the gas of interest from a large volume of air into a small volume. Nitrous oxide, for example, has been adsorbed onto molecular sieve 5A from an air stream passing through a chamber placed over the soil surface, and later released by adding water to the sieve in a sealed flask [26,66]. It has also been trapped out from 300 ml air samples on Porapak Q cooled to $-135°C$ [67].

The sampling of gases from incubation vessels in laboratory experiments is generally a straightforward operation. Samples can be withdrawn by syringe through a septum or serum cap and usually can be analyzed within a short time. However, major problems may be experienced in this type of experiment, due to a variety of causes. The creation of leaks, following repeated sampling through the same rubber closure, is a common problem, and adsorption of the gases of interest by the rubber may occur [68]. The latter effect can be minimized by using the "Mininert" type of bottle closure in which the underside of the rubber septum is exposed to the gases in the bottle only at the moment of sampling [69]. During the intervening periods, it is sealed off by a gas-tight Teflon plug. The repeated sampling of small incubation vessels can result in the removal of a significant fraction of the gases present, and this can have effects on the analytical results that are often ignored. The creation of a partial vacuum within the vessel will cause air to enter through any leaks by mass flow as well as by diffusion. A less obvious problem is that the gas in the sampling syringe will be at a reduced pressure, and on removal of the needle from the septum, air can enter the syringe. Thus, the presence of a small quantity of oxygen in the gas mixture injected into the chromatograph may provide false evidence of the presence of oxygen in the incubation vessel. The existence of a partial vacuum is often readily detectable when samples are withdrawn with a relatively large syringe. However, when 1 ml or smaller syringes with tightly fitting plungers are used, the friction of the plunger in the barrel is often quite sufficient to resist any inward movement under the influence of the pressure of the atmosphere; under these circumstances, it is possible for the operator to take samples that are considerably below atmospheric pressure without being aware of it. This contamination problem can be overcome by using syringes that can be closed by a valve. In analyses of the total gas mixture present (using a TCD), the occurrence of a reduced quantity of sample can then be detected by the reduced total area of the peaks in the chromatogram. However, in experiments in which trace gases only are being determined, this check cannot be made; but provided that the experimental system used is leak-tight, then the quantity removed on each successive sampling occasion can be calculated from the known volumes of the vessel and the syringe by applying Boyle's law. Alternatively, the use of low concentrations of a rare gas, e.g., krypton, as an internal standard in conjunction with a suitable detector [35] provides a satisfactory solution to the problem of estimating sample size.

B. Column-Detector Systems for Major Constituents

In many investigations, it is desirable to measure the concentrations of all the major constituents of the soil atmosphere: oxygen, nitrogen, argon, carbon dioxide, and sometimes, methane. In the course of studies of denitrification, nitrous oxide and acetylene may also be present, the latter having been introduced artificially into the soil to inhibit the reduction of N_2O to N_2 [70].

A number of multiple-column and/or multiple detector systems have been devised to permit the isothermal analysis of all these gases in a single sample. Several are illustrated in Fig. 6. The complete separation cannot be undertaken on a single column without the use of temperature programming or impractically long analysis times. Figure 7 shows three systems that have been used for the simultaneous measurement of trace gases (with a flame ionization or electron capture detector) and other permanent gases with a thermal conductivity detector. The gases measured in all these systems are listed in Table 2.

The system shown in Fig. 6a can be used to determine some or all of the following gases: oxygen (+ argon), nitrogen, carbon dioxide, nitrous oxide, methane, and acetylene. By appropriate adjustments of temperature, column lengths, and relative flow rates through the two columns, the relative positions of the gases from each column can be altered on the chromatogram. Thus, oxygen and nitrogen may be eluted before nitrous oxide and carbon dioxide [30], or after [52,70]. Alternatively, carbon dioxide may come between oxygen and nitrogen [55]. In the system used in the authors' laboratory, a 1.6 m × 4 mm ID Porapak Q column is connected in parallel with a 1.6 m × 1.5 mm ID molecular sieve 5A column, and the sample is split so that only a small fraction passes through the molecular sieve. A typical chromatogram of an air sample containing acetylene is shown in Fig. 8.

An alternative version of the two-column, one-detector system is shown in Fig. 6c; the columns may be at the same temperature, using different packings [64,71] or at different temperatures with the same packing [25,49,72].

An arrangement in which two parallel columns are connected to two separate thermal conductivity detectors (Fig. 6b) has been described [74]. This arrangement has been extended to include a third column coupled to a flame ionization detector to determine trace hydrocarbons (Fig. 7a) [15]. Another method for the simultaneous analysis of the main constituents of the soil atmosphere and trace hydrocarbons involves the use of a Porapak Q or R column with thermal conductivity and flame ionization detectors in series (Fig. 7b) [76].

Laskowski and Moraghan [77] used two columns and two detectors in series (Fig. 6d) to separate hydrogen, nitrogen, nitrous oxide, and methane. The system described by Frunzke and Zumft [78] (Fig. 6e) for the study of bacterial denitrification permits the separation of hydrogen, nitric oxide, and carbon monoxide, as well as oxygen, nitrogen, nitrous oxide, and carbon dioxide, in 2.5 min.

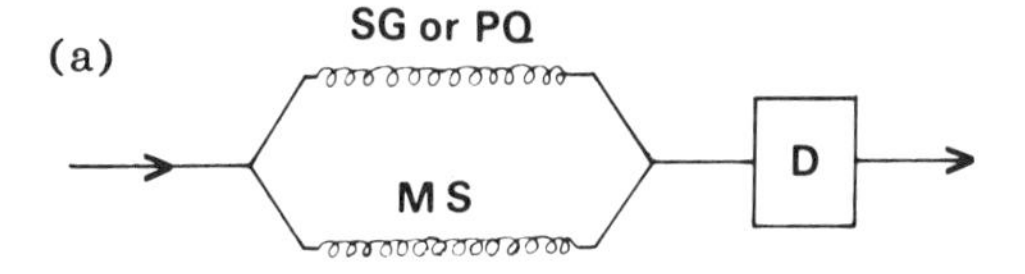

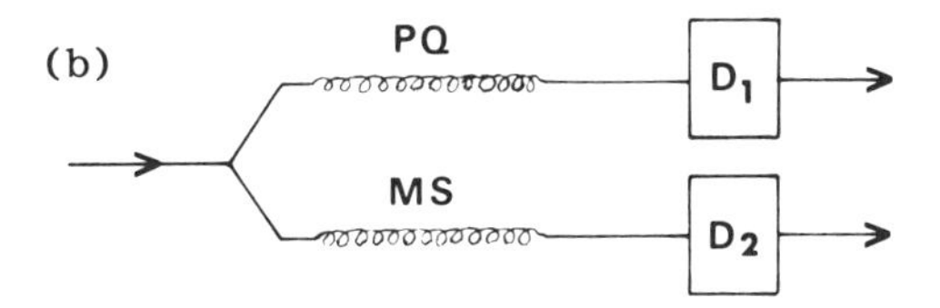

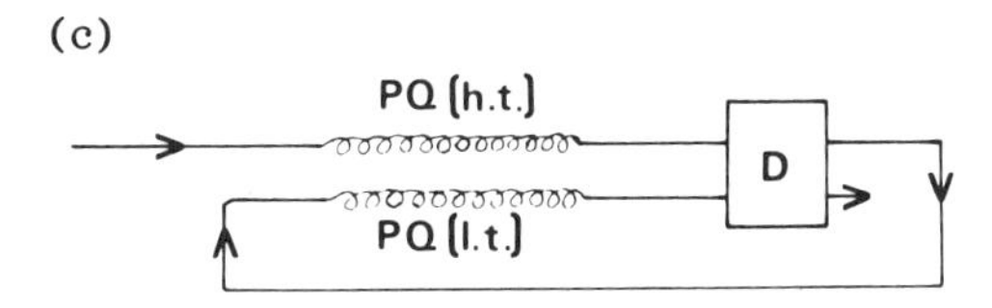

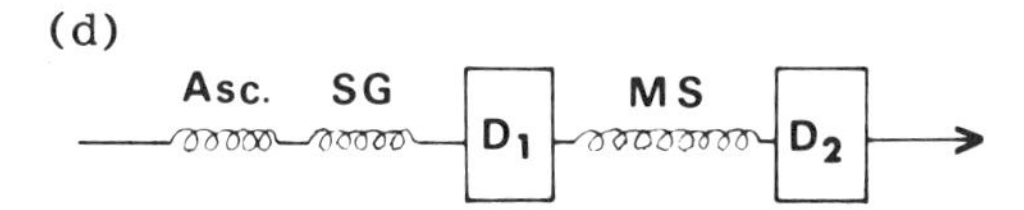

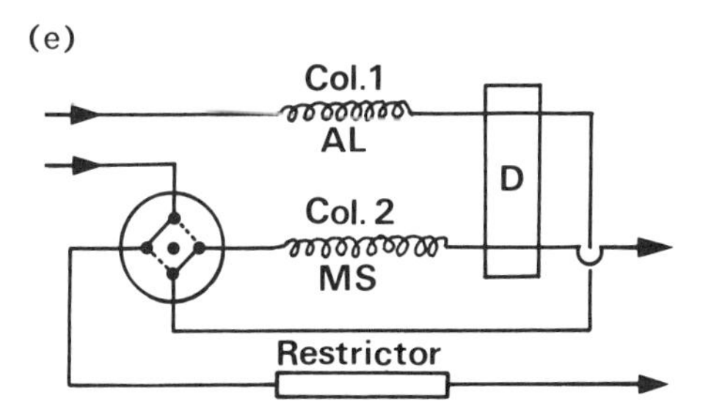

FIG. 6 Multiple-column/detector systems used for determination of the major constituents of the soil atmosphere. SG, silica gel; AL, alumina, MS, molecular sieve 5A or 13X; PQ, Porapak Q; Asc., Ascarite; h.t., high temperature; l.t., low temperature; D, detector. System c has also been used with different packings at the same temperature (see text and Table 2).

(a)

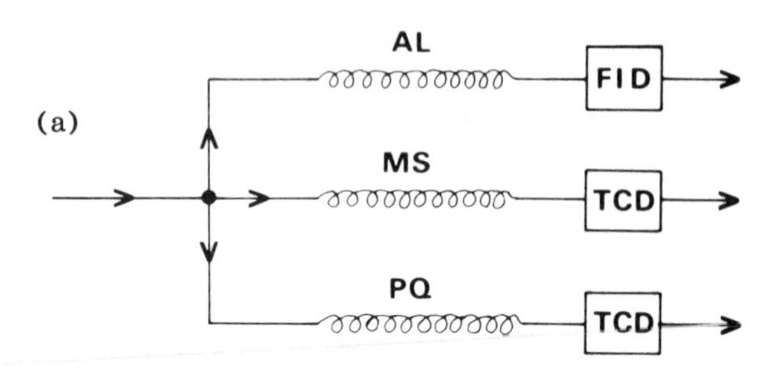

(b)

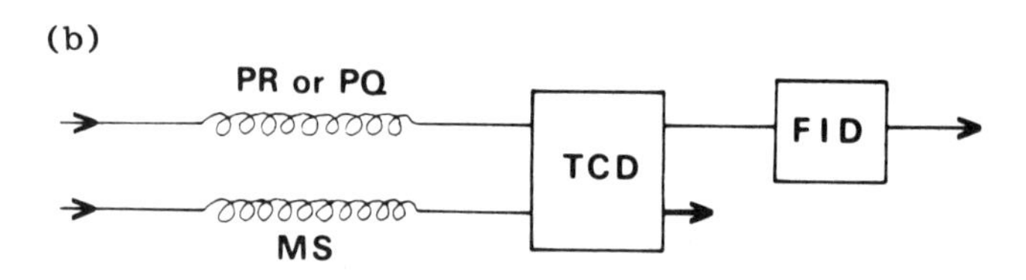

(c)

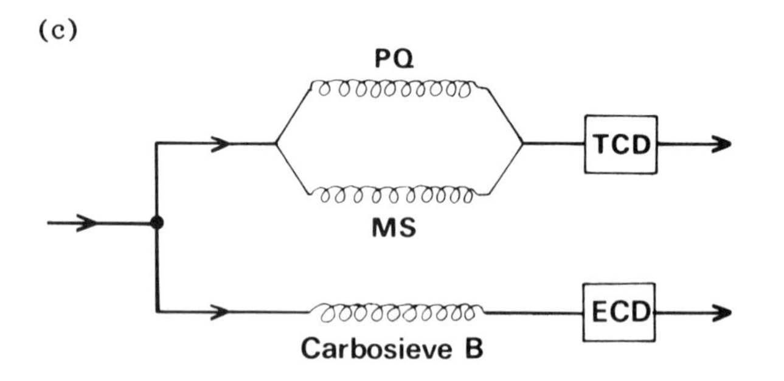

FIG. 7 Multiple-column/detector systems for simultaneous analysis of major and trace constituents in the soil atmosphere. AL, alumina deactivated with NaI; MS, molecular sieve 5A; PQ, Porapak Q; PR, Porapak R.

The light gases (H_2, O_2, N_2, NO, CO) are eluted as a composite peak from column 1 onto column 2, whereupon the valve is activated by a signal from a digital integrator to switch the columns from a series to a parallel arrangement. The light gases are then separated and detected before CO_2 and N_2O are eluted from column 1.

TABLE 2 Gases Measured with Multiple Column/Detector Systems Shown in Figs. 6 and 7

System	Columns	Gases	References
6a	silica gel, mol. sieve 5A	O_2, CO_2, N_2	[55]
	silica gel, mol. sieve 5A	O_2, N_2, N_2O, CO_2	[30]
	silica gel, mol. sieve 5A	CO_2, O_2, N_2, CH_4	[73]
	Porapak Q, mol. sieve 5A	CO_2, O_2, N_2	[61]
	Porapak Q, mol. sieve 5A	CO_2, C_2H_2, O_2, N_2	a
6b	Porapak Q, mol. sieve 5A	CH_4, CO_2, H_2O; O_2, N_2	[74]
6c	Porapak Q, Porapak Q	CH_4, CO_2, N_2O; N_2, O_2, Ar	[22,25, 49,72]
	silica gel, mol. sieve	CO_2, O_2, N_2	[64]
	HMPA, mol. sieve 13X	CO_2, O_2, N_2, CH_4, H_2S	[71]
	Porapak Q, mol. sieve 5A	CO_2, O_2, N_2	[75]
6d	Ascarite/silica gel, mol. sieve 5A	H_2, H_2, N_2O, CH_4	[77]
6e	alumina, mol. sieve 13X	H_2, O_2, N_2, NO, CO; CO_2, N_2O	[78]
7a	Porapak Q, mol. sieve 5A, alumina/NaI	CO_2, N_2O; O_2, N_2; CH_4, C_2H_6, C_2H_4, C_3H_8, C_3H_6, iso-C_4H_{10}, n-C_4H_{10}, C_4H_8	[15]
7b	Porapak R, mol. sieve 5A	C_2H_4, CO_2, O_2	[76]
7c	Porapak Q, mol. sieve 5A, Carbosieve B	O_2, N_2, CO_2, N_2O	[79]

[a] J. R. M. Arah and K. A. Smith, unpublished results.

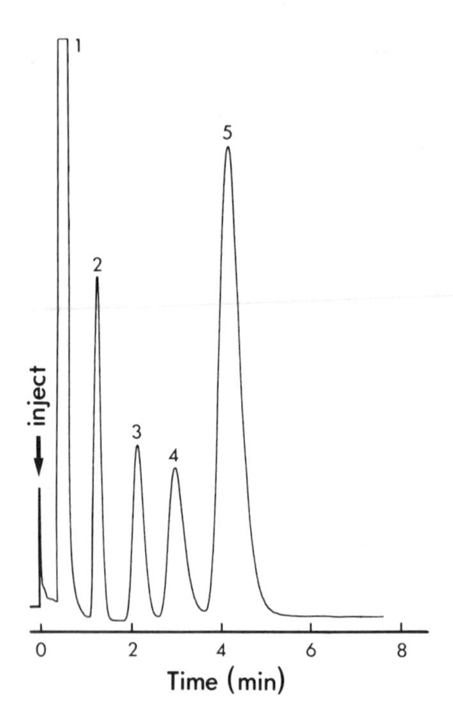

FIG. 8 Separation of gases on column system shown in Fig. 6a. Sample is split so that only a small fraction passes through MS 5A column. 1, O_2 + N_2 + Ar; 2, CO_2; 3, C_2H_2; 4, O_2 + Ar; 5, N_2.

1. *Measurement of Oxygen*

Satisfactory measurements of oxygen concentrations can be obtained normally in soil atmosphere investigations without having to resort to cryogenic separation from argon, because the argon contribution to the single peak obtained at ambient or elevated temperatures can be calculated to a sufficient level of accuracy, and the oxygen content obtained by difference. The nitrogen, oxygen, and argon contents of CO_2-free dry air are N_2, 0.7806; O_2, 0.2100; and Ar, 0.0094 ml ml^{-1}, respectively. The composition of a soil atmosphere sample will differ from this in that the carbon dioxide level will usually be above the 3×10^{-4} ml ml^{-1} found in unpolluted air, the oxygen may be depleted, and the sample will be almost saturated with water vapor (which contributes about 0.012 and 0.023 ml ml^{-1} at 10 and 20°C, respectively). The consumption of oxygen and evolution of carbon dioxide by respiring organisms causes a net reduction in pressure in the gaseous phase of the soil, because of the greater solubility of carbon dioxide. The result is an inward flow of air and, at equilibrium, an increase in the concentrations of nitrogen and argon. But none of these processes will change the ratio of these two gases to each other (83:1); the only way that such a change can be made is when nitrogen is evolved in the soil as the result of denitrification of nitrate. Calculations show that an unrealistically high rate of denitrification would have to occur in the field before the ratio of nitrogen to argon changed sufficiently to produce a significant error in the calculated concentration of argon, and thus in the calculation of oxygen in the composite oxygen-argon peak [15]. Table 3 shows the errors in the calculated oxygen concentration resulting from the extreme and unlikely situation in which 100 kg NO_3^--N ha^{-1} have been denitrified totally to N_2 before any has escaped from the profile. It is evident that in most circumstances, particularly for very variable field situations, such errors are quite acceptable, as they are likely to be well within the other errors associated with the experiment. To cite one example, Burford [80] has estimated that the maximum error in oxygen concentrations, using this approach in this study of the soil atmosphere after heavy applications of organic manure, was less than 0.002 ml ml^{-1}.

TABLE 3 Maximum Likely Errors in Determination of Oxygen Concentrations by Subtraction of Calculated Argon Contribution to Joint Peak

Concentration measured by GC (%)		Oxygen concentration		% error in O_2 value
O_2 + Ar	N_2	True	Calculated[a]	
15	80	14.12	14.04	0.57
10	85	9.07	8.98	0.99
5	90	4.01	3.92	2.24

[a]At 10% air-filled porosity (N_2 increased by denitrification of 100 kg NO_3^--N ha^{-1}).

C. Determination of Dissolved Gases

When studies of the composition of the soil atmosphere are carried out over extended periods, there may well be occasions when the soil is saturated, and any gases present in the profile are in solution, not in the gaseous phase. If the concentrations of those dissolved gases that are of interest can be determined in samples of soil water, and the corresponding equilibrium gas phase concentrations are calculated from Henry's law, it is possible to interrelate data obtained under waterlogged conditions with those obtained when the soil is drained [59]. This is of considerable value in investigations of the concentrations of physiologically active gases to which plant roots and microorganisms are exposed. Table 4 shows the solubilities of a number of gases at different temperatures.

A number of workers in several different scientific fields have successfully applied gas chromatographic methods to the determination of dissolved gases. In the method of Swinnerton et al. [81], a water sample (1 or 2 ml) was transferred to a glass chamber divided into two compartments by a fritted glass disk, through which the carrier gas could be diverted by operating a four-way switching valve. The passage of the gas through the fritted glass resulted in the formation of a stream of bubbles that removed the dissolved gases from solution. An alternative extraction method involves the equilibration of a water sample of known volume with a known volume of air, nitrogen, or helium in a sealed bottle by vigorous shaking

TABLE 4 Solubilities of Gases in Water at Different Temperatures

Temperature (°C)	Solubility (milliliters of gas at STP per milliliter of water)				
	O_2	N_2	CO_2	N_2O	C_2H_4
0	0.049	0.024	1.71	1.28	0.26
10	0.038	0.019	1.19	0.88	0.17
20	0.031	0.015	0.88	0.61	0.13
30	0.026	0.013	0.67	0.45	0.09

Source: *Handbook of Chemistry and Physics*, 37th ed., Chemical Rubber Publishing Co., Cleveland, Ohio, 1955.

and sampling the gas phase with a syringe. Concentrations of oxygen, nitrogen, nitrous oxide, carbon dioxide, and ethylene have been determined in this way [25,29,59,61,82]. However, shaking the sample with another gas creates problems of contamination with air when the method is applied to measurements of dissolved oxygen or nitrogen, and exceptional care has to be taken when flushing both the bottle and sampling syringe. These difficulties are not very great during the analysis of gases that are either very minor constituents of the atmosphere, or not normally present, e.g., carbon dioxide, nitrous oxide, and ethylene. The alternative methods that could be applied to oxygen analysis, apart from stripping with carrier gas, to which reference has already been made, include

1. Outgassing the sample under reduced pressure in a modified Van Slyke apparatus [83,84] or Toepler gauge [85] and transferring the evolved gases to a gas chromatograph
2. Pumping continuously an extractant gas through the water sample in a closed circuit by means of a peristaltic pump, followed by transfer of an aliquot of the gas to the chromatograph [86]
3. Direct injection of the water sample into the chromatograph, followed by separation of gases and water vapor

The first two methods should be suitable for soil water analysis; however, they require additional ancillary apparatus coupled to the gas chromatograph and may be time-consuming, thus limiting the amount of replication possible. The third method, which is as rapid as the analysis of a normal gaseous sample, was applied first to the determination of oxygen in soil water by Smith et al. [87]. The procedure involved the injection of 10 μl samples through an injection port heated to 120°C. The water vapor produced was adsorbed on a short Porapak T precolumn at room temperature, and the residual gases passed to a 1.8 m × 3 mm molecular sieve 5A column at 25°C, and then to a helium ionization detector. Operation of an eight-port switching valve allowed a second Porapak T precolumn to be used, while the first one was backflushed with helium to remove the accumulated water (Fig. 9). A very similar system, except for the use of a thermal conductivity detector, was developed independently by Yamaguchi and Komatsu [88], and Hall [89] described a version in which an electron capture detector replaced the HID. This has the advantage that the ECD does not respond to nitrogen, so nitrogen may be used as carrier gas. Figure 10 shows the chromatograms obtained with the HID and ECD systems.

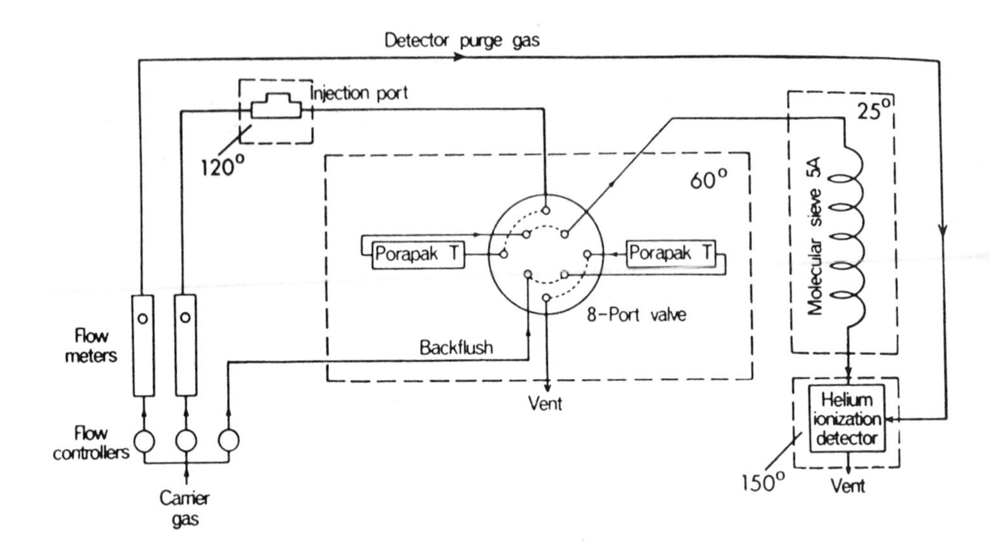

FIG. 9 Gas flow diagram for gas chromatographic measurement of dissolved oxygen using a helium ionization detector. (From Ref. 51.)

IV. APPLICATIONS

A. Soil Oxygen and Carbon Dioxide

The use of gas chromatographic methods such as those described in Sec. III.B has extended greatly the available information about the soil atmosphere under field conditions. In agricultural soils, the effects of drainage and irrigation [90], rainfall [91], cultivation and compaction by wheels [92], heavy applications of organic manures [80,93], and soil type [90] on the oxygen and carbon dioxide concentrations in the soil to which plant roots and microorganisms are exposed have all been investigated under field conditions. Similar studies have also been made of upland mineral soils and deep peat under afforestation [94,95].

The composition of the soil atmosphere varies greatly over short distances, and this has been attributed to the effects of soil structure on gaseous diffusion. In fine-textured, aggregated soils in particular, at high-moisture contents gaseous diffusion will occur primarily through the larger, interaggregate pores. If neighboring sampling probes are inserted (1) into the center of a wet aggregate and (2) close to a macropore between aggregates, respectively, there are likely to be major differences in the oxygen and

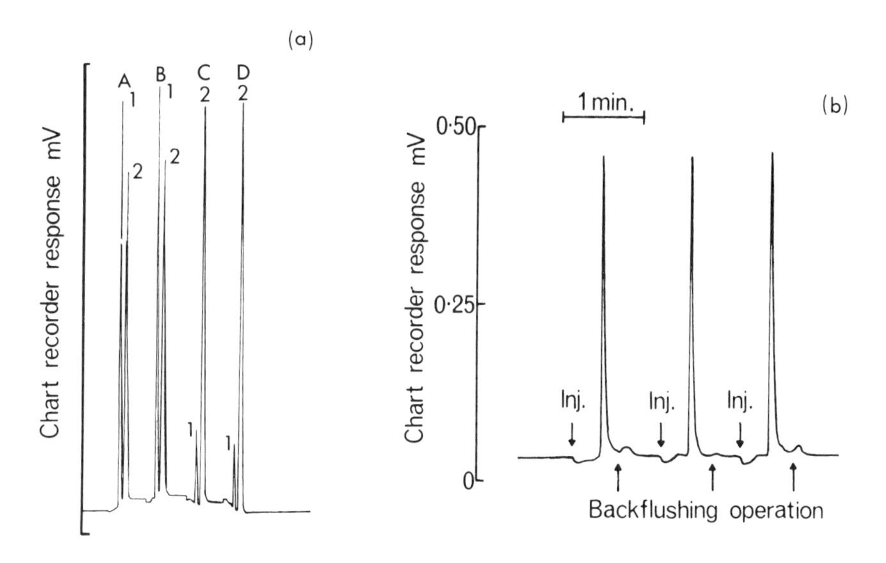

FIG. 10 Chromatograms obtained from successive injections of 10 μl water samples directly into a chromatograph. (a) System with helium ionization detector. A, B, water saturated with air; C, D, water equilibrated with a mixture of 2.8% O_2, 97.2% N_2; (1) O_2; (2) N_2 (from Ref. 51); (b) system with electron capture detector, showing O_2 peak only. (From Ref. 89, reproduced from the *Journal of Chromatographic Science*, by permission of Preston Publications, a division of Preston Industries Inc.)

carbon dioxide concentrations from the two probes. These effects make it necessary to maximize the number of replicate sampling locations. Usually, the upper limit is set by practical considerations such as the number of analyses that can be handled. Automatic injection techniques are particularly valuable in this context, because time lost between successive analyses can be kept to a minimum. Provided that the analysis time per sample can be kept down to a few minutes, then throughputs of the order of 100 samples per day are feasible.

Gas chromatographic analysis of oxygen and carbon dioxide has also found several useful applications in laboratory studies. An example is the measurement of carbon dioxide evolution in soil respiration [34,96]. A very different application is the measurement of soil carbonate content by determining the carbon dioxide liberated after adding hydrochloric acid in sealed vials [97]. Many investigations of soil microbiological processes require either strictly aerobic or strictly anaerobic conditions, and the ability to check rapidly the composition of the atmosphere in the incubation vessel has proved to be a major advantage when attempting to establish the necessary experimental conditions [98].

GC is not the only instrumental method used for soil aeration studies. Polarographic oxygen analyzers have been used by Patrick [99] and Blackwell [100]. In Patrick's method, 10 ml volumes of air were drawn from 75 ml reservoirs in the soil through a cell of <1 ml volume. Individual measurements could be made in less than 1 min, but such large sample volumes were likely to have been representative only of the largest pores. Working on a much smaller scale, Sexstone et al. [101] and Revsbech et al. [102] have measured the oxygen concentration profiles within individual soil crumbs by means of microelectrodes. This technique achieves a degree of spatial resolution that is unattainable by gas chromatographic methods.

B. Nitrous Oxide

A major application of GC to soil and environmental science in recent years has been in the study of the emission of nitrous oxide from soils, originating in the anaerobic microbial reduction of nitrate (denitrification) and also in the aerobic oxidation of ammonium to nitrate (nitrification). This work has been stimulated by the importance of the loss of fertilizer N by the former process, and also by the role of atmospheric nitrous oxide in the destruction of stratospheric ozone and the absorption of infrared radiation, the "greenhouse effect."

The inhibition of nitrous oxide reduction to molecular nitrogen by concentrations of acetylene above 10^{-3} ml ml^{-1} has been used widely in studies of denitrification, because the former gas is measured readily by GC (usually with an electron capture detector, but occasionally using thermal conductivity, ultrasonic or helium ionization detectors), and the atmospheric background is very low (0.3×10^{-6} ml ml^{-1}). In contrast, nitrogen gas from denitrification can only be distinguished easily from the nitrogen present naturally in the soil atmosphere when the former is labeled with ^{15}N and determined by optical emission or mass spectrometry (see Chaps. 10 and 11).

Examples of recent studies of denitrification employing gas chromatographic methods include the effects of inorganic nitrate fertilizer and organic manure on rates in agricultural soils [103]; spatial variability [104]; natural rates in unfertilized forests [105]; and the use of intact cores incubated in the presence of acetylene to assess rates in the field [106]. Among the studies of nitrous oxide release in the course of nitrification are those of Bremner and Blackmer [107] and Anderson and Levine [31]. The recent soil science and environmental science literature contains many more references that cannot be covered here.

C. Sulfur Gases

The release of sulfur-containing volatiles in soil has been studied following amendment with cruciferous plant residues [108], sulfur-containing amino acids [109], animal manures [110,111], and sewage sludge [110]. The effect of digestion by earthworms on the flux of sulfur gases from anaerobically digested sewage sludge has also been investigated [112]. In most of these studies, gas detection was by a flame photometric detector.

Surface soils have the capacity to adsorb considerable quantities of gaseous atmospheric pollutants, either by physicochemical processes or as the result of uptake by microorganisms. The rates of uptake may be studied conveniently by following the disappearance of a gas from the atmosphere above a soil in a closed system, by taking successive samples for gas chromatographic analysis. Experiments with sterilized and unsterilized soils have shown that sulfur dioxide, hydrogen sulfide, and methanethiol are taken up by physicochemical processes [113], whereas dimethyl sulfide, dimethyl disulfide, carbonyl sulfide, and carbon disulfide are partly taken up by microorganisms [114]. Sulfur dioxide has a much greater affinity for soil surfaces, wet or dry, than any of the other sulfur gases studied.

D. Hydrocarbons

The anaerobic decomposition of organic matter in waterlogged soil results in the formation of methane, or "marsh gas," in considerable quantities. The first systematic investigation of this process was as long ago as 1884 [115]. In recent years, interest in methane emission from soils has intensified with the recognition that this gas, like nitrous oxide, contributes to the greenhouse effect, and that its concentration in the atmosphere is increasing by $>1\%\ yr^{-1}$ [116]. Methane emission from rice paddies (considered to be the most important source [116]) has been investigated by several groups, including Seiler and his co-workers. They have explored the emission rates from soil surfaces in the laboratory [117] and in the field [118,119], and the effect on those rates of methane oxidation in the aerated surface layer of the soil. Methane was separated on a 1 m column of molecular sieve 13X at 70°C. Methane produced under anaerobic conditions, following heavy inputs of cattle manure, has also been monitored by GC [93].

With the flame ionization detector, it has also been possible to study other gaseous hydrocarbons occurring in trace concentrations in the soil atmosphere, in particular the physiologically active ethylene. Among the many ethylene-related studies using this sensitive detector, there have been investigations of the effect of microbial substrates on the rate of biosynthesis of the gas [120]; its possible role in the regulation of soil respiration [95]; and its release by plant roots when subjected to stress conditions in the soil [121]. Columns of activated alumina, partially deactivated alumina, and Porapaks N, Q, and R have all been used for ethylene separations in such studies.

The mixture of other gaseous hydrocarbons produced in soil simultaneously with ethylene has been investigated in field and laboratory studies [15,25]. Changes in relative and absolute concentrations of the C_1-C_4 gases with time in samples taken from a soil atmosphere sampling probe in the field are shown in Fig. 11.

An important area of soil science research in which GC is a valuable analytical tool is the study of biological nitrogen fixation. Since the discovery that the nitrogenase enzyme that reduces atmospheric nitrogen to ammonia can also reduce acetylene to ethylene [122], this reaction has become widely used to detect the presence of nitrogen-fixing activity in soil or soil-plant systems. The relative ease with which the production of ethylene from acetylene can be determined by GC contrasts with the more elaborate techniques, and usually the greater expense associated with the alternative method, namely, the use of ^{15}N tracer and its measurement either by mass spectrometry or optical emission spectrometry. However, considerable problems have been encountered by workers attempting to quantify nitrogen-fixing activity on the basis of results obtained with the acetylene reduction method. This subject is also discussed in Chaps. 10 and 11.

Ethylene may be separated readily from acetylene on columns of Porapak T [123], Porapaks N and R [124], and 20% ethyl, *N'*,*N'*-dimethyloxalamide on 100–200 mesh acid-washed firebrick [125]. On all these columns, ethylene is eluted before acetylene, and this means that even a very small trace of the former gas can be detected in the presence of a very large acetylene peak (concentrations of 0.1 ml ml^{-1} are used commonly in the reduction assay procedure). Propane has been used as an internal standard for the acetylene reduction assay; measurement of the propane peak, after the elution of ethylene and acetylene, provides a check on any leaks in the system [126].

In the low-cost portable GC system developed for use in the field by Mallard et al. [124], a 0.44 m column, half of which was packed with Porapak R, and the other half with Porapak N, was employed. The detector in this system is a Taguchi gas sensor, consisting of a SnO semiconductor molded around a small filament heater and connected in a bridge circuit. The limit of detection for ethylene is

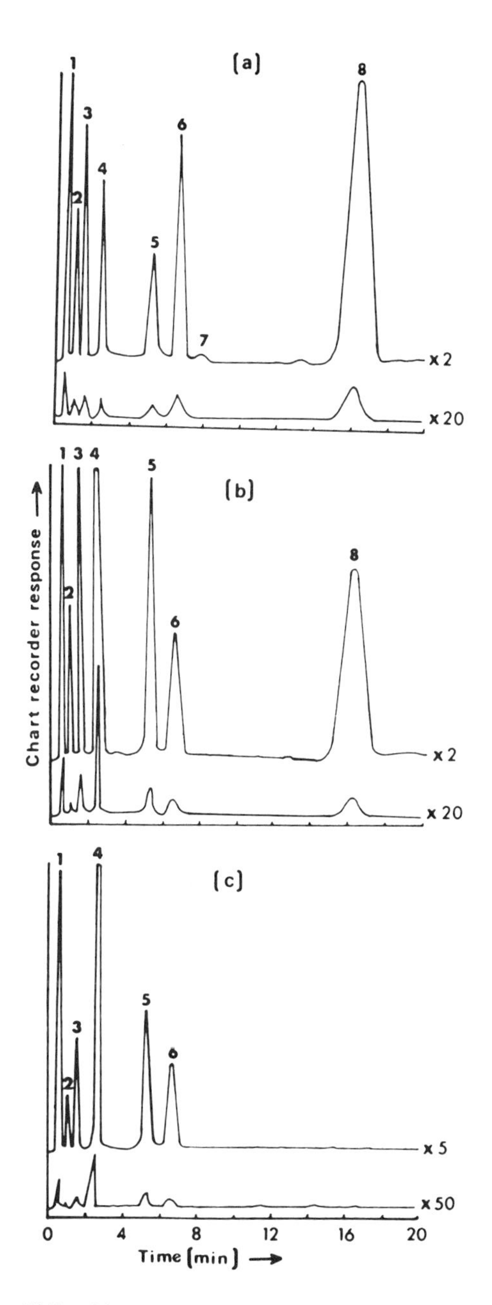

FIG. 11 Chromatograms of hydrocarbons in soil atmosphere samples taken on three occasions from the same probe at 300 mm depth. (a) April 26, (b) June 9, (c) June 21. (1) Methane (+ air), (2) ethane, (3) ethylene, (4) propane, (5) propene, (6) isobutane, (7) *n*-butane, (8), butene (mixture of isomers). Separation carried out on alumina partially deactivated with NaI. (From Ref. 15, reproduced from the *Journal of Chromatographic Science*, by permission of Preston Publications, a division of Preston Industries Inc.)

10^{-5} ml ml^{-1}. The original instrument was later improved [127,128] and has found application in remote locations where conventional laboratory facilities are lacking [128].

E. Other Carbon Compounds

Gas chromatographic methods have been used to determine a number of volatile organic substances in soil in addition to those already discussed. These substances include acetic and butyric acids [129, 130], and alcohols, aldehydes, ketones, and esters [131–133] resulting from the decomposition of plant residues.

The evolution from (and uptake by) soil of carbon monoxide, another gas implicated in environmentally significant atmospheric processes, has been measured by GC, using a molecular sieve column and helium ionization detector [113]. Other similar studies have used a nonchromatographic technique for this gas, based on the reduction of mercuric oxide to mercury vapor [134].

The distribution and persistence in soil of fumigants such as methyl bromide [135], 1,2-dibromoethane [136], and 1,2-dibromo-3-chloropropane (DBCP) [137] can be measured readily by GC. Less than 1 ng of compounds such as DBCP can be detected with an electron capture detector. A mass spectrometer has been used for the detection of 1,2-dibromoethane [136].

V. SUMMARY AND CONCLUSIONS

The development of gas chromatographic methods over the last 25 years has made possible the rapid and accurate measurement of permanent gases and volatile compounds present as major constituents or only as trace components of the soil atmosphere. For several years, the improvement in sensitivity over classical methods of analysis was vastly greater for organic compounds than for inorganic, because of the availability of the flame ionization detector. The development of the helium ionization and ultrasonic detectors then led to comparable sensitivity for all substances, and detectors such as the electron capture and flame photometric combine great sensitivity with selectivity for certain types of molecule, further extending the capabilities of the method.

It is possible, with currently available column packing materials, to separate all the gases of interest in the soil atmosphere, providing multiple-column systems or temperature programming procedures are used. One highly desirable development would be a column on which oxygen, nitrogen, and carbon dioxide could be separated isothermally in a reasonably short time. Another would be further improvement in the separation of argon from oxygen to make it readily achievable at ambient or elevated temperatures.

Naturally occurring soil gases that may be measured by GC include

1. Oxygen, nitrogen, and argon originating from the earth's atmosphere
2. Carbon dioxide produced by respiration of plant roots and microorganisms
3. Nitrous oxide, nitric oxide, and molecular nitrogen evolved by microbiological and chemical processes in the soil
4. Hydrogen sulfide, carbonyl sulfide, carbon disulfide, and organo-sulfur compounds produced by microbial decomposition of sulfur-amino acids and other sulfur compounds
5. Methane, ethylene, and other C_2-C_4 hydrocarbons produced by microbial activity in the soil, and also possibly present as a result of diffusion from underlying oil or gas-bearing strata
6. Carbon monoxide from chemical decomposition of soil organic matter; hydrogen resulting from microbial decomposition of organic matter under severely reducing conditions
7. The vapors of volatile fatty acids, aldehydes, alcohols, and ketones produced by biological activity

In addition, the fate of gases that have been deliberately or accidentally introduced into the soil may be monitored by GC. Applications include the measurement of residual quantities of gaseous fumigants such as brominated hydrocarbons; estimation of biological nitrogen fixation rates by the reduction of acetylene to ethylene; identification of leaks of natural and manufactured gas from underground pipes; measurement of the rate of sorption of atmospheric pollutant gases by soils; and measurement of gaseous diffusion rates.

REFERENCES

1. J. B. Boussingault and M. B. Lewy, *Ann. Chim. Phys.*, *37*:5 (1853).
2. J. C. Giddings, *Dynamics of Chromatography*, Marcel Dekker, New York, 1965.
3. L. Szepesy, *Gas Chromatography*, Iliffe, London, 1970.
4. P. G. Jeffery and P. J. Kipping, *Gas Analysis by Gas Chromatography*, Pergamon, Oxford, 1972.
5. C. J. Cowper and A. J. De Rose, *The Analysis of Gases by Chromatography*, Pergamon, Oxford, 1983.
6. A. J. P. Martin and R. L. M. Synge, *Biochem. J.*, *35*:1358 (1941).

7. A. T. James and A. J. P. Martin, *Biochem. J.*, *50*:679 (1952).
8. E. Glueckauf, *Trans. Faraday Soc.*, *51*:34 (1955).
9. J. J. van Deemter, F. J. Zuiderweg, and A. Klinkengerg, *Chem. Eng. Sci.*, *5*:271 (1956).
10. J. C. Giddings, *Anal. Chem.*, *35*:439 (1963).
11. E. Grushka, *J. Chromatogr. Sci.*, *13*:25 (1975).
12. F. L. Bayer, *J. Chromatogr. Sci.*, *24*:549 (1986).
13. O. L. Hollis, *Anal. Chem.*, *38*:309 (1966).
14. C. G. Scott, *J. Inst. Petroleum*, *45*:118 (1959).
15. K. A. Smith and R. J. Dowdell, *J. Chromatogr. Sci.*, *11*:655 (1973).
16. F. Bruner, A. Liberti, M. Possanzine, and I. Allegrine, *Anal. Chem.*, *44*:2070 (1972).
17. R. Kaiser, *Chromatographia*, *2*:453 (1969).
18. R. Kaiser, *Chromatographia*, *3*:38 (1970).
19. A. Zlatkis, H. R. Kaufman, and D. E. Durbin, *J. Chromatogr. Sci.*, *8*:416 (1970).
20. D. H. Bollman and D. M. Mortimore, *J. Chromatogr. Sci.*, *10*:523 (1972).
21. E. W. Lord and R. C. Horn, *Anal. Chem.*, *32*:878 (1960).
22. E. L. Obermiller and G. O. Charlier, *J. Gas Chromatogr.*, *6*:446 (1968).
23. G. E. Pollock, *J. Chromatogr. Sci.*, *24*:173 (1986).
24. T. B. Parkin, H. F. Kaspar, A. J. Sexstone, and J. M. Tiedje, *Soil Biol. Biochem.*, *16*:323 (1984).
25. K. A. Smith and S. W. F. Restall, *J. Soil Sci.*, *22*:430 (1971).
26. R. J. Dowdell and R. Crees, *Lab. Practice*, *23*:488 (1974).
27. J. C. Ryden, *J. Soil Sci.*, *33*:263 (1982).
28. S. Ghoshal and B. Larsson, *Acta Agric. Scand.*, *25*:275 (1975).
29. R. J. Dowdell, J. R. Burford, and R. Crees, *Nature (Lond.)*, *278*:342 (1979).
30. D. H. Smith and F. E. Clark, *Soil Sci. Soc. Am. Proc.*, *24*:111 (1960).
31. I. C. Anderson and J. S. Levine, *J. Geophys. Res.*, *92*:965 (1987).
32. C. W. Quinlan and J. R. Kittrel, *J. Chromatogr. Sci.*, *10*:691 (1972).
33. J. M. Trowell, *J. Chromatogr. Sci.*, *9*:253 (1971).
34. J. R. Burford and J. M. Bremner, *Soil Biol. Biochem.*, *4*:191 (1972).
35. J. R. Burford and J. M. Bremner, *Soil Biol. Biochem.*, *4*:489 (1972).
36. R. K. Stevens, J. D. Mulik, A. E. O'Keefe, and K. J. Krost, *Anal. Chem.*, *43*:827 (1971).
37. J. Macak, J. Kubat, V. Dobal, and J. Mizera, *J. Chromatogr.* *286*:69 (1984).

38. T. L. C. de Souza, *J. Chromatogr. Sci.*, *22*:470 (1984).
39. G. Wells and R. Simon, *J. Chromatogr.*, *256*:1 (1983).
40. P. L. Patterson, R. A. Gatten, J. Kolar, and C. Ontiveros, *J. Chromatogr. Sci.*, *20*:27 (1982).
41. F. Andrawes and R. Ramsey, *J. Chromatogr. Sci.*, *24*:513 (1986).
42. R. S. Brazell and R. A. Todd, *J. Chromatogr. Sci.*, *302*:257 (1984).
43. J. Connor, *J. Chromatogr.*, *210*:193 (1981).
44. W. E. Wentworth and R. R. Freeman, *J. Chromatogr.*, *79*:322 (1973).
45. I. C. Anderson and J. S. Levine, *Appl. Environ. Microbiol.*, *51*:938 (1986).
46. S. O. Farwell and C. J. Barinaga, *J. Chromatogr. Sci.*, *24*:483 (1986).
47. A. M. Blackmer and J. M. Bremner, *Soil Sci. Soc. Am. J.*, *41*:908 (1977).
48. D. R. Keeney, I. R. Fillery, and G. P. Marx, *Soil Sci. Soc. Am. J.*, *43*:1124 (1979).
49. D. A. Binstock, *Soil Biol. Biochem.*, *16*:287 (1984).
50. K. A. Smith and W. Harris, *J. Chromatogr.*, *53*:358 (1970).
51. P. D. Brooks and E. A. Paul, *Plant Soil*, *101*:183 (1987).
52. G. P. Robertson and J. M. Tiedje, *Plant Soil*, *83*:453 (1985).
53. S. Neglia and L. Favretto, in *Advances in Organic Geochemistry* (U. Colombo and G. D. Hobson, eds.), Pergamon, Oxford, 1964, p. 285.
54. E. J. Russell and A. Appleyard, *J. Agric. Sci. Camb.*, *7*:1 (1915).
55. J. L. Tackett, *Soil Sci. Soc. Am. Proc.*, *32*:346 (1968).
56. O. A. De Camargo, F. Grohmann, E. Salati, and E. Matsui, *Soil Sci.*, *117*:173 (1974).
57. J. R. Burford and R. J. Millington, *Transactions of the 9th International Congress of Soil Science*, Vol. 2, I.S.S.S./Angus and Robertson, Sydney, 1968, p. 505.
58. M. Yamaguchi, F. D. Howard, D. L. Hughes, and W. J. Flocker, *Soil Sci. Soc. Am. Proc.*, *26*:512 (1962).
59. R. J. Dowdell, K. A. Smith, R. Crees, and S. W. F. Restall, *Soil Biol. Biochem.*, *4*:325 (1972).
60. P. Colbourn, M. M. Iqbal, and I. W. Harper, *J. Soil Sci.*, *35*:11 (1984).
61. K. A. Smith, in *Advances in Chromatography*, Vol. 15 (J. C. Giddings, E. Grushka, J. Cazes, and P. R. Brown, eds.), Marcel Dekker, New York, 1977, p. 197.
62. J. R. Burford, *J. Chromatogr. Sci.*, *7*:760 (1969).
63. T. B. Parkin, *Soil Sci. Soc. Am. J.*, *49*:273 (1985).
64. M. E. Bednas and D. S. Russell, *J. Gas Chromatogr.*, *5*:592 (1967).

65. J. P. Slater, F. R. McLaren, D. Christensen, and D. Deneen, in *Proceedings of Conference on Characterization and Monitoring of Vadose (Unsaturated) Zone* (D. M. Nielsen and M. Curl, eds.), Nat. Well Water Association, Worthington, Ohio, 1983, p. 623.
66. G. M. Egginton and K. A. Smith, *J. Soil Sci.*, *37*:69 (1980).
67. J. M. Bremner and A. M. Blackmer, *Proceedings of Denitrification Seminar*, Fert. Inst., Washington, D.C., 1977.
68. E. P. Kavanagh and J. R. Postgate, *Lab. Practice*, *19*:159 (1970).
69. W. L. Banwart and J. M. Bremner, *Soil Biol. Biochem.*, *6*:113 (1974).
70. T. Yoshinari and R. Knowles, *Biochem. Biophys. Res. Comm.*, *9*:177 (1976).
71. M. Vilain and J.-P. Druelle, *Ann. Agron. (N.S.)*, *18*:507 (1967).
72. A. M. Blackmer, J. H. Baker, and M. E. Weeks, *Soil Sci. Soc. Am. Proc.*, *38*:689 (1974).
73. L. F. Elliot and T. M. McCalla, *Soil Sci. Soc. Am. Proc.*, *36*:68 (1972).
74. J. M. Bollag and S. T. Czlonkowski, *Soil Biol. Biochem.*, *5*:673 (1973).
75. B. T. Bunting and J. A. Campbell, *Can. J. Soil Sci.*, *55*:69 (1975).
76. I. S. Cornforth and R. J. Stevens, *Plant Soil*, *38*:581 (1973).
77. D. Laskowski and J. T. Moraghan, *Plant Soil*, *27*:357 (1967).
78. K. Frunzke and W. G. Zumft, *J. Chromatogr.*, *299*:477 (1984).
79. K. C. Hall and R. C. Dowdell, *J. Chromatogr. Sci.*, *19*:107 (1981).
80. J. R. Burford, *J. Sci. Fd. Agric.*, *27*:115 (1976).
81. J. W. Swinnerton, J. Linnenbom, and C. H. Cheek, *Anal. Chem.*, *34*:483 (1962).
82. E. A. Davidson and M. K. Firestone, *Soil Sci. Soc. Am. J.*, *52*:1201 (1988).
83. K. G. Ikels, *J. Gas Chromatogr.*, *3*:359 (1965).
84. E. Bjergbakke, in *Measurement of Oxygen* (H. Degn, I. Balslev, and R. Brook, eds.), Elsevier, Amsterdam, 1976, p. 1.
85. L. V. Betkovic, M. M. Kosanic, and J. G. Draganic, *Bull. Boris Kidric Inst. Nucl. Sci. (Belgrade)*, *15*:9 (1964).
86. H. A. C. Montgomery and C. Quarmby, *Lab. Practice*, *15*:538 (1966).
87. K. A. Smith, R. J. Dowdell, and K. C. Hall, in *Measurement of Oxygen* (H. Degn, I. Balslev, and R. Brooks, eds.), Elsevier, Amsterdam, 1976, p. 226.
88. M. Yamaguchi and Y. Komatsu, *Plant Soil*, *47*:265 (1977).
89. K. C. Hall, *J. Chromatogr. Sci.*, *16*:311 (1978).

90. K. A. Smith and R. J. Dowdell, *J. Soil Sci.*, *25*:217 (1974).
91. K. A. Smith, R. J. Dowdell, K. C. Hall, and R. J. Crees, *U.K. Agricultural Research Council Letcombe Laboratory Annual Report for 1973*, HMSO, London, 1974, p. 35.
92. R. J. Dowdell and R. Crees, *U.K. Agricultural Research Council Letcombe Laboratory Annual Report for 1975*, HMSO, London, 1976, p. 52.
93. F. A. Norstadt and L. K. Porter, *Soil Sci. Soc. Am. J.*, *48*: 783 (1984).
94. D. G. Pyatt and K. A. Smith, *J. Soil Sci.*, *34*:465 (1983).
95. J. A. King, K. A. Smith, and D. G. Pyatt, *J. Soil Sci.*, *37*: 485 (1986).
96. K. A. Smith, *Soil Boil. Biochem.*, *10*:269 (1978).
97. R. G. Amundson, J. Trask, and E. Pendall, *Soil Sci. Soc. Am. J.*, *52*:880 (1988).
98. G. Goodlass and K. A. Smith, *Soil Biol. Biochem.*, *10*:193 (1978).
99. W. H. Patrick, *Soil Sci. Soc. Am. J.*, *41*:651 (1977).
100. P. S. Blackwell, *J. Soil Sci.*, *34*:271 (1983).
101. A. J. Sexstone, N. P. Revsbech, T. B. Parkin, and J. M. Tiedje, *Soil Sci. Soc. Am. J.*, *49*:645 (1985).
102. N. P. Revsbech, L. P. Nielsen, P. B. Christensen, and J. M. Sorensen, *Appl. Environ. Microbiol.*, *54*:2245 (1988).
103. G. M. Egginton and K. A. Smith, *J. Soil Sci.*, *37*:69 (1986).
104. T. B. Parkin, *Soil Sci. Soc. Am. J.*, *51*:1194 (1987).
105. G. P. Robertson and J. M. Tiedje, *Soil Sci. Soc. Am. J.*, *48*: 383 (1984).
106. J. C. Ryden, J. H. Skinner, and D. J. Nixon, *Soil Biol. Biochem.*, *19*:753 (1987).
107. J. M. Bremner and A. M. Blackmer, *Sci.*, *199*:295 (1978).
108. J. A. Lewis and G. C. Papavizas, *Soil Biol. Biochem.*, *2*:239 (1970).
109. W. L. Banwart and J. M. Bremner, *Soil Biol. Biochem.*, *7*:359 (1975).
110. W. L. Banwart and J. M. Bremner, *Soil Biol. Biochem.*, *8*:439 (1976).
111. L. F. Elliott and T. A. Travis, *Soil Sci. Soc. Am. Proc.*, *37*: 700 (1973).
112. S. G. Horner and M. J. Mitchell, *Soil Biol. Biochem.*, *13*:367 (1981).
113. K. A. Smith, J. M. Bremner, and M. A. Tabatabai, *Soil Sci.*, *116*:313 (1973).
114. J. M. Bremner and W. L. Banwart, *Soil Biol. Biochem.*, *8*:79 (1976).
115. P. P. Deherain, *Ann. Agron.*, *10*:385 (1984).
116. R. E. Dickinson and R. J. Cicerone, *Nature (Lond.)*, *319*:109 (1986).

117. A. Holzapfel-Pschorn, R. Conrad, and W. Seiler, *FEMS Microbiol. Ecol.*, *31*:343 (1985).
118. W. Seiler, A. Holzapfel-Pschorn, R. Conrad, and D. Scharffe, *J. Atm. Chem.*, *1*:241 (1984).
119. A. Holzapfel-Pschorn and W. Seiler, *J. Geophys. Res.*, *91*:1803 (1986).
120. W. T. Frankenburger and P. J. Phelan, *Soil Sci. Soc. Am. J.*, *49*:1416 (1985).
121. M. B. Jackson and M. C. Drew, in *Flooding and Plant Growth* (T. T. Kozlowski, ed.), Academic Press, New York, 1984, p. 47.
122. M. J. Dilworth, *Biochim. Biophys. Acta*, *127*:285 (1966).
123. T. G. Baker and P. M. Attiwill, *Soil Biol. Biochem.*, *16*:241 (1984).
124. T. M. Mallard, C. S. Mallard, H. S. Holfeld, and T. A. LaRue, *Anal. Chem.*, *49*:1275 (1977).
125. A. B. Richmond, *J. Chromatogr. Sci.*, *7*:321 (1969).
126. P. J. Dart, J. M. Day, and D. Harris, in *Grain Legume Production Technical Booklet*, FAO/IAEA, Vienna, 1972.
127. H. S. Holfeld, C. S. Mallard, and T. A. LaRue, *Plant Soil*, *52*:595 (1979).
128. I. F. Grant, *Plant Soil*, *95*:435 (1986).
129. J. M. Lynch, *J. Appl. Bacteriol.*, *42*:81 (1977).
130. H. Okazawi and K. Nose, *Ann. Phytopath. Soc. Japan*, *52*:384 (1986).
131. J. A. Adamson, A. J. Francis, J. M. Duxbury, and M. Alexander, *Soil Biol. Biochem.*, *7*:45 (1975).
132. R. E. Holm, *Plant Physiol.*, *50*:293 (1972).
133. P. R. Herrington, J. T. Craig, and J. E. Sheridan, *Soil Biol. Biochem.*, *19*:509 (1987).
134. R. Conrad and W. Seiler, *Environ. Sci. Technol.*, *19*:1165 (1985).
135. M. J. Kolbezen and F. J. Abu-El-Haj, *Pesticide Sci.*, *3*:73 (1972).
136. B. L. Sawhney, J. J. Pignatello, and S. M. Steinberg, *J. Environ. Qual.*, *17*:149 (1988).
137. D. E. Johnson and B. Lear, *J. Chromatogr. Sci.*, *7*:384 (1969).

13

Determination of Pesticides by Gas Chromatography and High-Pressure Liquid Chromatography

DAVID J. EAGLE, JOHN L. O. JONES, EDWARD J. JEWELL, and ROGER P. PAXTON *Agricultural Development and Advisory Service, Ministry of Agriculture, Fisheries and Food, Cambridge, England*

I. INTRODUCTION

The rapid development in recent decades of pesticidal chemicals that are biologically active in minute concentrations has posed a formidable challenge to the analytical chemist, that of identifying and quantifying a wide variety of substances at nanogram or even picogram level in a highly complex organic matrix. The problems of the analyst are often complicated by an initial lack of information as to the pesticide he or she is to determine. Although screening methods are available for the main groups of chemicals, careful casework to obtain the maximum circumstantial evidence should always precede the work in the laboratory. Although no analytical method will detect all pesticides, the challenge of determining these materials has been met very largely by the advances that have been made in the technology of gas chromatography (GC). Of the hundreds of pesticides currently in use, the majority can be determined by GC, and for most of these it is the preferred method both for sensitivity and selectivity. More recently, high-pressure liquid chromatography (HPLC) has joined with GC as a valuable technique in the methodology of the residue analyst. It rapidly is being adopted for the determination of compounds for which GC is unsuitable and it is challenging established GC methods for others. It has the potential to equal or exceed GC in its importance.

The history of chromatography has been well documented in texts on the subject (e.g., [1]) and the major landmarks [2–7] are well known. Many books (e.g., [8–13]) are available that treat the theory and practice in greater depth than is possible here, but an outline of the basic theory is given in Chap. 12. Although a few pesticide substances have been separated by gas-solid chromatography (GSC), the overwhelming majority of GC separations have involved the use of columns containing liquid-coated supports as stationary phase, i.e., gas-liquid chromatography (GLC). In this chapter the principles involved in the application of GLC to pesticide analysis are discussed, together with developments in the use of HPLC, and selected applications to pesticide residue analysis in soil are given.

II. SAMPLING AND PREPARATION FOR CHROMATOGRAPHY

A. Sampling

Procedures for taking representative samples of soil are given in texts of agricultural chemistry and in a leaflet issued by the Agricultural Development and Advisory Service in England and Wales (ADAS) [14]. Sampling methods for pesticide residue analysis have been reported by Ford et al. [15]. When symptoms of crop damage are seen, samples from areas of differing severity can be helpful. When a sample can be obtained from an area of comparable soil known to be free of pesticide treatment, it can be valuable to the analyst in assessing the validity of the analysis. The depth is determined by the behavior of the compound to be determined. Many herbicides are adsorbed in the surface layer; sampling to greater than 50–75 mm dilutes the sample with uncontaminated material, raising the detection limit. For valid comparison of results the depth of sampling must always be known.

Unless metabolism of rapidly degraded chemicals is being studied, no special precautions are needed in transit of the sample to the laboratory other than protection from contamination. It should be noted that some packing materials, e.g., polythene [16] and cloth bags [17], may release interfering materials.

B. Sample Storage

It is often convenient to be able to store samples prior to analysis. However, few data are available on the stability of pesticides in stored samples and the analyst would be well advised to make his or her own investigation before undertaking a major program of work. Kawar et al. [18] examined the storage stability of 35 pesticides and found little evidence of breakdown under deep-freeze conditions other than for diazinon, malathion, parathion, captan, and dichlorvos, but at refrigerator temperatures a number of others showed some degradation. Webster and Reimer [19] have reported degradation of metribuzin in cold storage. Where evidence is lacking, storage at −20°C and retention for no more than 1–2 months are wise precautions.

C. Preparation and Extraction

Samples should be mixed thoroughly and stones and plant debris removed. When the pesticide is known to be stable during drying, the sample should be dried, as the presence of water may be undesirable in the extraction and cleanup. When the extractant is miscible with water and an aqueous phase is used in the first stage of cleanup, drying may be avoidable, but it should be remembered that even in these cases dilution by sample water may affect extraction efficiency. Drying at <30°C in a current of air is acceptable when stability data relating to these conditions indicate no problem of degradation. When data are lacking or preclude drying in this way, removal of moisture may be achieved by blending with anhydrous sodium sulfate. Freeze-drying is also worthy of consideration. An alternative approach is extraction with a succession of solvents. A typical scheme used acetone for the first extraction, then the residue is reextracted with acetone/dichloromethane mixture and finally with chloroform. On combination of the extracts and dilution, if necessary, with water the pesticide is obtained in the chloroform phase. With a judicious choice of solvents and volumes such a sequence may simultaneously achieve extraction, cleanup, and concentration. In all operations it should be noted that contamination can be introduced from the laboratory environment [16,20,21]; a blank run without sample to check the apparatus and chemicals should always be carried out.

Extraction may be by shaking or Soxhlet extraction. The latter is preferred when the pesticide is adsorbed by the soil or is of limited solubility, but its use may be restricted by the thermal stability of the compound. In specifying a procedure, the extraction time and rate of reflux should be stated. Condensers should allow the solvent to fall directly into the thimble.

Extraction of soil by shaking with methanol has been found to give satisfactory recoveries for the majority of pesticides. An advantage is that the extraction time is much shorter than that of the Soxhlet extraction method.

The first consideration in the choice of a solvent is solubility. Although residue levels are unlikely to exceed the solubility in all but a few solvents, the use of a reagent in which macro-quantities are soluble favors rapid and complete extraction. The next consideration is the ability to release adsorbed material. There are no generally applicable rules, but polar solvents tend to be more effective.

A further consideration is the solubility of unwanted organic matter. Miscibility with water is advantageous for the extraction of undried samples and acetone is used widely for this purpose. It is often more effective than other polar compounds, e.g., methanol, for adsorbed material but coextracts relatively large amounts of organic matter. Acetonitrile is used less widely on soils than on plants but the properties that make it popular for plant material are equally applicable to soil. Miscibility with water in most proportions makes it an efficient extractant for moist samples; it can be separated from water by dilution and addition of a salt, it is immiscible with a number of other solvents giving opportunities for partition cleanup, and it extracts relatively little organic matter. When the compund ionizes, an aqueous extractant may be advantageous, the pH being chosen to ensure it is entirely in the acid or salt form.

D. Cleanup

The extent of cleanup, i.e., the removal of unwanted constituents of the extract that may interfere with the subsequent analysis, is determined by the ability of the chromatographic column to separate the pesticide from these coextractives and by the specificity of the detector. With selective detectors many compounds may be determined with nothing more than concentration to small volume. For sensitive, nonspecific detectors, e.g., electron capture, cleanup is invariably needed.

1. Partition

Partition between immiscible solvents affords a rapid means of removing gross contamination. Apart from considerations of miscibility, a wide difference in polarity between the solvents is necessary to obtain a sufficiently favorable partition coefficient for the pesticide to be transferred completely to one phase in two to three extractions. Partition between aqueous and organic phases is most used but a system of polar and nonpolar organic phases, e.g., acetone and hexane, can be very useful, hexane being valuable for the removal of lipids. When the compound ionizes, adjustment of the pH will effect the transfer between aqueous and organic phases, but this can lead to troublesome emulsification with some solvents.

2. Column Chromatography

Various forms of silica, alumina, and carbon are used widely for column chromatographic cleanup. Their characteristics are dependent on their method of preparation and some batch-to-batch variation may be encountered. With silica and alumina the moisture content is critical to their separation characteristics and may change with storage or exposure to the atmosphere. It is always wise to check the performance of the material by passing a standard solution of the pesticide at the time of analysis rather than relying on published elution data or past experience. For ionizable compounds, ion-exchange resins may provide an effective cleanup material. The coarser HPLC packings have found some use as cleanup materials with relatively reproducible properties. HPLC itself, though primarily a determinative technique, should not be neglected as a possible preliminary to GC. Most detectors are nondestructive and with suitable fraction-collecting arrangements the eluate may be recovered, giving a very clean solution for GC and providing valuable confirmatory evidence of peak identity.

3. *Derivatization*

Derivatization is normally looked on as a confirmatory technique or a means of making the pesticide suitable in stability, retention characteristics, or detectability for GC. However, this alteration of the physical and chemical properties may also be exploited for cleanup. Few, if any, of the coextractives are altered similarly, so the pesticide is likely to become separable by means that were not applicable to the parent compound. Conversely but less commonly, a troublesome coextractive may be derivatized to enable its separation from the pesticide.

4. *Sweep Codistillation*

Sweep codistillation [22–24], in which the sample extract is distilled through a glass wool column in which the coextractives are retained on the glass wool and the pesticide recovered in the distillate, has found some favor as a means of cleanup and concentration.

Some degree of concentration may be achieved in the course of cleanup but it is usually necessary to remove excess solvent to obtain a solution that will give an acceptable detection limit when injected into the chromatograph. Rotary evaporation under reduced pressure is effective in most cases. Small amounts of volatile solvent may be removed by impinging a stream of nitrogen. When the solution has to be taken to dryness to be redissolved in another solvent, the addition of a few drops of a nonvolatile liquid as a "keeper" is advised, as many compounds are degraded or difficult to redissolve if held dry more than momentarily. Chiba and Morley [25] reported that in the preparation and cleanup of soil extracts the greatest losses are at the evaporation stage. Certain GC detectors require rigorous elimination of halogenated solvent, which may entail repeated evaporation to dryness.

The residue is diluted finally with or redissolved in a measured volume of the solvent in which it is to be injected into the chromatograph.

For HPLC this will always be the eluent. For GC any volatile solvent that is not retained by the stationary phase and does not saturate the detector may be used.

III. GAS-LIQUID CHROMATOGRAPHY

A. Columns

1. *Solid Supports*

Although sold under a variety of names, the majority of solid supports are derived from firebrick or diatomaceous earth, the latter being more common. Firebrick is preferred for strongly nonpolar compounds. An ideal support is inert; it serves only to keep the liquid in place and give it a large surface area. The base materials are therefore washed and treated to inactive surface hydroxyl groups. Ashby [26] has listed the main materials and their pretreatments. Particle size is important—the smaller the particles the more efficient the column, the limit being set by resistance to gas flow. Efficiency depends equally on close packing so uniformity of size and resistance to fracture are essential. For further discussion of support materials, the reviews by Ottenstein [27] and Palframan and Walker [28] will be found useful.

2. *Liquid Phases*

Many hundreds of compounds have been used as stationary phases and most suppliers offer a wide range. It is impracticable to find the most suitable by trial and error. Treatments of the factors governing their selection may be found in textbooks on GC, but a few simple guidelines are helpful to narrow down the choice. The most basic is the concept of polarity: A polar phase will best separate polar compounds, whereas a nonpolar phase is preferable for nonpolar analytes. Polarity is not directly quantifiable. Our knowledge of it is gained from the behavior of various compounds when chromatographed on different phases. From such studies comes the concept of retention index, based on the ability of a phase to retain benzene compared with the least polar—squalene. This concept has been reviewed by Ettre [29]. Rohrschneider [30] used the retention times of benzene, ethanol, methyl ethyl ketone, nitromethane, and pyridine to determine numerical indexes, and McReynolds [31], developing this work further, determined the behavior of 10 compounds on over 200 phases. Within a group of phases of similar polarity the Rohrschneider index may be used to predict separations from the structure of the compounds of interest, e.g., a high index number for methyl ethyl ketone relative to that for ethanol would be chosen if it were desired to elute an electron attractor compound (e.g., alcohols or acids) before an electron donor (e.g., ketones or ethers). An extensive list of McReynolds constants has been published by Supina [32]. The latter also discussed the effect of temperature and stationary phase loading on column efficiency and the effect of column packing.

In listing the properties of over 200 stationary phases, McReynolds sought to show the close similarity between many of them. His view that the number of phases had proliferated unnecessarily is shared by many chromatographers. A list has been published [33] of six preferred phases for pesticide analysis, with a further 24 to be tried if none in the short list is suitable. The chromatographer would do well to make an initial choice from these lists, exploring others only if the separation is unsatisfactory. Some analysts go further and use one column almost exclusively for a wide range of determinations. Although this approach is unlikely to succeed for multiresidue analysis, the analyst who determines one particular compound in a sample, and has only to separate it from coextractives, may achieve considerable success with this approach and avoid the unproductive use of time that is a consequence of frequent column changing.

3. *Coating and Packing*

The efficiency of a column depends on the care and skill of its preparation. The preparation is not difficult and with a little practice it should be possible to equal or better the performance of commercially packed columns. The essential requirements are uniformity of coating, avoidance of particle damage, and no chemical degradation.

About 7 g of material is required to pack a 1.5 × 4 mm column. Many liquid phases are viscous liquids or gums at ambient temperature and difficult to weigh out exactly. It is, therefore, best to weight the coating material first and then the appropriate amount of support to give the required loading. Suppliers' lists usually indicate a suitable solvent. Sufficient should be used to wet the support and form a slurry. The slurry (in a shallow basin) is placed over a gentle source of heat and kept continuously stirred. Thorough stirring is especially important as the slurry reaches a pasty state and particular attention should be given to the side of the basin where, if material is left undisturbed, there is a tendency for the coating to be deposited on the basin rather than the support. Once a friable state is reached, stirring may be relaxed and the basin put aside in a well-ventilated place for the last traces of solvent to evaporate.

Some chromatographers favor the use of a rotary evaporator to keep the slurry in motion. However, this is more liable to damage the particles. Another method is to prepare a liquid suspension of the support in a solution of the stationary phase. The procedure of Cook et al. [34] is typical. The suspension is poured into a Buchner funnel or wide-bore tube and the excess solution drained away. The wetted support is removed and spread in a thin layer to dry. The loading achieved is calculated from the difference in volume between the initial solution and that drained off. Some practice is needed to judge the concentration of solution required. The support may be expected to retain about its own volume of solution but this can vary by a factor of two. This method is the best where uniformity of coating and avoidance of particle damage are critical.

It is advisable to acid-wash and silanize glass columns and glass wool plugs. Pressure or vacuum may be used to assist close packing. The latter is more convenient, but with columns exceeding 1 m in length the flow becomes barely sufficient in the later stages. The packing is introduced a little at a time with constant vibration and each lot allowed to pack down until there is no further movement before the next is added. Final settlement may be achieved conveniently by connecting the column to the injection port of the chromatograph and tapping the column throughout its length with maximum carrier pressure applied. The space remaining at the inlet should be such that a syringe needle just reaches the top of the packing. This space is filled with silanized glass beads and finally a small porous plug.

To condition the column it is placed in the chromatograph with carrier flow but no connection to the detector. The temperature is set 20–30°C higher than that at which the column will be used, but not exceeding the maximum for the liquid phase. This temperature is maintained at least overnight to remove last traces of solvent and volatile impurities and ensure stable behavior when the column is put into use.

4. *Column Efficiency*

It is important to be able to measure the efficiency of a column, so that if the separation is unsatisfactory, it will be known whether the choice of liquid phase or the preparation of the column is at fault. The number of theoretical plates n is given by

$$n = \frac{16V_R}{w} \tag{1}$$

where V_R is the retention volume (or, more conveniently, retention time) of an eluted peak and w is the base width of a triangle formed by extrapolating the sides of the peak to the baseline.

It should be noted that V_R should be measured properly from the elution of an unretained peak, not from the injection. If an air peak is discernible, this may be taken as the reference point; otherwise, for maximum accuracy it is necessary to measure two peaks and solve the equation

$$\frac{V_{R_1} - V_a}{w_1} = \frac{V_{R_2} - V_a}{w_2} \quad (2)$$

to obtain the retention volume V_a of the air peak and then the true values of the retention volume for the measured peaks. The number of theoretical plates is a measure of the ability of the column to separate but not of analysis time. A poor 2 m column may have the same n as a good 1 m column but take twice as long to achieve the same separation. Efficiency is given by n/L, where L is the length of the column. Values of up to 3000 per meter are attainable; 1500 is satisfactory. Efficiency may also be expressed as the height equivalent to a theoretical plate (HETP), h, where h is the number of plates divided by the column length in millimeters.

5. *Capillary Columns*

Capillary columns are glass or metal tubing, <1 mm in diameter and 25–100 m in length. They contain no support materials, the liquid phase being coated on the wall of the tube. They are comparable with packed columns in efficiency per unit length, but by virtue of the great length that is possible without undue resistance to gas flow, they can achieve very high efficiencies. Whereas 1000–2000 plates are usual for a packed column, 100,000 are normal for capillary columns and up to 1,000,000 have been achieved. The small amount of liquid phase that can be deposited on the wall of the capillary makes the sample size admissible without overloading very small. As 0.2 μl is the least that can be injected reproducibly, a sample splitter is needed to further reduce the amount reaching the column. It is difficult to design a splitter that gives a constant ratio for all components of a mixture, so careful calibration is needed. The development of support-coated capillaries, in which the tube is coated first with support material to which the liquid may be bonded more permanently and uniformly, has allowed greater loadings than were formerly possible and made the technique more attractive. Though costly and short-lived, capillary columns are worthy of consideration when a rapid separation is required of a particularly complex mixture.

A recent development of the capillary column is the wide-bore capillary column. This is a flexible tube of about 0.55 mm internal diameter made of inert fused silica glass. A thin film of stationary phase is bonded chemically to the support. This type of column is as easy to handle and use as a conventional packed column.

The column operates at flow rates and a loading capacity similar to those of a normal packed column. It is capable of better resolution and more rapid analyses than the packed column and does not need a complex injection system such as the split/splitless injector.

B. Mobile Phase

In GC the mobile phase, the gas, plays little part in the separation. The choice is governed by the characteristics of the detector. Thermal conductivity detectors require a gas with high heat conductivity and therefore low molecular weight; helium is usual. Argon or nitrogen may be used with electron capture detectors. With alkali ionization detectors nitrogen is usual, although some forms of this detector have been reported to give higher sensitivity with helium. The gas must always be purified by passage through a tube containing a molecular sieve to remove moisture and impurities. Moisture degrades many liquid phases, particularly the more polar ones. Oxygen must be removed when an electron capture detector is used [35].

C. Detectors

The choice of detector is influenced by the compounds to be determined and the sensitivity required, with reliability and cost as further considerations. There is much to be gained by using a detector that is selective for a particular element. Cleanup may be much reduced and some contributory evidence is afforded as to the identity of the compound detected. Often, however, such detectors may be less sensitive and less reliable than a nonspecific method such as electron capture. A number of useful comparative reviews have been published [36–42] that will guide the chromatographer as to the best choice for particular needs.

1. Flame Ionization Detector

Frequent reference to the flame ionization detector will be found in early papers on GC of pesticides, but for analyses at residue level it has been superseded by more sensitive and selective methods. Its operation is based on a combustion reaction that may be summarized as

$$\text{Organic sample} + H_2 + O_2 \rightarrow CO_2 + H_2O + \text{+ve ions} + \text{−ve ions} + e^-$$

It responds to all carbon compounds, so cleanup is invariably necessary. When cleanup is not difficult and sensitivity not demanding, the detector has some merit for its reliability and wide linear range. A description of the principle and operation is given in Chap. 12.

2. *Thermal Conductivity Detector*

Like flame ionization, the thermal conductivity detector, also known as the katharometer or hot-wire detector, was much used in the early days of GC of pesticides but has been replaced by more sensitive and selective techniques. Dependent on a temperature effect, it is slow to stabilize and sensitive to the fluctuation of the detector oven temperature and carrier gas flow. Interruption of carrier flow will damage the filaments by overheating. The detector and its operation are described more fully in Chap. 12.

3. *Electron Capture Detector*

The development of the electron capture detector (ECD) [43] opened up a new order of sensitivity in residue analysis and did much to establish GC as the method of choice in the field. It remains the most sensitive detector for a wide range of compounds. The detector contains a beta-emittig radioactive source, usually cylindrical, and a central collector electrode (Fig. 1). Early models used tritium contained in metal foil or strontium-90. Both a volatile, limiting the safe operating temperature to 225°C or less. They have been superseded largely by nickel-63, which allows temperatures up to 400°C. Collisions of beta particles with the carrier gas cause ionization

$$N_2 + \beta \rightleftharpoons N_2^+ + e^-$$

A potential applied between the source and collector causes current to flow that may be amplified and recorded. When an electrophilic compound, e.g., one containing a halogen, is introduced, one of two reactions may take place

$$e^- + AB \rightarrow AB^-$$

or

$$e^- + AB \rightarrow A + B^-$$

In either case the fast-moving electron is "captured" and replaced by a slow-moving negative ion that has a much greater chance of recombining with a positive ion. Less negative charges reach the collector and the current is reduced.

The detector may be operated in one of three modes. In the dc mode a constant potential of 2–30 V is applied. This mode is the

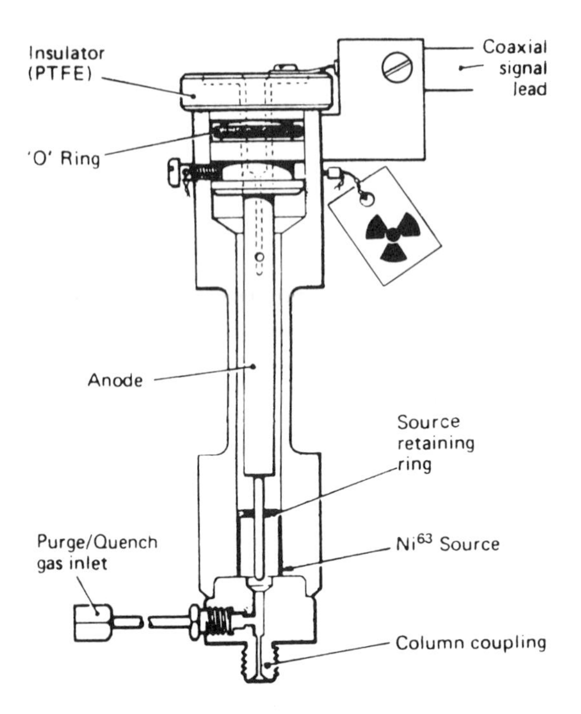

FIG. 1 Electron capture detector. (From Ref. 26.)

least sensitive, as electrons are removed constantly and negative ions are most likely to reach the anode and lead to a buildup of contamination. In the pulsed mode a potential of 50 V is applied in pulses of 0.5–1 μs at intervals of 50–500 μs. The interval allows a buildup of electrons available to be captured, leading to increased sensitivity. The migration of negative ions is reduced greatly, lessening contamination. One of the limitations of both modes is the very limited range of concentration in which the response is linear. Beyond the concentration that gives 20% reduction in standing current, the change for a given concentration increment is reduced progressively. The increase of the pulse interval may be expected to improve sensitivity, but by reducing the standing current, it lowers the threshold of nonlinearity. The pulse modulated mode [44] attempts to obtain the best compromise. The pulse interval is adjusted electronically to maintain constant detector current, the change of interval being monitored to give the measured sample signal. A linear range of four decades of concentration is attainable and it is claimed that contamination is reduced, but the authors have found it still to be a problem, particularly at higher column temperatures.

Although particularly sensitive to halogenated compounds, the ECD should not be regarded as specific to this class of compounds, let alone discriminating between them. The mistaken identification as the DDT of peaks actually due to polychlorinated biphenyls is well known, but many less obvious pitfalls await the overconfident chromatographer. Cleanup is, therefore, invariably necessary. If the high sensitivity of the ECD is not essential, a more selective detector may obviate much time-consuming preparator work. When the object is screening to eliminate negative samples rather than identification of positive findings, the lack of specificity may be an advantage.

The detector is slow to stabilize and very slow to recover from overload. Thus, halogenated solvents, which cause great problems in this respect, must be excluded rigorously from sample solutions. After a stable baseline has been achieved, there may be further drift in sensitivity, and injection of a reference material at intervals after starting up or recovery is advisable. If not abused, the detector is one of the most stable and reliable, requiring little maintenance. Intervals between cleaning can be lengthened if it is operated so as to avoid condensation of contaminants. The temperature should be appreciably higher than that of the column but, although heating to the maximum working temperature is used as a means of cleaning, the use of maximum temperature under routine conditions may be disadvantageous, as some contaminants may bake onto the electrodes rather than volatilize. Some silicone stationary phases are troublesome in this respect. Cleaning by any other means than heating should not be attempted, except by personnel familiar with the regulations and precautions applicable to radioactive material. Otherwise, the detector should be returned to the manufacturer for cleaning. If the necessary expertise is available, some cleaning may be undertaken by the user, following the manufacturer's recommended procedure. Examination of the collector electrode will reveal the extent of cleaning necessary. Any discoloration is a sign that cleaning will be beneficial; a varnish-like deposit is indicative of gross contamination requiring action to prevent its recurrence.

4. *Alkali Ionization Detector*

The alkali ionization detector is a great asset to the residue analyst, enabling many analyses to be performed without cleanup. In its simplest form, it resembles a flame ionization detector to which is added some means of introducing an alkali metal. It has been produced in many configurations and given a variety of names. Aue [45] proposed the term "alkali flame detector" to avoid mechanistic implications, but subsequent developments have made "flame" a misnomer for some types. The alkali source is in the form of an independently heated bead above the jet. They were first described by

Kolb et al. [46,47] in 1974 but have some similarities to much earlier dual flame detectors [48,49]. Four variants are now marketed and further advances may be expected. This type of detector results from the recognition that the gas mixture that is optimal for conversion of the sample to the ions necessary to the reaction is not necessarily that which will heat the alkali sufficiently for vaporization. The position of the alkali relative to the jet is much less critical than with other types, allowing rapid differentiation between nitrogen and phosphorus by an alteration of operating conditions that does not involve mechanical adjustment.

In the form described by Kolb et al., operation in the phosphorus mode uses a gas mixture sufficient to heat the alkali bead. The bead is polarized negatively and the jet grounded.

Electrons from combustion are prevented from reaching the collector by the polarity of the bead, whereas those from the reaction of phosphorus at the bead surface are captured by the collector. In the phosphorus-nitrogen mode, bead and jet are polarized and hydrogen flow is reduced tenfold, the bead being heated electrically. It is postulated that a partial pyrolysis occurs, giving cyanide radicals. At the bead these capture electrons form alkali atoms, giving alkali ions that return to the bead and cyanide ions that either move to the collector or react in the oxidizing zone to give electrons, which are captured. Compounds containing nitrogen in the forms

$$-CO-NH-CO- \quad \text{or} \quad -CO-NH_2 \quad \text{or} \quad -O-NH_2$$

are not detected, as they are incapable of forming cyanide radicals unless alkylated. A paper by Kolb et al. [50] further describes the reaction mechanism and other workers have reviewed the performance of the detector [51–53]. Further commercial developments of the heated bead type of detector have been described by Burgett et al. [54] and by Patterson and Howe [55]. The detector of Burgett et al. is responsive to compounds not detected by the configuration of Kolb et al., so a mechanism other than the formation of cyanide radicals must be postulated.

5. *Flame Photometric Detector (FPD)*

The first flame photometric detector (FPD) to be developed commercially was described by Brody and Chaney [56]. Eluted compounds are burnt in a hydrogen-rich flame and the emitted light observed through a narrow bandpass filter by a sensitive photomultiplier. Use of a 526 nm filter gives selectivity for phosphorus; 393 nm is selective for sulfur. Krost et al. [57] have reported flame photometric detection of nitrogen but with a detection limit two orders higher than for sulfur. Detectors are available that simultaneously monitor phosphorus and sulfur emission. Response to phosphorus is linear; mechanistic considerations suggest a sulfur response proportional to the square of the concentration. In practice, responses ranging from linear to square have been reported [58], exponents between 1.5 and 2.0 being usual. The variations are attributed to differences in burner design and gas flows [59], sample concentration, and the oxidation state of the sulfur. Grob [11] has cited reports of the ratio of the phosphorus response to the square root of the sulfur response being a good indicator of their relative amounts, independent of concentration, column temperature, and retention time, and of the exponent of the sulfur response being used as an indicator of its oxidation state. In the sulfur mode, selectivity over phosphorus is 10^3–10^4; the phosphorus mode is much less discriminating against sulfur. Electronic devices for linearizing the sulfur response have been described (e.g., [60]) but Burnett et al. [61] have drawn attention to potential errors that may arise from their use.

Early designs of the detector were prone to extinction of the flame by elution of the solvent. Burgett and Green [62] developed a detector that gave improved performance and overcame extinction, but there remained a major downward disturbance of the baseline that would interfere with quantitation of rapidly eluted peaks and could cause false baseline assignment by some types of integrator. Zehner and Simonaitis [63] attempted to offset this by the addition of a volatile phosphorus compound. The response of the detector is quenched by the simultaneous elution of nonsulfur compounds [64]. The selective response cannot, therefore, be used to quantify compounds not resolved by the column.

With response characteristics critically dependent on operating parameters, careful optimization is essential. Greenhalgh and Wilson [65] have reviewed the factors affecting optimization of the Pye Unicam detector. Recently Patterson et al. [66] have described a dual flame detector in which a hydrogen-rich flame is used to decompose the sample and a second flame to produce light emission. A more uniform response to concentration and a high solvent tolerance is reported. Patterson [67] has compared its performance with that of single flame operation and found the dual flame mode to give less quenching but the single flame to be more sensitive.

6. *Microcoulometric Detector*

The microcoulometric detector described by Coulson et al. [68,69] in 1960 was the first successful approach to element-specific detection. They tackled the problem of detection of halogens by pyrolysis of the sample to halide ions and reaction with a solution containing electrochemically generated silver ions. The current passed to restore the silver ion concentration was measured. By using reducing conditions to form H_2S the principle was also applied to sulfur. More detailed papers on the theory and practice of the system and its application to pesticides subsequently appeared [70,71]. Burchfield et al. [72–74] reported on further developments in the system and extended its application to phosphorus. Martin [75] determined nitrogen by catalytic reduction to ammonia that was titrated with coulometrically generated hydrogen ions. Guiffrida and Ives [76] described a miniaturized furnace for use with the detector. Sensitivity tends to be poor—detection limits at the microgram level are usual although nanogram limits are attainable under favorable conditions. Careful optimization due to substantial changes in response with gas flow is generally considered to be necessary. However, Westlake [77] reported the Dohrman microcoulometric detector to be insensitive to gas flow, in contrast with earlier types. The later developments in the microcoulometric detector were paralleled by the development of the electrolytic conductivity detector, which has to a considerable extent superseded it.

7. *Electrolytic Conductivity Detector*

Developed by the same workers as the microcoulometric detector, the electrolytic conductivity detector is also based on the oxidation or catalytic reduction of the column effluent and its dissolution. The ionic species is, however, measured by its conductivity rather than by titrimetry. The original detector was described by Coulson [78–80] and its performance compared with microcoulometric and electron capture detectors. A detailed description of its application to nitrogen has been given by Cassil et al. [81], and Westlake [82] has reported on its use for bromine and iodine. Patchett [83] described improvements to the detector and evaluated its performance. Hall [84] introduced an improved design that overcame many of the limitations of the Coulson detector and achieved substantially better sensitivity. The two models have been compared by Wilson and Cochrane [85].

8. *Mass Spectrometry*

The qualitative significance of data obtained from GC is discussed in Sec. VI. Suffice it to say here that the chromatographic behavior of a compound, although providing valuable evidence, is by no means conclusive proof of its identity. Almost the sole exception to this is mass spectrometric detection. The technique requires a large capital outlay and considerable operator skill. The reward is data that not only quantify a compound but identify it. The coincidence of retention time with the appearance of the characteristic ions with the appropriate relative intensities demonstrates the presence of the compound more conclusively than any other means available to the analyst.

In its most sophisticated form the complete mass spectrum of the compound is scanned and compared with the mass spectra of known compounds. Such a mode of operation requires a complex data system. It is also insensitive—on a par with flame ionization detection—as the time taken for a complete scan limits the number of possible in the time the compound is eluted. The more usual mode is to monitor only two or three of the most intense ions. Their relative intensities provide identification with sufficient certainty for most purposes and this mode is much more sensitive. Although capable of giving qualitative data, mass spectrometry should not be looked on as a means of elucidating the composition of unresolved mixtures. Although in experienced hands it can be used to identify components of partially resolved peaks, it should not be treated as a substitute for good chromatography. A typical mass spectrum of a pesticide is shown in Fig. 2, and mass chromatograms (showing peak height as a function of time) at selected mass/energy ratios are illustrated in Fig. 3.

Discussion of the types of mass spectrometer and the practical aspects of their operation is beyond the scope of this chapter. For further reading on the technique the works of Brooks and Middleditch [86], Jenden and Cho [87], and McFadden [88] will be found useful, and for its application to pesticides those by Vander Velde and Ryan [89] and Ryan [90] should be consulted.

D. Operating Parameters and Technique

A number of matters of technique already have been referred to, e.g., optimization of detectors. Before concluding this section,

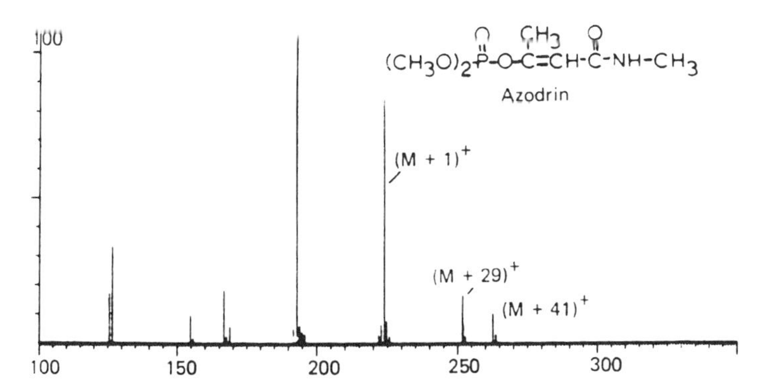

FIG. 2 Chemical ionization (CH_4) mass spectrum of Azodrin. (From Ref. 89, reproduced from the *Journal of Chromatographic Science* by permission of Preston Publications Inc.)

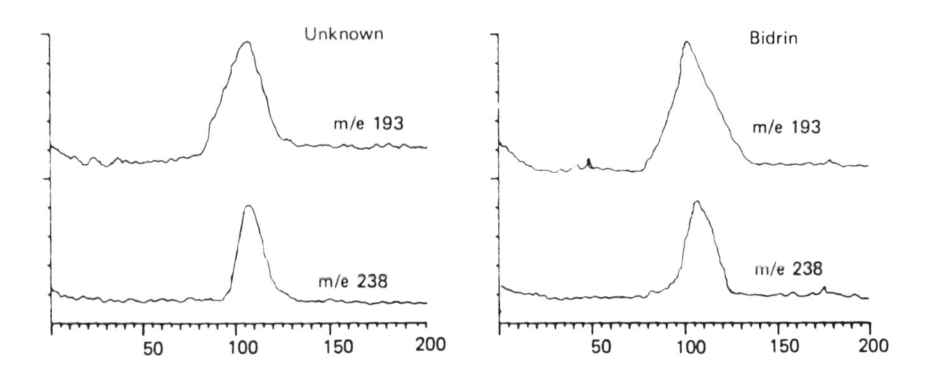

FIG. 3 Mass chromatograms of *m/e* 193 and 238, Bidrin and unknown extract. (From Ref. 89, reproduced with permission from the *Journal of Chromatographic Science* by permission of Preston Publications Inc.)

some others should be mentioned that are of importance in ensuring a good chromatogram. Introduction of the sample is normally by syringe. With care good reproducibility is not difficult. Between samples the syringe should always be rinsed with solvent—it is good practice to do so at least 10 times—and finally with the sample to be injected. Checks should be made that there is no leakage; bubbles are indicative of a leak between needle and barrel, upward movement of the liquid as the septum is penetrated shows a loose fit of the plunger. Septa should be checked frequently for leakage, the most common cause of drift and noise. Injection heaters should be set a few tens of degrees above the column temperature to ensure rapid vaporization of the sample.

Column temperature is determined by the analysis time that is acceptable for the slowest eluting peak, by its effect on separation and, in some cases, by the susceptibility of the analyte to thermal degradation. As a general rule, increasing the temperature improves peak shape but tends to bunch rapidly eluting peaks. The increase of retention time as the temperature is lowered is accompanied by peak broadening so the effect on resolution tends to be small, but when peaks are partially resolved, a change of temperature can sometimes have a beneficial effect.

Drift, noise, and distorted peaks are problems encountered by all chromatographers. They can arise from a variety of causes; leaks and bleeding of the liquid phase are the most common. With experience much can be learned from symptoms revealed by the chromatogram. A helpful guide to the operating problems of gas chromatographs and their causes and cure has been published by one of the instrument manufacturers [91], and a booklet by Swan [92] gives flowcharts for problem solving with the ECD, alkali ionization detector and FPD, as well as concise theoretical and constructional descriptions of these detectors.

IV. HIGH-PRESSURE LIQUID CHROMATOGRAPHY

Whereas gas chromatographic apparatus is supplied usually as a complete unit, in liquid chromatography it is more usual for the individual parts—pump, injection system, column, detector, and data system—to be purchased separately. The absence of a column oven—most separations are performed at ambient temperature—facilitates the combination of components from more than one manufacturer, enabling a system to be assembled that meets the specific needs of the chromatographer at least cost. There is, however, a trend toward complete systems and this may be expected to increase as the equipment becomes more sophisticated and greater automation is introduced.

A. Pumps

Pumps are of two types: constant pressure, powered by pressure from a gas cylinder, or constant flow, electrically driven. In the former type the pressure may be applied directly to the liquid. Such pumps are simple and reliable; the pressure is pulse-free and easily controlled. For routine repetitive analysis they are convenient but, when frequent changes of solvent are called for, other types are preferable. The pressure attainable is limited by the cylinder pressure. Most analytical separations can be achieved below 10 MPa (100 bar), so this limitation is not serious. It may be overcome by means of a pneumatic amplifier, a piston system in which air pressure is applied to a large piston drives a small piston pressurizing the liquid. With reciprocating pistons a wide range of flow rates is attainable but pulsing is introduced. Constant pressure pumps can give constant flow only if the impedance of the column and plumbing remain constant. As detector signals are normally recorded in relation to time and retention volumes are assumed to be proportional, this is an important consideration and again points to constant pressure pumps being most useful in routine analysis where operating parameters are infrequently altered.

Constant volume pumps employ either a large piston driven by a stepping electric motor or a small piston driven by a cam. The piston may be in contact with the liquid directly or via a diaphragm. Small pistons are normally mounted in pairs, the second maintaining the flow while the first is making its return stroke. Even with this arrangement they are subject to pulsing but with careful design of the timing of the strokes and, in some cases, with feedback to detect and compensate for pulses, the problem is very much reduced. Large piston pumps are pulse free but, as with direct gas driven pumps, their running time is limited by the capacity of the reservoir. Figure 4 illustrates schematically the main types of pump.

When compounds of widely differing polarity have to be separated, a single eluent may not separate them in a reasonable time. A means of mixing the eluents in varying proportion as the run proceeds may be needed. These solvent gradients may be formed on either the low- or high-pressure side of the pump. The most simple but least versatile system is to form a step-wise gradient by changing reservoirs at intervals during the run. A sudden change of eluent is liable to disturb the baseline so the changes should be timed to avoid peaks of interest. A continuous gradient can be formed at low pressure by using a proportioning valve to feed from two reservoirs to a mixing chamber and then to the pump. Time proportioning at high pressure is feasible but time accuracy and speed become critical as does efficient mixing. The more usual way to form a gradient at high pressure is to use two constant volume pumps. For separation development this may be the most satisfactory system but costly for routine applications. High accuracy of flow at low flow rates is necessary to ensure reproducibility at the extremes of the gradient.

B. Injection Devices

Injection with a syringe onto the column poses greater problems than in GC because of the pressure and effects of solvents on septum materials. These problems are overcome by using an injection valve.

The sample solution is introduced into a loop in the valve with a syringe and the valve then turned so that the flow to the column passes through the loop. Fixed-volume loops give superior precision to syringe injection but the volume is difficult to change quickly. Some loops can be filled partially with a syringe. The volume is, therefore, variable but the precision of a fixed loop is lost and there is greater dilution of the sample before it reaches the column. In most valve injectors the column end fitting does not allow the sample to reach the top of the column coaxially, as it would form a syringe needle, so efficiency is impaired by edge effects.

C. Columns

Stainless steel is the usual material for columns. Unless the system can be operated in such a manner that the sample material does not reach the walls, the interior finish is of considerable importance.

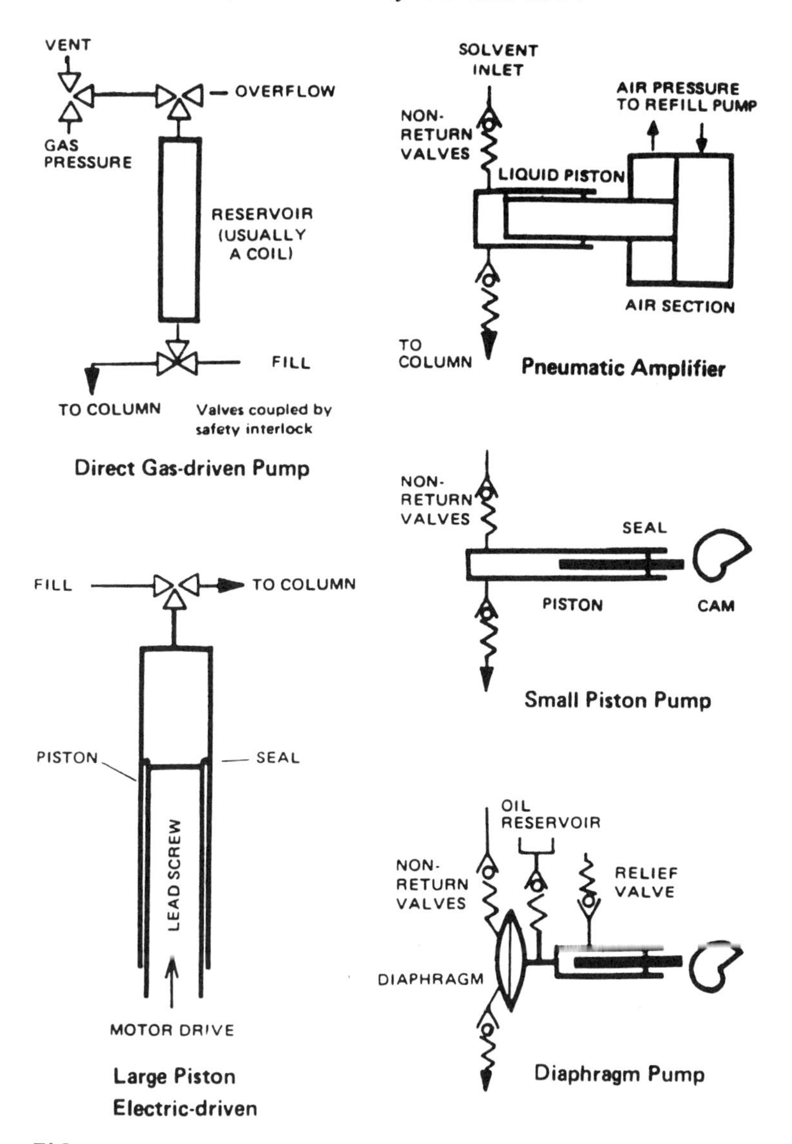

FIG. 4 Principal types of HPLC pump. (From Ref. 13.)

Glass has been used for its smoothness and resistance to corrosion but has serious pressure limitations. The lateral spread of the sample as it passes through the column can be calculated and it has been shown that if it is introduced coaxially and is of negligible diameter at the point of entry, it will not reach the wall if $Ld_p/d_c < 0.4$, where L is the length of the column, d_p the diameter of the particles, and d_c the internal diameter of the column. If the column diameter satisfies this condition, the adsorptive properties of the wall and the effect of surface irregularities on the flow pattern are of no consequence. An appreciably greater diameter will needlessly dilute the sample.

Packing materials have not proliferated to the same extent as GC packings. Porous silica is used widely, a pore size of 5–10 nm being most useful. Its adsorptive properties result from surface hydroxyl groups; polar solutes are, therefore, most strongly adsorbed. The various silica materials differ in their particle size and shape. The trend now is toward small (3 μm) spherical particles and short columns (10 cm). With these small particles, uniformity of size is of equal or greater importance than shape in obtaining close packing. These short columns have excellent separation characteristics and also operate at low flow rates (0.2 ml/min), which means a reduction in the use of expensive solvents as mobile phase.

The strongly adsorptive properties of silica can produce excessive retention times for some solutes. To reduce the time taken for diffusion, packings have been developed in which nonporous glass beads are coated with the adsorbing material. This concept also allows a greater variety of active materials to be used. The disadvantage of these packings is that considerable column volume is occupied by inert material, leading to peak broadening, which reduces their resolving power and gives greater sample dilution. This, together with their very low sample capacity, has limited their application. Bonded-layer materials are now generally preferred.

Packings of lower polarity than silica have been produced by chemically bonding an organic molecule to the hydroxyl of the silica surface. The technique offers the possibility of an infinite variety of materials, but the difficulties of preparing them reproducibly has kept their number sensibly restrained. Attachment of the organic molecule is achieved usually by the reaction of di- or tri-chlorosilanes with the silica, the Si—O—Si—C linkage being more stable and easy to prepare than a direct Si—C bond. The most common radical is an alkyl group, notably octadecyl, producing a nonpolar stationary phase. Chromatography with such materials commonly is referred to as "reversed phase," as a polar solvent is used with a nonpolar packing, in contrast to the practice with silica. The bonded layer may be a single molecule thick, but if moisture is present in the preparation, cross-linkage of the chlorosilanes gives a polymerized layer. Polymerized layers have greater capacity but are less efficient.

HPLC columns are often purchased ready packed, but with some practice and patience it is well within the capability of a user in a chromatography laboratory to fill his or her own. Particles of 30 μm or over and spherical materials down to 10 μm may be packed dry, but packing as a slurry that is pumped into the column generally is preferred. Typical procedures are given by Bristow [13]. The solvent for preparing the slurry is chosen for low viscosity, flocculation or dispersion, and density to control settlement. There is conflicting evidence as to whether a flocculated or dispersed slurry is preferable. For silica, polar solvents should be used if flocculation is to be avoided; for nonpolar bonded phases a nonpolar solvent is needed. Acetone will give a dispersed suspension of both types.

After filling, the top 5 mm of packing is removed and a mesh disk is pushed in to allow removal of septum and sample debris without disturbing the packing. This is followed by glass beads and a PTFE sinter that is trimmed off about 1 mm proud of the column tube. After assembling the end fittings, the column may be attached to the chromatograph and eluting solvent pumped to wash out the slurry solvent. If the column is to be used in the "infinite diameter" mode, a top fitting that will introduce the eluent coaxially is desirable. Testing and subsequent care of the column are described in a booklet issued by one of the column suppliers [93].

D. Mobile Phase

In GC the mobile phase serves as a carrier and plays virtually no role in the separation. By contrast, in HPLC the choice of solvents is crucial. As with GC phases, there have been many theoretical attempts to predict suitable solvents, with some degree of success, but in the present state of knowledge the most practical approach is still one of trial and error with generalized theoretical considerations pointing the direction. Snyder and Kirkland [94] have published a practical guide to solvent selection and Saunders [95] has provided practical conclusions from theoretical work. The principal points of their studies have been summarized by Bristow [13]. The complexities of solvent selection are intensified by the possibilities of using mixtures. A binary mixture greatly aids the attainment of suitable polarity. Use of more than two major components is not often practiced, but it is not unusual for minor amounts of other solvents to be added to achieve a particular selectivity. When the solutes to be separated cover a wide range of polarity, gradient elution can be helpful in achieving a separation in reasonable time. Even when the compounds of interest can be separated isocratically, i.e., with a single eluent, a change of solvent after they have been eluted may be beneficial to clear strongly adsorbed coextractives from the column. The guidelines of Saunders [95] are helpful in assessing the mixtures. For chromatography on silica a small concentration of a polar solvent may have a very marked effect compared with none. Doubling of the concentration approximately halves retention time. Alteration of the nonpolar component has relatively little effect.

E. Detectors

HPLC is demanding in detector design. Peak volumes as low as 100 μl may have to be detected so much of the resolution achieved by chromatography is lost unless the detection volume is small. A rapid response is required—on short columns of small particles peak duration may be as little as 1 sec. Photometric detectors are well capable of detecting such peaks, others less so. The recorder response may in some cases be limiting. The types of detector available for liquid chromatography have been described by Scott [96].

1. Photometric Detectors

Light absorption, usually in the ultraviolet, is by far the most widely used means of detection, both generally and in pesticide residue analysis. A choice has to be made between a fixed-wavelength instrument using a monochromatic source or a more versatile but more expensive variable-wavelength instrument. Fixed-wavelength instruments usually operate at the 254 nm mercury emission wavelength. Variable wavelength allows detection at the peak of the absorption spectrum, offering potentially greater sensitivity and selectivity. When, however, a number of peaks with differing absorption maxima have to be quantified on a chromatogram, a compromise wavelength may have to be chosen to the detriment of sensitivity. Variable wavelength offers the possibility of obtaining some qualitative evidence by stopping the flow and observing the wavelength of maximum absorption. A good detector should have a cell volume not greater than a tenth of the smallest peak volume likely to be encountered and it should be so designed that any bubbles formed are cleared quickly. A restriction downstream helps to prevent their formation. The available types of photometric and fluorimetric detectors have been compared by Baker et al. [97].

2. Fluorimetric Detectors

Many applications have been reported in the pesticide residue field for fluorimetric detectors and the technique has considerable potential for sensitivity and selectivity if suitable fluorescent derivatives can be prepared. Derivatization may be carried out prior to chromatography or, if the conditions for the reaction can be satisfied, post-column. Suitable reactions have been described by Udenfriend et al. [98] and Katz and Pitt [99]. Methods specifically for pesticides have been reported by Duenges [100], Moye et al. [101], and Seiler and Demisch [102].

Instruments designed specifically for HPLC use electronics to process the detector output. This enables flow cells of very low (0.3 μl) volume to be used so that excellent peak resolution is obtained. Absorption by the fluorescent compound or any other eluate at the existing wavelength seriously affects linearity of response, so it is prudent to simultaneously monitor this absorption; it should not exceed 0.1 absorbance unit.

3. *Photodiode Array Detector*

A limitation of most variable-wavelength HPLC detectors is that they operate at only one wavelength at a time, which makes confirmation difficult. A scan of the compound is only possible if the flow is stopped, making quantification impossible. The photodiode array detector is based on a series of light-sensitive cells etched onto a silicon chip. This series or photodiode array can monitor all wavelengths from 190–600 nm simultaneously. A scan is performed in about 10 millisec so that the flow does not need to be stopped. The result is that no chromatographic data are lost.

Each diode produces 25 data points each second, so that a very sophisticated and high-speed data processing network is needed to make full use of the detector's capability. This detector gives a much higher degree of confirmation for a compound than the normal UV detector.

4. *Photoconductivity Detector*

This detector is based on photochemical post-column reactions and the detection of any resultant ionic species in the conductivity cell. Effluent emerging from the HPLC column splits into two streams, analytical and reference. The analytical effluent passes through a quartz reaction coil where it is irradiated with intense UV light. On the reference side, the balance of the effluent passes into the reference delay coil in which the flow and volume are matched carefully to that of the reaction coil and then into the reference half of the conductivity cell. Measurement of the difference in cell conductivities gives the change in sample stream conductance.

5. *Other Detectors*

Refractometers have found application in HPLC, mainly in permeation chromatography. For pesticide residue analysis they do not have the necessary sensitivity and selectivity. They are highly sensitive to temperature and pressure.

Atomic absorption spectrometry has been applied to HPLC eluates. Its application to pesticides is limited, as few contain a metal. If the technique gains wider acceptance, extension of the range of application by derivatization may be worthy of consideration.

Dolphin et al. [103] have described the use of electron capture to detect organochlorine pesticides. A commercial instrument is available and the range of application is being extended to nitrogen and organophosphorus compounds. The range of solvents that are compatible is restricted and exclusion of dissolved oxygen is essential.

Attempts have been made (e.g., [104]) to separate the solvent from the eluate and transfer the solute to a flame ionization detector. At present, recovery of the solute presents formidable problems and response is too slow for adequate resolution with efficient columns. If the separation mechanism can be perfected, the concept holds promise for the future, particularly if linked to other types of GC flame detectors to give selectivity.

Highly specific identification information can be obtained by mass spectrometry of eluted peaks from HPLC. The technique has been assessed by Scott [105] in a review of various aspects of HPLC. With wide-bore columns it is not difficult to collect sufficient material. Linked HPLC-MS systems have been described, but as with HPLC-FID, a breakthrough in solvent/solute separation is needed to establish this technique. Like GC-MS it will always be costly and can only be justified when there is a high throughput and certainty of identification is paramount.

V. DATA HANDLING

The devices available for recording the output of chromatographs range from simple chart records to complex systems capable of quantifying the data, comparing it with standards, and calculating the final result. If a recorder is used, the choice must be made between measurement of peak height or area. Much has been written on their relative merits. If the peak is of the ideal Gaussian shape and well resolved, height measurement is legitimate and easier; less perfectly shaped peaks may be measured accurately in this way if the shape is reproducible. It should always be the aim of the chromatographer to produce such peaks. Where it cannot be done, area measurement may be more accurate. Areas may be measured manually by triangulation or by multiplying the height by the width at half height. Both are subject to some degree of error, which becomes significant and broad peaks and depends on regular peak shape. The shape problem can be overcome by cutting out the peak from the chart paper and weighing it but it is tedious, the assignment of the baseline remains subjective, and the permanent record of the separation is destroyed.

The simplest integrators provide a cumulative measure of all the area below the trace. Their use is, therefore, limited to chromatograms with a steady, level baseline. More sophisticated instruments assign the baseline and assess partially resolved peaks according to predetermined criteria, removing the subjective element. The accuracy of their assessment is only as good as the criteria programmed into them. The possibility of an unusual chromatogram being wrongly interpreted should be kept in mind and the recorder trace examined before results are accepted. An integrator is an adjunct to the chart recorder, not a replacement. Electronic data handling methods, including interfacing with computers, have been reviewed by Pierson and Steible [106].

VI. IDENTIFICATION AND CONFIRMATION

The coincidence of retention time of an unknown with that of a standard provides some evidence as to its identity. It is no more than a probability. If the chromatographer has extensive experience of a compound and the type of sample, in which no "clean" samples have shown the peak and all positive findings have been confirmed, the balance of probability may be sufficient for some purposes. If the sample is rerun on a column of substantially different characteristics the probability is improved but it is still not conclusive. To be convincing a confirmatory technique must employ a different fundamental principle or modify the compound of interest so that its chromatographic behavior is different. Chemical derivatization is, therefore, a very valuable means of confirmation. A number of reviews of derivatization have been published [107–111]. On a routine basis it is convenient to derivatize an aliquot of the sample extract and chromatograph it immediately following the initial determination of the pesticide.

A. Triazines

The chloro-*s*-triazines can be derivatized by the following general reactions.

1. Methylation

The residue is heated with dimethyl sulfoxide, sodium hydride, and methyl iodide [112,113] to give the reaction

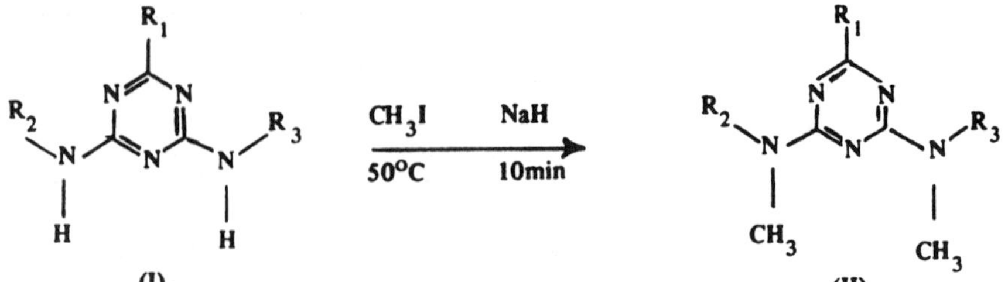

$$\text{(I)} \xrightarrow[\text{50°C} \quad \text{10min}]{CH_3I \quad NaH} \text{(II)}$$

$R_1 = Cl, OCH_3, SCH_3$

$R_2, R_3 = C_2H_5, CH_3.CH.CH_3$

Figure 5 [113] shows chromatograms of atrazine before and after alkylation.

2. Methoxylation

The reaction involves the replacement of the labile chlorine atom of the triazine with the methoxyl radical [112].

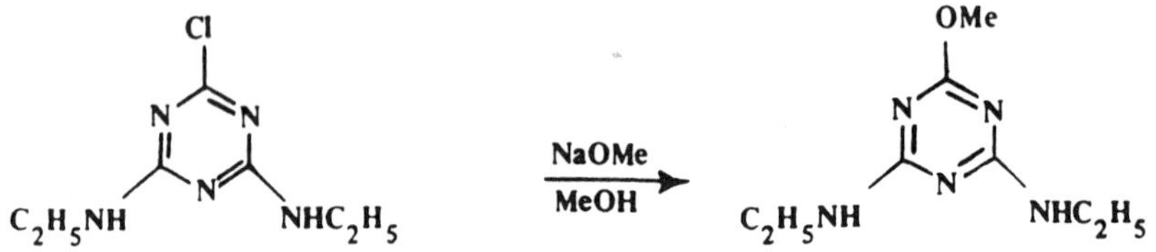

$$\text{Simazine (III)} \xrightarrow[MeOH]{NaOMe} \text{Simetone (IV)}$$

3. Hydrolysis and 2,4-Dinitrofluorobenzene

Hydrolysis with *M* HCl at 150°C is followed by conversion to the dinitrophenyl derivatives by reaction with 2,4-dinitrofluorobenzene in alkaline solution [112].

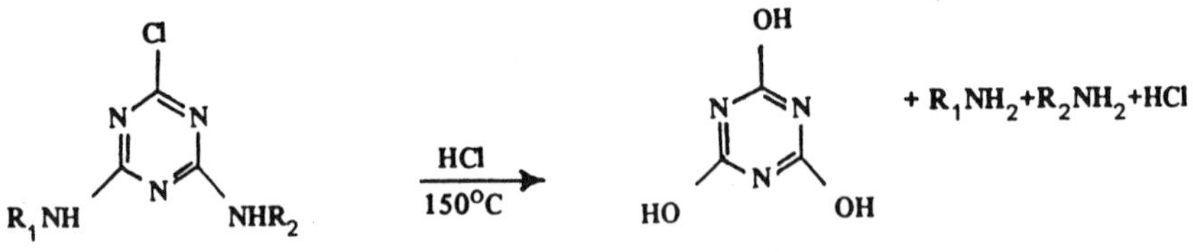

$$\text{(V)} \xrightarrow[150^\circ C]{HCl} \text{(VI)} + R_1NH_2 + R_2NH_2 + HCl$$

$$R_1NH_2 + R_2NH_2 + 2\,\text{(VII)} \xrightarrow[85^\circ C]{pH9} \text{(VIII)} + \text{(IX)}$$

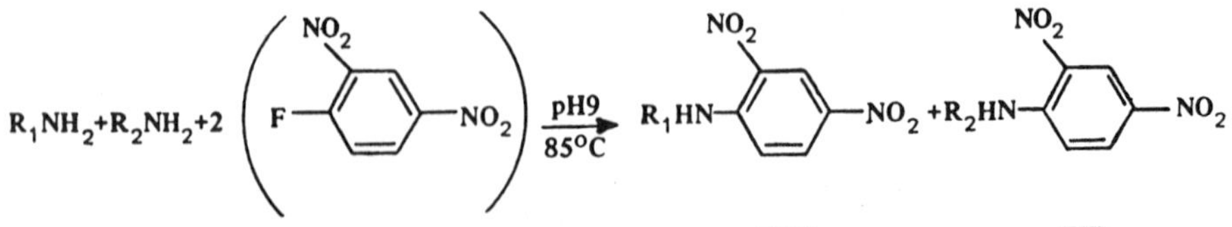

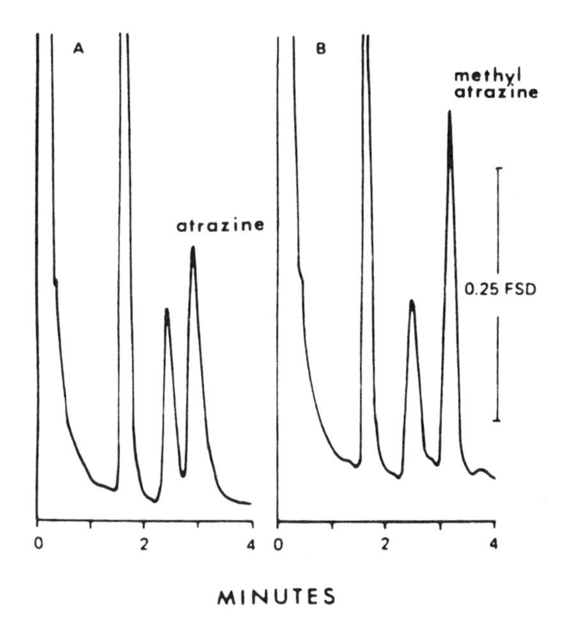

FIG. 5 Gas chromatograms of crude methanol extract of organic soil (0.26 μg g^{-1} of atrazine) before and after alkylation. A, crude soil extract; B, alkylated soil extract. [From Ref. 113, reprinted with permission from the *Journal of Agricultural and Food Chemistry, 23*:325 (1975). Copyright © by the American Chemical Society, Washington, D.C.]

Lawrence [112] concluded that the methoxylation reaction is the simplest and fastest method. To confirm the presence of methoxy-*s*-triazines, methylation with methyl iodide should be used. The hydrolysis-DNFB reaction is useful for characterization of the triazines.

B. Phenoxyalkanoic Acids

Phenoxyalkanoic acid herbicides can be derivatized using the following reagents.

1. Pentafluorobenzyl Bromide

The acids are converted to pentafluorobenzyl bromide (PFB) esters by reacting with PFB in alkaline acetone solution overnight at room temperature and extracted into iso-octane [114]. The product is separated by column chromatography.

2. *Dicyclohexylcarbodiimide Reagent*

2-Chloro esters are prepared by reaction with dicyclohexylcarbodiimide (DCC) and 2-chloroethanol overnight at 60°C [114]. The product is separated from the reaction mixture by partition and column chromatography. The pentafluorobenzyl bromide reaction is simpler to perform than the dicyclocarbodiimide but less specific The GC retention times are twice as long for the PFB esters as for the 2-chloro esters under the same conditions. The PFB esters have a greater sensitivity than the 2-chloro esters when determined by electron capture.

C. Organophosphorus Compounds

1. *Hydrolysis and Pentafluorobenzyl Bromide Reagent*

Organophosphorus pesticides are hydrolyzed with 10% methanolic potassium hydroxide and the product separated from the acidified reaction mixture by partition into benzene. Derivatization with pentafluorobenzyl bromide is carried out in alkaline solution and the product, partitioned into iso-octane, purified by column chromatography [115].

$(C_2H_5O)_2P(=S)-O-C_6H_4-NO_2 \xrightarrow[\text{Hydrolysis}]{OH} HO-C_6H_4-NO_2$

Parathion (X) p · Nitrophenol (XI)

$\xrightarrow{\text{Pentafluorobenzyl bromide}} C_6F_5-CH_2-O-C_6H_4-NO_2$

(XII)

2. *Chemical Reduction*

Metal chlorides, especially chromous chloride, have been used to confirm the identity of organophosphorus pesticides [116]. Their reactivity toward heptachlor was found to decrease in the order Cr^{2+}, Pd^{2+}, Ti^{2+}, Co^{2+}, and Fe^{2+}. Chromous chloride can be used to confirm organophosphorus pesticides containing a nitro group. The suggested pathway of the reduction of parathion in acetone to yield *O,O*-diethyl-*O*-(isopropylidineamino)-phosphorothioate is

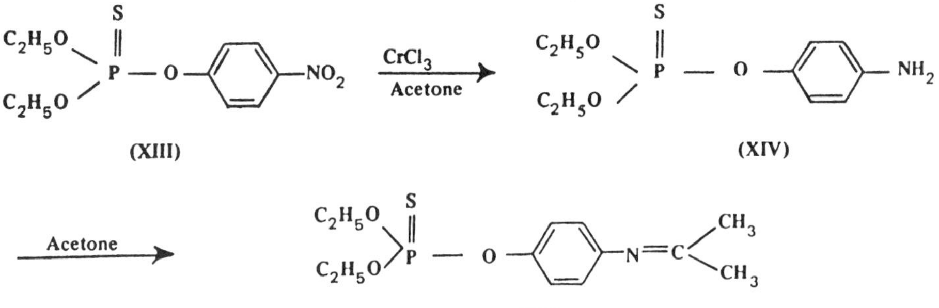

$$(C_2H_5O)_2P(S){-}O{-}C_6H_4{-}NO_2 \ \xrightarrow[\text{Acetone}]{CrCl_3}\ (C_2H_5O)_2P(S){-}O{-}C_6H_4{-}NH_2$$

(XIII) (XIV)

$$\xrightarrow{\text{Acetone}}\ (C_2H_5O)_2P(S){-}O{-}C_6H_4{-}N{=}C(CH_3)_2$$

(XV)

3. *Neutral Sodium Hypochlorite*

Organothiophosphorus compounds are oxidized by neutral sodium hypochlorite solution [117]. Generally, the compounds are oxidized to their corresponding oxons

$$>P{=}S \rightarrow\ >P{=}O$$

4. *Alkylation (Base-Catalyzed) Using Sodium Hydride-Methyl Iodide-Dimethyl Sulfoxide*

This reaction is applicable to organophosphorus pesticides that have NH or NH_2 moieties [113]. This includes phosphoroamidothioates, $R_1R_2P(S)\text{-}NHR_3$, and phosphoroamidates, $R_1R_2(S)\text{-}NHR_3$.

It has been shown that these compounds in the presence of a strong base react with various alkylating agents to give the methyl derivative, e.g.,

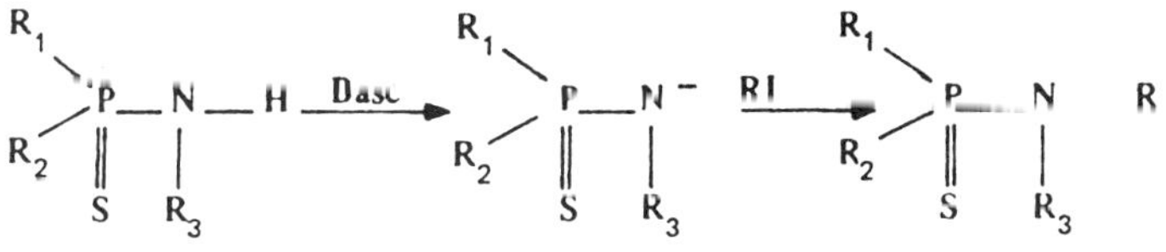

$$R_1R_2P(S){-}N(R_3){-}H \ \xrightarrow{\text{Base}}\ R_1R_2P(S){-}N^{-}(R_3) \ \xrightarrow{RI}\ R_1R_2P(S){-}N(R_3)\ R$$

(XVI) (XVII) (XVIII)

The yield of this reaction is related to the polarity of the solvent.

5. *Methylation Using Trimethylanilinium Hydroxide*

Trimethylanilinium hydroxide (TMAH) reacts with most organophosphorus compounds to form short-chain trialkyl phosphates that can be separated and quantified by GC [118]. A solution containing the organophosphate pesticide and TMAH is injected into a gas chromatograph equipped with a flame ionization detector. The efficiency of the on-column reaction varies from 61–100%, depending on the compound under test.

D. Carbamates and Ureas

These two classes will be considered together because of the similarity of their structure

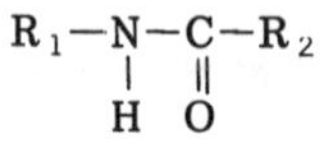

$$R_1-\underset{\displaystyle H}{\underset{|}{N}}-\underset{\displaystyle O}{\underset{\|}{C}}-R_2 \qquad\qquad R_1-\underset{\displaystyle H}{\underset{|}{N}}-\underset{\displaystyle O}{\underset{\|}{C}}-\underset{\displaystyle R_3}{\underset{|}{N}}-R_2$$

Carbamates (XIX) 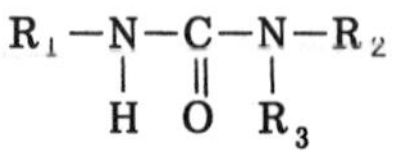 Ureas (XX)

1. *Alkylation (Base-Catalyzed)*

This is applicable to pesticides that have NH or NH_2 moieties. The residue applicable is heated in a sealed tube with sodium hydroxide, dimethyl sulfoxide, and methyl iodide [113]. The product is extracted from the reaction mixture with ether.

Carbamates decompose if heated at 50°C for 10 min at the alkylation stage. A single product is obtained if the reaction is carried out at room temperature for 5 min.

2. *Esters of Sulfonic Acids: 2,5-Dichloro-benzene Sulfonyl Chloride*

Esters of sulfonyl chlorides can be prepared by the reaction of carbamate pesticides with halogenated benzene sulfonyl chlorides [119]. These derivatives respond well to electron capture detection. The reaction is carried out in a sealed tube at 80°C. Figure 6 shows a typical chromatogram of the derivatives.

3. *Perfluoroacetylation*

N-Methylcarbamate insecticides react rapidly and quantitatively with trifluoroacetic, pentafluoropropionic, and heptafluorobutyric anhydrides [120,121].

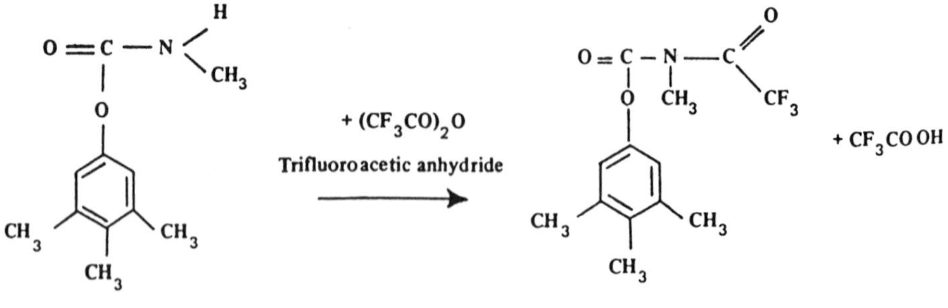

O=C—N(H)(CH₃) on O on trimethylphenyl ring + $(CF_3CO)_2O$ (Trifluoroacetic anhydride) → O=C(—N(CH₃)—C(=O)CF₃) on O on trimethylphenyl ring + CF_3COOH

Methiocarb (XXI) (XXII)

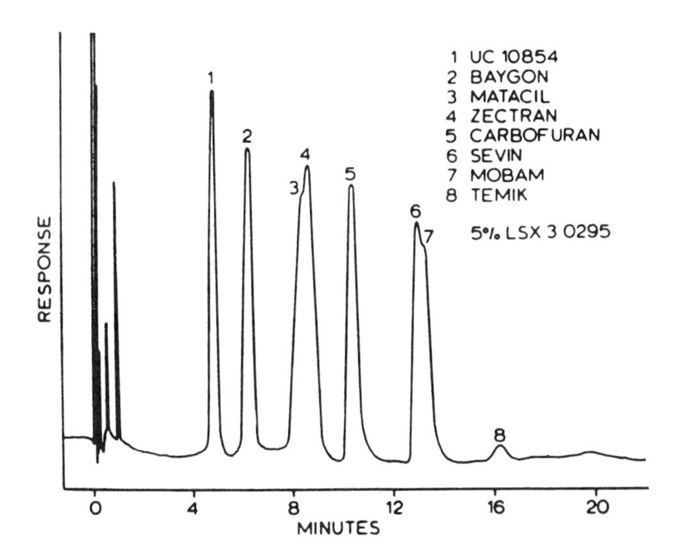

FIG. 6 Chromatogram of 1 ng each of seven carbamates and two carbamoyl oximes after derivatization with 2,5-dichlorobenzenesulfonyl chloride. [From Ref. 119, reprinted with permission from the *Journal of Agricultural and Food Chemistry*, *23*:416 (1975). Copyright © by the American Chemical Society, Washington, D.C.]

GC examinations of these derivatives are nearly identical. Relative electron capture responses increase, e.g., twofold with carbaryl and fivefold for carbofuran on passing from the trifluoroacetyl to the heptafluorobutyryl derivatives. The use of pyridine as a catalyst enables the condensation of pentafluoropropionic anhydride to proceed at room temperature.

4. 2,4-Dinitrophenyl Derivatives

Methylcarbamate residues are treated with 1-fluoro-2,4-dinitrobenzene by heating for 20 min at 100°C in alkaline sodium borate solution [122]. The ether derivative is extracted into iso-octane and analyzed by electron capture GC. The derivatives of carbamates are stable compounds.

5. 2,4-Dinitroaniline Derivatives

Methyl- or dimethylcarbamates are hydrolyzed with alkali and the liberated amines are reacted with 1-fluoro-2,4-dinitrobenzene to form dinitroaniline derivatives that are determined using electron capture detection [123]. This procedure is not specific.

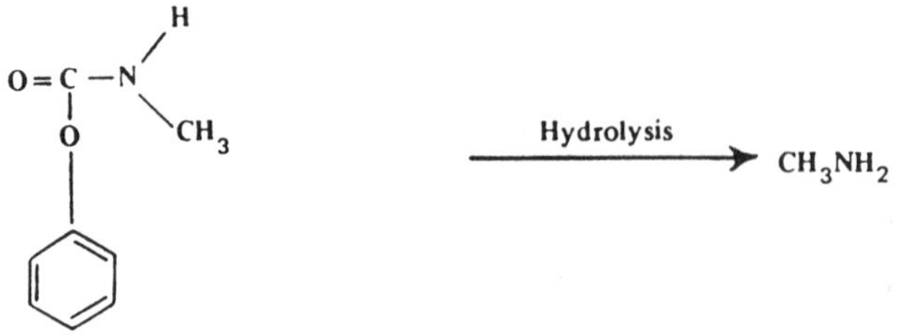

(XXIII)

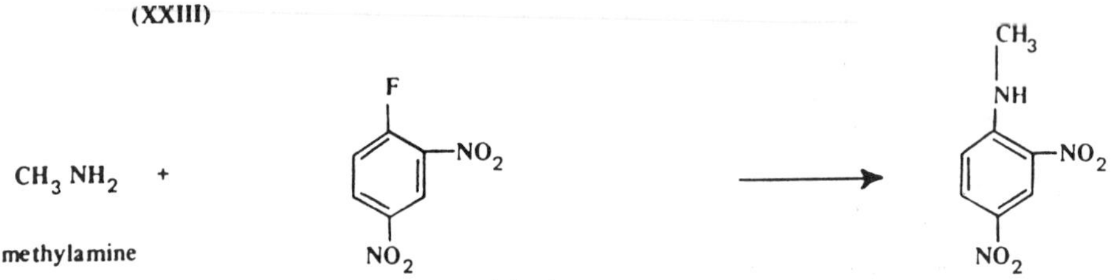

Hydrolysis → CH_3NH_2

CH_3NH_2 + 1 fluoro - 2, 4 - dinitrobenzene → 2, 4 - dinitro - N - methylaniline

methylamine

(XXIV)

6. *Pentafluorobenzyl Bromide Reagent*

N-Methylcarbamates are hydrolyzed in a 10% methanolic potassium hydroxide solution to form phenolic products. These are derivatized with pentafluorobenzyl bromide to produce ether derivatives that are cleaned up on a silica gel microcolumn [124]. They are determined by GC using an electron capture detector.

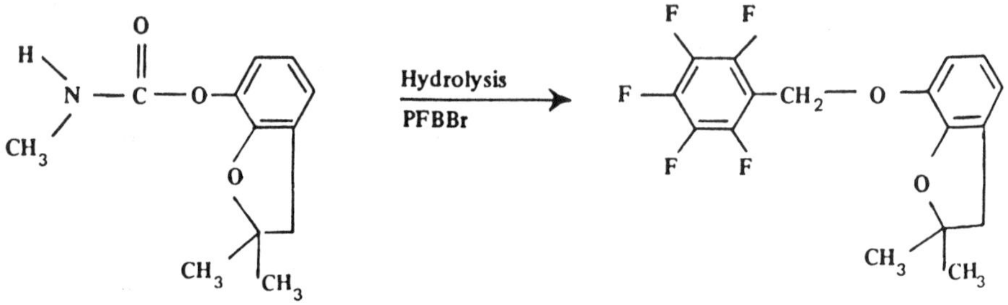

Hydrolysis PFBBr →

Carbofuran (XXV)

Pentafluorobenzyl derivative (XXVI)

VII. METHODS FOR THE DETERMINATION OF PESTICIDES IN SOIL

Gas-liquid chromatography is currently the main analytical technique used for the determination of pesticide residues in soil. High-pressure liquid chromatography has been applied to the assay of a number of herbicides including the substituted ureas, triazines, and uracils. This technique is particularly applicable to thermally labile compounds that suffer some decomposition at the temperature necessary for gas-liquid chromatography or when problems occur in the preparation of a volatile stable derivative.

In general, the sensitivity and specificity of the detectors coupled to gas chromatographs are superior to the UV detector used with high-pressure liquid chromatographs. The post-column fluorimetric labeling of some carbamate insecticides [125] has enabled nanogram quantities of these compounds to be detected. Chromatographic data published by Lawrence and Turton [126] detail the application of high-pressure liquid chromatography to the determination of 166 pesticides in standard solutions and a range of biological materials, including some soil samples. The methods of detection are based, in the main, on measurement in the UV, but some pesticides are determined by electron capture or by the production of a fluorophore and monitored with a fluorescence detector.

Pesticides are extracted from soil by end-over-end shaking, Soxhlet refluxing, or a leaching procedure. Comparative studies have been published for some residues in soil. Solvent systems commonly employed include hexane, acetone, methanol, acetonitrile, and chloroform, used either singly or in a mixture of two or more. Methods for the cleanup of soil extracts prior to chromatography involve the use of florisil, alumina, or liquid partition. This may not always be necessary if a selective detector is used and the soil extract is clean.

From the available solid support materials for gas chromatography, Chromosorb W-HP and Gas Chrom Q are used widely. The stationary phases SE-30, QF-1, Carbowax 20M, and the OV series are suitable for a range of residues. High pressure liquid chromatography columns for adsorption chromatography are available with 5 μm size silica or alumina particles. Bonded phase columns with both polar and nonpolar phases enables the operator to utilize their selectivity with high efficiency. Solvent programming may be used to decrease analysis time.

Confirmation of the result of a pesticide residue analysis should always be obtained by one of the standard procedures. The recovery of a pesticide added to a soil at residue level should genrally be better than 75%.

A selection of published methods for the determination of the more commonly used pesticides in soil is given in Table 1. Figure 7 is typical of chromatograms obtainable by GC and HPLC, respectively.

TABLE 1 Methods for the Determination of Pesticides in Soil

Compound	Soil extractant cleanup	Method	Detector	Ref.
A. Substituted ureas				
Chlorotoluron	methanol/silica gel	HPLC on Merckosorb SI-60; eluting solvent *n*-hexane/2-propanol (85:15)	UV (240 nm)	[127]
Chlorotoluron	methanol/silica gel	GLC		[128]
Diflubenzuron	acetonitrile/water	HPLC, Pak-CH	UV (254 nm)	[129]
Diflubenzuron	acetonitrile/water Celite/Florisil-alumina-silica gel	HPLC, μ-Bondapak C-18 or μ-Porasil	UV (254 nm)	[130]
Fenuron	methanol	GLC of methyl derivative; 4% SE-30 + 6% QF-1 on Gas Chrom Q	alkali ionization (RbCl)	[131]
Linuron		GLC of 3,4-dichloroaniline formed by alkaline hydrolysis; 7.5% SE-30 on silanized Chromosorb W	electron capture	[132]
Linuron	methanol/florisil	GLC, 1.5% XE-60 on Chromosorb W-HP	electron capture	[133]
11 Substituted ureas	acetone	GLC of brominated anilines; 5% high vacuum silicone grease on Gas Chrom Q (typical chromatogram Fig. 7)	electron capture	[134]
8 Substituted phenylureas	methanol	HPLC, Spherisorb ODS	UV (240 nm)	[135]

B. Carbamates				
Asulam	$Ca(OH)_2$ solution	colorimetric, reaction with sodium nitrate and *N*-(1-naphthyl)ethylenediamine dihydrochloride	spectrophotometer (540 nm)	[136]
Carbofuran	chloroform/diethyl ether (1:1)	TLC, silica gel G; mobile phase ether/hexane (3:1) colorimetric, hydrolysis and diazotization	spectrophotometer (490 nm)	[137]
Carbofuran	chloroform	colorimetric, reaction with diazotized aniline	spectrophotometer (460 nm)	[138]
Methomyl, carbofuran, and metabolite	acidified ammonium acetate	GLC, on-column transesterification; Porapak P	alkali ionization (Rb_2SO_4)	[139]
Bendiocarb	ethylacetate	GLC, 2% OV-101 on Ultra-bond 20M	alkali ionization	[187]
Carbofuran	methanol, partition with hexane-methanol-water	HPLC, 25 cm column RP-8	UV, 280 nm	[191]
Carbamates (general)	acetone	GLC of dichlorobenzene sulfonyl chloride derivatives (typical chromatogram Fig. 6)	electron capture or FPD (sulfur)	[119]
C. Triazines				
Atrazine and metabolites	methanol/acidic alumina	GLC of methyl derivative; 3% Carbo wax 20M on Chromosorb W-HP	alkali ionization (RbCl)	[141]
Atrazine		GLC of methyl derivative; 4% SE-30 and 6% QF-1 on Gas Chrom Q	alkali ionization	[131]
Atrazine and hydroxyatrazine	methanol/acidic alumina	GLC, 3% SE-30 or 3% Carbowax 20M on Chromosorb W-HP	alkali ionization (CsBr)	[142]

(continued)

TABLE 1 (Continued)

Compound	Soil extractant/ cleanup	Method	Detector	Ref.
Simazine and prometryne	methanol	GLC, 2% neopentyl glycol succinate on Chromosorb W-HP	alkali ionization (RbCl)	[143]
7s-Triazines	acetone/alumina	GLC, 5% Reoplex or 3% Carbowax 20M or 3% OV-17 on Chromosorb W-HP	alkali ionization (CsBr)	[144]
D. Organochlorine insecticides				
Aldrin		GLC, 10% Apiezon L on Chromosorb W AW-DCMS	electron capture	[145]
Chlordane	hexane/2-propanol (1:1)	GLC, 2% SE-30 + 6% on QF-1 Chromosorb W-HP	electron capture	[146]
Dieldrin metabolites	methanol	GLC, 1% OV-1 on Chromosorb W AW-DCMS. TLC, on silica gel, mobile phase cyclohexane/acetone (4:1)	electron capture and FID	[147]
Dieldrin and heptachlor	hexane/2-propanol (3:1)/alumina	GLC, 5% DC-200 and 7.5% QF-1 or 3% DC-200 on Gas Chrom Q	electron capture	[148]
Hexachlorobenzene	hexane/acetone (1:1)/Florisil	GLC, 5% DC-200 and 7.5% QF-1	electron capture	[149]
Hexachlorobenzene and dichloran	petroleum ether/ acetone (1:1)/ Florisil	GLC, 2% OV-225 or 3% mixture of OV-17 and OV-210 (3:22) on Gas Chrom Q	electron capture	[150]

11 Organochlorine pesticides	hexane/acetone (1:1)/Florasil	GLC, 10% DC-200 or OV-101 on Chromosorb W-HP	electron capture	[152]
E. Organophosphorus insecticides				
Bromofenvinfos	acetone/light petroleum (3:7)	GLC, 5% DC-200 on Varaport	alkali ionization	[153]
Dichlorvos, trichlorphon, and isofenfos	acetone/benzene (19:1); Nuchar C-190/low temperature	GLC, 3% SE-30 on Gas Chrom Q	alkali ionization (Rb_2SO_4)	[139]
Fonofos	hexane/acetone (1:1)	GLC, 0.8% Carbowax 20M and 8% SE-30 on Gas Chrom Q	alkali ionization (RbCl)	[154]
Methyl parathion	incubation with sodium acetate, pH 7	colorimetric determination of the *p*-nitrophenol produced	spectrophotometer (405 nm)	[155]
Staufer N-2596 [O-ethyl-S-(4-chlorophenyl)ethanephosphorodithioate]	benzene	GLC, 10% OV-17 on Gas Chrom Q	FPD (phosphorus)	[156]
Isofenfos and metabolite	ethyl acetate/Florisil, silica gel, alumina	GLC, 5% DEGS on Gas Chrom Q	FPD (phosphorus)	[151]
Parathion, phorate	acetone and ethyl acetate/Celite 545, MgO, Norit SG	GLC, 3% OV-1 on Gas Chrom Q	FPD (phosphorus)	[157]

(continued)

TABLE 1 (Continued)

Compound	Soil extractant/ cleanup	Method	Detector	Ref.
F. Halo-alkanoic acids				
Dalapon	methanolic NaOH	GLC of 1-butyl ester; 2% OV-17 on Chromosorb W-HP	electron capture	[158]
G. Uracils				
Bromacil, lenacil, and terbacil	$Ca(OH)_2$ solution/ chloroform	GLC, 5% high vacuum silicone grease on Gas Chrom Q	alkali ionization (RbCl)	[159]
Bromacil		quantitative bioassay		
Terbacil and metab-olites	chloroform/liquid–liquid partition	GLC of silyl derivative; 5% XE-60 and 0.2% Epon 1001 on Gas Chrom Q	microcoulometer	[160]
H. Heterocyclic compounds				
Bentazone and metab-olites	leachate/methylene chloride	TLC, on silica gel coated sheets with a fluorescent indicator; mo-bile phase chloroform/methanol/ ammonia (69:30:1)	UV light	[161]
Metribuzin and metab-olites	acetonitrile/water/ silica gel-Florisil	GLC, 5% OV-225 on Chromosorb W-HP	electron capture	[162]
Metribuzin and metab-olites	water/methanol (1:9)/chloroform	GLC, 3% Silar 5 CP on Chromosorb W AW-DCMS	alkali ionization	[19]
Metribuzin and metab-olites	20% aq. methanol/ chloroform	GLC, 3% OV-225 or Silar 5 CP on Chromosorb W-HP; 1% Reoplex 400 on Gas Chrom Q	electron capture, Coulson conduc-tivity, FPD (sul-fur)	[163]

Oxadiazon	benzene/ethyl acetate/acetone (1:1:2) and 10% aq. methanol	TLC, on silica gel plates (F-254); mobile phases hexane/actone (60:40), methylene chloride/methanol (90:10 and 85:15), benzene	UV light	[164]
Oxadiazon	10% aq. methanol	GLC of methyl derivative; 3% OV-17 on Gas Chrom Q	electron capture	[164]
Picloram	10% aq. KOH, and $Ca(OH)_2$ solution	GLC of methyl exter; 1.5% XE-60 on Chromosorb W-HP	electron capture	[165]
I. Aromatic acids				
2,4-D, dicamba and mecoprop	acidified acetone/Florisil	GLC of methyl ester or brominated derivative; 11% OV-17/QF-1 on Chromosorb W-HP	electron capture	[166]
2,4-D and dicamba		isotope dilution GC/MS technique		[188]
J. Pyridines				
Clopyralid	NaCl solution, pH 7–8, 5% ethanol chloroform	GLC of methyl ester; 3% OV-17 on Chromosorb W-HP	electron capture	[167]
Clopyralid	water	colorimetric	UV spectrophotometer (280 nm)	[168]
Clopyralid	0.25 *M* NaOH	GLC of methyl derivative, wide-bore capillary column, methyl silicon fluid	electron capture	[185]

(continued)

TABLE 1 (Continued)

Compound	Soil extractant/ cleanup	Method	Detector	Ref.
Cyperquat	H_2SO_4 solution; hydrogenation/ methylene chloride	GLC, 3% Carbowax 20M + KOH on Carbosorb W-HP	alkali ionization (RbCl)	[169]
K. Anilides				
Benzoylprop ethyl and metabolites	methanol/water/ ethyl acetate/ acetic acid (40:40:20:1)/ alumina H	GLC of methyl derivatives; 2% OV-225 on Gas Chrom Q	electron capture	[170]
L. Nitroanilines				
Trifluralin	methylene chloride	GLC, 10% DC-200 on Gas Chrom Q	Coulson conductivity (nitrogen mode)	[171]
Cryzalin	methanol/alumina	GLC of methyl derivative; 5% XE-60 on Chromosorb W-HP	electron capture	[172]
M. Nitrophenols				
Dinoterb	ethanol	GLC of methyl derivative; 4% SE-30 on Gas Chrom Q	electron capture	[173]
N. Amides and nitriles				
Dichlobenil	ethyl acetate	GLC, 10% DC-200 on Gas Chrom Q	electron capture	[174]

Dichlobenil	20% acetone in hexane/TLC on alumina H	GLC, 10% silicone oil and 0.5% Epikote resin 1001 on Celite; 2% NPGS or PDES on Celite	electron capture	[175]
Diphenamid	methylene chloride	GLC, 10% DC-200 on Gas Chrom Q	Coulson conductivity (nitrogen mode)	[167]
N-methylcarbamates	acidified ammonium acetate/silica gel	GLC of PFB derivative; 3% OV-225 on Chromosorb W-HP and 3.6% OV-101 + 5.5% OV-210 on Chromosorb W-HP	electron capture	[140]
Propyzamide	sulfuric acid/methanol/Florisil	GLC of methyldichlorobenzoate; 10% OV-17 on Gas Chrom Q	electron capture	[176]
O. Oxime derivates				
Aldicarb		GLC, 5% Reoplex 400 on Gas Chrom Q	FPD (sulfur)	[177]
Aldicarb, sulfoxide and sulfone	aqueous leachate; chloroform, silica gel	GLC, 8% Carbowax 20M on Anakrom ABS	FPD (sulfur)	[178]
Glyphosate and metabolites	Dilute NH_4OH	TLC, on microcrystalline cellulose plates; two-dimensional; mobile phases: phenol/water/acetic acid (84:16:1), isobutyric acid/water/1-propanol/conc. ammonia/2-propanol/1-butanol (500:95:70:20:15:15)	UV light/Hanes reagent, ninhydrin	[179]

(continued)

TABLE 1 (Continued)

Compound	Soil extractant/ cleanup	Method	Detector	Ref.
Oxamyl	ethyl acetate	GLC, after alkaline hydrolysis to oximino fragment, 10% SP-1200/ 1% H_3PO_4 on Chromosorb W-AW	FPD (sulfur)	[180]
Oxamyl	acetone/dichloromethane/Florisil	GLC, on-column reaction with trimethylphenylammonium hydroxide; 0.5% Carbowax 20M and 5% SE-30 on Chromosorb W	FPD (sulfur)	[181]
P. Sulphonyl ureas				
Chlorosulfuron		enzyme immunoassay (ELISA)		[195]
Q. Aromatic ethers				
Diclofop-methyl	methanol/water/ ethylacetate/ phosphoric acid (40:40:19:1)	GLC of methyl esters, 3% OV-17 on Chromosorb G AW-DMCS	electron capture	[197]
Diclofop	acetonitrile/water/ acetic acid (80:20:2.5) partition by aq. $NaHCO_3$	GLC of methyl esters, 2% Apiezon L on Chromosorb W-HP or Ultrabond 20M	electron capture	[198]
R. Fungicides and Acaricides				
Benomyl	acetone/M aqueous ammonium chloride (1:1); solvent partition	colorimetric	UV (282 nm)	[182]

Benomyl	benzene; solvent partition	GLC of trifluoroacetyl derivative; 5% SE 30 on Chromosorb R	electron capture	[183]
Quintozene	petroleum ether; acetone (1:1)	GLC 2% OV-225 or 3% of mixture of OV-17, OV-210 (3:22) on Gas Chrom Q	electron capture	[150]
Iprodione	acetone solvent partition	GLC 5% SE 30 on Chromosorb W-HP		[184]
Vinclozolin	acetone solvent partition	GLC 3% OV-1 Chromosorb W-HP		[184]
Amitraz	base hydrolysis; steam distillation; acid-base partition	GLC of heptafluorobutyranilide derivative, 3% OV-17 on Gas Chrom Q	electron capture	[186]
Propiconazole	methanol/water (4:1) partitioned into methylene chloride	GLC	alkali ionization	[194]
S. Multiresidue methods				
10 Acidic herbicides	acetone/hexane, partitioned with acidified water	GLC of pentafluorobenzyl esters	electron capture	[192]
7 Natural herbicides	acetone, partitioned with 2% $KHCO_3$ dichloromethane	GLC	Electron capture or alkali ionization	[193]

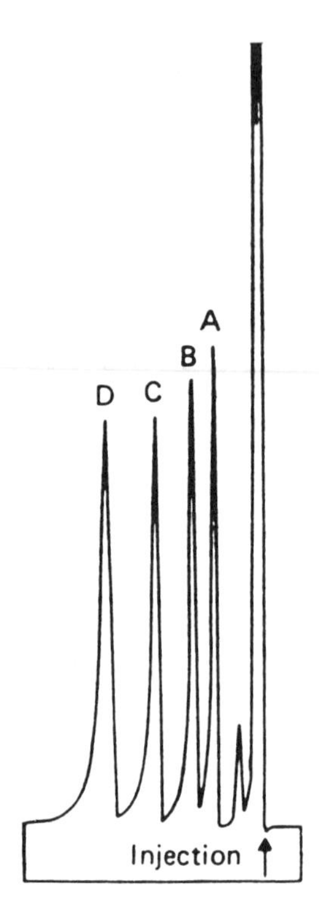

FIG. 7 Gas chromatographic response of a standard mixture of brominated anilines derived from: A, monolinuron; B, fenuron; C, linuron; and D, chlorobromuron (R_t = 14 min). Amount applied 0.5 ng. (From Ref. 134.)

ACKNOWLEDGMENTS

We thank Dr. Keith Smith for affording us the opportunity to contribute to this volume. We acknowledge gratefully the permission of the Ministry of Agriculture, Fisheries and Food to draw on experience gained in our employment.

REFERNCES

1. R. L. Grob, in *Modern Practice of Gas Chromatography* (R. L. Grob, ed.), Wiley, New York, 1977, p. 3.
2. M. Tswett, *Ber. Deut. Bot. Ges.*, *24*:316 (1906).

3. M. Tswett, *Ber. Deut. Bot. Ges.*, *24*:384 (1906).
4. M. Tswett, *Ber. Deut. Bot. Ges.*, *25*:71 (1907).
5. A. J. P. Martin and R. L. M. Synge, *Biochem. J.*, *35*:1358 (1941).
6. A. T. James and A. J. P. Martin, *Analyst (Lond.)*, *77*:915 (1952).
7. A. T. James and A. J. P. Martin, *Biochem. J.*, *50*:679 (1952).
8. R. L. Grob (ed.), *Chromatographic Analysis of the Environment*, Marcel Dekker, New York, 1975.
9. L. S. Ettre and A. Zlatkis (eds.), *The Practice of Gas Chromatography*, Wiley-Interscience, New York, 1967.
10. D. W. Grant, *Gas Liquid Chromatography*, Van Nostrand Reinhold, New York, 1971.
11. R. L. Grob (ed.), *Modern Practice of Gas Chromatography*, Wiley, New York, 1977.
12. C. F. Simpson, *Practical High Performance Liquid Chromatography*, Heyden, London, 1976.
13. P. A. Bristow, *Liquid Chromatography in Practice*, HETP, Wilmslow, Cheshire, England, 1976.
14. ADAS, *Sampling Soil for Analysis*, Leaflet 655, MAFF, Pinner, Middlesex, England, 1979.
15. J. H. Ford, C. A. McDaniel, F. C. White, R. E. Vest, and R. E. Roberts, *J. Chromatogr. Sci.*, *13*:291 (1975).
16. D. R. Rourke, W. F. Mueller, and S. H. Yang, *J. Assoc. Off. Anal. Chem.*, *60*:233 (1977).
17. I. Levi and T. W. Nowicki, *Bull. Environ. Contam. Toxic.*, *7*:133 (1972).
18. N. S. Kawar, G. C. deBatista, and F. A. Gunther, *Residue Rev.*, *48*:45 (1973).
19. G. R. B. Webster and G. J. Reimer, *Pestic. Sci.*, *7*:292 (1976).
20. I. Levi and T. W. Nowicki, *Bull. Environ. Contam. Toxic.*, *7*:193 (1972).
21. J. A. Singmaster and D. G. Crosby, *Bull. Environ. Contam. Toxic.*, *16*:291 (1976).
22. R. R. Watts and R. W. Storherr, *J. Assoc. Off. Anal. Chem.*, *48*:1158 (1965).
23. R. W. Storherr and R. R. Watts, *J. Assoc. Off. Anal. Chem.*, *51*:662 (1968).
24. R. W. Storherr and R. R. Watts, *J. Assoc. Off. Anal. Chem.*, *48*:1154 (1965).
25. M. Chiba and H. V. Morley, *J. Assoc. Off. Anal. Chem.*, *51*:55 (1968).
26. J. D. Ashby, *An Introduction to Gas Chromatography*, Pye Unicam Ltd., Cambridge, England, 1977, p. 25.
27. D. M. Ottenstein, *Advances in Chromatography*, Vol. 3, Marcel Dekker, New York, 1966, p. 137.

28. J. F. Palframan and E. A. Walker, *Analyst (Lond.)*, *92*:71 (1967).
29. L. Ettre, *Anal. Chem.*, *36*:31A (1964).
30. L. Rohrschneider, in *Advances in Chromatography*, Vol. 4 (J. C. Giddings and R. A. Keller, eds.), Marcel Dekker, New York, 1967, p. 333.
31. W. O. McReynolds, *J. Chromatogr. Sci.*, *8*:685 (1970).
32. W. R. Supina, *The Packed Column in Gas Chromatography*, Supelco, Belafonte, Pa., 1974.
33. Anonymous, *J. Chromatogr. Sci.*, *13*:115 (1975).
34. R. F. Cook, R. P. Stanovick, and C. C. Cassil, *J. Agric. Fd. Chem.*, *17*:277 (1969).
35. H. J. Van de Wiel and P. Tommassen, *J. Chromatogr.*, *71*:1 (1972).
36. D. J. David, *Gas Chromatographic Detectors*, Wiley, New York, 1974.
37. J. Sevcik, *Detectors in Gas Chromatography*, Elsevier, Amsterdam, 1975.
38. D. F. S. Natusch and T. M. Thorpe, *Anal. Chem.*, *45*:1185A (1973).
39. W. A. Aue and S. Kapila, *J. Chromatogr. Sci.*, *11*:255 (1973).
40. W. A. Aue, *J. Chromatogr. Sci.*, *13*:329 (1975).
41. W. P. Cochrane and R. Greenhalgh, *Chromatographia*, *9*:255 (1976).
42. J. F. Taylor, *Proc. Soc. Anal. Chem.*, *13*:168 (1976).
43. J. E. Lovelock and S. R. Lipsky, *J. Am. Chem. Soc.*, *82*:431 (1960).
44. R. J. Maggs, P. L. Joynes, A. J. Davies, and J. E. Lovelock, *Anal. Chem.*, *43*:1966 (1971).
45. W. A. Aue, in *Pesticide Identification at the Residue Level* (F. J. Biros, ed.), Am. Chem. Soc., Washington, D.C., 1971, p. 43.
46. B. Kolb and J. Bischoff, *J. Chromatogr. Sci.*, *12*:625 (1974).
47. B. Kolb, M. Linder, and B. Kempen, *Appl. Chromatogr.* (Perkin Elmer), *21E*:1 (1974).
48. A. Karmen, *Anal. Chem.*, *36*:1416 (1964).
49. K. Abel, K. Lannéau, and R. K. Stevens, *J. Assoc. Off. Anal. Chem.*, *49*:1022 (1966).
50. B. Kolb, M. Auer, and P. Popisil, *J. Chromatogr. Sci.*, *15*:53 (1977).
51. J. A. Lubkowitz, J. L. Glajch, B. P. Semonian, and L. B. Rogers, *J. Chromatogr.*, *133*:34 (1977).
52. B. P. Semonian, J. A. Lubkowitz, and L. B. Rogers, *J. Chromatogr.*, *151*:1 (1978).
53. M. J. Hartigan, J. E. Purcell, M. L. McConnell, and M. L. Lee, *J. Chromatogr.*, *99*:339 (1974).

54. C. A. Burgett, D. H. Smith, and H. B. Bente, *J. Chromatogr.*, *134*:57 (1977).
55. P. L. Patterson and R. L. Howe, *J. Chromatogr. Sci.*, *16*:275 (1978).
56. S. S. Brody and J. E. Chaney, *J. Gas Chromatogr.*, *4*:42 (1966).
57. K. J. Krost, J. A. Hodgeson, and R. K. Stevens, *Anal. Chem.*, *45*:1800 (1973).
58. A. I. Mizany, *J. Chromatogr. Sci.*, *8*:151 (1970).
59. S. O. Farwell and R. A. Rasmussen, *J. Chromatogr. Sci.*, *14*:224 (1976).
60. J. G. Eckhardt, M. B. Denton, and J. L. Moyers, *J. Chromatogr. Sci.*, *13*:133 (1975).
61. C. H. Burnett, D. F. Adams, and S. O. Farwell, *J. Chromatogr. Sci.*, *15*:230 (1977).
62. C. A. Burgett and L. E. Green, *J. Chromatogr. Sci.*, *12*:356 (1974).
63. J. M. Zehner and R. A. Simonaitis, *Anal. Chem.*, *47*:2485 (1975).
64. T. Sugiyama, Y. Suzuki, and T. Takeuchi, *J. Chromatogr.*, *80*:61 (1973).
65. R. Greenhalgh and M. A. Wilson, *J. Chromatogr.*, *128*:157 (1976).
66. P. L. Patterson, R. L. Howe, and A. Abu-Shumays, *Anal. Chem.*, *50*:339 (1978).
67. P. L. Patterson, *Anal. Chem.*, *50*:345 (1978).
68. D. M. Coulson, L. A. Cavanagh, J. E. De Vries, and ?. ?. Walther, *J. Agric. Fd. Chem.*, *8*:399 (1960).
69. D. M. Coulson and L. A. Cavanagh, *Anal. Chem.*, *32*:1245 (1960).
70. D. M. Coulson, *Advan. Pest Control Res.*, *5*:153 (1962).
71. C C. Cassil, *Residue Rev.*, *1*:37 (1962).
72. H. P. Burchfield, D. E. Johnson, J. W. Rhoades, and R. J. Wheeler, *J. Gas Chromatogr.*, *3*:28 (1965).,
73. H. P. Burchfield, J. W. Rhoades, and R. J. Wheeler, *J. Agric. Fd. Chem.*, *13*:511 (1965),.
74. H. P. Burchfield and R. J. Wheeler, *J. Assoc. Off. Anal. Chem.*, *49*:651 (1966).
75. R. L. Martin, *Anal. Chem.*, *38*:1209 (1966).
76. L. Guiffrida and N. F. Ives, *J. Assoc. Off. Anal,. Chem.*, *52*:541 (1969).
77. W. E. Westlake, in *Pesticide Identification at the Residue Level* (F. J. Biros, ed.), Am. Chem. Soc., Washington, D.C., 1971, p. 78.
78. D. M. Coulson, *J. Gas Chromatogr.*, *3*:134 (1965).

79. D. M. Coulson, *J. Gas Chromatogr.*, *4*:285 (1966).
80. D. M. Coulson, in *Advances in Chromatography*, Vol. 3 (J. C. Giddings and R. A. Keller, eds.), Marcel Dekker, New York, 1966, p. 197.
81. C. C. Cassil, R. P. Stanovick, and R. F. Cook, *Residue Rev.*, *26*:63 (1969).
82. W. E. Westlake, in *Pesticide Identification at the Residue Level* (F. J. Biros, ed.), Am. Chem. Soc., Washington, D.C., 1971, p. 79.
83. G. G. Patchett, *J. Chromatogr. Sci.*, *8*:155 (1970).
84. R. C. Hall, *J. Chromatogr. Sci.*, *12*:152 (1974).
85. B. P. Wilson and W. P. Cochrane, *J. Chromatogr.*, *106*:174 (1975).
86. C. J. W. Brooks and B. S. Middleditch, *Clin. Chem. Acta*, *34*:145 (1971).
87. D. J. Jenden and A. K. Cho, *Ann. Rev. Pharmacol.*, *13*: 371 (1973).
88. W. McFadden, *Techniques of Combined Gas Chromatography/ Mass Spectrometry*, Wiley-Interscience, New York, 1973.
89. G., Vander Velde and J. F. Ryan, *J. Chromatogr. Sci.*, *13*: 322 (1975).
90. J. F. Ryan, in *Analytical Methods for Pesticides and Plant Growth Regulators, Spectroscopic Methods of Analysis*, Vol. 9, (G. Zweig and J. Sherma, eds.), Academic Press, New York, 1978, p. 1.
91. Anonymous, *Logical Troubleshooting*, Hewlett-Packard, Inc., Avondale, Pa., 1970.
92. D. F. K. Swan, *Three Selective Detectors*, Pye Unicam Ltd., Cambridge, England, 1978.
93. Anonymous, *Care and Use of Partisil Columns for HPLC*, Whatman Inc., Clifton, N.J., 1975.
94. L. R. Snyder and J. J. Kirkland, *Introduction to Modern Liquid Chromatography*, Wiley, New York, 1974.
95. D. L. Saunders, *Anal. Chem.*, *46*:470 (1974).
96. R. P. W. Scott, *Liquid Chromatography Detectors*, Journal of Chromatography Library, Vol. 11, Elsevier, Amsterdam, 1977.
97. D. R. Baker, R. C. Williams, and J. C. Steichen, *J. Chromatogr. Sci.*, *12*:499 (1974).
98. S. Udenfriend, S. Stein, P. Böhlen, W. Dairman, W. Leimgruber, and M. Weigele, *Sci.*, *178*:871 (1972).
99. S. Katz and W. W. Pitt, *Anal. Lett.*, *5*:177 (1972).
100. W. Duenges, UV Spectrom. Group Bull, No. 5, 1977, p. 38.
101. H. A. Moye, S . J. Scherer, and P. A. St. John, *Anal. Lett.*, *10*:1439 (1977).

102. N. Seiler and L. Demisch, in *Handbook of Derivatives for Chromatography* (K. Blau and G. S. King, eds.), Heyden, London, 1977.
103. R. J. Dolphin, F. W. Willmott, A. D. Mills, and L. P. J. Hoogeveen, *J. Chromatogr.*, *122*:259 (1976).
104. R. P. W. Scott and J. G. Lawrence, *J. Chromatogr. Sci.*, *8*:65 (1970).
105. R. P. W. Scott, *Analyst (Lond.)*, *103*:37 (1978).
106. H. L. Pierson and D. J. Steible, in *Modern Practice of Gas Chromatography* (R. L. Grob, ed.), Wiley, New York, 1977.
107. J. C. Cavagnol and W. R. Betker, in *The Practice of Gas Chromatography* (L. S. Ettre and A. Zlatkis, eds.), Wiley-Interscience, New York, 1967.
108. M. Beroza and W. R. Betker, in *The Practice of Gas Chromatography* (L. S. Ettre and A. Zlatkis, eds.), Wiley-Interscience, New York, 1967.
109. W. P. Cochrane, *J. Chromatogr. Sci.*, *13*:246 (1975).
110. Anonymous, *Handbook and General Catalog*, Pierce Chemical Co., Rockford, Ill., 1979.
111. K. Blau and G. S. King (eds.), *Handbook of Derivatives for Chromatography*, Heyden, London, 1977.
112. J. F. Lawrence, *J. Agric. Fd. Chem.*, *22*:936 (1974).
113. R. Greenhalgh and J. Kovacicova, *J. Agric. Fd. Chem.*, *23*:325 (1975).
114. H. Agemian and A. S. Y. Chau, *J. Assoc. Off. Anal. Chem.*, *60*:1070 (1977).
115. J. A. Coburn and A. S. Y. Chau, *J. Assoc. Off. Anal. Chem.*, *57*:1272 (1974).
116. M. A. Forbes, B. P. Wilson, R. Greenhalgh, and W. P. Cochrane, *Bull. Environ. Contam. Toxic.*, *13*:141 (1975).
117. J. Singh and M. R. Lapointe, *J. Assoc. Off. Anal. Chem.*, *57*:1285 (1974).
118. J. W. Miles and W. E. Dale, *J. Agric. Fd. Chem.*, *26*:480 (1978).
119, H. A. Moye, *J. Agric. Fd. Chem.*, *23*:415 (1975).
120. J. N. Sieber, *J. Agric. Fd. Chem.*, *20*:443 (1972).
121. S. C. Lau and R. L. Marxmiller, *J. Agric. Fd. Chem.*, *18*:413 (1970).
122. E. R. Holden, *J. Assoc. Off. Anal. Chem.*, *56*:713 (1973).
123. E. R. Holden, W. M. Jones, and M. Beroza, *J. Agric. Fd. Chem.*, *17*:56 (1969).
124. J. A. Coburn, B. D. Ripley, and A. S. Y. Chau, *J. Assoc. Off. Anal. Chem.*, *59*:188 (1976).
125. R. T. Krause, *J. Chromatogr. Sci.*, *13*:281 (1978).
126. J. F. Lawrence and D. Turton, *J. Chromatogr.*, *159*:207 (1978).

127. A. E. Smith and K. A. Lord, *J. Chromatogr.*, *107*:407 (1975).
128. E. Allan and K. Lord, *J. Chromatogr.*, *106*:409 (1975).
129. C. H. Schaefer and E. F. Dupras, *J. Agric. Fd. Chem.*, *25*: 1026 (1977).
130. S. J. Diprima, R. D. Cannizzaro, J. Roger, and C. D. Ferrell, *J. Agric. Fd. Chem.*, *26*:968 (1978).
131. R. Greenhalgh and J. Kovacicova, *Bull. Environ. Contam. Toxic.*, *14*:47 (1975).
132. L. B. Dmitriev and L. P. Yudina, *Izv. Timiryazev. Sel-khoz Akad.*, *2*:215 (1974).
133. S. U. Khan, R. Greenhalgh, and W. P. Cochrane, *Bull. Environ. Contam. Toxic.*, *13*:602 (1975).
134. D. J. Caverly and R. C. Denney, *Analyst (Lond.)*, *103*:368 (1978).
135. D. S. Farrington, R. G. Hopkins, and J. H. A. Ruzicka, *Analyst (Lond.)*, *102*:377 (1977).
136. A. E. Smith and A. Walker, *Pestic. Sci.*, *8*:449 (1977).
137. K. Venkateswarlu, T. K. Siddarame Gowda, and N. Sethunathan, *J. Agric. Fd. Chem.*, *25*:533 (1977).
138. J. R. Rangaswamy, Y. N. Vijayashankar, and S. Prakash, *J. Assoc. Off. Anal. Chem.*, *59*:1276 (1976).
139. P. T. Holland, *Pestic. Sci.*, *8*:354 (1977).
140. J. A. Coburn, B. D. Ripley, and A. S. Y. Chau, *J. Assoc. Off. Anal. Chem.*, *59*:188 (1976).
141. S. U. Khan and P. B. Marriage, *J. Agric. Fd. Chem.*, *25*: 1408 (1977).
142. S. U. Khan, R. Greenhalgh, and W. P. Cochrane, *J. Agric. Fd. Chem.*, *23*:430 (1975).
143. A. Walker, *Pestic. Sci.*, *7*:41 (1976).
144. S. U. Khan and P. Purkayastha, *J. Agric. Fd. Chem.*, *23*: 311 (1975).
145. D. L. Struble, *Bull. Environ. Contam. Toxic.*, *11*:231 (1974).
146. F. Tafuri, M. Businelli, L. Scarponi, and C. Marucchini, *J. Agric. Fd. Chem.*, *25*:353 (1977).
147. I. Weisgerber, D. Bieniek, J. Kohli, and W. Klein, *J. Agric. Fd. Chem.*, *23*:682 (1975).
148. H. P. Freeman, A. W. Taylor, and W. M. Edwards, *J. Agric. Fd. Chem.*, *23*:1101 (1975).
149. A. R. Isensee, E. R. Holden, E. A. Woolson, and G. E. Jones, *J. Agric. Fd. Chem.*, *24*:1210 (1976).
150. W. Dejonckheere, W. Steurbaut, and R. H. Kips, *Bull. Environ. Contam. Toxic.*, *13*:720 (1975).
151. M. J. Brown and I. H. Williams, *Pestic. Sci.*, *7*:545 (1976).
152. Anonymous, *Official Methods of Analysis*, 12th ed., Assoc. Off. Anal. Chem., Washington, D.C., Sec. 29.013.

153. A. Chmiel, R. Knapek, and T. Utracki, *Pr. Inst. Przem. org.*, *6*:147 (1974).
154. S. U. Khan, H. A. Hamilton, and E. J. Hogue, *Pestic. Sci.*, *7*:553 (1976).
155. F. M. Kishk, T. El-Essawi, S. Abdel-Chafar, and M. B. Abou-Donia, *J. Agric. Fd. Chem.*, *24*:305 (1976).
156. R. J. Bussey, M. A. Christenson, and M. S. O'Connor, *J. Agric. Fd. Chem.*, *25*:993 (1977).
157. A. M. Kadoum and D. E. Moch, *J. Agric. Fd. Chem.*, *26*:45 (1978).
158. E. G. Cotterill, *J. Chromatogr.*, *106*:409 (1975).
159. D. J. Caverly and R. C. Denney, *Analyst (Lond.)*, *102*:576 (1977).
160. R. F. Holt and H. L. Pease, *J. Agric. Fd. Chem.*, *25*:373 (1977).
161. E. W. Stoller, L. W. Wax, L. C. Haderlie, and F. W. Slife, *J. Agric. Fd. Chem.*, *23*:682 (1975).
162. J. S. Thornton and C. W. Stanley, *J. Agric. Fd. Chem.*, *25*:380 (1977).
163. G. R. B. Webster, S. R. Macdonald, and L. P. Sarna, *J. Agric. Fd. Chem.*, *23*:74 (1975).
164. D. Ambrosi, P. C. Kearney, and J. A. Macchia, *J. Agric. Fd. Chem.*, *25*:868 (1977).,
165. C. E. McKone and E. G. Cotterill, *Bull. Environ. Contam. Toxic.*, *11*:233 (1974).
166. S. U. Khan, *J. Assoc. Off. Anal. Chem.*, *58*:1027 (1975).
167. A. J. Pik and G. W. Hodgson, *J. Assoc. Off. Anal. Chem.*, *59*:264 (1976).
168. A. J. Pik, E. Peake, M. T. Strosher, and G. W. Hodgson, *J. Agric. Fd. Chem.*, *25*:1054 (1977).
169. S. U. Khan and K. S. Lee, *J. Agric. Fd. Chem.*, *24*:684 (1976).
170. A. N. Wright and B. L. Mathews, *Pestic. Sci.*, *7*:339 (1976).
171. W. R. Payne, J. D. Pope, and J. E. Benner, *J. Agric. Fd. Chem.*, *22*:79 (1974).
172. R. F. Sieck, W. S. Johnson, A. F. Cockerill, D. N. B. Mallen, D. J. Osborne, and S. J. Barton, *J. Agric. Fd. Chem.*, *24*:617 (1976).
173. W. H. Dekker and H. A. Selling, *J. Agric. Fd. Chem.*, *23*: 1013 (1975).
174. G. G. Briggs and J. E. Dawson, *J. Agric. Fd. Chem.*, *18*: 97 (1970).
175. K. I. Benyon, L. Davies, K. Elgar, and A. N. Wright, *J. Sci. Fd. Agric.*, *17*:151 (1966).
176. I. R. Adler, C. F. Gordon, L. D. Haines, and J. P. Wargo, *J. Assoc. Off. Anal. Chem.*, *55*:802 (1972).

177. Y. Iwata, W. E. Westlake, J. H. Barkley, G. E. Carman, and F. A. Gunther, *J. Agric. Fd. Chem.*, *25*:933 (1977).
178. M. Leistra, J. H. Smelt, and T. M. Lexmund, *Pestic. Sci.*, *7*:471 (1976).
179. M. L. Rueppel, B. B. Brightwell, J. Schaefer, and J. T. Marvel, *J. Agric. Fd. Chem.*, *25*:517 (1977).
180. R. F. Holt and H. L. Pease, *J. Agric. Fd. Chem.*, *24*:263 (1976).
181. R. H. Bromilow, *Analyst (Lond.)*, *101*:982 (1976).
182. D. J. Austin and G. G. Briggs, *Pestic. Sci.*, *7*:201 (1976).
183. J. P. Rouchaud and J. R. Decallonne, *J. Agric. Fd. Chem.*, *22*:259 (1974).
184. A. Walker, P. A. Brown, and A. R. Entwistle, *Pestic. Sci.*, *17*:183 (1986).
185. M. Galoux, A. Bernes, and J. C. Van Damme, *J. Agric. Fd. Chem.*, *33*:965 (1985).
186. R. Hornish, M. Clasby, J. L. Nappier, and G. Hoffman, *J. Agric. Fd. Chem.*, *32*:1219 (1984).
187. S. Szeto, A. Wilkinson, and M. Brown, *J. Agric. Fd. Chem.*, *32*:78 (1984).
188. A. Lopez-Avila, P. Hirata, S. Kraska, and J. Taylor, *J. Agric. Fd. Chem.*, *34*:530 (1986).
189. L. N. Lundgren, *J. Agric. Fd. Chem.*, *34*:535 (1986).
190. M. Åkerblom and G. Alex, *J. Assoc. Off. Anal. Chem.*, *67*:653 (1984).
191. D. R. Lauren, *J. Assoc. Off. Anal. Chem.*, *67*:655 (1984).
192. H. B. Lee and A. S. Y. Chau, *J. Assoc. Off. Anal. Chem.*, *66*:1023 (1983).
193. H. B. Lee and A. S. Y. Chau, *J. Assoc. Off. Anal. Chem.*, *66*:1322 (1983).
194. Q. Y. Bai and C. W. Liu, *J. Assoc. Off. Anal. Chem.*, *68*:602 (1985).
195. M. Kelly, E. Zahnow, W. C. Peterson, and S. Toy, *J. Agric. Fd. Chem.*, *33*:962 (1985).
196. P. H. Bennett and P. R. de Beer, *Pestic. Sci.*, *15*:425 (1984).
197. P. K. Johnstone, I. R. Minchinton, and R. J. Truscott, *Pestic. Sci.*, *16*:159 (1985).
198. A. E. Smith, R. Grover, A. J. Cessna, S. R. Shewchuk, and J. H. Hunter, *J. Environ. Qual.*, *15*:234 (1986).
199. Anonymous, *Pesticide Analytical Manual*, Food and Drug Administration, Washington, D.C., 1983.

14

Analysis of Functional Groups in Soil by Nuclear Magnetic Resonance Spectroscopy

MICHAEL A. WILSON *CSIRO, North Ryde, New South Wales, Australia*

I. INTRODUCTION

Although soil organic matter has been studied extensively by spectroscopic methods, only recently have investigators turned to nulear magnetic resonance (NMR) techniques to understand structure. This is due largely to the fact that in the late 1960s and early 1970s problems associated with the low sensitivity of ^{13}C NMR spectroscopy were overcome. The technique of pulse Fourier transform spectroscopy uses multichannel excitation to allow many scans to be recorded and averaged in a comparatively short time. Much higher signal-to-noise ratios may be obtained than with conventional methods and this has allowed otherwise unobtainable spectra of complex mixtures to be recorded.

NMR is now useful in measuring a range of structural groups in soils and soil extracts, both in solution and the solid state. Nor are studies confined to the ^{13}C nucleus. ^{1}H, ^{29}Si, ^{27}Al, ^{15}N, and ^{31}P can be studied in soils (Table 1) and information on functional group content can be obtained readily.

However, operation of a new-generation nuclear magnetic resonance spectrometer requires a high degree of skill and training. The soil chemist does not normally have this background and will require the NMR specialist to produce a spectrum that he or she can interpret. Without prior instructions the spectroscopist will have to undertake a considerable amount of research before producing an acceptable spectrum, and unless the soil chemist can tell

TABLE 1 Nuclei That Have Been Studied Adequately in Soils by NMR

Nucleus	Special techniques and problems: Solution	Special techniques and problems: Solid state
^{1}H	Deuterated solvents necessary; HOD or H_2O peak needs to be irradiated.	Broad line with no chemical structural information.
^{13}C	Large numbers of scans required, short pulse delays, small pulse angles, large line broadening.	Large numbers of scans required, low-frequency ($\leq$ 25 MHz) instrumentation and rapid spinning speeds necessary, cross-polarization normally needed.
^{29}Si	Large numbers of scans required, small pulse angles, large line broadening.	Large numbers of scans required, sometimes long pulse delays needed.
^{27}Al	Some ^{27}Al lines so broad they are not observed.	Some ^{27}Al lines so broad they are not observed. Large (>7 T, ^{1}H > 300 MHz) fields needed.
^{15}N	Large numbers of scans needed, usually suitable for isotopic tracer studies only.	Large numbers of scans needed, usually suitable for isotopic tracer studies only.
^{31}P		Spinning sidebands present in all spectra.

the spectroscopist how the NMR experiment is to be performed, he or she is likely to be disappointed with the result.

The purpose of this chapter is to bridge the gap in knowledge between spectroscopist and soil chemist. Emphasis is placed on the nature of the instrumentation, not on interpretation that has been reviewed elsewhere [1–4]. An in-depth treatment of NMR instrumentation would cover several volumes and is not within the scope of this chapter. The reader is referred to two excellent texts on this subject [5,6]. Instead, the basic components of the NMR spectrometer are described and the various instrumental parameters discussed. The chapter concentrates on those instrumental factors that affect the spectrum of a sample of soil or soil-derived material. A sample containing a combination of organic macromolecules, inorganic aluminosilicates, and paramagnetics may never have passed between the poles of the spectroscopist's magnet before, and he or she will need guidance from someone who has a background in instrumentation as well as soil science.

II. BASIC PRINCIPLES

A. Nuclei Studied in Soils by NMR

When a sample containing suitable nuclei is placed in a magnetic field, the spin energy levels of the nucleus are split. The size of the splitting depends on the strength of the magnetic field H_0 (in tesla, T) and characteristics of the nucleus. Transitions (resonance) between the new energy levels can be brought about by electromagnetic radiation ν_0 in the rf range (MHz). The strength of the magnetic field and frequency of irradiation are related by $2\pi\nu_0 = \gamma H_0$ where γ is the gyromagnetic ratio and is characteristic of the nucleus. Thus, irradiating a sample at different frequencies can identify various elements. However, NMR can do much more than this. The electrons around a nucleus shield the nucleus from the magnetic field so that nuclei of the same element with different electron distributions resonate at different frequencies. Thus, carbon nuclei in aromatic structures resonate at a different frequency than those of carboxylic, methylene, or methyl carbon and this can be used to detect these functional groups in soils.

It should be understood that not all nuclei undergo nuclear magnetic resonance. The number of energy states created in the presence of a magnetic field depends on the spin quantum number I of the nucleus investigated. For ^{1}H and ^{13}C, $I = 1/2$ and the number of energy states is $2I + 1 = 2$. For a nucleus with $I = 0$, such as ^{13}C, there is only $2I + 1 = 1$ energy level, and hence, it is not possible to induce transitions by irradiating this nucleus. Thus, if the structure of carbon compounds is to be studied, the NMR experiment is carried out on the ^{13}C nucleus.

Theoretically, it should be possible to study any element that is present in a soil by NMR. However, there are a number of limitations. Some nuclei are intrinsically insensitive and if present in only trace concentrations cannot be detected. This is particularly troublesome for elements for which rare nuclides have to be studied. A further problem is that the width of the lines in the NMR spectra can be greater than the differences between individual resonance frequencies so that whatever the structure, only one broad line is obtained. As already noted, to date only a few nuclei (^{1}H, ^{13}C, ^{27}Al, ^{29}Si, ^{15}N, ^{31}P) have been studied successfully in soils, but others are almost certain to be exploited in the future.

B. Signal Averaging

There are two ways of observing the NMR phenomenon. The obvious method (continuous-wave NMR) is to sweep the rf over a range of values until resonance is observed. This method is quite adequate for simple proton NMR because the nucleus is so sensitive that one experiment is sufficient to obtain a spectrum of adequate signal to noise. However, if a rare nucleus such as ^{29}Si were to be studied by this method, the signal would not be detectable above the background noise. Instead, a large number of scans must be recorded and added (averaged) together so that the signal becomes proportionally larger than the background noise. Usually, sweeping the spectrum takes many seconds, so that a more efficient way of collecting the spectral data is to pulse the sample with rf irradiation causing all the nuclei to resonate at the same time. The signal detected in this way (Fig. 1a) is known as a free induction decay (FID) since it decays with time. Fourier transformation converts the data into the form of a continuous-wave spectrum (Fig. 1b). The advantage of Fourier transform NMR over continuous-wave NMR is that a large number of scans can be collected in a very short period of time.

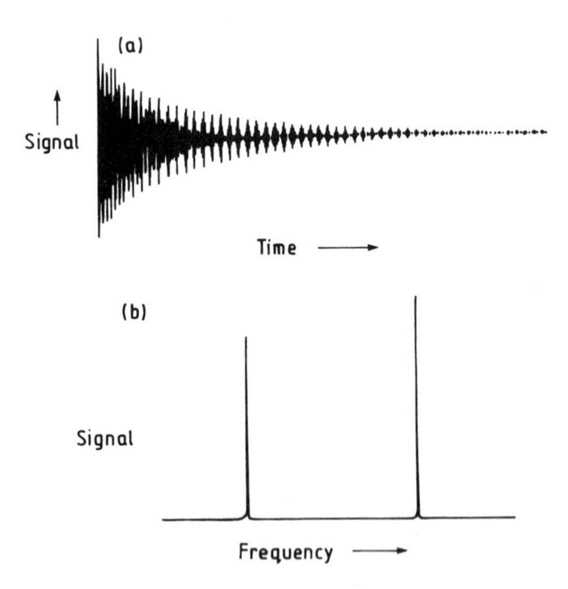

FIG. 1 Two forms of NMR spectrum. (a) Free induction decay; this is a plot of signal strength against time. (b) Continuous-wave spectrum or spectrum of (a) after Fourier transformation.

Large numbers of scans (10^6) must be collected to obtain a ^{13}C or ^{29}Si spectrum of a soil or a soil component such as a humic acid. Even for a readily observable nucleus such as a proton, it is desirable to collect at least a few hundred scans to obtain a spectrum of a humic acid. Continuous-wave NMR is therefore of little use to the soil chemist and only Fourier transform methods will be discussed below.

A prerequisite for adequate signal-averaging experiments is that a sufficient length of time be left for nuclei to relax back to their original equilibrium distribution after pulsing. To ensure complete relaxation, about three times the relaxation time (spin lattice relaxation time T_1) is left between pulses. However, because signal-to-noise ratios are often poor in experiments on soil organic matter extracts, it is necessary to pulse more rapidly than this. Hence, some structural groups that relax slowly may be diminished in intensity relative to others. Fortunately, many structures in soil extracts relax at the same rate, so that although complete relaxation is not obtained, data are still quantitative.

C. Solution Versus Solid-State NMR

Solid-state NMR has always been popular with physicists because it provides information on molecular motion, but only recently has it become possible to obtain the sort of structural information useful to chemists that can be gleaned from spectra consisting of more than one broad line. In contrast, solution NMR spectroscopy has been used to obtain structural information for over 30 years. High-resolution spectra can now be obtained in the solid state by a technique known as magic angle spinning. By rapidly spinning the sample at 54.7° to the magnetic field, the interactions between nuclei that give rise to broad lines in solid-state spectra are averaged to zero, so that in effect, the material behaves as if it were a liquid.

Today, the soil chemist must decide whether he or she wishes to obtain a spectrum in the solid or solution state. If the soil or soil component is not soluble in a suitable NMR solvent [$CDCl_3$, CCl_4, $(CD_3)_2SO$, NaOD, D_2O, $(CD_3)_2CO$], no decision is necessary: The experimentalist must use solid-state techniques. If a decision is to be made, it may be based on cost. Older broad line solid-state instruments without magic angle spinning facilities are unsuitable, and more recent instruments are expensive. On the other hand, solution instruments are widely available, although the best results in solution will be obtained on high field machines (e.g., 11.7 T or 500 MHz frequency for protons) and these are equally expensive. Adequate results in solution can be obtained at 2.3 T (100 MHz frequency for protons). As a general rule, it is preferable to use solution rather than solid-state techniques, but for ^{13}C and ^{29}Si sensitivity is paramount and thus the enhancement gained by exploiting the greater concentration of atoms between the poles of the magnet in the solid-state technique is significant. In addition, special techniques (which have some drawbacks) can give additional enhancements to signal intensities. Some of the special conditions for solid-state and solution work on soil components are listed in Table 1.

D. Sample Preparation

Since paramagnetic ions can cause broadening of NMR spectral features, ideally the sample should first be treated to remove these ions. For soils this is almost an impossible prerequisite, and spectra are obtained usually on samples containing considerable amounts of Fe^{3+} ions. It should be realized, however, that the presence of paramagnetic ions can affect quantitation and for some samples spectral lines become so broad that spectral features are obscured altogether. In these cases, it is necessary to remove Fe^{3+} or reduce it to Fe^{2+}. Dithionite appears to be a useful reagent for this purpose [7]. Complete demineralization may even be necessary for some studies.

The amount of sample needed for an NMR experiment depends greatly on the type and concentration of the nuclide under study. Typical requirements for nuclides studied in soils to date are listed in Table 2. In solution experiments the greater the quantity of material that can be dissolved in a suitable solvent the better, provided

TABLE 2 Amounts of Sample Required to Obtain Spectra of Soils or Soil Components

Nucleus	Solution	Solid state (mg)
^{1}H	$\geq$ 10 mg	—
^{13}C	As much as possible, preferably 1–2 g	300
^{29}Si	As much as possible, preferably 1–2 g	300
^{27}Al	$\geq$ 10–50 mg	5–300
^{15}N	As much as possible, preferably 1–2 g	300
^{31}P	$\geq$ 10–50 mg	5–300

the solution does not become viscous. In the solid state enough material to fill the sample holder, normally a "mushroom-" or "bullet-" shaped spinning device (rotor), is required. Most rotors in use take about 300 mg but new designs may hold more (see Sec. III). Obviously, the more sample used, the fewer the number of scans needed to obtain a spectrum. For solid-state studies the sample should be crushed to a fine powder and should be dry for adequate decoupling or cross-polarization to be achieved (see Sec. III.G).

Because of minor fluctuations in field strength and because the differences between the frequencies of resonances are so small compared with the magnet field strength, it is desirable to include in the sample an internal standard from which frequency differences (chemical shifts) can be measured. This internal standard (reference) is normally tetramethylsilane (TMS) for solution studies on ^{29}Si, ^{13}C, and ^{1}H but other standards are used for other nuclei (Table 3). In the solid state internal standards can be used, but more frequently a spectrum of a sample of known chemical shift is obtained immediately before and after the spectrum of the soil or soil component and used to set the chemical shift scale. If the standard used in solution is a liquid, the standard used in the solid state must be different. However, if the chemical shift of the solid standard in solution is known, the chemical shifts in the solid state can be corrected. Thus, the chemical shifts quoted are with respect to the chemical shifts for internal standards used in solution. If the solid-state standard is not an internal standard, the chemical shifts are not as accurate as solution chemical shifts

TABLE 3 Reference Standards for Chemical Shifts

Nucleus	Standard
^{1}H	$Si(CH_3)_4$ aqueous solutions $(CH_3)_3Si(CH_2)_3SO_3^-Na^+$, $(CH_3)_3Si(CH_2)_2CO_2^-Na^+$
^{13}C	$Si(CH_3)_4$ aqueous solutions, dioxane, or $(CH_3)_3Si(CH_2)_2COO^-Na^+$
^{29}Si	$Si(CH_3)_4$
^{27}Al	$AlCl_3 \cdot 6H_2O$
^{15}N	NH_3
^{31}P	H_3PO_4 (85% in H_2O)

since the magnetic field may change slightly between experiments. It should also be noted that chemical shifts are field-dependent and hence are normally quoted as a dimensionless ratio in parts per million rather than frequency units. Thus, chemical shift is defined as

$$\text{chemical shift } (\delta) = \frac{\text{frequency of sample} - \text{frequency of standard}}{\text{frequency of standard}} \times 10^6$$

Some special problems arise with obtaining chemical shifts of protons in extracts of soil components such as humic acids. In aqueous sodium deuteroxide or hydroxide, $(CH_3)_3Si(CH_2)_2SO_3^-Na^+$ is used usually as an internal standard. The CH_2 resonance can be distinguished readily from resonances of the sample. However, in spectra of humic or fulvic acids there are problems because of overlap with the broad lines. TMS, $[Si(CH_3)_4]$, is insoluble in sodium deuteroxide or hydroxide and hence cannot be used as an internal standard. However, a sealed capillary of TMS inserted into the sample is a suitable external reference. Calibration of small differences in chemical shift between external and internal reference can be achieved by measurements on compounds for which the chemical shifts are known.

III. SPECTROMETER

The principal components of a Fourier transform NMR spectrometer are outlined in Fig. 2. The physical arrangement is shown pictorially in Fig. 3. The spectrometer consists of a magnet; rf generating source; probe that contains the sample; detector; pulse programmer that controls timed events, and computer with an analog-to-digital converter. Also needed is a decoupler for irradiating one nucleus while observing another. A decoupler is necessary because the spins of nuclei may couple to create additional energy states that result in linesplitting or broadening.

The overall procedure used to obtain an NMR spectrum is summarized in Fig. 4.

A. Magnet

There are three types of magnets currently in use in modern NMR spectrometers. These are the permanent, electro, and superconducting (cryogenic) magnets. The permanent magnet is used mainly in cheap spectrometers unsuitable for soils work. The electromagnet (Fig. 5) has to be applied continuously from a stable power

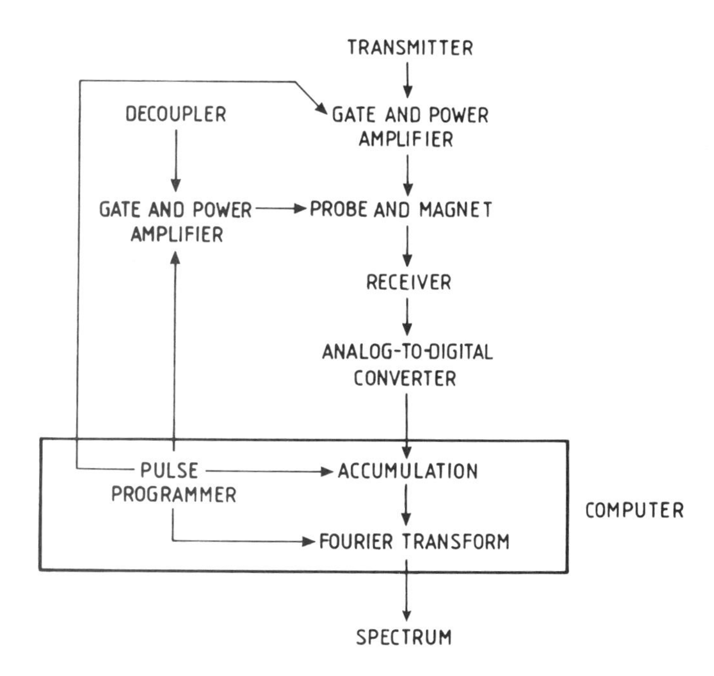

FIG. 2 Diagrammatic layout of a NMR spectrometer.

supply and excess heat removed by water cooling. About 2.3 T (100 MHz for protons) is the upper limit of the field of the electromagnet in NMR spectrometers.

The third type of magnet used is the superconducting magnet. This consists of a solenoid coil in a dewar of liquid helium. Within the coil a large electric current circulates continuously inside a jacket of cooling liquid nitrogen. The field at which superconducting magnets can be produced is continuing to increase but about 14.09 T (600 MHz frequency for protons) is the current limit.

The fields produced by the magnets described above are not sufficiently homogeneous to produce a good spectrum. Homogeneity is improved by spinning the sample and by using shim coils, a series of coils that generate small local field gradients. These gradients are of equal magnitude and opposite to the main magnet field gradients and therefore can homogenize the field.

The higher the field strength of the magnet, the better the ratio of signal to noise in the NMR spectrum. In most applications, high field magnets are more desirable than low field ones. This is

FIG. 3 300 MHz spectrometer operated by CSIRO at North Ryde, Sydney, Australia. (a) Transmitter; (b) superconducting cryogenic magnet, the probe is inside; (c) re-deiver display; (d) computer; (e) decoupler; (f) high-power source for solid-state de-coupling.

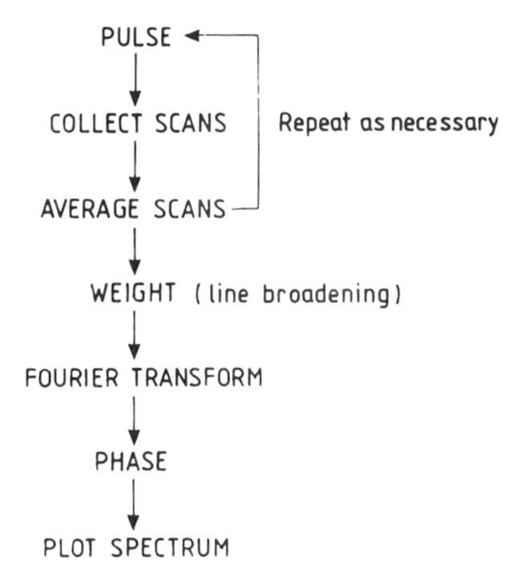

FIG. 4 Stepwise procedure for the NMR experiment.

particularly true for solution studies. Nevertheless, some relaxation mechanisms depend on the field strength used, so that with some very large molecules that reorientate slowly in solution, problems can be encountered with relaxation at high fields. Many soil organic substances may be of sufficient size to relax by field-dependent mechanisms, but no work has been done on soil solutions to compare relaxation at different field strengths.

For solids the choice of magnetic field strength depends on the nucleus under investigation. In obtaining a high-resolution solid-state spectrum, the principal aim is to remove interactions that give rise to line broadening. Some nuclei, such as ^{29}Si, have weak interactions that can be removed at low field strength. In these cases it is desirable to go to the highest field strength for the greatest sensitivity and resolution. In others, such as ^{13}C, interactions cannot be removed adequately at high field strength and hence lower field strengths are desirable. In contrast, for nuclei with quadrupole moments, e.g., ^{27}Al, the interactions decrease with field strength so that the low magnetic fields suitable for ^{13}C are inadequate.

The effects of field strength on solid-state spectra of some soil components and related materials are shown in Figs. 6 and 7. Note that at 7.05 T (75.5 MHz frequency) in the ^{13}C spectrum of a humic acid (Fig. 6b), the ^{13}C COOH signal is split into three resonances: one major signal and two spinning sidebands. However, at the lower field strength (2.11 T or 22.5 MHz frequency) there

FIG. 5 90 MHz electromagnet operated by CSIRO at North Ryde, Sydney, Australia.

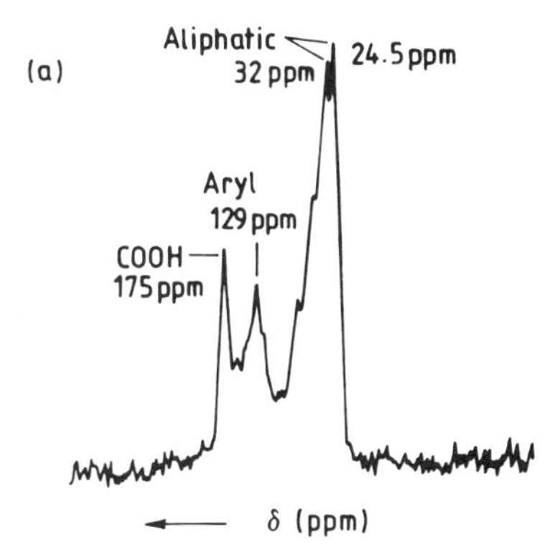

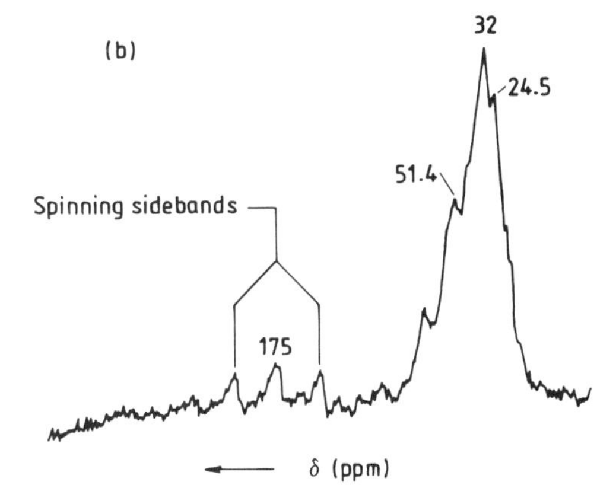

FIG. 6 ^{13}C solid state spectra of a humic acid obtained at two different field strengths. (a) 2.11 T, 22.5 MHz frequency; (b) 7.046 T, 75.5 MHz frequency. The peaks marked in (b) are spinning sidebands of the COOH resonance.

is only one resonance. In contrast, in ^{27}Al spectra (Fig. 7) the sidebands (SSB) are smaller and more clearly resolved at 11.74 T (Fig. 7b) than at 7.05 T (Fig. 7e) so that spectra are interpreted more easily at the higher field strength. Table 4 lists the most suitable field strengths for experiments on the major nuclei studied in soils.

B. Probe

The probe holds the sample and also the coils for transmitting and receiving the NMR signal. Usually, a single coil design is used for

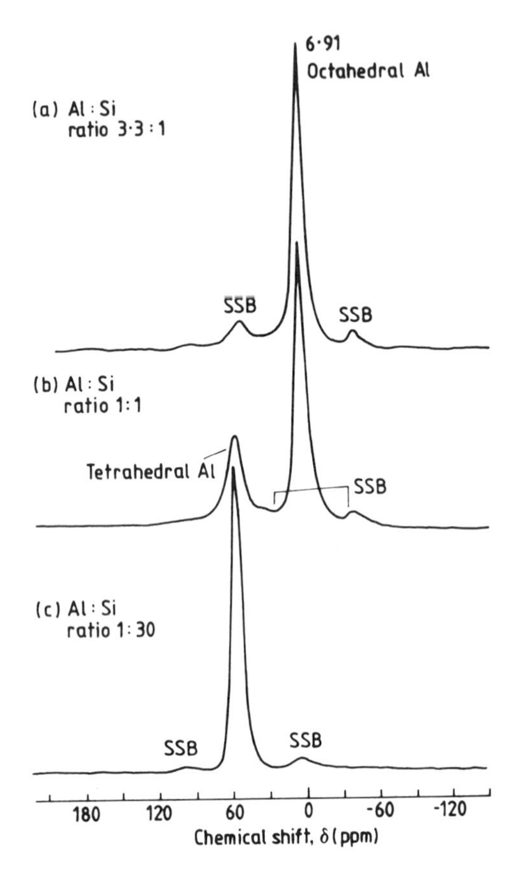

FIG. 7 ^{27}Al solid-state spectra of aluminosilicates obtained at (a–c) 11.744 T, 130.2 MHz frequency and (d, e) 7.046 T, 78.2 MHz frequency. (In part from Ref. 29.)

both purposes. An additional coil may also be provided for decoupling experiments.

To obtain a spectrum, it is necessary to have a very stable excitation frequency and magnetic field but this is not possible in nonsuperconducting magnets. To overcome this problem, a field/frequency lock that detects minor fluctuations with respect to a reference is used. This lock works by applying correcting signals to the magnet when fluctuations are detected in the field observed between the lock and a stable reference frequency. The lock material can be incorporated with the material under investigation (internal

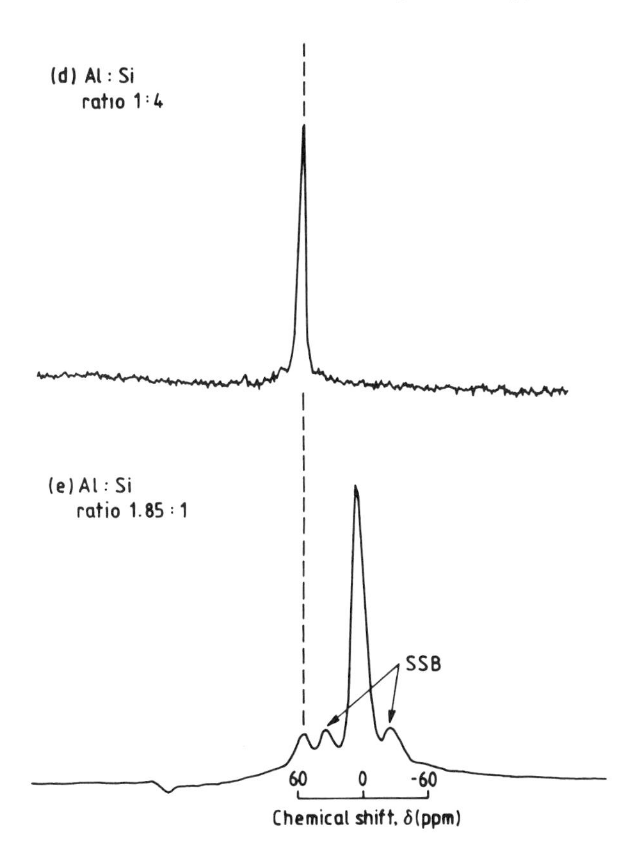

FIG. 7 (Continued)

lock) or externally (external lock). Deuterium in a deuterated solvent is used normally as an internal lock, whereas lithium is popular as an external lock.

The probe design for high-resolution solid-state studies is quite different from that used in solution studies. As already noted, nuclei in a single functional group experience a range of interactions in the solid state due to coupling with protons and also due to the fact that molecules have different orientations in the magnetic field. That is, the nuclei experience slightly different magnetic field strengths because of their various orientations (termed chemical shift anisotropy). Magic angle spinning at 54.7° removes chemical shift anisotropy. Speeds needed vary depending on field strength: About 3600 cycles/sec are needed at 2.11 T (22.5 MHz

TABLE 4 Suitable Magnetic Fields (Expressed as Frequency Needed for ^1H Irradiation) for Solid-State Studies of Nuclei of Interest in Soil Science

Nucleus	Suitable field strength (T)	Frequency at field strength quoted (MHz)	Notes
^1H	not established	—	At present, multiple pulse sequences needed to narrow lines
^{13}C	2.11	22.6 ($^1H \leq 90$ MHz)	Sidebands increase at higher fields.
^{29}Si	2.11	17.9 ($^1H \geq 90$ MHz), higher the better	
^{27}Al	7.05	78.2 ($^1H > 300$ MHz)	Linewidths and sidebands decrease with an increase in field strength.
^{15}N	2.11	>9.11 ($^1H >$ 22.5 MHz)	Some interactions are field-dependent.
^{31}P	not established		Sidebands at all field strengths.

^{13}C frequency). In older spectrometer probes the sample is placed in a mushroom-design rotor (Fig. 8a) that fits into a stator (Fig. 8b), but a considerable amount of research has gone into developing rotors with double bearings to increase stability, and these are now the rotors of choice.

It is essential to choose rotors that do not give background signals. Thus, the more readily available delrin and deuteroplexiglass rotors are not suitable for ^{13}C studies on soils. We have adopted a design based on a Kel-F tip and boron nitride base for routine work. Most rotors hold only about 300 mg of sample, although in our laboratory we have also used mushroom rotors made from a zirconium ceramic that have very thin walls and can take up to 600 mg. Solution studies are carried out usually in glass tubes but these are not always suitable for soil extracts. When the ^{29}Si nucleus is to be examined, plastic or some other suitable container that does not contain silicon must be used.

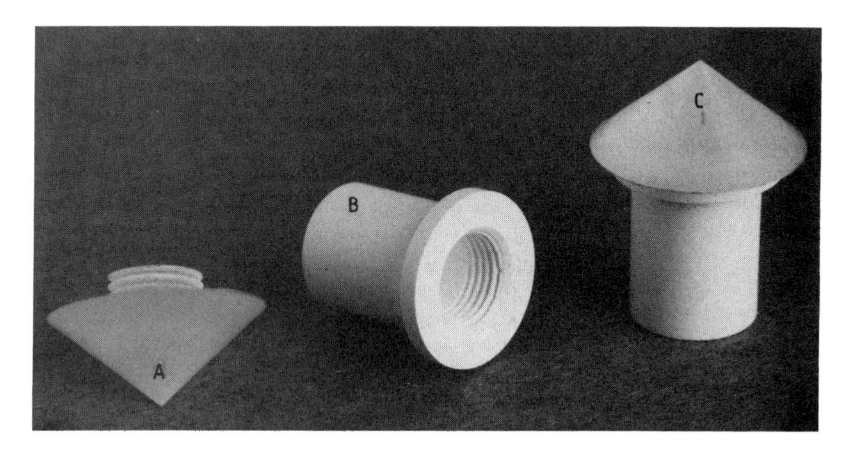

(a)

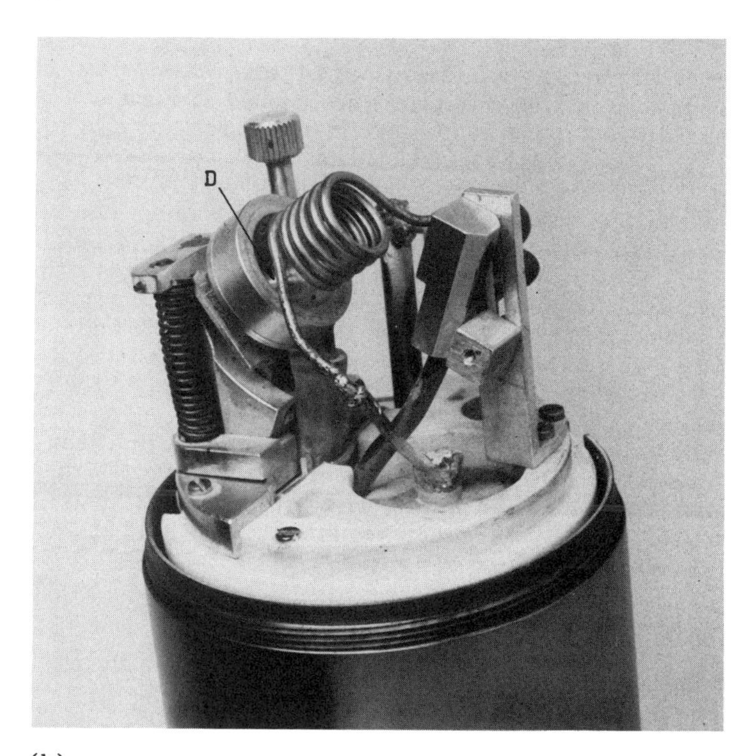

(b)

FIG. 8 (a) Rotor for Andrews–Lowe spinner. The design shown, which was developed at the University of Utah by Drs. Zilm, Pugmire, and Grant, has a Kel-F tip and boron nitride base. A, rotor tip; B, rotor base; C, assembled rotor. (b) Stator for A–L spinner. Air jets D provide an air bearing on which the rotor shown in Fig. 8a spins.

C. Setting Spectral Width, Acquisition Time, and Pulse Width

Before the NMR experiment is started and scans are acquired, it is necessary to set a number of spectral parameters. Since acquisition time and frequency are related by Fourier transformation, the spectral width of the spectrum obtained after Fourier transformation is determined by the time for which the free induction decay is acquired (the acquisition time). The spectrometer computer automatically sets the spectral width from acquisition time and vice versa. Sufficient spectral width must be chosen to observe all the lines in the NMR spectrum and to obtain a straight baseline on either side of these peaks. Normally, the irradiation frequency is set such that the lines of the spectrum are offset about the center of the irradiation pulse. To sample the FID for Fourier transformation, a number of data points must be used. This is preset in the computer as 4, 8, 16, or 32K.

If two resonances are close in frequency, then small spectral widths and large numbers of data points are needed to resolve the peaks. For example, Table 5 shows that to obtain resolution of two peaks 0.003 Hz apart, 32,768 (32K) data points are needed with a spectral width of 50 Hz. On the other hand, if only 4.88 Hz resolution is needed, the spectral width can be 10,000 Hz and 4096 (4K) data points can be used. Small spectral widths require long acquisition times, and to sample a large number of data points, the computer also needs long acquisition times. Thus, the spectroscopist must set an acquisition time to suit both the spectral width and number of data points (Table 5). In practice, the spectral width and data points are first set manually and the computer sets the acquisition time.

For soil organic matter, spectral lines are broad so acquisition times can be short and spectral widths wide. Whereas most high-resolution spectra are obtained in 8192 data points or more, only 4096 (4K) points are usually necessary to obtain a humic acid spectrum.

D. Choice of Pulse Width

The length of time for which the pulse is left on during an NMR experiment is the pulse width. During this time the nuclei are excited by transmission of the electromagnetic radiation. Pulse widths are also expressed in terms of pulse angles. To understand the concept of pulse angle, it is necessary to describe the NMR experiment by vector diagrams.

When a nucleus is placed in a magnetic field, the magnetic moment μ aligns with the direction z of the magnetic field H_0; but the angular momentum of the nucleus makes it precess about z, much as

TABLE 5 Relationship Between Spectral Width, Acquisition Time, and Number of Data Points

Spectral width (Hz)	Acquisition time (sec)			Resolution (Hz)		
	4K	8K	32K	4K	8K	32K
50	40.960	81.920	327.680	0.02	0.01	0.003
100	20.480	40.960	163.840	0.05	0.02	0.006
500	4.096	8.192	32.768	0.24	0.12	0.03
1000	2.048	4.096	16.384	0.49	0.24	0.06
5000	0.410	0.819	3.277	2.44	1.22	0.30
10,000	0.205	0.410	1.638	4.88	2.44	0.61

a spinning top precesses about a point (Fig. 9a). It is normal to change the reference coordinates of the magnetic moment from the laboratory frame x, y, z to a frame rotating in x', y', z' at the precessional frequency. This is not such an unusual step as one might expect, since the laboratory frame in itself is a rotating frame that is rotating about the center of the earth at 72.7×10^{-6} rad/sec, i.e., once a day. In the new frame of reference (x', y', z'), μ is static. Since a large number of nuclei are present in a real sample, the net magnetization M, representing the sum of all the moments μ at different points around the cone of precession (Fig. 9a), is static and along z' (Fig. 9b).

How then does the magnetic field of the electromagnetic radiation used to produce resonance interact with the moment of the nuclei? At resonance the field due to the electromagnetic radiation (H_1) also rotates at the frequency of the rotating frame and hence can be regarded as static in the rotating frame along x' (Fig. 9c). Now when the rf field H_1 is applied, it will operate on M so that M will change its orientation with respect to the static field. The amount by which M is moved will depend on the intensity and duration of the applied field H_1. If the applied field turns M by 90° (Fig. 9c), then a 90° pulse is said to be applied. After H_1 is turned off, the magnetization will return to the z' direction. The spins may exchange spin energy with each other during this process (spin-spin relaxation T_2), but eventually, they realign along the z' direction by interacting with the surroundings (spin lattice relaxation T_1).

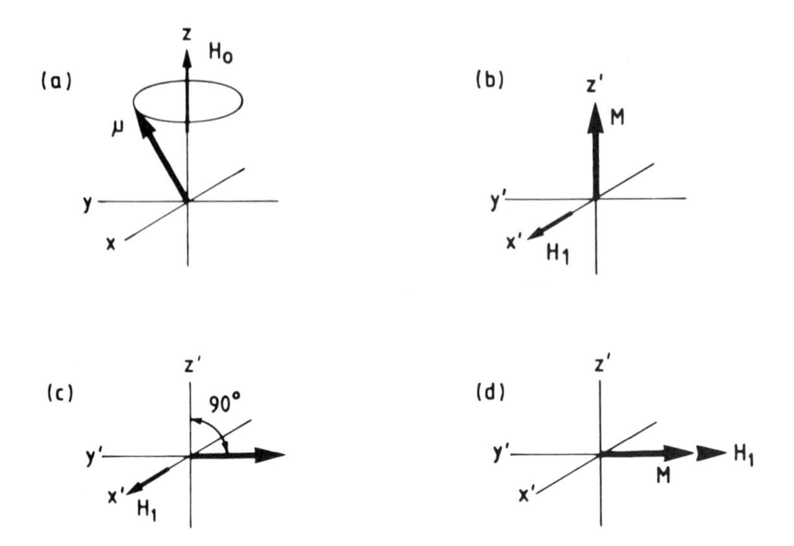

FIG. 9 Vector diagrams to describe the behavior of the net magnetization of a single μ or collection M of nuclear spins in a static magnetic field H_0. (a) μ precesses around laboratory frame coordinate in the direction of the static magnetic field; (b) corrdinates x, y rotated at rate of precession of M; (c) after applying a rf-induced magnetic field H_1 along rotating coordinate x' to produce a 90° pulse; (d) spin locking.

It is not necessary to give a sample a 90° pulse to obtain an NMR spectrum. Nor does a 90° pulse always produce the most signal intensity. If the nuclei are not allowed to relax fully between scans, then a pulse less than 90° is desirable. The relationship between pulse angle, relaxation time T_1, pulse delay time T_{delay}, and signal intensity is shown in Fig. 10a. Equation (1) gives the best criterion to decide on the optimum pulse angle α and is shown graphically in Fig. 10b.

$$\alpha = \cos^{-1} [\exp(-T_{delay}/T_1)] \tag{1}$$

It can be seen that when the ratio of the pulse delay to spin lattice relaxation time is short (0.02), a small pulse angle (10°) pulse is optimum.

Setting an optimum pulse angle requires knowledge of the relaxation time of the substance being studied. This requires extensive

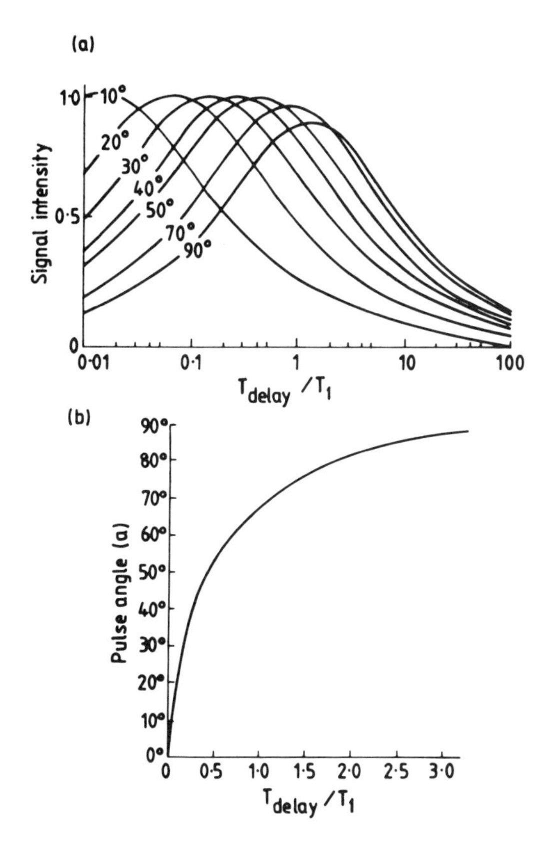

FIG. 10 (a) Relationship between pulse angle, signal intensity, and the ratio of pulse delay time T_{delay} and spin-lattice relaxation time (T_1). (b) Optimum angle (Ernst angle) as a function of T_1 and pulsing delay.

investigation. Moreover, different structural groups in a sample will have different relaxation times. Fortunately, work by a number of groups [8–15] has demonstrated clearly that the relaxation times of humic substances are much shorter than those of pure compounds, so short pulse delays can be used. For example, Table 6 lists relaxation times for resonances of the fulvic and humic acids investigated by Preston and Blackwell [11].

Reasons for the short spin lattice relaxation times of humic substances have not been established. An obvious candidate is paramagnetic ions that are known to relax samples rapidly, although

TABLE 6 Chemical Shifts (δ), Spin-Lattice Relaxation Times (T_1), and Nuclear Overhauser Enhancements (η) for Selected Peaks of the ^{13}C Spectra of Humic and Fulvic Acids

δ (ppm)		η		T_1 (sec)	
HA	FA	HA	FA	HA	FA
176	180	0.13	0.05	1.7	0.05
132	103	0.39	0.42	0.4	0.42
76	74	0.54	0.51	0.5	0.51
65	64	0.59	0.77	0.8	0.71
59	58	0.63	0.63	0.8	0.63
41	52	0.62	0.26	0.6	0.26
37	41	—	0.75	0.6	0.75
32	25	0.60	0.45	0.6	0.45
27	20	—	1.0	0.5	1.0
25		1.13		0.3	
21		0.77		0.5	
15		0.44		—	

Newman and Tate [15] have found no relationship between the paramagnetic (Fe^{3+}) content of humic extracts and relaxation time.

Because so many transients are needed to obtain a humic spectrum of adequate signal to noise, the pulse delay left between scans is usually very short, much shorter than the relaxation time. Therefore, for maximum sensitivity, short pulse lengths of 45° or less are used normally in solution studies of soil organic matter. Similar arguments can be applied to solid-state studies, except that when cross-polarization techniques are used (see Sec. III.G), at 90° pulse must be applied. Some nuclei such as ^{27}Al can be excited to a number of energy levels. It is the energy level transition $I = 1/2 \rightarrow -1/2$ that produces high-resolution spectra, and to ensure that only this transition is excited, very short pulse times (1–2 μsec) which produce small pulse angles (<20°) are used.

E. Receiver and Transmitter

Once the NMR experiment is started, the rf irradiation is fed from the frequency generator to a transmitter whose function is to amplify the signal and apply the signal to the transmitter coil. The transmitter also contains a gate to turn the rf radiation on or off as required so that the pulse width and pulse angle can be varied. The precession of the nuclei after excitation induces a voltage in the receiver coil. This voltage is amplified by a preamplifier that is usually as close to the receiver as possible; in many cases it is part of the probe. After further amplification in the spectrometer console, the signal is detected by two phase-sensitive detectors as a free induction decay (Fig. 1a). As already noted, the time for which the data are acquired, the acquisition time, depends on the number of data points used and the spectral width.

F. Computer, Pulse Programmer, and Interfaces

The pulse programmer is responsible for controlling timed events such as the pulse times, turning the decoupler off and on (see Sec. III.G), and the delay between pulses, and is thus coupled to the computer. In a typical experiment the required number of scans to be collected is preset in the spectrometer computer. After starting the program, i.e., running the experiment, the required number of scans are collected and, on completion, the free induction decay (FID) is Fourier-transformed after conversion from analog-to-digital form. At the expense of resolution, the computer may be used to improve the signal to noise by preweighting the FID by multiplying it by an exponential function (termed line broadening or the window function). In ^{13}C spectra of soil samples, the signal to noise is poor and it is necessary to use line broadenings of 50 Hz or greater, at least 50 times greater than that used for simple organic substances. Insufficient line broadening is a common cause of poor spectra in soil chemical investigations.

After Fourier transformation, some of the resonances will appear as peaks above the baseline of the spectrum and some as peaks below. This is a result of phase shifts introduced by the Fourier transform process and they can be adjusted by the spectrometer computer, by the process of phasing, to produce an acceptable plot of signal intensity versus frequency. In some advanced methods using complex pulse sequences, involving turning the decoupler on and off at specific times, additional structural information can be gleaned from phasing information.

G. Decoupler and Its Use in Spectral Editing

Many nuclei, e.g., ^{13}C nuclei, are in close proximity to protons. The spins of these protons can be aligned with or against those of the nucleus under study. Thus, each energy level of the nucleus is further split due to the orientation of the adjacent proton spins, so that more than one spectral line is obtained for a given nucleus. In a liquid this multiplicity can be quite simple and an aid to assignment. For example, if one proton is adjacent to a ^{13}C nucleus, then two lines for the ^{13}C nucleus should result since the proton may be aligned with or against the spin of the ^{13}C nucleus. This is called J-coupling. In a solid the situation is much more complicated because of the various orientations of the protons and lack of translational motion of the molecules. The energy levels are also split because of the strong dipolar forces across the C—H bond. This is called dipolar coupling.

To overcome these problems, it is essential to irradiate all the protons in a solid sample to produce single lines for each nucleus in different structural groups. In practice, for complex materials like soils or soil extracts broad lines are obtained even with irradiation, because there are so many different types of nuclei in a given region of the spectrum.

The process by which the interaction with protons is removed is called decoupling and is essential to obtain a simple spectrum of the substance. Decoupling involves irradiating the protons in the sample so that all the protons are, on average, at the same energy state and hence the carbons experience the presence of only one type of proton spin. Normally, all protons are irradiated (the process is called broadband proton decoupling), but it is also possible in solution studies to decouple just one type of proton from ^{13}C or any other nucleus (heterodecoupling) or even from the other protons (homodecoupling).

Decoupling has one important side effect. When protons are irradiated, the populations of the carbon energy levels are also changed. Those carbons in close proximity to protons are affected more greatly so that the signal intensity from these carbons is greater than that from carbons remote from protons. This effect is termed the nuclear Overhauser effect (NOE) and the enhancement of signal intensity is termed the nuclear Overhauser enhancement (η). Because some carbon signals are enhanced by proton decoupling, and others are not, NOE can affect quantitation. In order to reduce the enhancement effect, the decoupler may be turned off during the pulse delay and used only when collecting the FID. This process is termed inverse gated decoupling. Gated decoupling is the reverse, i.e., when the decoupler is turned off only during data acquisition.

Nuclear Overhauser enhancements in NMR experiments on humic substances have been studied by a number of groups [8,11,15]. The most comprehensive work is that of Preston ahd Blakewell [11]. All samples were observed to experience an NOE, and this can contribute to an increase in aliphatic and aromatic resonance signal relative to carboxyl signal. However, there was little effect on the aromatic to aliphatic ratio, so for this measurement suppression of

NOE is not necessary. This result is surprising because NOEs can often be used to make assignments of the degree of protonation. Clearly protonated carbons should experience significant NOEs and nonprotonated carbons should not. The unusual behavior of humic substances may be related to their molecular motion in solution. The observed enhancements are all less than the theoretical maximum η value of 2, and this may be because of slow motion in solution (as experienced by macromolecules) or because some relaxation mechanism that does not involve interaction with protons is also operating.

The second type of decoupling experiment that is useful in studying soil organic substances in solution is proton homonuclear decoupling. For example, this has allowed the identification of lactic acid units in some humic extracts [16] (Fig. 11). In these experiments one particular proton species is irradiated to determine which type of proton is coupling with another. The lactic acid [$CH_3CH(OH)COOH$] ^1H spectrum contains resonances only from CH_3 and CH because the OH and COOH protons exchange with solvent. The CH_3 peak is split into a doublet because the adjacent H spin of the CH group can be aligned with or against the methyl protons. Likewise, the

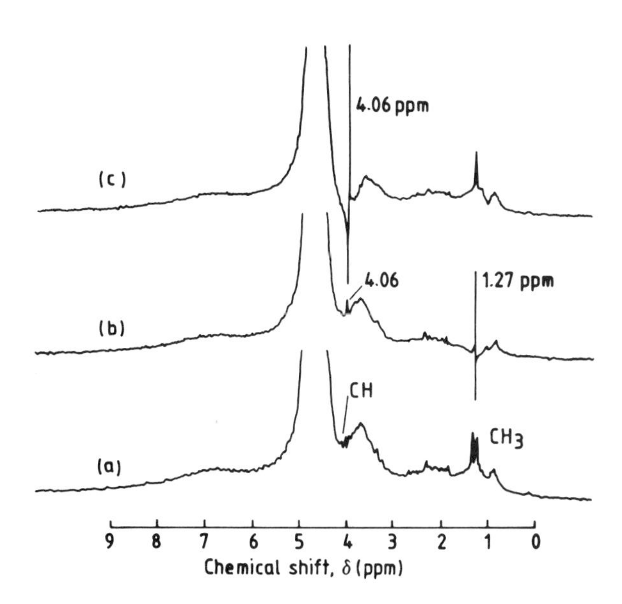

FIG. 11 ^1H NMR spectrum of humic extract containing lactic acid groups. (a) Spectrum without homonuclear decoupling; (b) irradiation at 1.27 ppm; (c) irradiation at 4.06 ppm.

CH proton consists of four lines because the CH_3 protons can be coupled in a total of four different ways. Figure 11 shows that when the CH proton is decoupled, the CH_3 groups reduce to a singlet. Likewise, the CH peak becomes visible as a singlet when the CH_3 peak is irradiated. Thus, on irradiation at 4.06 ppm (Fig. 11c) the signal at 1.27 ppm reduces to a singlet and on irradiation at 1.27 ppm a singlet appears at 4.06 ppm (Fig. 11b).

Homonuclear decoupling is also useful to remove water peaks from proton spectra. Unfortunately, there are so many protons in the sodium deuteroxide solutions used to obtain spectra from humic substances that the spectra can be hidden by the large water peak [17]. Irradiating the protons reveals spectral features (Fig. 12). Indeed,

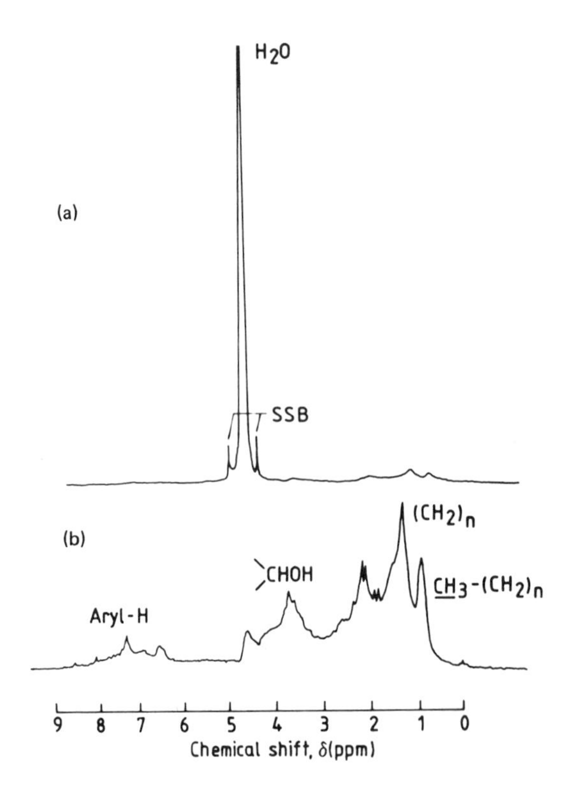

FIG. 12 (a) Attempted nondecoupled ^1H spectrum of humic acid. Only a water peak plus spinning sidebands (SSB) are detectable. (b) ^1H spectrum of humic acid after irradiation of the water peak, spectral features become visible.

resolution of small peaks can be lost because of problems with the analog-to-digital conversion of all the spectral information if the water peak is not irradiated.

Decoupling is also essential in the solid state. However, the decoupler is not truly broadband and the power may drop off greatly on either side of the irradiation frequency. Generally, there is sufficient power to decouple all protons, but methylene (CH_2) carbon is coupled strongly and unless the decoupler resonance is set carefully to the center of the proton spectrum, the polymethylene resonances become broad. Resolution may also be lost (see Fig. 13; note that

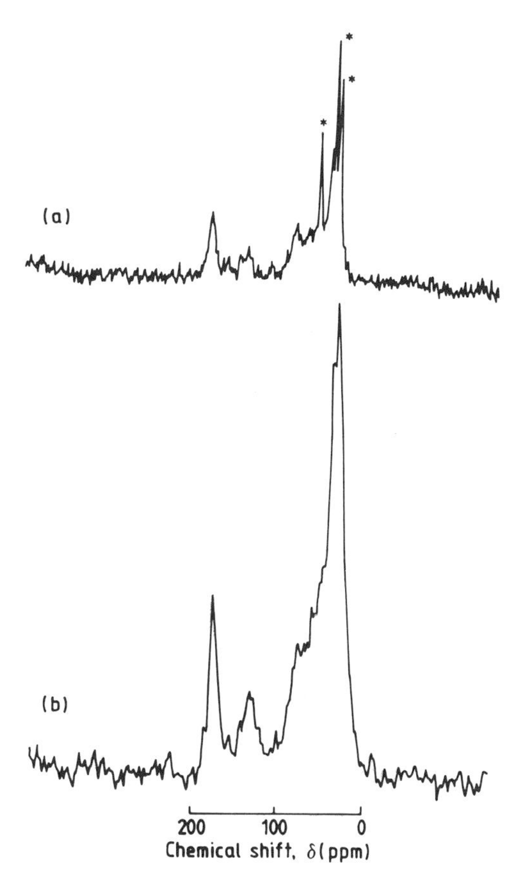

FIG. 13 Solid-state high-resolution ^{13}C NMR spectra of a humic preparation using different offsets. (a) ^{1}H decoupler on resonance; (b) 5000 Hz offset.

off-resonance decoupling results in the loss of resolution of the peaks marked *).

The decoupler is also used to transfer polarization from protons to other nuclei in the solid state. This technique, called cross-polarization, is used to gain signal intensity enhancements (i.e., better signal-to-noise ratios) and also to allow carbons to relax at a much faster rate, close to the intrinsic proton relaxation rate.

The cross-polarization technique is based on energy transfer from protons to carbons. If a sample containing both ^{1}H and ^{13}C nuclei is placed in a magnetic field, energy levels are split, but the difference depends on a constant that is characteristic of the nucleus (the gyromagnetic ratio). In the same magnetic field the energy levels of the protons and carbons, which are different because of the different gyromagnetic ratios, separately undergo magnetic resonance and are excited by different rf's. Normally, energy is not transferred between the protons and carbons. If, however, the energy of the two systems can be equalized, energy can be transferred from protons to carbons and vice versa.

Equalization is achieved by creating a variable magnetic field on the protons. The first step is to apply a 90° pulse on the protons along y' (Fig. 9c) by a rf field H_1. The direction of the proton irradiation is now changed so that it is along y' (Fig. 9d). In effect, the proton polarization is now trapped along y' much as it was along z' before the 90° pulse. If the strength of the field H_1 is now adjusted, the energy levels can be made similar to those for carbons and the polarization is transferred to the carbons. There is a signal enhancement (the ratio of the gyromagnetic ratios), but more important, the carbons relax back through protons. This can be 1000 times faster than in a simple carbon experiment without cross-polarization.

Transferring polarization from protons to carbons takes a particular time (termed the cross-polarization time T_{CH}), and carbons remote from protons, e.g., quaternary carbons, take longer to cross-polarize than carbons close to protons. It is essential to allow all carbons to cross-polarize effectively before data acquisition, i.e., leave them in contact for a sufficient time, the contact time, but unfortunately some carbons may relax through protons faster than other carbons can cross-polarize. Hence, quantitative data cannot be obtained unless corrections are made for relaxation [3,16]. This effect is illustrated by Fig. 14, which shows the spectra of humic and fulvic acids obtained at several different contact times. The carbohydrate resonance (72 ppm) clearly is diminished in intensity at long contact times (5 msec, Fig. 14a) than at shorter contact times (0.8 msec, Fig. 14b). In Fig. 14c, obtained at 0.5 msec, the carbohydrate resonance is even bigger, but the other carbons have not fully cross-polarized because the contact time is not long enough. To obtain the correct proportions of functional groups, it is necessary to extrapolate the relative signal intensities back to a contact time of zero [2,16]. In some cases, relaxation may be so fast that some carbon is not observed at all. It is clear that paramagnetics are particularly important in selectively relaxing functional groups in cross-polarization experiments. In particular, those structures that can bond to paramagnetics (e.g., OH in carbohydrate) seem to be most affected [13,18]. Nevertheless, if only a single spectrum is to be obtained, a 1 msec contact time is optimum for most humic substances.

Polarization transfer has also been used successfully in solution. When two different types of nuclides are irradiated, polarization transfer can occur from one nucleus to another through spin-spin interactions (i.e., the J-coupling). This has become a valuable technique for enhancing carbon signal intensities but can also be used for distinguishing protonated and quaternary carbon in solution spectra. Quaternary carbons do not have J-coupling and therefore do not cross-polarize. (This is different from the solid state in which dipole coupling is involved.) The most established technique has been called INEPT ("insensitive nuclear enhancement by polarization transfer") [19,20]. An improved version, DEPT (distortion-free nuclear enhancement by polarization transfer) has also been developed [21]. INEPT or DEPT can be used to distinguish CH, CH_2, and CH_3 carbons as well as quaternary carbons. After cross-polarization, a series of pulses (Fig. 15a) can be used to distinguish the carbon types. As an example of the way INEPT works, consider the case of a methine CH group in a system in which both C and H are precessing at their resonance frequencies (i.e., the chemical shifts). For protons the coupling of the carbon produces two vectors (Fig. 16a) in the rotating frame, and if the irradiation frequency is between the chemical shift of the two resonances, one vector will be rotating slightly faster than the rotating frame and one slightly slower. The difference between the frequencies is the coupling constant J, so that one vector rotates at $J/2$ faster than the irradiation frequency (i.e., the rotating frame) and one vector at $J/2$ slower than the irradiation frequency. The difference between the vectors is J Hz/sec or $2\pi J$ rad/sec. After a 90° pulse, these vectors will be in the x',y' plane (Fig. 16b). After a period of time τ the faster-moving vector will become 180° out of phase with the slower-moving vector (Fig. 16c). (Usually, a 180° pulse on x' is applied in the middle of dephasing to refocus spins, but details of this are not necessary for a general understanding of the method.) To travel through 180° or π radians, the time taken will be $\pi/(2\pi J) = 1/(2J)$ sec. If a second 90° pulse is now applied on y' (Fig. 16d), the two vectors are rotated so that they lie along the z' axis. However, one vector is inverted with respect to the

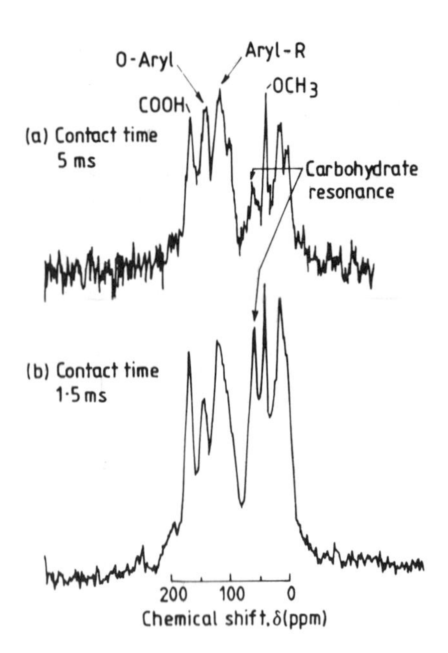

FIG. 14 Cross-polarization magic angle spinning ^{13}C NMR spectra of (a, b) humic acid and (c, d) fulvic acid from Georgia at different contact times. (From Ref. 16.)

other. This polarization of the protons is transferred by spin exchange to the carbons. It is important to realize that two carbon vectors are created and thus the coupling information of the protons is transferred to the carbons. It is from this information that assignments to the CH, CH_2, CH_3, or C carbon can be made [22–27] after various further pulses and dephasing (Figs. 16f and 16g).

There are other sequences, SEFT (spin echo Fourier transform [24,25]), PCSE (partial coupled spin echo), and GASPE (gated spin echo [22,23,26]), which give the same information, but the polarization of carbon is direct and not from protons. Thus, signals from quaternary carbons are also seen. In principle, these sequences and INEPT and DEPT are the same in that they rely on the fact that refocusing the carbon vectors containing coupling information means leaving a time period for refocusing that depends on J. For GASPE the actual signal intensities at time 2τ are given by [26]

$$\text{quaternary carbon } M = M_0 \exp(-2\tau_1/T_2) \tag{2}$$

(The simple NMR exponential decay equation)

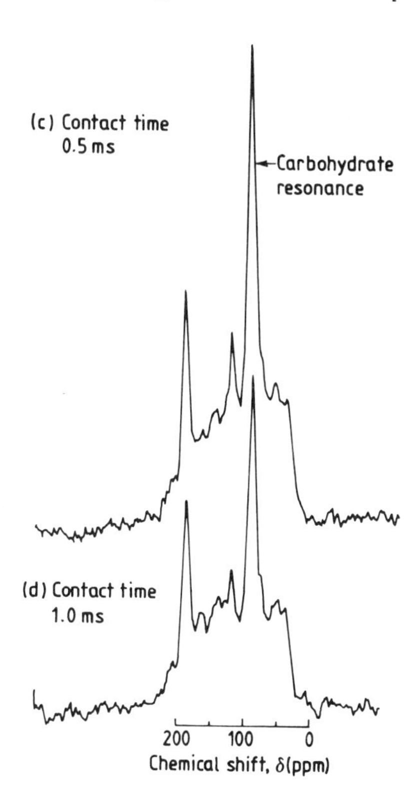

FIG. 14 (Continued)

$$\text{CH carbon } M = M_0 \,(\exp{-2\tau_1/T_2})\cos \pi\tau_1 J \quad (3)$$

$$\text{CH}_2 \text{ carbon } M = M_0 \,(\exp{-2\tau_1/T_2})(0.5 + 0.5 \cos 2\pi\tau_1 J) \quad (4)$$

$$\text{CH}_3 \text{ carbon } M = M_0 \,(\exp{-2\tau_1/T_2})\,(0.75 \cos \pi\tau_1 J + 0.25 \cos 3\pi\tau_1 J) \quad (5)$$

where τ is the delay period with the decoupler off before, and on after, a 180° pulse on x' (Fig. 16b). It is clear from these equations that the observed magnetization M for the different structural groups will be different at a given value of τ. For example, at $\tau = 1/(2J)$ then $\cos\pi\tau J = \cos\pi/2$ (rad) = 0, $M = 0$, and there is no signal from the CH carbon. Quaternary carbon will, however, still be in the spectrum and can thus be distinguished from the CH carbon.

Delay settings for INEPT and GASPE are different because in INEPT the two ^{13}C vectors begin 180° out of phase to each other,

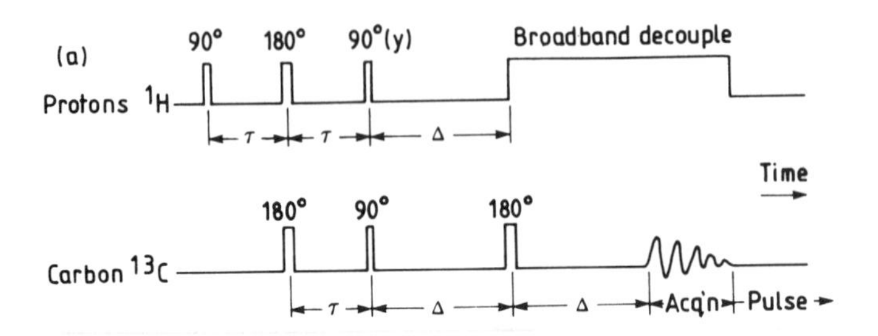

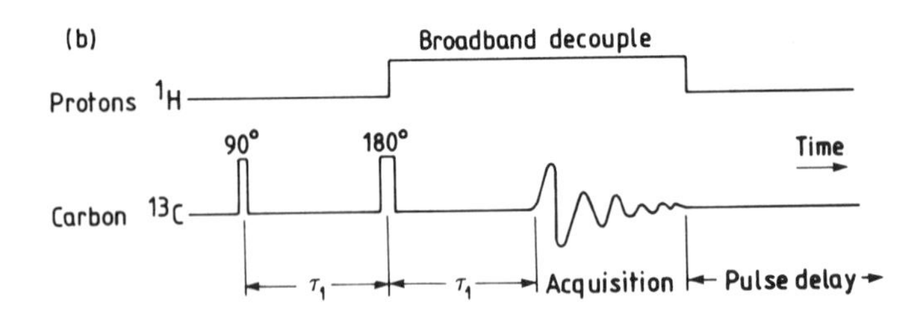

FIG. 15 (a) INEPT and (b) GASPE pulse sequences to distinguish the number of protons directly bound to carbons. Pulse sequence (a) is only one example from the range of INEPT sequences now available.

whereas in GASPE they dephase after an initial 90° pulse on carbons. The delays for these two sequences to obtain spectra edited for different functional groups are listed in Table 7. Some resonances can have negative values of *M* and hence are inverted. Preston and Blackwell [11] have used spectral editing techniques to assign C, CH, CH_2, and CH_3 resonance observed in some humic acids and fulvic acids. Typical results are shown in Fig. 17.

In solids, J-coupling is not seen because of the larger dipolar coupling. Thus, INEPT and GASPE or similar sequences cannot be used to assign the degree of protonation of carbon atoms. Moreover, assignments cannot be made from NOEs because the decoupler must always be gated off during pulse delay; otherwise, the high power needed to decouple protons in solids would destroy the probe. However, the decoupler can be used in an additional way to distinguish CH and CH_2 carbons from nonprotonated carbons and methyl groups. The method is to turn off the decoupler for a short period after cross-polarization, but before data acquisition, so that the CH and CH_2 carbons relax rapidly and thus are not observed. This technique, termed dipolar dephasing, has been used to distinguish protonated and nonprotonated carbons in soils and humic

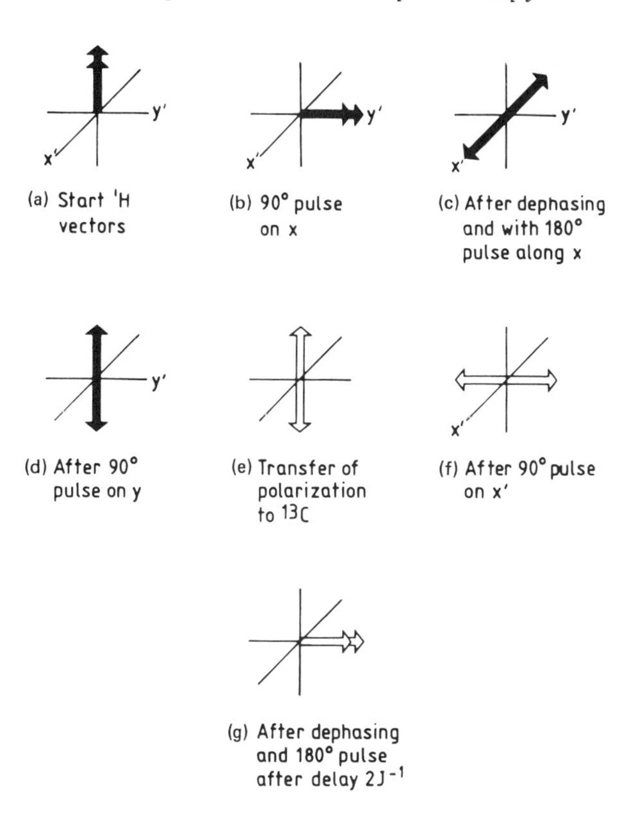

FIG. 16 Vector diagram to show the INEPT experiment. (a) Proton vectors in z' direction; (b) after 90° pulse on protons; (c) after 180° pulse on x' and dephasing; (d) after 90° pulse on y'; (e) spin transfer, ⇨ carbon vectors; (f) after 90° pulse on carbons; (g) after selected pulse delay $(2J)^{-1}$.

substances [28] (Fig. 18). If relaxation effects are taken into account, it can be made quantitative [3,16].

Decoupling experiments can also be useful in studying other nuclei. The dipolar interaction between protons bonded through oxygen to silicon is weaker, but strong enough to cause significant line broadening. Thus, if soil or a soil fraction is studied by ^{29}Si NMR with the decoupler off or on, then the proximity of protons can be investigated. Whereas the ^{29}Si resonance of soils containing protomogolite or imogolite is broadened by turning off the decoupler, that from montmorillonite is not [29]. The ^{29}Si in imogolite will also cross-polarize faster [30].

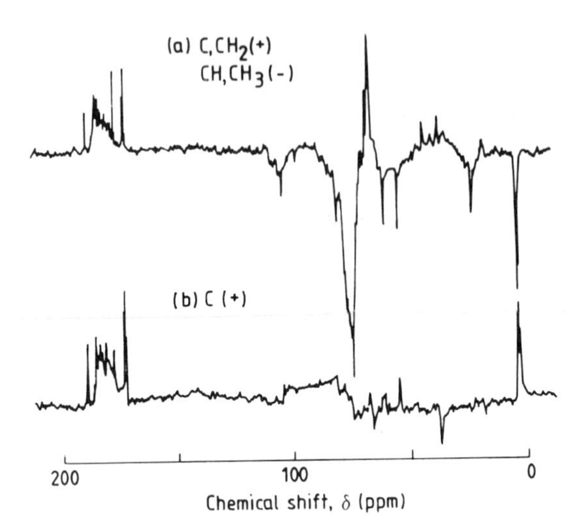

FIG. 17 J-modulated spectra of fulvic acid from Bainville, Canada. (a) Quaternary carbons and CH_2 peaks (+); CH and CH_3 peaks (−); (b) quaternary carbons only. (From Ref. 11.)

TABLE 7 Effect of Varying Delay Time on Signal Intensities in INEPT and GASPE Pulse Sequences

Delay time T_1 (sec)	Signal intensity[a]			
	C	CH	CH_2	CH_3
INEPT				
$(2J)^{-1}$	0	+1	0	0
$3(4J)^{-1}$	0	+	−1	+
GASPE				
$(2J)^{-1}$	+1	0	0	0
J^{-1}	+1	−1	+1	−1

[a]+1 = maximum positive, −1 = maximum negative, + = positive.

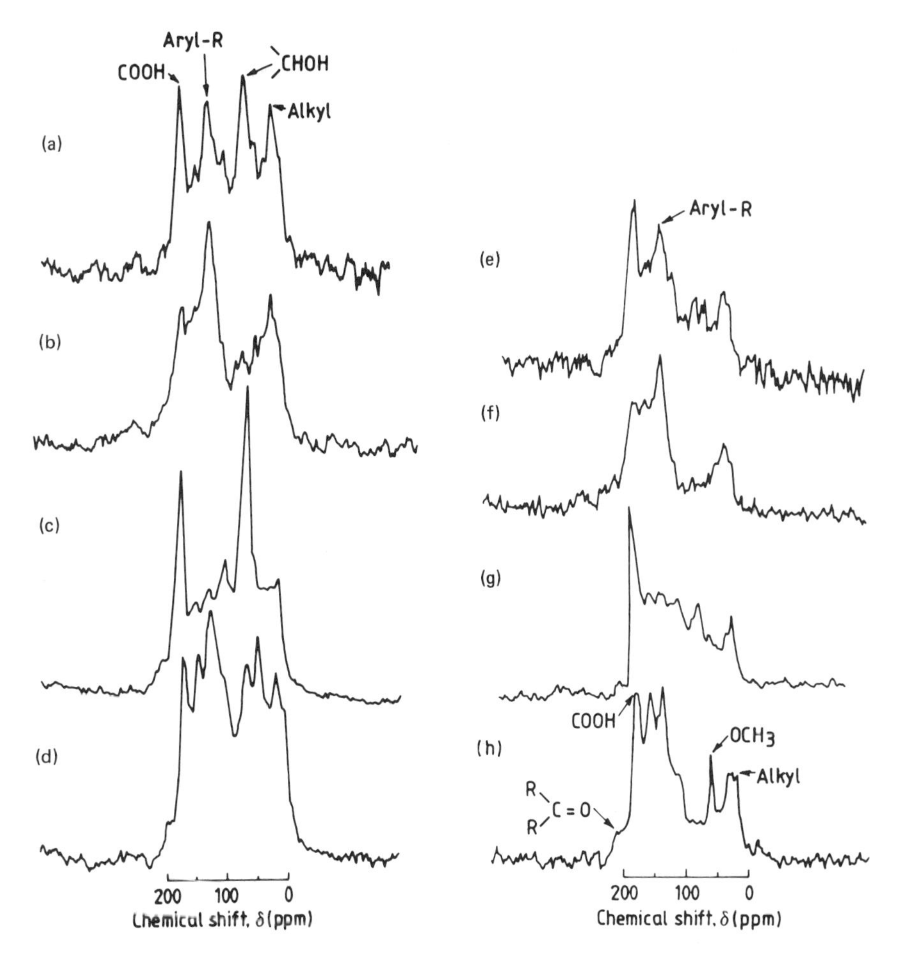

FIG. 18 (a–d) Cross-polarization magic angle spinning ^{13}C spectra of humic and fulvic acid samples. (e–h) Corresponding samples after dipolar dephasing to reveal methyl, methoxy, and quaternary resonances, respectively.

IV. INTERPRETATION

Interpretation of spectra and any subspectra obtained to establish the proximity of other nuclei is the final aim of the NMR experiment. Assignments can be supported by quantitative data or relaxation experiments, and the soil chemist will also have data from other sources

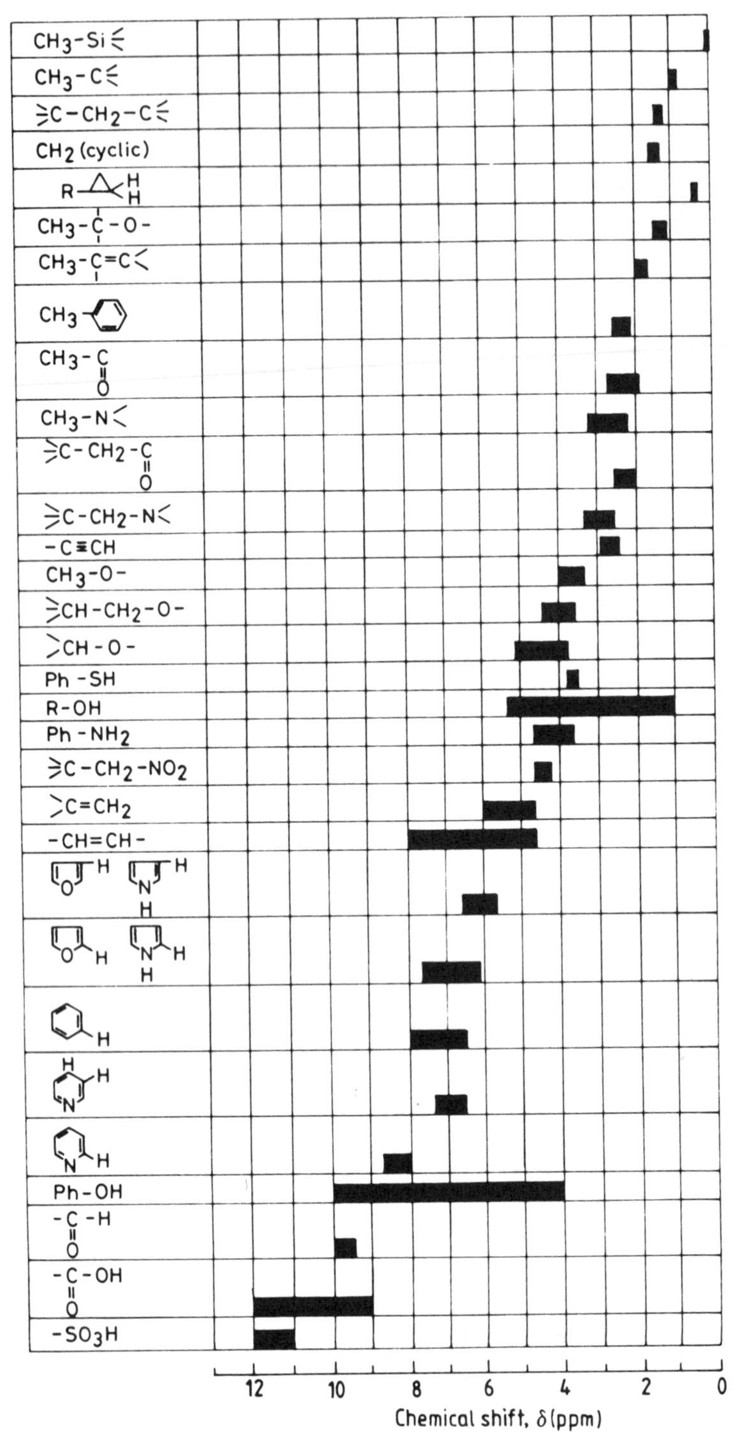

FIG. 19 Proton chemical shifts (^{1}H).

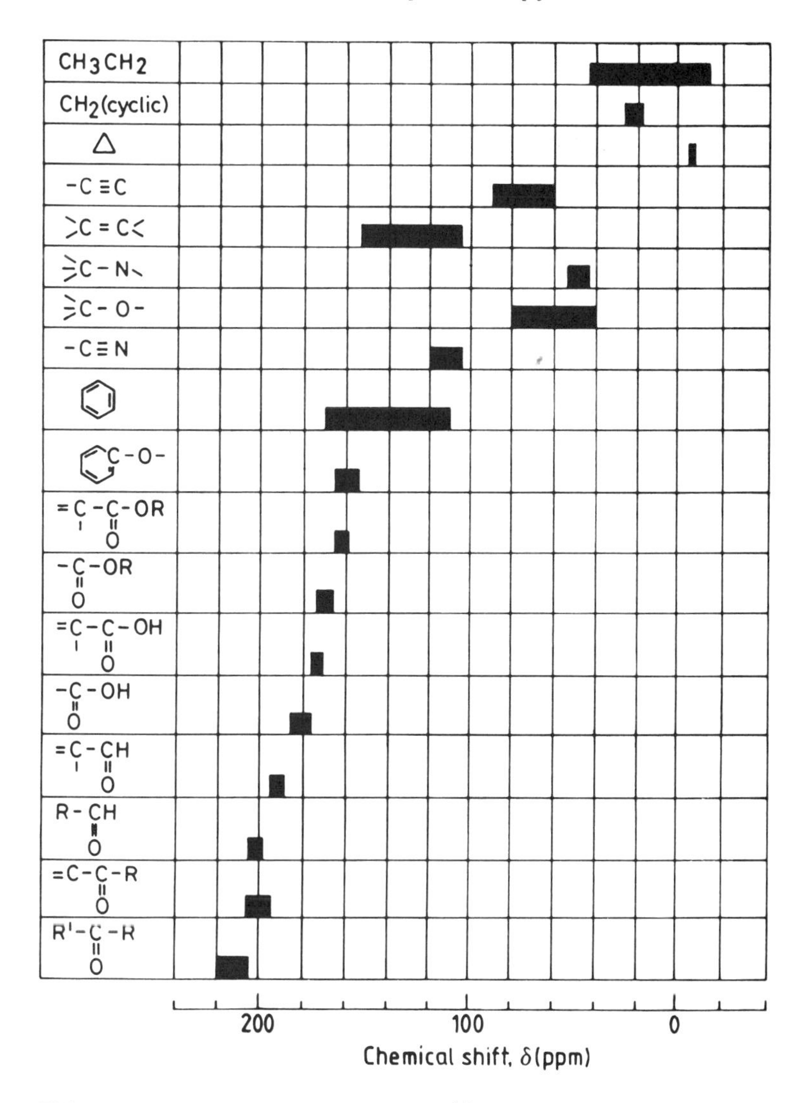

FIG. 20 Carbon chemical shifts (^{13}C).

that will be of assistance. The first step is to consult standard tables of chemical shifts. The chemical shifts of various structural groups that may be found in soils are listed in Figs. 19–24, but the researcher must consult both the original soil and NMR literature for definitive assignments. Further discussion of assignments

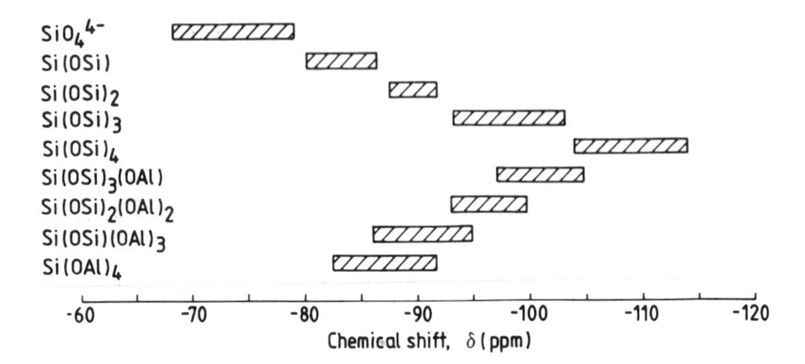

FIG. 21 Silicon chemical shifts (^{29}Si).

is not within the frame of reference of this chapter and the reader is referred elsewhere [1–4,31].

V. APPLICATIONS IN SOIL CHEMISTRY

Although applications of NMR in soil chemistry have been extensively reviewed elsewhere [1–4,31], it is worthwhile summarizing the major applications of ^{1}H, ^{13}C, ^{29}Si, ^{27}Al, ^{15}N, and ^{31}P NMR to soil analysis.

A. ^{1}H NMR

1. From spectra such as those shown in Fig. 12b, the fraction of protons that are aromatic in the sample can be measured by integrating the area of signal from ~6.4–8.4 ppm. Resonances from

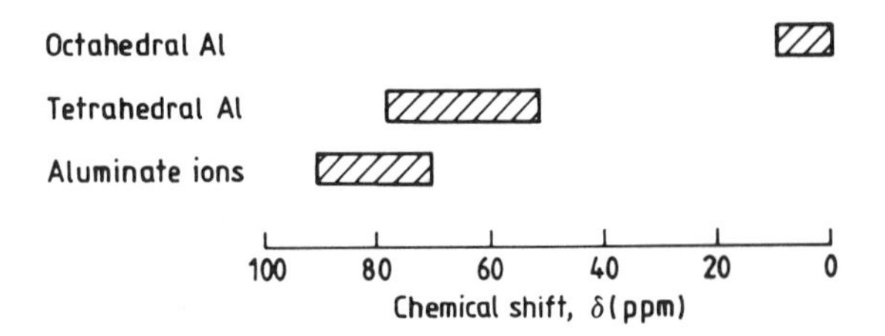

FIG. 22 Aluminum chemical shifts (^{27}Al).

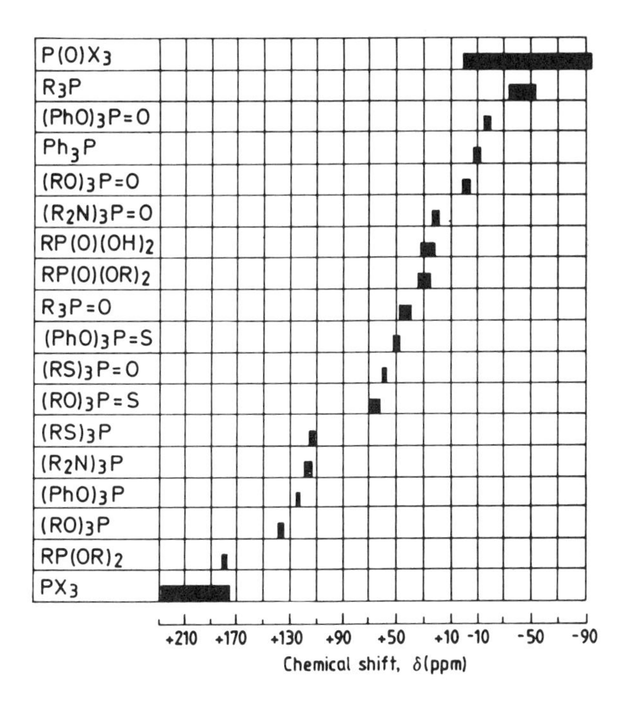

FIG. 23 Phosphorus chemical shifts, X = halogen (^{31}P).

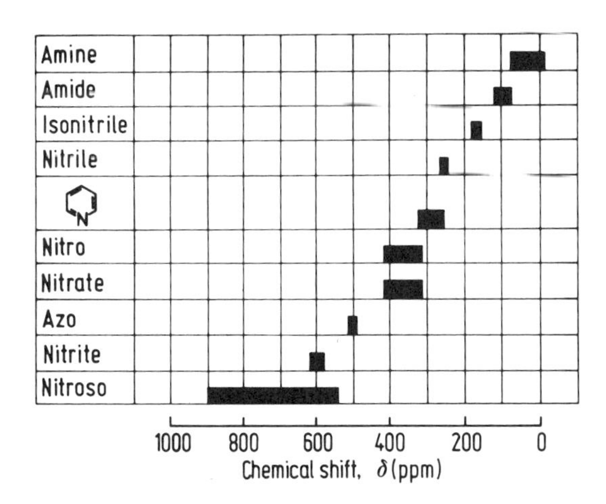

FIG. 24 Nitrogen chemical shifts (^{15}N).

formate at ~ 8.4 ppm should not be included. It should be appreciated that protons that exchange with the solvent are not observed.

2. Some measure of the length of alkyl chains in the extracts can be determined by measuring the area of the resonance at 0.9 ppm (CH_3) and the area at about 1.25–1.29 ppm [$(CH_2)_n$]. The spectrum in Fig. 12b clearly shows that the alkyl chains in this sample are fairly short. It should be appreciated that some branched-chain methine resonances may also appear in the 1.25–1.29 ppm region.

3. Likewise, a measure of the carbohydrate protons can be made by integrating the area at or about 3–6 ppm. Some amino acid protons will also appear in this region.

4. Discrete individual compounds in the extracts can be observed. These include acetate, methanol, formate, aryl methoxy carboxylic acids, succinate, and lactate. The chemical shifts of these resonances are solvent-dependent and hence they can only be identified by adding small amounts to the humic solution and observing an increase in the size of the observed resonance [31].

B. ^{13}C NMR

1. The aromaticity, i.e., the fraction of carbon that is aromatic f_a, can be estimated from the ratio of the integrated area of the resonances at ~107–165 ppm to total area (e.g., see Fig. 18a). However, some overlap with other functional groups can occur. A range of relaxation experiments is necessary to ensure that the data are quantitative.

2. The fraction of carbohydrate carbon can be measured from the sum of resonances at ~72–76 ppm and ~104 ppm. There are some carbohydrate resonances outside these regions, and some other materials may also resonate in this region. It is preferable to place integration cut-off limits at natural valleys in the spectra and not be bound by specific conventions.

3. Cut-off limits can be set at 0–~50 ppm to measure alkyl carbon, or 165–190 ppm to measure COOH carbon and 190–220 ppm to measure ketone and aldehyde carbon. Methoxy carbon resonates with sharply defined limits of 55–57 ppm and can also be estimated. Amino acid carbon can also resonate in this region (50–60 ppm) and care should be taken not to make misassignments. Again, it should be stressed that measurements are comparative and not absolute. Integration limits for one series of samples may not be applicable to another. Dipolar dephasing (solid-state) or INEPT, DEPT, GASPE, or other multiple pulse sequences can be used to further identify the degree of protonation of various structures. A full relaxation study is necessary to obtain quantitative data. Dipolar dephasing is also useful to distinguish methoxy from amino acid carbon.

C. ^{29}Si, ^{27}Al, ^{15}N, and ^{31}P NMR

1. The degree of coordination of silicon to aluminum and other silicon atoms through oxygen can be determined by ^{29}Si NMR (Fig. 25). It is worthwhile removing iron from samples to increase the signal-to-noise ratio (Fig. 25c). It may also reduce or remove spinning sidebands that can hinder assignments and quantitation.

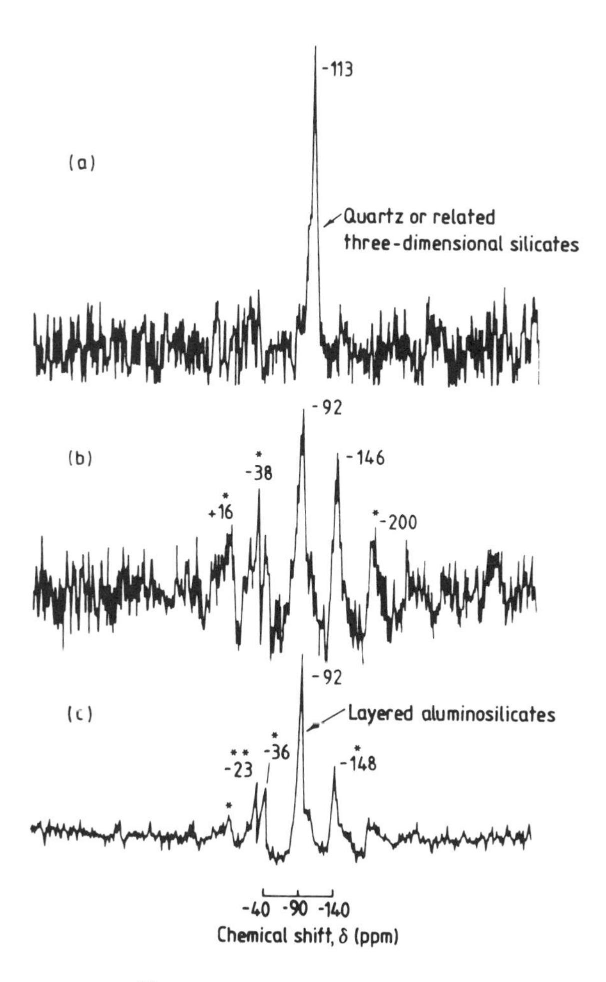

FIG. 25 ^{29}Si solid-state NMR spectrum of sized fractions from Arboretum permanent pasture, South Australia. (a) 2–20 μm size fraction; (b) clay (< 0.2 μm) fraction; (c) clay (< 0.2 μm) fraction after dithionite reduction. *, Spinning sideband; **, artifact.

2. ^{27}Al NMR can be useful in determining the degree of tetrahedral and octahedral coordination of aluminum. However, it should be borne in mind that aluminum in highly distorted environments may not be observed [3,31].

3. ^{31}P NMR can be used to identify a wide range of inorganic and organic phosphates in solution (Fig. 26), including polyphosphate, orthophosphate esters, inorganic orthophosphate, and phosphonates. In the solid state similar assignments apply, but spinning sidebands can interfere with assignments.

4. ^{15}N NMR is useful in identifying amides in ^{15}N-labeled samples (Fig. 27). Other nitrogen-containing functional groups may be identified in the near future by ^{15}N NMR.

VI. FINAL REMARKS

Some comment should be made regarding the future. A wide range of techniques exist that are now almost routinely used for measuring the structural parameters of simple molecules. Most of these techniques have still to be exploited on humic substances or other soil components. In particular, chemical shift information is not the only parameter that gives structural information from NMR measurements. Relaxation times, nuclear Overhauser enhancements, and coupling constants all yield complementary information that is sometimes more valuable. Only Preston [11] and Newman [15] have attempted to make use of these parameters to characterize soils and soil components.

The concept of the Fourier transformation can be used to relate any set of data in the time dimension to the frequency dimension.

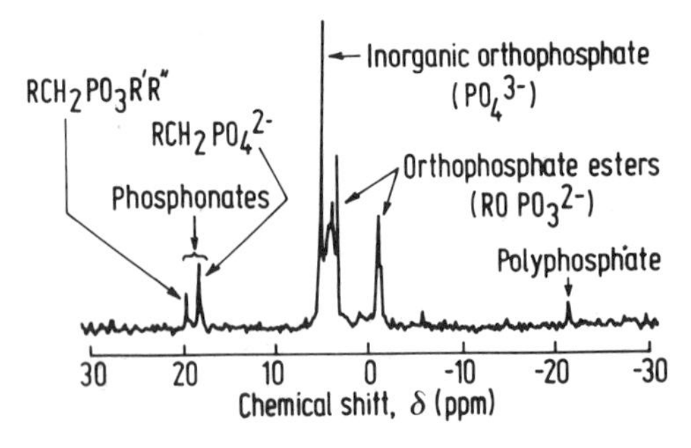

FIG. 26 ^{31}P solution NMR spectrum of soil extract (McKerrow, New Zealand).

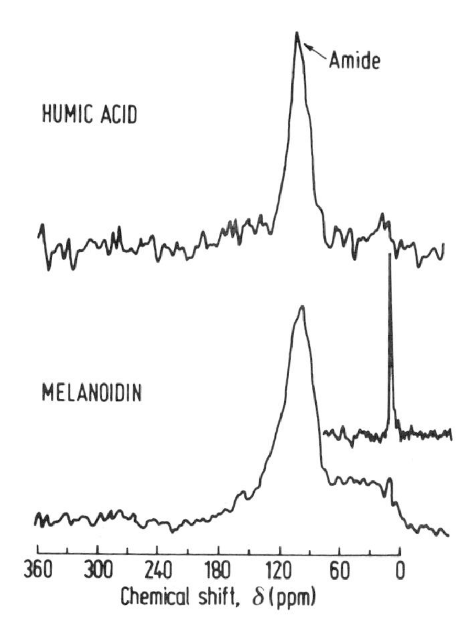

FIG. 27 Cross-polarization magic angle spinning ^{15}N NMR spectrum of ^{15}N labeled humic acid and synthetic melanoidin. Chemical shifts are measured with respect to $^{15}NH_4NO_3$.

Thus, data obtained as a function of the period when the decoupler is turned off can be Fourier-transformed to give a spectrum of signal intensity versus frequency. Coupled with data obtained as a function of time of acquisition (the normal free induction decay), two-dimensional spectra of whole soils or solid soil components can be obtained [33]. These techniques also need to be explored further. Likewise, solution NMR techniques have been developed that relate resonances in proton spectra to those in carbon spectra. Even homonuclear interactions can be explored. None of these solution techniques have yet been exploited in soil systems.

Finally, mention should be made of NMR imaging [34], which has become a useful adjunct to X-ray information in medicine. Such techniques are also applicable to soils, particularly in mapping soil water. The future will also see the development of spectrometers with higher fields and greater sensitivity. This should extend the range of nuclei that can be studied adequately in soils. ^{15}N, ^{31}P, and ^{29}Si soil spectra are just beginning to appear in the literature but other nuclei may soon follow. The soil chemist may find innovative ideas by studying applications in other areas of geochemistry, and so as a final gesture, I refer the reader to my book [31] that covers the applications of NMR to soil science and other areas of geochemistry in more detail.

REFERENCES

1. R. L. Wershaw, in *Humic Substances in Soil, Sediment and Water* (G. R. Aiken, D. M. McKnight, R. L. Wershaw, and P. MacCarthy, eds.), Wiley, New York, 1985, p. 561.
2. M. A. Wilson, in *Humic substances II, in search of structure* (M. Haynes, R. Swift, and R. Malcolm, eds.), Wiley, New York, 1989, pp. 309–339.
3. M. A. Wilson, in *Humic Substances in Soil and Crop Sciences*, Am. Soc. Agron. Public. (in press).
4. M. A. Wilson, *J. Soil Sci., 32*:167 (1981).
5. E. Fukushuma and S. B. W. Roeder, *Experimental Pulse NMR—A Nuts and Bolts Approach*, Addison-Wesley, Reading, Mass., USA, 1981.
6. D. Shaw, *Fourier Transform N.M.R. Spectroscopy*, 2nd ed., Elsevier, Amsterdam, 1981.
7. M. Oades, A. M. Vassallo, A. Waters, and M. A. Wilson, *Aust. J. Soil Res.*, 25, 71 (1987).
8. R. H. Newman, K. R. Tate, P. F. Barron, and M. A. Wilson, *J. Soil Sci., 31*:623 (1980).
9. M. A. Wilson, P. F. Barron, and A. H. Gillam, *Geochim. Cosmochim. Acta, 45*:1743 (1981).
10. M. A. Wilson and K. M. Goh, *J. Soil Sci., 34*:305 (1983).
11. C. M. Preston and B. A. Blackwell, *Soil Sci., 139*:88 (1985).
12. C. M. Preston and J. A. Ripmeester, *Can. J. Spectrosc., 27*:99 (1982).
13. C. M. Preston, R. L. Dudley, C. A. Fyfe, and S. P. Mathus, *Geoderma, 33*:245 (1984).
14. C. M. Preston and M. Schnitzer, *Soil Sci. Soc. Am. J., 48*:305 (1984).
15. R. H. Newman and K. R. Tate, *J. Soil Sci., 35*:47 (1984).
16. M. A. Wilson, A. M. Vassallo, M. Purdue, and H. Reuter, *Anal. Chem., 59*:551 (1987).
17. M. A. Wilson, P. J. Collin, and K. R. Tate, *J. Soil Sci., 34*:297 (1983).
18. P. E. Pfeffer, W. V. Gerasimowicz, and E. G. Piotrowski, *Anal. Chem., 56*:734 (1984).
19. G. A. Morris and R. Freeman, *J. Am. Chem. Soc., 101*:760 (1979).
20. M. R. Bendall, D. M. Doddrell, and D. T. Pegg, *J. Am. Chem. Soc., 103*:4603 (1981).
21. D. M. Doddrell, D. T. Pegg, and M. R. Bendall, *J. Magn. Reson., 48*:323 (1982).

22. C. E. Snape, *Fuel, 61*:1164 (1982).
23. C. E. Snape, *Fuel, 62*:621 (1983).
24. D. W. Brown, T. T. Nakeshima, and D. L. Rabenstein, *J. Magn. Reson., 43*:302 (1984).
25. S. L. Patt and J. N. Shoolery, *J. Magn. Reson., 46*:535 (1982).
26. D. J. Cookson and B. E. Smith, *Org. Magn. Reson., 16*:111 (1981).
27. D. J. Cookson and B. E. Smith, *Fuel, 62*:39 (1983).
28. M. A. Wilson, R. J. Pugmire, and D. M. Grant, *Org. Geochem., 5*:121 (1983).
29. M. A. Wilson and S. A. McCarthy, *Anal. Chem., 57*:2733 (1985).
30. P. F. Barron, M. A. Wilson, A. S. Campbell, and R. L. Frost, *Nature, 299*:616 (1982).
31. M. A. Wilson, *Technique and Applications of N.M.R. Spectroscopy in Geochemistry and Soil Science*, Pergamon Press, Oxford, 1987.
32. R. H. Newman and K. R. Tate, *Commun. Soil Sci. and Plant Analysis, 11*:835 (1980).
33. M. A. Wilson, *J. Soil Sci., 35*:209 (1984).
34. P. Mansfield and P. Morris, *N.M.R. Imaging in Biomedicine*, Academic Press, New York, 1982.

Index